NATO ASI Series

Advanced Science Institutes Series

A series presenting the results of activities sponsored by the NATO Science Committee, which aims at the dissemination of advanced scientific and technological knowledge, with a view to strengthening links between scientific communities.

The Series is published by an international board of publishers in conjunction with the NATO Scientific Affairs Division

A Life Sciences B Physics	Plenum Publishing Corporation London and New York
C Mathematical and Physical Sciences D Behavioural and Social Sciences E Applied Sciences	Kluwer Academic Publishers Dordrecht, Boston and London
F Computer and Systems Sciences G Ecological Sciences H Cell Biology	Springer-Verlag Berlin Heidelberg New York London Paris Tokyo Hong Kong

Series H: Cell Biology Vol. 32

The ASI Series Books Published as a Result of Activities of the Special Programme on CELL TO CELL SIGNALS IN PLANTS AND ANIMALS

This book contains the proceedings of a NATO Advanched Research Workshop held within the activities of the NATO Special Programme on Cell to Cell Signals in Plants and Animals, running from 1984 to 1989 under the auspices of the NATO Science Committee.

The books published as a result of the activities of the Special Programme are:

Molecular Biology of Neuroreceptors and Ion Channels

Edited by

Alfred Maelicke

Max-Planck-Institut für Ernährungsphysiologie
Rheinlanddamm 201, 4600 Dortmund, FRG

Springer-Verlag Berlin Heidelberg New York
London Paris Tokyo Hong Kong
Published in cooperation with NATO Scientific Affairs Division

Proceedings of the NATO Advanced Research Workshop on Molecular Biology of Neuroreceptors and Ion Channels held on the Island of Santorini, Greece, October 2-7, 1988

ISBN-13:978-3-642-74157-9 e-ISBN-13:978-3-642-74155-5
DOI: 10.1007/978-3-642-74155-5

Library of Congress Cataloging-in-Publication Data. NATO Advanced Research Workshop on Molecular Biology of Neuroreceptors and Ion Channels (1988 : Thera Island, Greece). Molecular biology of neuroreceptors and ion channels / edited by Alfred Maelicke. p. cm.—(NATO ASI series. Series H, Cell biology ; vol. 32) "Proceedings of the NATO Advanced Research Workshop on Molecular Biology of Neuroreceptors and Ion Channels held on the Island of Santorini, Greece, October 2–7, 1988"—T.p. verso. Includes indexes. ISBN-13:978-3-642-74157-9 (U.S.)
1. Neurotransmitter receptors—Congresses. 2. Ion channels—Congresses. 3. Molecular biology—Congresses. 4. Gene expression—Congresses. 5. Genetic regulation—Congresses. I. Maelicke, Alfred. II. North Atlantic Treaty Organization. Scientific Affairs Division. III. Title. IV. Series.
[DNLM: 1. Ion Channels—physiology—congresses. 2. Molecular Biology—congresses. 3. Receptors, Cholinergic—metabolism—congresses. 4. Receptors, GABA-Benzodiazepine—metabolism—congresses. 5. Receptors, Sensory—physiology—congresses. WL 102.9 N2787m 1988] QP364.7.N362 1988
591'.188—dc20 DNLM/DLC for Library of Congress 89-11460

Softcover reprint of the hardcover 1st edition 1989

2131/3140-543210 – Printed on acid-free-paper

Contents

III. Molecular Biology of Aminoacid Receptors

IV. Molecular Biology of Voltage-Gated Ion Channels

V. Molecular Biology of Other Neuroreceptors and Ion Channels

VI. Structure-Function Relationships of Ion Channel Proteins

VI. Gene Expression, Gene Regulation and Development

VII. Clinical Aspects

PREFACE

This workshop was the second of this series held on the island of Santorini in the Cycladic Sea. The first one ("Mechanism of Action of the Nicotinic Acetylcholine Receptor", NATO ASI Series H, vol. 10) took place in May 1986 and focused on what was at the time the best studied of all neuroreceptors. This second one, held only two years later, demonstrates the immense progress achieved since then in the field of neuroreceptors and ion channels. Molecular cloning techniques have now made available the primary structures of a whole array of ion channel proteins, and this in turn has shed light on some general principles of the structure-function relationships of these central elements of intercellular communication.

The purpose of this workshop was to explore the common elements in gene and protein structure of already cloned ion channel proteins, and to assess the status of other cloning projects in progress. It explicitly focused on very recently published and unpublished results. All participants kept to these goals thereby demonstrating the very value of such workshops for the progress of science.

As might be expected, these proceedings cannot truly reflect the full content of presentations and discussions at the workshop. This is not so much due to the limits in space allocated to each contribution but rather to matters of publication policy followed by the various laboratories. In the case of one central contribution to the workshop, this has even lead to first delaying and finally withdrawing from providing a manuscript for this volume. I consider such practice as unfair to those other colleagues who have willingly participated not only in the advantages but also in the burdens associated with such a workshop. Disregarding these limitations, this volume should provide an almost complete overview of the present state of this field.

I have organized the manuscripts of these proceedings following only marginally the conference program. The first four chapters provide an overview of the Molecular Biology of neuroreceptors and ion channels. The fifth chapter deals with structure-function relationships and models of receptor and channel structures, the final ones with gene expression, gene regulation and development, and clinical applications. Now that gene cloning has left infancy, it will be applied increasingly to solving general biological and medical problems.

The workshop was again made possible by a generous award from the NATO Special Programme on Cell-to-Cell Signals. (Regretfully, this program has recently been terminated.) I particularly thank the Program Director, Dr. Alain H. Jubier, for his continuous interest and support in staging this comparatively large workshop at a location which is not quite as easily manageable as others are. (I hope to have him convinced though of the advantages an exciting location provides to the atmosphere of a scientific gathering.) Further support has come from the Paul-Martini-Foundation, Hannover, F.R.G., and the companies Boehringer Mannheim, F.R.G. and Latoxan, Rosans, France.

The workshop took place in Thira, the island's capital. We are indebted to the Metropolitan of the Greek Orthodox Church who allowed us to use the church's seminar building as conference hall. Much of the local arrangements including transportion from and to Athens, hotels, meals and excursions were again handled to our satisfaction by Stelios Petropoulos from Damigos Tours.

We are particularly grateful to Professor C.G. Doumas, Director of the Excavations at Akrotiri who delivered a most fascinating lecture on "The Archaeological Treasures of Santorini". This was a marvelously clear and concise account of the discovery, excavation and archaeological and cultural value of the remains of this important Minoan city. It was given in a very personal and exciting manner giving rise to an enchanted discussion in the lecture hall and later at the bar in Selene's restaurant.

This workshop would not have been possible without the help of many colleagues and friends. Heinrich Betz, Heidelberg, helped in selecting speakers and discussants, Christine Riemer was an excellent conference secretary, and my colleagues Ralf Seidel and Bernhard Hermsen acted as projectionists and helpers-for-all-and-everything. Most of the work and success is owed to Reinhard Heiermann who handled all financial matters including travel and hotel arrangements for the workshop. Due to our sincere shortage in funds and some (expected) difficulties in making definite arrangements for travel and hotels in Greece, this was an exhausting and time-consuming task.

Finally, I wish to thank all participants for the exciting scientific atmosphere they created during this workshop. Free flowing discussions inside and outside of the lecture hall, readiness of sharing unpublished results, and collective enjoyment of the beauty of this island high above the sea, all contributed to a feeling of community which often is lost in this age of highly competitive science.

January 1, 1989 Alfred Maelicke

Symposium Participants

AAS, PAL, Norwegian Defence Research Establishment, Kjeller, Norway

BALLIVET, MARC, Dept. de Biochimie, Université de Gènève, Gènève, Switzerland

BARKAS, TOM, Neurobiology Laboratory, Lausanne, Switzerland

BARON, MIRON, Dept. of Medical Genetics, New York State Psychiatric Institute and Columbia University, New York, USA

BAUMANN, ARND, Abteilung Biochemie, Ruhr-Universität Bochum, Bochum, Federal Republic of Germany

BECKER, CORD-MICHAEL, Zentrum für Molekulare Biologie, Universität Heidelberg, Heidelberg, Federal Republic of Germany

BECKH, SYNNÖVE, Präklinische Forschung, Firma Sandoz AG, Basel, Switzerland

BETZ, HEINRICH, Zentrum für Molekulare Biologie, Universität Heidelberg, Heidelberg, Federal Republic of Germany

BIRDSALL, NIGEL, MRC National Institute for Medical Research, London, England

BLACHLY-DYSON, ELISABETH, Oregon Health Science University, Portland, Oregon, USA

BLOUNT, PAUL, Dept. of Pharmacology, Washington University School of Medicine, St. Louis, Mo., USA

BOHEIM, GÜNTHER, Abteilung Zellphysiologie, Ruhr-Universität Bochum, Bochum, Federal Republic of Germany

BREER, HEINZ, Institut für Zoophysiologie, Universität Hohenheim, Stuttgart, Federal Republic of Germany

BRENNER, HANS, Physiologisches Institut, Universität Basel, Basel, Switzerland

BUNZOW, JIM, Oregon Health Sciences University, Portland, Oregon, USA

BUONANNO, ANDRES, Dept. of Pharmacology, Washington University School of Medicine, St. Louis, Mo., USA

BURDEN, STEVE, Dept. of Biology, Massachusetts Institute, Cambridge, MA, USA

CATTERALL, WILLIAM, Dept. of Pharmacology, University of Washington, Seattle, WA, USA

CHANGEUX, JEAN-PIERRE, Laboratory of Molecular Neurobiology, Institut Pasteur, Paris, France

CIVELLI, OLIVIER, Oregon Health Sciences University, Portland, Oregon, USA

CLAUDIO, TONI, Dept. of Physiology, Yale University School of Medicine, New Haven, C.T., USA

CLEMENTI, FRANCESCO, Cattedra di Farmacologia, Universitá di Milano, Milano, Italy

CONTI-TRONCONI, BIANCA, Dept. of Biochemistry, University of Minnesota, St. Paul, MN, USA

COUTURIER, SABINE, Dept. de Biochimie, Université de Gènève, Gènève, Switzerland

DARLISON, MARK, MRC Molecular Neurobiology Laboratory, Cambridge, England

DAVIS, HEATHER-LYNN, Dept. of Anatomy, McGill University, Montreal, Canada

DE SAINT-MICHEL, A., Laboratoire des Toxines Animales, Rosans, France

DEMENEIX, BARBARA, Institut de Physiologie et de Chimie Biologique, Strasbourg, France

DENERIS, EVAN, Molecular Neurobiology Laboratory, The Salk Institute, San Diego, CA, USA

ETEROVIC, VESNA, Dept. of Biochemistry, Universidad Central del Caribe, Cayey, Puerto Rico, USA

FLEMING, JOHN T., MRC Molecular Neurobiology Laboratory, Cambridge, England

FONTAINE, BERTRAND, Laboratory of Molecular Neurobiology, Institut Pasteur, Paris, France

FORNASARI, DIEGO, Dept. of Medical Pharmacology, Universitá di Milano, Milano, Italy

FUCHS, SARA, Dept. of Chemical Immunology, The Weizmann Institute of Science, Rehovot, Israel

GANETZKY, BARRY, Laboratory of Genetics, University of Wisconsin, Madison, Wisconsin, USA

GERSHONI, JON, Dept. of Biophysics, The Weizmann Institute of Science, Rehovot, Israel

GUERRINI, LUISA, Dept. of Molecular Biology, ENICHEM, Monterotondo, Italy

GUNDELFINGER, ECKART, Zentrum für Molekulare Biologie, Universität Heidelberg, Heidelberg, Federal Republic of Germany

HADJIDAKIS, IOANNIS, Dept. of Chemistry, University of Ioannina, Ioannina, Greece

HARIRI, NURAN, Dept. of Physiology, Ege University, Bornova, Izmir, Turkey

HEIERMANN, REINHARD, Max-Planck-Institut für Ernährungsphysiologie, Dortmund, Federal Republic of Germany

HEINEMANN, STEPHEN F., Molecular Neurobiology Laboratory, The Salk Institute, San Diego, CA, USA

HERMSEN, BERNHARD, Max-Planck-Institut für Ernährungsphysiologie, Dortmund, Federal Republic of Germany

HOCH, WERNER, Zentrum für Molekulare Biologie, Universität Heidelberg, Heidelberg, Federal Republic of Germany

HUCHO, FERDINAND, Institut für Biochemie, Freie Universität Berlin, Berlin, Federal Republic of Germany

KAWASHIMA, ERIC, Glaxo Institute for Molecular Biology, Varouge, Géneva, Switzerland

KLARSFELD, ANDRE, Laboratory of Molecular Neurobiology, Institut Pasteur, Paris, France

KOSOWER, EDWARD, M., Sackler Faculty of Exact Sciences, Tel-Aviv University, Ramat-Aviv, Tel-Aviv, Israel

KOSOWER, NECHAMA, Dept. of Chemistry, Tel-Aviv University, Tel-Aviv, Israel

LANGOSCH, DIETER, Zentrum für Molekulare Biologie, Universität Heidelberg, Heidelberg, Federal Republic of Germany

LOEFFLER, J. PH., Institut de Physiologie et de Chimie Biologique, Strasbourg, France

LOUTRARI, HELENI, Laboratory of Biochemistry, Hellenic Pasteur Institute, Athens, Greece

LÜBBERT, HERMANN, Klinische Forschung, Firma Sandoz AG, Basel, Switzerland

MAELICKE, ALFRED, Max-Planck-Institut für Ernährungsphysiologie, Dortmund, Federal Republic of Germany

MALHERBE, PARI, Dept. of Pharmaceutical Research, Hoffmann La Roche, Basel, Switzerland

MEYERINK, JAN H., Microbiological Laboratories, Organon International BV, Oss, The Netherlands

MOSCKOVITZ, RACHEL, Dept. of Biophysics, The Weizmann Institute of Science, Rehovot, Israel

NAKANISHI, SHIGETADA, Faculty of Medicine, Kyoto University, Kyoto, Japan

NEF, PATRICK, Dept. of Biophysics, The Weizmann Institute of Science, Rehovot, Israel

NODA, MASAHARU, Faculty of Medicine, Kyoto University, Kyoto, Japan

OHANA, BELLA, Dept. of Biophysics, The Weizmann Institute of Science, Rehovot, Israel

PANAYOTOVA, MARIANNE, Max-Planck-Institut für Ernährungsphysiologie, Dortmund, Federal Republic of Germany

PAPADOULI, IRENE, Laboratory of Biochemistry, Hellenic Pasteur Institute, Athens, Greece

PEROZO, EDUARDO, Marine Biology Laboratory, Woods Hole, MA, USA

PONGS, OLAF, Abteilung Biochemie, Ruhr-Universität Bochum, Bochum, Federal Republic of Germany

PRITCHETT, DOLAN, Zentrum für Molekulare Biologie, Universität Heidelberg, Heidelberg, Federal Republic of Germany

REED, JUTA K., Dept. of Chemistry, University of Toronto, Mississauga, Ont., Canada

REINHARDT-MAELICKE, SIGRID, Institut für Tumorforschung, Universität Essen, Federal Republic of Germany

SCHLOSS, PATRICK, Zentrum für Molekulare Biologie, Universität Heidelberg, Heidelberg, Federal Republic of Germany

SCHOEPFER, RALF, Receptor Biology Laboratory, The Salk Institute, San Diego, CA, USA

SCHROEDER, WERNER, Institut für Biochemie, Freie Universität Berlin, Berlin, Federal Republic of Germany

SCHWARZ, THOMAS L., Medical Institute, University of California, San Francisco, CA, USA

SEEBURG, PETER, Zentrum für Molekulare Biologie, Universität Heidelberg, Heidelberg, Federal Republic of Germany

SEIDEL, RALF, Max-Planck-Institut für Ernährungsphysiologie, Dortmund, Federal Republic of Germany

SHAW, CHRIS, Dept. of Ophthalmology, The University of British Columbia, Vancouver, B.C., Canada

SIEGHART, WERNER, Psychiatrische Universitätsklinik, Wien, Austria

SOPHIANOS, DEMETRIS, Laboratory of Biochemistry, Hellenic Pasteur Institute, Athens, Greece

STROSBERG, A. DONNY, Dept. of Biotechnology, Institut Pasteur, Paris, France

SUMIKAWA, KATUMI, Dept. of Psychobiology, University of California, Irvine, CA, USA

THOMAS, LEO, Zentrum für Molekulare Biologie, Universität Heidelberg, Heidelberg, Federal Republic of Germany

TZARTOS, SOCRATES, Laboratory of Biochemistry, Hellenic Pasteur Institute, Athens, Greece

VAN HUIZEN, FRANS, Dept. of Ophthalmology, The University of British Columbia, Vancouver, B.C., Canada

VOGEL, ZVI, Dept. of Neurobiology, The Weizmann Institute of Science, Rehovot, Israel

WITZEMANN, VEIT, Max-Planck-Institut für Biophysikalische Chemie, Göttingen, Federal Republic of Germany

WRIGHT, MARIANNE, Institute for Medical Biochemistry, Oslo, Norway

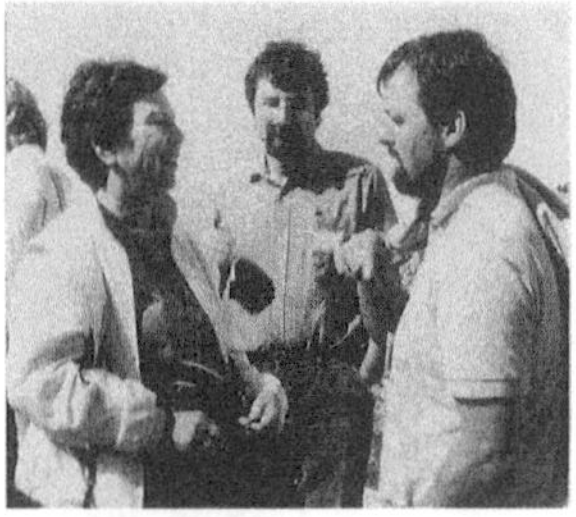

First row: A. Baumann, B. Ganetzky, S. Burden,
R. Schoepfer, J. Gershoni, J. Reed;
M. Noda, O. Pongs; S. Fuchs, H. Brenner,
V. Witzemann

Second row: N. Birdsall, M. Wright; P. Seeburg,
S. Nakanishi, M. Ballivet, A. Maelicke;
M. Darlison, R. Heiermann

Third row: R. Schoepfer, T. Claudio; B. Ganetzky,
P. Nef, H. Betz, Z. Vogel; J.-P. Changeux

Fourth row: H. Betz, G. Boheim; O. Pongs, C. Riemer,
P. Blount, B. Ganetzky, O. Civelli,
S. Reinhardt-Maelicke; S. Petropoulos

COMMON STRUCTURAL PRINCIPLES OF ION CHANNEL PROTEINS

Alfred Maelicke
Max-Planck-Institut für Ernährungsphysiologie,
Rheinlanddamm 201, D-4600 Dortmund 1, FRG

Hydropathy profiles of the available sequences of ion channel proteins (Fig. 1) suggest that ligand-gated and voltage-gated channels each form a structural family. Ligand-gated channel proteins consist of several subunits a combination of which must be inserted into the membrane to provide ligand-gated conductance (Mishina et al., 1984). In contrast, functional voltage-gated sodium channels are already obtained when only the large subunit is expressed (Noda et al., 1986).

Ligand-gated ion channels.

Common features in hydropathy profiles of subunits of ligand-gated ion channels are a large hydrophilic (probably extracellular) domain followed by a cluster of three hydrophobic (α-helical) domains, M1-M3, another large hydrophilic domain (probably cytosolic), and a fourth hydrophobic (α-helical) domain, M4, close to the carboxy terminus of the sequence (Fig. 1). Much of the topological assignment (extracellular or cytosolic) of hydrophilic domains comes from studies of nicotinic acetylcholine receptors (nAChR). Thus, epitope mapping studies have provided evidence for the sidedness of N-terminus and C-terminus (Ratnam et al., 1986; Young et al., 1985); further determinants of the N-terminal extracellular domain are the main immunogenic region (Barkas et al., 1987), the N-glycosylation site, and a pair of adjacent cysteine residues

NATO ASI Series, Vol. H 32
Molecular Biology of Neuroreceptors and
Ion Channels. Edited by A. Maelicke

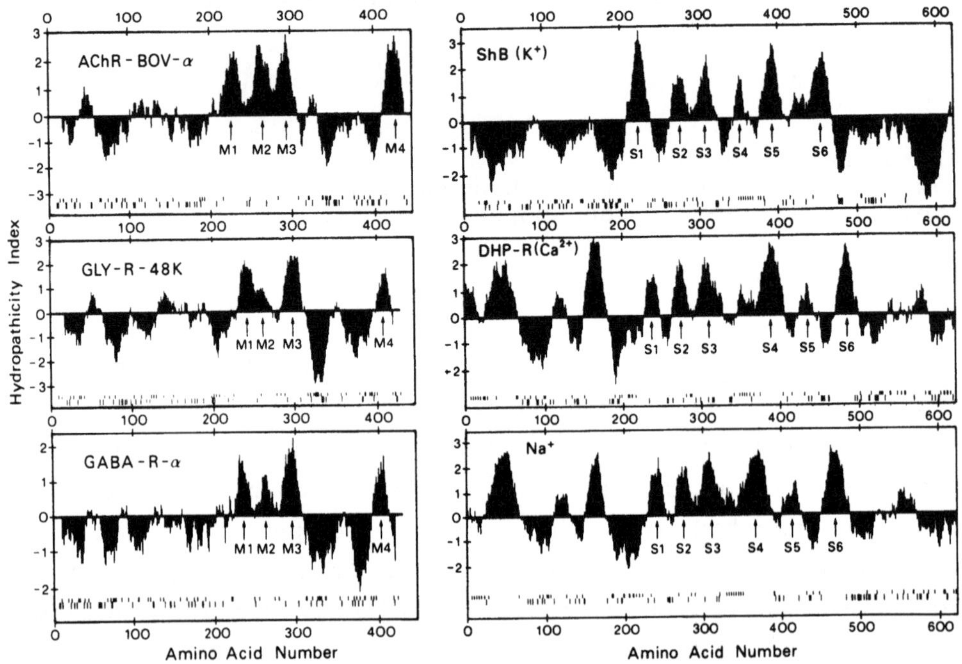

Fig. 1: Aligned hydropathy profiles of subunit sequences of representative ligand-gated and voltage gated ion channels.

(a) Ligand-gated channels: Sequence of α-subunit of nAChR from bovine skeletal muscle (AChR-BOV-α), of 48K subunit of glycine receptor (GLY-R-48K), and of α-subunit of GABA receptor (GABA-R-α) are aligned according to their putative transmembrane domains M2.
(b) Voltage-gated channels: Hydropathy plots of repeat I of rat brain Na^+ channel (Na^+), and of repeat I of DHP-receptor (DHP-R (Ca^{2+})) are aligned according to transmembrane segment S2 of Shaker B protein (ShB(K^+)).
Positions of positively charged residues (Arg, Lys) and of negatively charged residues (Asp, Glu) are indicated by upward and downward vertical lines, respectively, below each hydropathy profile.
Hydropathy was calculated according to Ref. 31 employing a program developed for MS-DOS computers by Thomas Karsch, Lehrstuhl für Allgemeine Botanik, Ruhr-Universität, D-4630 Bochum, FRG.

"close to" the transmitter binding site at the α-subunit (Karlin et al., 1986). All putative phosphorylation sites are located in the hydrophilic loop just in front of M4.

Patch-clamp studies of chimeric nAChRs expressed in Xenopus oocytes (Imoto et al., 1986), and affinity labelling with local anesthetics (Giraudat et al., 1986; Oberthür et al., 1986), both show that putative transmembrane domain M2 is involved in the control of transmitter-activated ion flux. As all subunits of nAChRs have very similar hydrophobicity profiles, it has been suggested that transmembrane domains M1-M3 of all subunits are arranged in a pseudosymmetric fashion around a central axis defining the position of the receptor-integral channel, with the M2 domains providing the inner lining of the channel.

Voltage-gated ion channels.

The main polypeptide of voltage-gated sodium channels (Noda et al., 1984) and of a putative calcium channel (Tanabe et al., 1987) all contain four repeated units of strong sequence homology. In contrast, the putative potassium channel from the nervous system of *Drosophila* may be composed of several polypeptides each related to one of the repeated units of large channel proteins (Baumann et al., 1987) (Fig. 1b). Each of the repeats (Na^+, Ca^{2+}) or channel proteins (K^+) consists of six hydrophobic (α-helical) segments S1-S6 one of which (S2) being negatively charged, another one (S4) (or its environment) exhibiting an ordered pattern of positive charges (Fig. 1b). In analogy to the structure of ligand-gated channels, the four repeated units (or the set of K^+-channel proteins) are proposed to also being oriented in a pseudosymmetric fashion around a central axis (channel). The four S2 segments are suggested to form the inner wall, the S4 segments are suggested to be voltage sensors and to be situated within the clusters of other segments.

Structural models of ion channel proteins.

It is important to keep in mind that the above assignments are based on limited structural information (primary sequences of a few channel proteins) and rather unsophisticated functional requirements. Furthermore, a number of earlier structural suggestions has already been shown to be either incorrect or not of general significance. As a prominent example, earlier proposals of nAChR structure assumed an additional (amphipathic) α-helical domain M5 in each nAChR subunit which was suggested to forming instead of M2 the inner lining of the channel (Finer-Moore and Stroud, 1984; Kosower, 1987). A homologous sequence is neither found in glycine receptor (Grenningloh et al., 1987) nor in GABA receptor (Barnard and Darlison, 1987), two ligand-gated anion channels. In the case of nAChR, epitope mapping and site-directed mutagenesis studies also argue against the existence of M5 (Maelicke et al. 1989; Lindstrom, 1986).

Another example refers to the position of ligand binding sites at ligand-gated ion channels. The similarity of hydropathy profiles (Fig. 1a) suggests that these may reside in an identical area of the large extracellular loop preceding transmembrane domains M1-M3. Originally a consensus sequence found in all nAChR α-subunits (between two cysteines in approximate positions 128 and 142) was suggested to contain essential elements of this site (Noda et al., 1982; Smart et al., 1984). After Karlin's group identified cysteines 192 and 193 in *Torpedo* receptor as target sites of affinity alkylation with bromoacetylcholine (Karlin et al., 1986), and a pair of adjacent cysteines in front of M1 was found also in the ligand-binding subunit of glycine receptor (Betz, 1987), a general motif for a ligand binding domain was apparently discovered. Soon later it was found, however, that the GABA receptor does not display this motif which therefore cannot be considered a general one for ligand-gated ion channels (Barnard and Darlison, 1987).

It still is a popular assumption that gene structure (exon/ intron distribution) and function are correlated, i.e. that a functional domain normally is encoded by a single exon. Evidence against this assumption is mounting.

How accurate are these models?

How much then can assignments of functional sites be trusted when based on sequence homology or hydropathy considerations only? Or, in other words, do the gross structural features detected in this way already correlate to specific function or are they of more basic nature?

Part of the answer depends on the importance of interactions of amino acids from *separate* sequence portions of the protein under consideration. Such interloop interactions are expected to occur in particular in extramembrane (hydrophilic) regions. As binding affinities and pharmacological discrimination of ligands (e.g. agonist versus antagonist action) have been shown to sensibly depend on nativity of receptor structure (Maelicke 1988), functional assignments of structural motifs of hydrophilic sequence portions appears particularly risky. Noteworthy in this context is the lack of indication of common structural features consistent with an acetylcholine binding site in the sequences of nicotinic and muscarinic acetylcholine receptors (Kubo et al., 1986), and of acetylcholine esterase (Schumacher et al., 1986).

With only very limited ultrastructural information on integral membrane proteins available (Deisenhofer et al., 1985), little is known about their general structural requirements, i.e. *below* the level of specific function. Thus, different clustering of transmembrane domains in ligand-gated as compared to voltage-gated ion channels may just reflect different means of membrane anchoring rather than manifesting (expected) differences in channel structure. Furthermore, it is debatable whether water-filled pores (as the nAChR channel) require a hydrophilic channel lining or not, whether the net value of charge clusters at the entrance and/or exit of a channel define it as cationic or anionic, how voltage sensors should be structured and where they should be located. With so many basic questions still open, any functional assignment of transmembrane domains appears quite ambiguous. As a case in point, similar transmembrane topology as observed for rhodopsins (Nathans et al., 1986) and β-adrenergic receptor (Lefkowitz et al., 1986) (seven similarly spaced hydrophobic domains) might reflect common principles of interaction with other membrane proteins (G proteins), it certainly does not reflect a similar mechanism of signal reception (chemical messages versus light quanta).

Except for the classical approach of active sites location (applicable for determination of ligand binding sites and enzyme controlled regulatory sites), there does not exist a general strategy for detecting functional domains in integral membrane proteins. Elucidation of their structure-function relationship, therefore, is more likely to occur by indirect means. As an example, the difference in size of essential polypeptides of ligand-gated channels (several subunits each consisting of some 500 amino acid residues) as compared to voltage-gated sodium and calcium channels (a single subunit of some 2000 amino acids) provides an interesting functional correlation: Ligand-gated channels are developmentally regulated by exchange of one or more subunits (Mishina et al., 1986), a mechanism not applicable to channels formed by a single large polypeptide.

Structural models of ion channel proteins should be considered preliminary.

With the above in mind, I propose more cautious handling of gross sequence homologies and hydropathy profiles. Besides suggesting common ancestral genes (which may even have coded for proteins with unrelated function), gross structural features such as exposed by hydropathy calculations may reflect properties of structural design *more basic* than functional specialization. Thus, clusters of three (ligand-gated ion channels), six (voltage-gated ion channels) or seven (rhodopsin, β-adrenergic receptor) transmembrane domains may just reflect alternative means of anchoring a protein in phospholipid bilayers. Such basic structural requirements will also lead to comparable hydropathy profiles and sequence homologies. This does not exclude but reduces the emphasis on functional similarity: The degree of membrane anchoring is likely to affect properties such as the lateral diffusion coefficient, interaction with other membrane components or protein turnover, and these properties in turn may directly or indirectly contribute to control of function. At our present level of knowledge of the structural features of membrane proteins in general, however, it seems premature to link functional specialization to such gross structural features as clusters of hydrophobic domains. 'Families of proteins' should therefore be defined more restrictively on the basis of true consensus sequences which are more likely to reflect common (conserved) principles of structure and function.

As a case in point, the glycine receptor from spinal cord and the GABA receptor from bovine brain are much more related to each other than either of them to nicotinic acetylcholine receptors. In addition to similar hydropathy profiles of their subunit sequences (Fig. 1a), these two ligand-gated chloride channels share a consensus sequence of ten conserved amino acids in the putative transmembrane domain M2 (Grenningloh et al. 1987; Barnard and Darlison, 1987). On the functional level, they exhibit a similar order of ion conductivities and similar channel kinetics including subconductance states (Hamill et al., 1983; Bormann et al., 1987). Together with their similar estimated channel diameter (Bormann et al., 1987), it is thus conceivable that the chloride channels of glycine and GABA receptor are governed by similar, if not identical, structural principles.

In contrast, the structural similarity of the two cloned chloride channels with the cation channel of the nAChR appears restricted to their hydrophobicity profiles (Fig. 1a) and their total number of polypeptides per receptor monomer. Their sets of clustered hydrophobic domains M1-M3 do not exhibit significant sequence homology. The same stochiometry of subunits per monomer of glycine receptor and *Torpedo* nAChR ($\alpha_3\beta_2$ (Langosch et al., this volume), and $\alpha_2\beta\gamma\delta$ (Maelicke, 1988), respectively) may be a consequence of functional optimization during evolution; alternatively, four to five hydrophobic domains of comparable structure (e.g. M2 or S2) may be a minimal requirement for the formation of stable membranes pores (Noda et al., 1984; Tanabe et al., 1987; Baumann et al., 1987; Barnard and Darlison, 1987) with the resulting channel pore size merely being a circumstantial correlate. As these examples should demonstrate, different levels of structural homology exist, and their correlation to function specialization may not necessarily be direct.

Acknowledgement

Work from my laboratory is supported by grants from the Deutsche Forschungsgemeinschaft and the Fonds der Chemischen Industrie.

References:

Barkas T, Mauron A, Roth B, Alliod C, Tzartos SJ, Ballivet M (1987) Mapping the main immunogenic region and toxin-binding site of the nicotinic acetylcholine receptor. Science 235:77-80

Barnard EA, Darlison MG, Seeburg P (1987) Molecular biology of the $GABA_A$ receptor: the receptor/channel superfamily. Trends Neurosci 10:502-509

Baumann A, Krah-Jentgens I, Müller R, Müller-Holtkamp F, Seidel R, Kecskemethy N, Casal J, Ferrus A, Pongs O (1987) Molecular organization of the maternal effect region of the *Shaker* complex of *Drosophila*: charakterization of an I_A channel transcript with homology to vertebrate Na^+ channel. EMBO J 6:3419-3429

Betz H (1987) Biology and structure of the mammalian glycine receptor. Trends Neurosci 10:113-117

Bormann J, Hamill OP, Sakmann B (1987) Mechanism of anion permeation through channels gated by glycine and γ-aminobutyric acid in mouse cultured spinal neurones. J Physiol 385:243-286

Deisenhofer J, Michel H, Huber R (1985) The structural basis of photosynthetic light reactions in bacteria. Trends Biol Sci 10:243-248

Finer-Moore J, Stroud RM (1984) Amphipathic analysis and possible formation of the ion channel in an acetylcholine receptor. Proc Natl Acad Sci USA 81:155-159

Giraudat J, Dennis M, Heidmann T, Chang J-Y, Changeux J-P (1986) Structure of the high-affinity binding site for noncompetitive blockers of the acetylcholine receptor: Serine-262 of the δ subunit is labeled by [^{3}H]chlorpromazine. Proc Natl Acad Sci USA 83:2719-2723

Grenningloh A, Rienitz A, Schmitt B, Methfessel C, Zensen M, Beyreuther K, Gundelfinger ED, Betz H (1987) The strychnine-binding subunit of the glycine receptor shows homology with nicotinic acetylcholine receptors. Nature 328:215-220

Hamill OP, Bormann J, Sakmann B (1983) Activation of multiple-conductance state chloride channels in spinal neurones by glycine and GABA. Nature 305:805-808

Imoto K, Methfessel C, Sakmann B, Mishina M, Mori Y, Konno T, Fukuda K, Kurasaki M, Bujo H, Fujita Y, Numa S. (1986) Location of a δ-subunit region determining ion transport through the acetylcholine receptor channel. Nature 324:670-674

Karlin A, Kao PN, DiPaola M (1986) Molecular pharmacology of the nicotinic acetylcholine receptor. Trends Pharmacol Sci 7:304-308

Kosower EM (1987) A structural and dynamic model for the nicotinic acetylcholine receptor. Eur J Biochem 168:431-449

Kubo, T., Fukuda, K., Mikami, A., Maeda, A., Takahashi, H., Mishina, M., Haga, T., Haga, K., Ichiyama, A., Kangawa, K., Kojima, M., Matsuo, H., Hirose, T. & Numa, S. (1986) Cloning, sequencing and expression of complementary DNA encoding the muscarinic acetylcholine receptor. Nature 323:411-416

Kyte J, Doolittle RJ (1982) A simple method for displaying the hydropathic character of a protein. J Mol Biol 157:105-132

Langosch D, Grenningloh G, Schmieden V, Prior P, Malosio M-L, Schmitt B, Betz H (1989) The postsynaptic glycine receptor - a member of the neurotransmitter-gated channel protein family. This volume, pp. 125-130

Lefkowitz RJ, Benovic JL, Kobilka B, Caron MG (1986) β-Adrenergic receptors and rhodopsin: shedding new light on an old subject. Trends Pharmacol Sci 7:444-448

Lindstrom J (1986) Probing nicotinic acetylcholine receptors with monoclonal antibodies. Trends NeuroSci 9:401-407

Maelicke A (1988) In: Whittaker VP (ed) Structure and Function of the Nicotinic Acetylcholine Receptor. Springer, Berlin Heidelberg New York, pp 267-313

Maelicke A, Plümer-Wilk R, Fels G, Spencer SR, Engelhard M, Veltel D, Conti-Tronconi BM (1989) Epitope mapping employing antibodies raised against short synthetic peptides: A study of the nicotinic acetylcholine receptor. Biochemistry 28:1396-1405

Mishina M, Kurosaki T, Tobimatsu T, Morimoto Y, Noda M, Yamamoto T, Terao M, Lindstrom J, Takahashi T, Kuno M, Numa S (1984) Expression of functional acetylcholine receptor from cloned cDNAs. Nature 307:604-608

Mishina M, Takai T, Imoto K, Noda M, Takahashi T, Numa S, Methfessel C, Sakmann B (1986) Molecular distinction between fetal and adult forms of muscle acetylcholine receptor. Nature 321:406-411

Nathans J, Thomas D, Hogness DS (1986) Molecular genetics of human color vision: the genes encoding blue, green, and red pigments. Science 232:193-202

Noda M, Ikeda T, Suzuki H, Takeshima H, Takahashi T, Kuno M, Numa S, (1986) Expression of functional sodium channels from cloned cDNA. Nature 322:826-828

Noda M, Shimizu S, Tanabe T, Takai T, Kayano T, Ikeda T, Takahashi H, Nakayama H, Kanaoka Y, Minamino N, Kangawa K, Matsuo H, Raftery MA, Hirose T, Inayama S, Hayashida H, Miyata T, Numa S (1984) Primary structure of *Electrophorus electricus* sodium channel deduced from cDNA sequence. Nature 312:121-127

Noda M, Takahashi H, Tanabe T, Toyosato M, Furutani Y, Hirose T, Asai M, Inayama S, Miyata T, Numa S (1982) Primary structure of alpha-subunit precursor of *Torpedo californica* acetylcholine receptor deduced from cDNA sequence. Nature 299:793-797

Oberthür W, Muhn P, Baumann H, Lottspeich F, Wittmann-Liebold B, Hucho F (1986) The reaction site of a non-competitive antagonist in the ω-subunit of the nicotinic acetylcholine receptor. EMBO J 5:(8)1815-1819

Ratnam M, Le Nguyen DL, Sargent PB, Lindstrom J (1986a) Transmembrane topology of nicotinic acetylcholine receptor: Immunochemical tests contradict theoretical predictions based on hydrophobicity profiles. Biochemistry 25:2633-2643

Sanchez JA, Dani JA, Siemen D, Hille B (1986) Slow permeation of organic cations in acetylcholine receptor channels. J Gen Physiol 87:985-1001

Schumacher M, Camp S, Manlet Y, Newton M, MacPhee-Quigley K, Taylor SS, Friedmann T, Taylor P (1986) Primary structure of *Torpedo californica* acetylcholinesterase deduced from its cDNA sequence. Nature 319:407-409

Smart L, Meyers H-W, Hilgenfeld R, Saenger W, Maelicke A (1984) A structural model for the ligand-binding sites at the nicotinic acetylcholine receptor. FEBS Letters 178:64-68

Tanabe T, Takeshima H, Mikami A, Flockerzi V, Takahashi H, Kangawa K, Kojima M, Matsuo H, Hirose T, Numa S (1987) Primary structure of the receptor for calcium channel blockers from skeletal muscle. Nature 328:313-318

Young EF, Ralston E, Blake J, Ramachandran J, Hall ZW, Stroud RM (1985) Topological mapping of acetylcholine receptor: evidence for a model with five transmembrane segments and a cytoplasmic COOH-terminal peptide. Proc Natl Acad Sci USA 82:626-630

BRAIN AND MUSCLE NICOTINIC ACETYLCHOLINE RECEPTORS: A GENE FAMILY

Steve Heinemann, Jim Boulter, Evan Deneris, John Connolly, Paul Gardner, Etsuko Wada, Keiji Wada, Robert Duvoisin, Marc Ballivet #, Larry Swanson*, and Jim Patrick

Molecular Neurobiology Laboratory, The Salk Institute, P.O. Box 85800, San Diego, CA 92138 / USA

#Department of Biochemistry, University of Genèva, CH-1211 Genèva, Switzerland

*Neural Systems Laboratory and Howard Hughes Medical Institute, San Diego, CA 92138 / USA

The synapse plays a key role in the nervous system and it is likely that biochemical changes at the synapse underlie some aspects of higher brain function. Most plausible theories of learning, pattern recognition and memory depend upon changes in the efficiency of chemical synapses and changes in the ion channels involved in altering and maintaining the membrane potential, Cajal, 1911, Hebb,1949, Hopfield 1982. It seems unlikely that these theories will be testable until we have learned more about the structure, function and regulation of receptor and ion channel molecules. It is also now known that receptors can be directly implicated in human disease. Myasthenia gravis is an auto-immune disease involving the production of antibodies against the nicotinic acetylcholine receptor present in skeletal muscle,Patrick and Lindstrom 1973. Degenerative diseases of the brain such as Parkinson's, Huntington's and Alzheimer's disease may involve a breakdown in one or more transmitter systems, Perry et al, 1987. Most of the therapeutic drugs that have proven to be effective in treating mental illnesses, (such as depression and schizophrenia), are known to act by altering receptor function or by affecting

NATO ASI Series, Vol. H 32
Molecular Biology of Neuroreceptors and Ion Channels. Edited by A. Maelicke

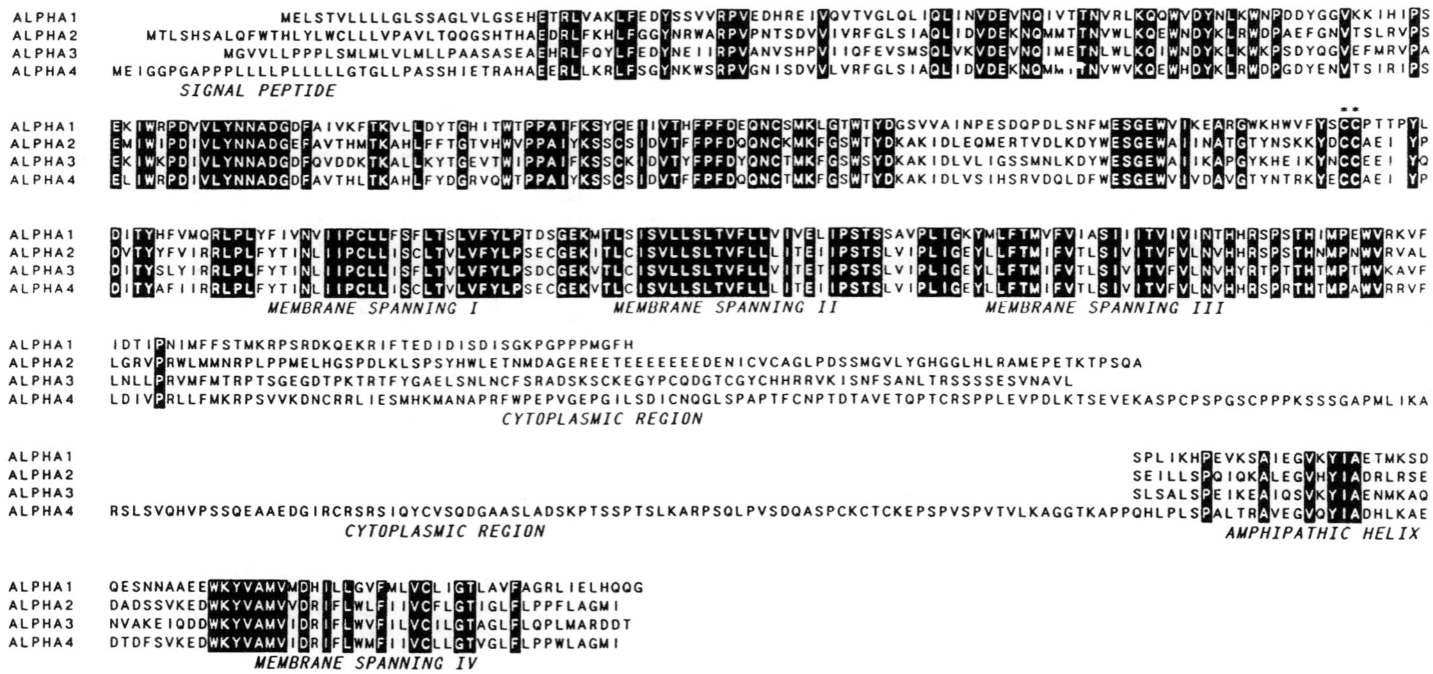

Fig. 1: Alignment of the amino acid sequences of mouse muscle α-subunit (alpha1) and rat neuronal α-subunits (alpha2, alpha3 and alpha4). Amino acids conserved in all four α-subunits are shown on a black background. The two cysteine residues that are thought to be close to the acetylcholine binding site are indicated by asterisks. Signal peptide, putative membrane-spanning and cytoplasmic regions and the proposed amphipatic helix are indicated below the aligned sequences. The mature alpha2 protein has 49, 57 and 67% amino acid sequence identity with the mature alpha1, alpha3 and alpha4 proteins, respectively. The percentages of sequence identity were calculated by dividing the number of identical residues by the number of residues in the shorter of the two compared sequences.

NEURONAL NICOTINIC ACETYLCHOLINE RECEPTOR SUBUNITS

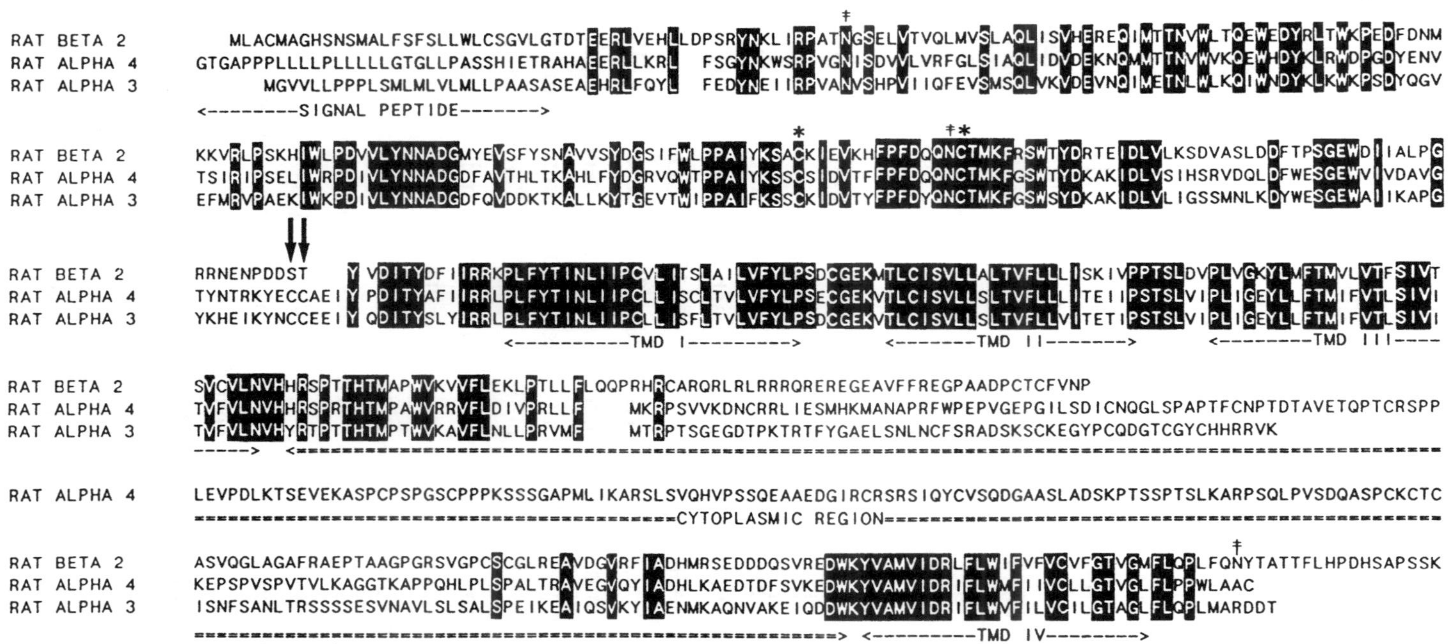

Fig. 2: Amino-acid of the beta2 subunit with the rat neuronal alpha subunits. Aligned with the beta2 subunit are the neuronal alpha4-1 and alpha3 subunits. Dark background highlights sequence identity among each of the subunits. Double daggers mark potential N-linked glycosylation sites, asterisks mark cysteine residues conserved in each member of the neurotransmitter-gated ion-channel subunit superfamily, arrows mark conserved cysteine residues in the nicotinic alpha subunits that are different in the beta2 subunit. Putative trans-membrane domains, (TMD I-IV), predicted using the algorithm of Kyte and Doolittle (1982) and a cytoplasmic domain are identified below the aligned sequences.

transmitter metabolism or transmitter uptake. Thus, understanding the receptor systems is crucial for the understanding of basic brain function as well as the rational treatment of disease.

Because of the complexity of the brain and the low amounts of receptor and ion channel protein it has been difficult to study these molecules by conventional biochemical approaches. Thus,for many years progress in studying these molecules,so important for brain function, has been slow. In addition pharmacological and, more recently, molecular genetic studies have demonstrated the general rule that the nervous system makes use of large families of closely related receptor and channel subtypes which are difficult to resolve by conventional biochemical procedures. Thus, a number of laboratories have taken another approach in recent years and used molecular genetics to identify receptor and channel genes. The genes or corresponding cDNAs are then sequenced to determine the primary structures of the proteins.

The subject of this paper is the family of nicotinic acetylcholine receptors that we have discovered in the past few years through the use of molecular genetics, Boulter et al, 1985,1986ab, Goldman et al. 1986,1987, Boulter et al,1987, Deneris et al, 1988, Wada et al, 1988a. Recently the GABA and glycine receptors have also been cloned and shown to be structurally related to the nicotinic receptor, Grenningloh et al,1987, Schofield et al, 1987, Levitan et al, 1988. Thus, the major inhibitory receptors in the nervous system are similar in structure to the nicotinic receptor, an excitatory receptor. This was an unexpected finding and raises the exciting possibility that all ligand gated channels are members of one super family of related proteins. Thus, it is reasonable to expect that knowledge about one receptor will be useful in understanding the role and function of other ligand gated channel receptors. It is also plausible to suggest that the glutamate receptors are related to the nicotinic receptor. If this is correct, this will have important implications since the NMDA class of glutamate receptor has been shown to play a role in the phenomenon of Long-Term-Potentiation which may underlie learning and memory. The NMDA receptors may also be involved in the brain damage that occurs during ischaemia and epilepsy.

We first cloned the genes coding for the nicotinic acetylcholine receptor expressed in skeletal muscle which is the target of the autoimmune response in patients with myasthenia gravis, Ballivet et. al, 1981,1982, Claudio et al 1983, Boulter et al 1985,1986a. We then used these muscle genes as probes and discovered a family of genes coding for nicotinic acetylcholine receptors expressed in the brain, Boulter et al,1986b,Goldman et al. 1986,1987, Boulter et al,1987, Deneris et al, 1988, Wada et al, 1988. It is likely that some, if not all, of the members of the receptor gene family that we have discovered code for the nicotine binding sites that have been mapped in the brain, Clarke et al, 1985, for review see Martin 1986. These receptors are probably responsible for the many diverse pharmacological effects that nicotine has on the central nervous system.

The best understood synapse is the neuromuscular junction because of its accessibility to biochemical and electrophysiological techniques. Thus, the neuromuscular junction serves as a model for synapses in the brain ,(for recent reviews see Maelicke 1984; Popot and Changeux 1984; Stroud and Finer-Moore 1985; Heinemann et al, 1986, Karlin et al 1986;). Developmental studies have demonstrated that the acetylcholine receptor found in skeletal muscle is regulated in an interesting way which may involve mechanisms unique to the nervous system. When synaptic activity is blocked either by surgical denervation or by a pharmacological block of the acetylcholine receptor, the synthesis of receptor is dramatically increased (Miledi 1960 ab; Cohen and Fischbach 1973; Brockes and Hall 1975b). Newly made receptors are incorporated over the whole extent of the muscle membrane and are responsible for making denervated muscle supersensitive to acetylcholine. This demonstrates that the level of electrical activity in the muscle can regulate the synthesis of macromolecules important for synaptic function. We have extended this work and have shown that the level of receptor mRNA is regulated by electrical activity, Goldman et al, 1988. Our working model is that the activity of the promoters of the receptor subunits are themselves regulated by electrical activity. In order to test this idea we have isolated a genomic fragment that contains at least part of the promoter of the gamma subunit, Gardner et al, 1987. Studying how these mechanisms work in muscle should provide insight into how similar mechanisms might work in the brain where it is known that electrical activity is important for the normal development of specific nerve networks,Hubel et al.,1977.

The isolation of cDNA clones coding for the muscle nicotinic receptor has made it possible to extend these studies in an important new direction. Both the peripheral and central nervous systems are known to contain nicotinic acetylcholine receptors at synapses between nerve cells. These receptors have been identified by functional assays and binding studies using cholinergic ligands such as acetylcholine and nicotine, (for example see Clarke et al 1985). Neuronal receptors have been difficult to study because of their low abundance and lack of high affinity probes to aid in their purification. Indirect studies indicate that neuronal nicotinic receptors can be either presynaptic or postsynaptic, Briggs and Cooper 1982; Brown et al.,1984; Clarke et.al,1986. We have used the cDNA clones coding for the muscle nicotinic receptor to screen rat and mouse cDNA and genomic libraries for related mRNAs or genes. The assumption behind this approach is that the neural nicotinic receptors are evolutionarily related to the muscle receptors and that this is reflected at the genetic level by a significant level of nucleotide sequence homology. This method has resulted in the isolation of cDNA and genomic clones coding for peptides which have about 50% amino acid sequence homology with the muscle alpha subunit of the nicotinic receptor. At the present time we have identified seven genes that are expressed in the brain and code for proteins with homology to the nicotinic receptor subunits. Four subunits have been identified that have structural features found only in the muscle ligand binding subunit. We have named these putative ligand binding neuronal subunits; alpha2, alpha3, alpha4 and alpha5, figure 1. We have identified three additional genes that code for proteins that are expressed in the brain and that have structural features that suggest that they are nicotinic receptor subunits. We have called these genes beta2, beta3 and beta4, (in this terminology the alpha1 and beta1 subunits are expressed in skeletal muscle). We have named these subunits beta subunits because they are missing the adjacent cysteines analogous to the cysteines 192,193 found in the alpha subunits of the Torpedo fish nicotinic receptor and all other alpha subunits, see figures 1 and 2. Furthermore, the beta2 subunit can functionally substitute for the muscle beta1 subunit to form a functional muscle-type nicotinic receptor, Boulter et al, 1986b, Goldman et al. 1986,1987, Boulter et al,1987, Deneris et al, 1988, Wada et al, 1988a.

The primary structures of the brain nicotinic receptor subtypes have been deduced from the sequences of the cDNA clones. Analysis of the hydrophobicity profiles of the brain nicotinic receptor subunits suggests to us that they fold through the membrane in an identical manner to the *Torpedo* fish receptor. Each subunit contains four stretches of about 25 amino acids that are very hydrophobic and we have proposed that they are membrane spanning regions. Thus, we have proposed that each subunit spans the membrane four times placing the N-terminal and C-terminal outside the cell, Claudio et al 1983. This model places about half the protein bulk outside the cell, consistent with structural studies on the *Torpedo* receptor, Stroud, R.M. and Finer-Moore, J. 1985, Karlin, et al 1986. This model also predicts that a number of potential phosphorylation sites are cytoplasmic and in one case phosphorylation at one of these sites is known to regulate receptor function, Huganir et al 1986, Safran et al 1986. The brain alpha subunits all contain adjacent cysteines at homologous positions to the Torpedo alpha subunit cysteines 192 and 193 which are known to be near the acetylcholine binding site, Karlin, et al 1986, for review see Heinemann et al 1986.

The availability of cDNA clones coding for the neuronal nicotinic receptors has made it possible to study their function. Expression studies in *Xenopus* oocytes have shown that the following combinations of subunits form functional nicotinic receptors: alpha2/beta2; alpha3/beta2 and alpha4/beta2, Boulter et al,1987., Wada et al 1988a. Thus, beta2 is a promiscuous subunit which can combine with three different alpha subunits to form a functional receptor. This is consistent with its wide distribution of expression in the brain. Each of the three functional combinations of subunits forms a pharmacologically distinct subtype which is activated by acetylcholine and nicotine and is resistant to alpha-bungarotoxin. Alpha3/beta2 and alpha4/beta2 are blocked by *Bungarus* toxin 3.1 consistent with a ganglionic nicotinic type pharmacology, figure 3. However the alpha2/beta2 receptor is resistant to *Bungarus* toxin 3.1 and is a new

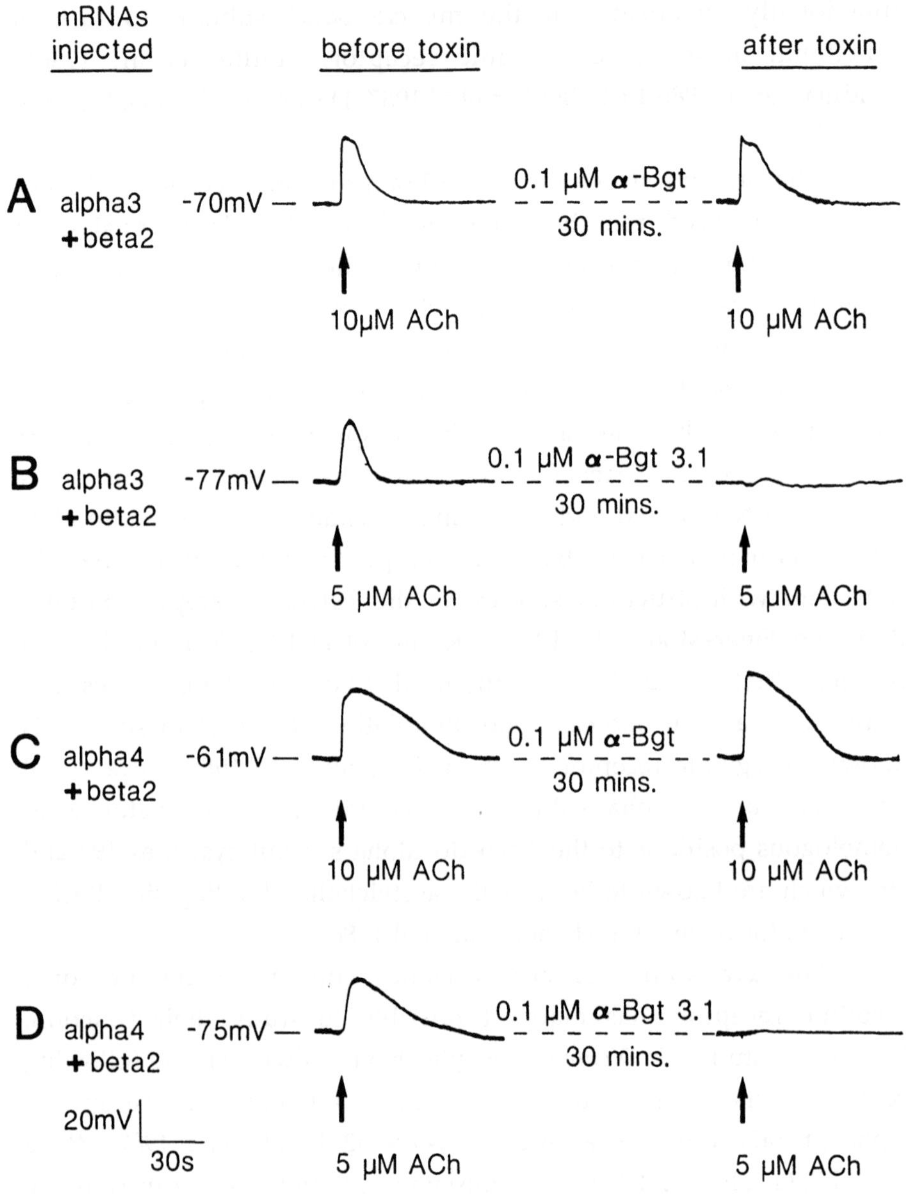

Figure 3. This figure shows the effect of two different neurotoxins on the activation by acetylcholine of two neuronal nicotinic acetylcholine receptor subtypes. The voltage tracing on the left shows the response before application of the toxin and the voltage tracing on the right shows the response following a brief washing and a 30 minute incubation in the indicated concentrations of the two toxins.

receptor subtype with a pharmacology that has not been observed previously, Boulter et al,1987, Wada et al 1988a.

In order to determine where the nicotinic receptors are expressed in the brain we utilized the method of in situ hybridization to visualize the distribution of expression of the mRNA coding for each nicotinic receptor subtype. These experiments demonstrated that each of the three alpha subunit mRNAs has a unique distribution of expression in the brain, consistent with the proposal that they are part of three independent nicotinic receptor systems, figure 4. The beta2 transcript is widely distributed throughout the brain consistent with the hypothesis that it is a common subunit used to form at least three different receptor subtypes, figure 5. In general the distribution of mRNA coding for the nicotinic receptor family parallels the map of nicotine binding, Boulter et al,1986b,Goldman et al. 1986,1987, Boulter et al,1987, Deneris et al, 1988, Wada et al, 1988ab.

The discovery of a family of genes coding for nicotinic acetylcholine receptors expressed in the brain brings for the first time the full power of molecular biology to the study of this important receptor system. The existence of multiple subtypes and the fact that they are expressed throughout the brain suggests that the nicotinic receptor is a major excitatory system.The anatomical distribution of these receptors can now be studied using specific antibodies to localize the receptor subtypes in the brain. In situ hybridization methods can be used to localize the cell bodies that synthesize the receptor protein. The regulation of receptor gene expression can be studied under various experimental conditions. Understanding the promoters of these receptor genes will provide insight into how the brain builds, regulates and maintains specific neural networks. At present there is little reliable pharmacological data on the properties of the nicotinic acetylcholine receptor in the brain. The receptors can now be expressed in the oocyte system where rigorous pharmacology can be performed without the complications that are inherent in pharmacological studies in the brain. By engineering mammalian cell lines it will be possible to study these receptors in a number of convenient cell systems to ask specific questions about their function. The availability of the primary structure of the brain and muscle receptors makes it possible to build models that relate structure to pharmacology and

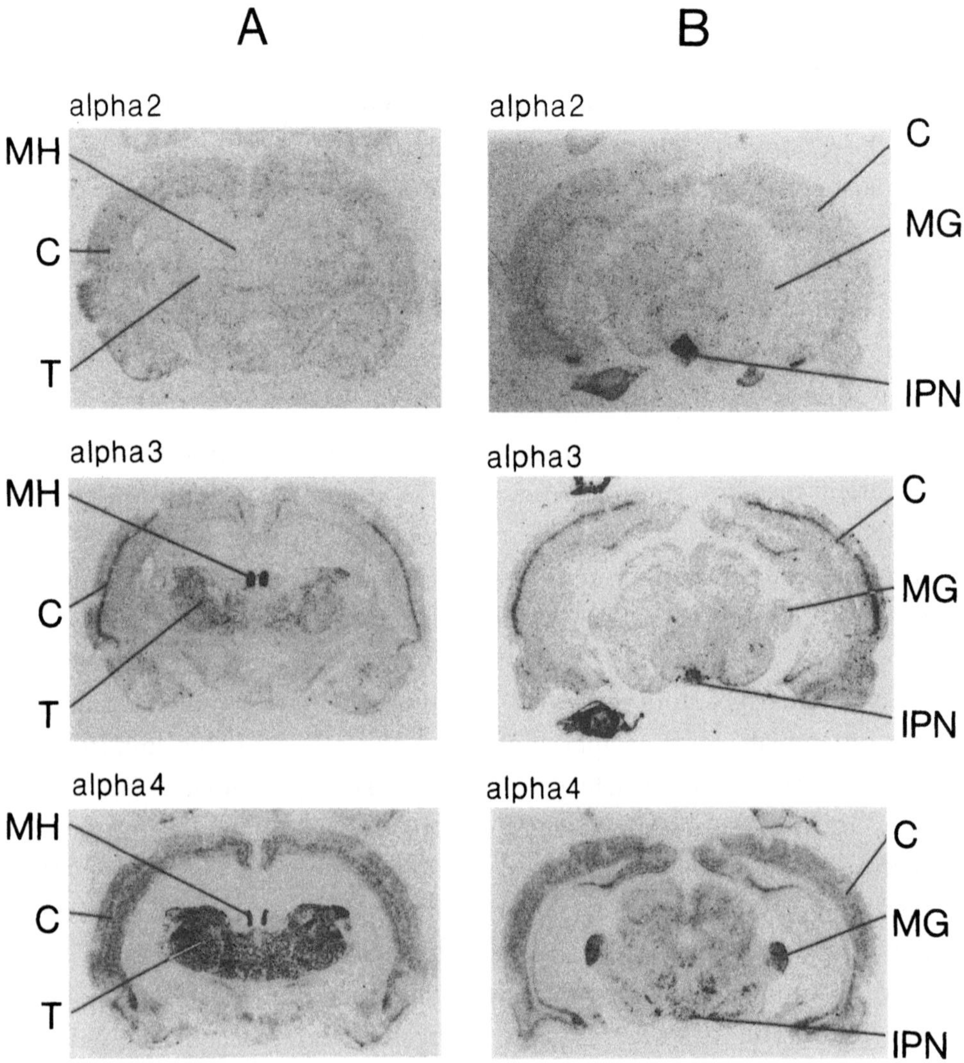
A
B
alpha2
MH
C
T
alpha2
C
MG
IPN
alpha3
MH
C
T
alpha3
C
MG
IPN
alpha4
MH
C
T
alpha4
C
MG
IPN

Figure 4. Comparison of the distribution of alpha2, alpha3 and alpha4 transcripts by *in situ* hybridization histochemistry. Serial coronal sections through the medial habenula (A) and the interpeduncular nucleus (B) were hybridized with the probes for alpha2, alpha3 and alpha4. In (B), slides contain sections of the trigeminal ganglion. Abbreviations: C, cortex; IPN, interpeduncular nucleus; MH, medial habenula; MG, medial geniculate nucleus; T, thalamus. Tissue preparation and hybridization were performed with minor modifications. Briefly, paraformaldehyde-fixed rat brain sections(25μm) were mounted on poly-L-lysine-coated slides, digested with proteinase K(10μg/ml, 37°C, 30 minutes), acetylated, and dehydrated. Hybridization with ^{35}S-radiolabelled RNA probe (5-10 x 10^6cpm/ml) was performed at 55°C for 12-18 hrs in a solution containing 50% formamide, 0.3M NaCl, 10mM Tris(pH 8.0), 1mM EDTA, 0.05% tRNA, 10mM DTT, 1x Denhardt's solution and 10% dextran sulfate. Because of the high sequence similarities in the protein coding regions of the cDNAs, 3' untranslated sequences were used to make probes. The *Eco*RI/3' end, *Bal*I/3' end and *Bgl*I/3' end fragments derived from C183, PCA48 and α4-2 cDNA clones, respectively, were subcloned into the plasmid, pSP65 and used to synthesize antisense RNA probes *in vitro*. After hybridization, sections were treated with RNaseA (20μg/ml, 37°C, 30minutes) and washed in 0.1 x SSC at 55°C. Dehydrated slides were exposed to X-ray films for 3-16 days at 4°C. A RNA probe coding the sense strand of C183 clone was used as a control.

function. These models can be tested by site specific mutagenesis. At present the only ligand gated channel that can be isolated in milligram amounts is the *Torpedo* nicotinic receptor. Thus, it is the only candidate for structural studies leading to a complete high resolution three dimensional structure. Because of the structural homologies between the ligand gated channels it should be possible to apply the results of the structural studies of the *Torpedo* nicotinic receptor to the muscle and brain nicotinic receptors and even to the glycine and GABA receptors.

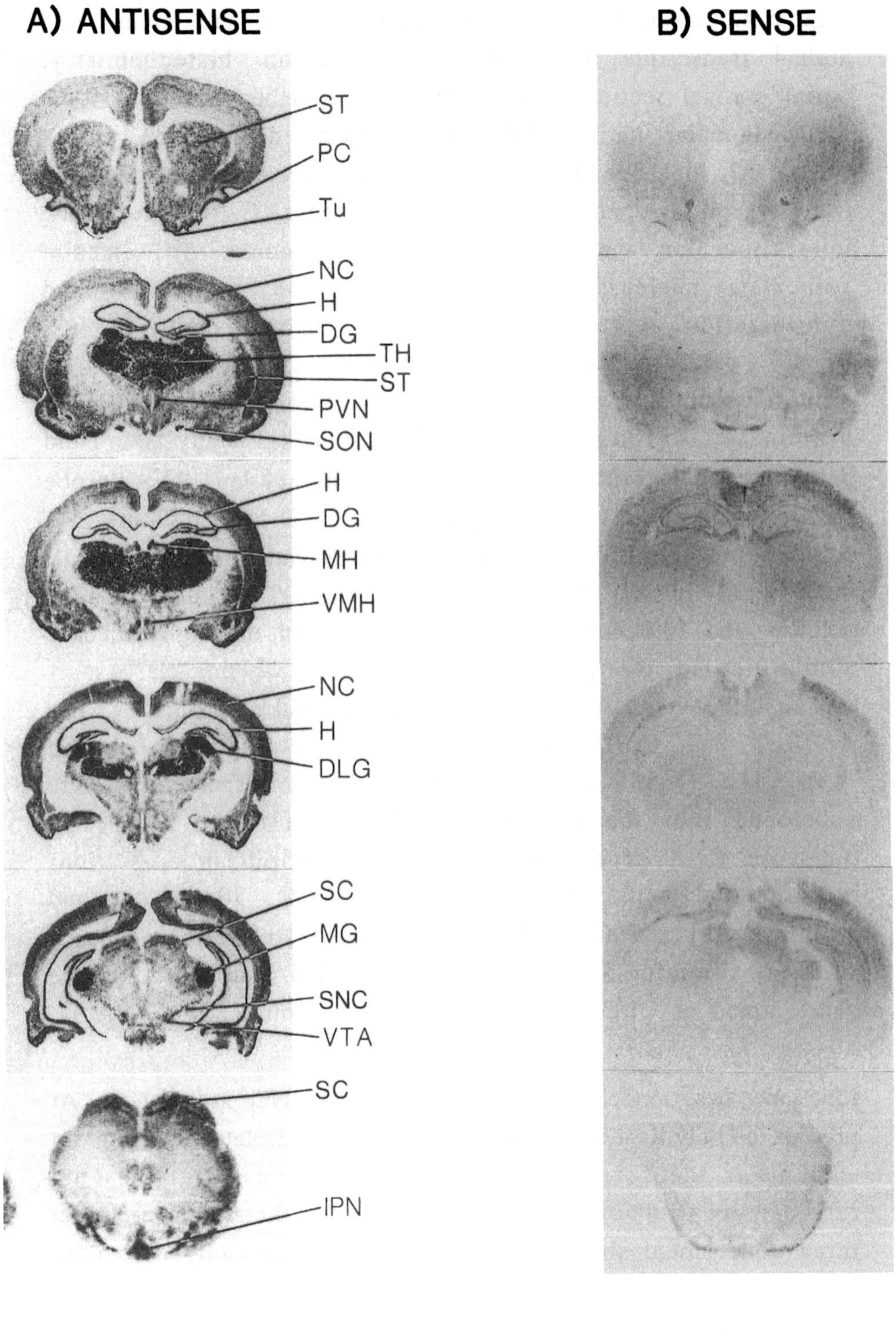
A) ANTISENSE
B) SENSE
ST
PC
Tu
NC
H
DG
TH
ST
PVN
SON
H
DG
MH
VMH
NC
H
DLG
SC
MG
SNC
VTA
SC
IPN

Figure 5. *In situ* hybridization analysis. Rat forebrain and midbrain sections were probed with [^{35}S]-radiolabeled antisense (A) or sense (B) beta2 RNA transcribed *in vitro* using a plasmid into which a 571bp Pstl/EcoR1 fragment of PCX49 was subcloned. Abbreviations are: DLG, lateral geniculate nucleus (dorsal part); DG, dentate gyrus; H, Ammon's horn (hippocampus); IPN, interpeduncular nucleus; MG, medial geniculate nucleus; MH, medial habenular nucleus; NC, neocortex; PC, piriform cortex; PVN, paraventricular hypothalamic nucleus; SON, supraoptic hypothalamic nucleus; SNC, substantia nigra, pars compacta; SC, superior colliculus; ST, striatum; TH, thalamus; Tu, olfactory tubercle; VTA, ventral tegmental area; VMH, ventromedial hypothalamic nucleus.

Interesting mutations can be introduced into the receptors and expressed in trans-genic mice and the mutant mice studied in order to gain insight into the function of the nicotinic receptors.The diversity of structure and function in this receptor system can now be studied with the long range goal of understanding the functional roles that the nicotinic receptors play in the nervous system.

At present little is known about the role of the nicotinic receptor system in the brain. However, nicotine is one of the most widely consumed and addictive drugs. Many behavioral effects have been observed which include mood changes, effects on learning and memory and weight loss, Wellman et al,1986, for review see Clarke et al 1988. Under some conditions nicotine is a relaxant and under other conditions nicotine is a stimulant. Many of these effects may be beneficial and this raises the possibility of designing better and more specific nicotinic drugs, see Luyten and Heinemann, 1987.

The nicotinic receptor system has been implicated in a number of serious health problems. Nicotine is a widely consumed and addictive drug and is a major factor in the present smoking epidemic . Behavioral studies have linked the nicotinic receptor system to learning and memory which is intriguing given the finding that Alzheimer's patients have a deficiency in memory and cortical nicotinic receptors, Perry et al, 1987. The existence of receptor subtypes and the availability of cDNA clones may make it possible to design new drugs that are subtype specific and which will prove useful in the battle against the smoking and Alzheimer's epidemics. The known ability of nicotine to effect mood suggests that new nicotinic drugs may also prove useful in the treatment of various mood disorders.

Acknowledgements

The work reviewed in this paper was supported by the National Institutes of Health, USA, National Institute of Neurological and Communicative Disorders and Stroke, (Steve Heinemann and Jim Patrick)., and the Muscular Dystrophy Association of America, (Steve Heinemann and Jim Patrick).

References

Ballivet, M., Patrick, J., Lee, J. and Heinemann, S. (1981) Progress in Cloning of the Acetylcholine Receptor cDNAs and Genes. Cold Spring Harbor Symposium. Molecular and Cellular Control of Muscle Development. Cold Spring Harbor Press.

Ballivet, M., Patrick, J., Lee, J. and Heinemann, S. (1982) Molecular cloning of cDNA coding for the gamma subunit of the *Torpedo* acetylcholine receptor. *Proc. Nat. Acad. Sci.* **79**:4466-4470.

Boulter, J., Luyten, W., Evans, K., Mason, P., Ballivet, M., Goldman, D., Stengelin, S., Martin, G., Heinemann, S. and Patrick, J. (1985) Isolation of a clone coding for the alpha subunit of a mouse acetylcholine receptor. *J. Neuroscience.* **5**:2545-2552.

Boulter J., Goldman, D., Evans, K., Martin, G., Mason, P., Stengelin, S., Heinemann, S. and Patrick, J. (1986a) Isolation and sequence of cDNA clones coding for the precursor to the gamma-subunit of mouse muscle acetylcholine receptor. *J. Neurosci. Res.* **16**:37-49.

Boulter, J., Evans, K., Goldman, D., Martin, G., Treco, D., Heinemann, S. and Patrick J. (1986b) Isolation of a cDNA clone coding for a possible neural nicotinic acetylcholine receptor alpha subunit. *Nature* **319**:368-374.

Boulter, J., Connolly, J., Deneris, E., Goldman, D., Heinemann, S. and Patrick, J. (1987) Functional expression of two neuronal nicotinic acetylcholine receptors from cDNA clones identifies a gene family. *Proc. Natl. Acad. Sci.*, **84**:7763-7767.

Briggs, C.A. and Cooper, J.R. (1982) Cholinergic modulation of the release of [^{3}H]acetylcholine from synaptosomes of myenteric plexus. *J. Neurochem.* **38**:501-508.

Brockes, J.P. and Hall, Z.W. (1975) Acetylcholine receptors in normal and denervated rat diaphragm muscle. II. Comparison of junctional and extrajunctional receptors. *Biochemistry* **14**:2100-2106.

Brown, D.A., Docherty, R.J. and Halliwell, J.V. (1984) The action of cholinomimetic substances on impulse conduction in the habenulointerpeduncular pathway of the rat *in vitro*. *J. Physiol.* **353**:101-109.

Cajal, S. R. (1911), Histologie du Systeme Nerveux de l'Homme et des Vertebres, Vol II (Paris: Moloine), pp 993

Clarke, P.B.S., Hamill, G.S., Nadi, N.S., Jacobowitz, D.M., and Pert, A. (1986) ^{3}H-nicotine- and ^{125}I-alpha-bungarotoxin-labeled nicotinic receptors in the interpeduncular nucleus of rats. II. Effects of habenular deafferentation. *J. Comparative Neurol.* **251**:407-413.

Clarke, P.B.S., Schwartz, R.D., Paul S.M., Pert, C.B. and Pert A. (1985) Nicotinic binding in rat brain: autoradiographic comparison of [^{3}H] acetylcholine, [^{3}H] alicotine, and [^{125}I]-alpha-bungarotoxin. *J. Neurosci.* **5**:1307-1315.

Clarke, P.B.S. (1988) The central pharmacology of nicotine: electrophysiological approaches. In: Actions and Medical Implications. (eds. I.P. Stolerman, S. Wonnacott and M.A.H. Russell). Oxford University Press. pp 2-55.

Claudio, T., Ballivet, M., Patrick, J. and Heinemann, S. (1983) Nucleotide and deduced amino acid sequences of *Torpedo* californica acetylcholine receptor gamma subunit. *Proc. Nat. Acad. Sci.* **80**:1111- 1115.

Cohen, S.A. and Fischbach, G.D. (1973) Regulation of muscle acetylcholine sensitivity by muscle activity in cell culture. *Science* **181**:76-78

Deneris, E.S., Connolly, J., Boulter, J., Wada, E., Wada, K., Swanson, L., Patrick, J. and Heinemann, S. (1988) Primary structure and expression of beta2: a novel subunit of neuronal nicotinic acetylcholine receptors. *Neuron* **1**:45-54.

Gardner, P., Heinemann, S., and Patrick, J. (1987) Transcriptional regulation of nicotinic acetylcholine receptor genes: identification of control elements of a γ-subunit gene. *Molecular Brain Research* **3**:69-76.

Goldman, D., Simmons, D., Swanson, L.W., Patrick, J. and Heinemann S. (1986) Mapping brain areas expressing RNA homologous to two different acetylcholine receptor alpha-subunit cDNAs. *Proc. Natl. Acad. Sci.* **83**:4076-4080.

Goldman, D., Deneris, E., Kochhar, A., Patrick, J. and Heinemann, S. (1987) Members of a nicotinic acetylcholine receptor gene family are expressed in different regions of the mammalian central nervous system. *Cell* **48**:965-973.

Goldman, D., Brenner, H.R., and Heinemann, S., Acetylcholine Receptor Alpha, Beta, Gamma, and Delta-Subunit mRNA Levels Are Regulated by Muscle Activity. *Neuron* , 1 329-333, (1988).

Grenningloh, G., Rienitz, A., Schmitt, B., Methfessel, C., Zensen, M., Beyreuther, K., Gundelfinger, E.D., and Betz, H. (1987). The strychnine-binding subunit of the glycine receptor shows homology with nicotinic acetylcholine receptors. *Nature* **328**:215-220.

Heinemann, S. Asouline, G., Ballivet, M., Boulter, J., Connolly, J., Deneris, E., Evans, K., Evans, S., Forrest, J., Gardner, P., Goldman, D., Kochhar, A., Luyten, W., Mason, P., Treco, D., Wada, K. and Patrick, J. (1986) Molecular biology of the neural and muscle acetylcholine receptors. In: Molecular Neurobiology (J. Patrick and S. Heinemann eds.) Plenum Press, New York, NY.

Hebb, D. O., (1949). The Organization of Behavior(New York: John Wiley and Sons, Inc.)

Hopfield, J.J. (1982) Neural networks and physical systems with emergent collective computational abilities. *Proc. Natl. Acad. Sci.* USA **79**:2554-2558.

Hubel, D.H., Wiesel, T.N., and LeVay, S. (1977) Plasticity of ocular dominance columns in monkey striate cortex. *Philos. Trans. R. Soc. Lond., Biol. Sci.* **278**: 377-409.

Huganir,R.L., Delcour,A.H.,Greengard P., and Hess,G.P., (1986), Phosphorylation of the nicotinic acetylcholine receptor regulates its rate of desensitization. Nature 321 , 774-776

Karlin, A., DiPaola, M., Kao, P.N. and Lobel, P. (1986) Functional sites and transient states of the nicotinic acetylcholine receptor. In: Proteins of Excitable Membrane (B. Hille and D.M Fambrough, eds.) John Wiley Inc.

Levitan,E.S.; Schofield,P.R.; Burt,D.R.; Rhee,L.M.; Wisden,W.; Kohler,M.; Fujita,N; Rodriguez,H.F.; Stehenson A.; Darlison,M.G.; Barnard,E.A.; Seeburg,P.H., (1988) Structural and functional basis for $GABA_A$ receptor heterogeneity. Nature 335 76-79

Luyten, H.W.M.L. and Heinemann, S.F. (1987) Molecular cloning of the nicotinic acetylcholine receptor: new opportunities in drug design? *Reports Medicinal Chem.* **22**:281-291.

Maelicke, A. (1984) Biochemical aspects of cholinergic excitation. *Agew. Chem. Int. Ed. Engl.* **23**:195-221.

Martin, B.R. (1986) Nicotine receptors in the central nervous system. *The Receptors* Vol. III. Academic Press, Orlando pp. 393-415.

Maelicke, A. (1984) Biochemical aspects of cholinergic excitation. *Agew. Chem. Int. Ed. Engl.* **23**:195-221.

Miledi, R. (1960a) The acetylcholine sensitivity of frog muscle fibers after complete or partial denervation. *J. Physiol.* **151**:1-23.

Miledi, R. (1960b) Junctional and extrajunctional receptors in skeletal muscle fibers. *J. Physiol.* **151**:24-30.

Patrick, J. and Lindstrom, J. (1973) Autoimmune response to acetylcholine receptor. *Science* **180**:871-872.

Perry, E.K., Perry, R.H., Smith, C.J., Dick, D.J., Candy, J.M., Edwardson, J.A., Fairbairn, A. and Blessed, G. (1987) Nicotinic receptor abnormalities in Alzheimer's and Parkinson's diseases. *J. Neurology, Neurosurgery, and Psychiatry* **50**:806-809.

Popot, J.L. and Changeux, J.-P. (1984) The nicotinic receptor of acetylcholine: structure of an oligomeric integral membrane protein. *Physiol. Rev.* **64**:1162-1239.

Safran, A., Neumann,D., and Fuchs,S., Analysis of acetylcholine receptor phosphorylation sites using antibodies to synthetic peptides and monoclonal antibodies. EMBO J., 5, 3175-3178

Schofield, P.R., Darlison, M.G., Fujita, N., Burt, D.R., Stephenson, F.A., Rodriguez,H., Rhee, L.M., Ramachandran, J., Reale, V., Glencorse, T.A., Seeburg, P.H., and Barnard, E.A. (1987) Sequence and functional expression of the $GABA_A$ receptor shows a ligand-gated receptor super-family. *Nature* **328**:221-227.

Stroud, R.M. and Finer-Moore, J. (1985) Acetylcholine receptor structure, function and evolution. *Ann. Rev. Cell Biol.* **1**:369-401.

Wada, W., Ballivet, M., Boulter, J., Connolly, J., Wada, E., Deneris, E.S., Swanson, L.W., Heinemann, S. and Patrick, J. (1988a) Functional expression of a new pharmacological subtype of brain nicotinic acetylcholine receptor. *Science*, 240 330-334

Wada,E.; Wada, K.; Boulter,J.; Deneris,E.; Heinemann,S.; Patrick, J.; Swanson, L.W. The Distribution of alpha2, alpha3, alpha4 and beta2 Neuronal Nicotinic Receptor Subunit mRNAs in the Central Nervous System. A Hybridization Histochemical Study in the Rat. (1988b) Journal of Comparitive Neurology *in press* .

Wellman, P.J., Marmon, M.M., Reich, S. and Ruddle, J. (1986) Effects of nicotine on body weight, food intake and brown adipose tissue thermogenesis. *Pharm. Biochem. and Behav.* **24**:1605-1609.

THE GENE FAMILY ENCODING NEURONAL ACETYLCHOLINE RECEPTORS: STRUCTURE OF THE AVIAN ALPHA2 AND ALPHA3 GENES.

Sabine Couturier, Patrick Nef, Carole Oneyser, and Marc Ballivet
Department of Biochemistry, University of Geneva, Quai Ansermet 30,
CH 1211 Geneva 4, Switzerland

INTRODUCTION

The neurotransmitter-gated ion channels whose primary structures have been established (i.e. the nicotinic acetylcholine receptor, and the GABAa and glycine receptors) each assemble from several transmembrane subunit proteins that are closely related in sequence. Moreover, there is definite, albeit weaker, homology between the subunits of the different receptors, all of which present conserved motifs and a characteristic pattern of aliphatic stretches thought to span the postsynaptic membrane. Thus, it appears that most, and perhaps all, neurotransmitter-gated ion channels have evolved from a single ancestral gene, as if the making of a transmembrane structure allowing gated ion permeation had been a bottleneck in the evolution of ionic signalling.

In addition, neurotransmitter-gated ion channels of a given specificity appear likely to be of several types. Thus, the muscle endplate nicotinic acetylcholine receptor (nAChR) of vertebrates has recently been shown to have many sibs, all of them expressed in central or in peripheral neural tissue. At least six distinct genes encoding neuronal nAChR subunits have been isolated (in birds, Nef, 1986; Nef et al., 1988; in rodents, Boulter et al., 1986; Goldman et al., 1987; Wada et al., 1988; Deneris et al., 1988 and unpublished results from the Salk Institute and Geneva laboratories), in addition to the five genes needed to encode the mature and embryonic forms of the mammalian endplate receptor (Takai et al., 1985; Witzemann et al., 1987). These neuronal subunits belong to two classes: the alpha-like subunits (alpha2, -3, -4 and -5) possessing a pair of vicinal cysteines at positions 192,193 (endplate alpha subunit numbering) of their extracellular domain, and therefore likely to be ACh binding subunits (Kao and Karlin, 1986), and the non-alpha subunits (non-alpha and non-alpha3 in our terminology) that have no extracellular vicinal cysteines and are thus expected to participate in receptor assembly and function but not to bind the ligand.

It is now a major goal in several laboratories to establish the subunit composition, stoichiometry and properties of these neuronal receptor types, to determine their sites of synthesis in the adult and in the developing nervous system and to characterise them physiologically either in vivo or in reconstitution experiments. The current working hypothesis is that in vertebrates a given neuronal alpha subunit defines a neuronal nAChR type and assembles with a particular non-alpha subunit in a stoichiometry of two neuronal

NATO ASI Series, Vol. H 32
Molecular Biology of Neuroreceptors and
Ion Channels. Edited by A. Maelicke

alphas and three neuronal non-alphas. Hence, a particular neuronal AChR type would be a pentameric assembly of only two different subunits, a radical departure in composition, but not in stoichiometry, from the endplate nAChR. Direct experimental evidence has recently been provided for this model by the purification from chick and rat brain of a receptor consisting of the alpha4 and non-alpha subunits and the demonstration that alpha4 could be specifically labelled by nicotinic affinity reagents (Whiting et al., 1987; Whiting and Lindstrom, 1987).

How many neuronal nicotinic receptors can we expect to discover? It is possible that the six neuronal alpha subunits identified so far may be found to assemble in vivo with any of the two identified non-alpha subunits and perhaps also with yet undiscovered non-alphas. The repertoire of possible neuronal receptor types could therefore be quite large and present major difficulties of developmental and functional analysis. Reconstitution experiments (Boulter et al., 1987; Wada et al., 1988) in the Xenopus oocyte system have indeed demonstrated that functional, ACh-operated cation conductances result from the assembly of rat alpha2, -3 and -4 subunits with beta2 (the rat subunit corresponding to avian non-alpha). The avian alpha4:non-alpha receptor has been characterized physiologically in voltage clamp and patch clamp studies of oocytes injected with the appropriate cDNAs (Ballivet et al., 1988), but the physiology of the other neuronal receptors has yet to be established. There is mounting evidence, however, that many of the neuronal nAChR subunit genes are closely linked genetically (Jim Boulter and S.C., unpublished). If neuronal alpha and non-alpha subunit genes were generally linked pairwise in the same relationship found for the delta and gamma subunit genes of endplate receptor (Nef et al., 1984), it could then be argued that a given pair of such genes probably encodes a particular receptor type because the transcription of linked genes is easily co-regulated and because linked genes must nearly always reside in a segment of chromatin having an homogeneous transcriptional activity. Thus, linkage would ensure that a given receptor type be specifically expressed at certain developmental stages or in certain neuronal cell-types.

In this communication, we present the protein sequences of two avian neuronal nAChR subunits, alpha2 and alpha3, as derived from their respective genes (Nef et al., 1988) and compare them to the corresponding subunits from the rat (Wada et al., 1988; Boulter et al., 1986). In addition, we describe and discuss differences found between the known avian neuronal subunits, some of which may be of functional significance.

PRIMARY STRUCTURE OF THE ALPHA2 SUBUNIT.

When faced with sets of homologous genes from two species, it is not always easy to determine which genes encode functionally equivalent subunits. In the case of the neuronal cholinoceptors, however, the task is rather simple because the extracellular domain of a given subunit is essentially invariant and the cytoplasmic domain, although quite variable between species, retains patches of homologies not found in any other subunit. As seen in figure 1 below, the chicken and rat alpha2 sequences can easily be aligned to maximise identity. Moreover, no pairwise alignement of any other combination of subunits yields nearly as good an identity score.

```
               .         .         .         .         .         .         .         .
c a2  REQKQPHGFAEDRLFKHLFTGYNRWSRPVPNTSDVVIVKFGLSIAQLIDVDEKNQMMTTNVWLKQEWSDYKLRWNPEDFD
r a2    QQGSHTHAEDRLFKHLFGGYNRWARPVPNTSDVVIVRFGLSIAQLIDVDEKNQMMTTNVWLKQEWNDYKLRWDPAEFG
         *** **          *     *            *                            *      * ** *

c a2  NVTSIRVPSEMIWIPDIVLYNNADGEFAVTHMTKAHLFSNGKVKWVPPAIYKSSCSIDVTYFPFDQQNCKMKFGSWTYDK
r a2  NVTSLRVPSEMIWIPDIVLYNNADGEFAVTHMTKAHLFFTGTVHWVPPAIYKSSCSIDVTFFPFDQQNCKMKFGSWTYDK
          *                                 ** * *                *

c a2  AKIDLENMEHHVDLKDYWESGEWAIINAIGRYNSKKYDCCTEIYPDITFYFVIRRLPLFYTINLIIPCLLISCLTVLVFY
r a2  AKIDLEQMERTVDLKDYWESGEWAIINATGTYNSKKYDCCAEIYPDVTYYFVIRRLPLFYTINLIIPCLLISCLTVLVFY
            *  **                 * *         *     * *        -----------------------

c a2  LPSDCGEKITLCISVLLSLTVFLLLITEIIPSTSLVIPLIGEYLLFTMIFVTLSIIITVFVLNVHHRSPSTHTMPHWVRS
r a2  LPSECGEKITLCISVLLSLTVFLLLITEIIPSTSLVIPLIGEYLLFTMIFVTLSIVITVFVLNVHHRSPSTHNMPNWVRV
      -  *    --------------------              --------------*--------       *  *   *

c a2  FFLGFIPRWLFMKRPPLLLPAEGTTGQYDPPGTRLSTSRCWLETDVDDKWEEEEEEEEEEEEEEEEEKAYPSRVPSGGSQ
r a2  ALLGRVPRWLMMNRPLPPMELHGS      PDLKLSPSYHWLETNMDAGEREETEEEEEEEDENICVCAGLPDSSMGVLY
      **  **    * *  ******* *       ***  * **    ** ****  *       * ***** ******* ***

c a2  GTQCHYSCERQAGKASGGPAPQVPLKGEEVGSDQGLTLSPSILRALEGVQYIADHLRAEDADFSVKEDWKYVAMVIDRIF
r a2  GHGGLHLRAMEPETKTPSQASEIL             LSPQIQKALEGVHYIADRLRSEDADSSVKEDWKYVAMVVDRIF
       ****************** ****                *  **     *    *  *    *            *  --

c a2  LWMFIIVCLLGTVGLFLPPYLAGMI   505
r a2  LWLFIIVCFLGTIGLFLPPFLAGMI   484
      --*-----*---*----  *
```

Figure 1. A comparison between the mature avian (c a2; Nef et al., 1988) and rat (r a2; Wada et al., 1988) alpha2 subunits. Asterisks mark residue differences between the two proteins. Underlined, aliphatic regions. Overall identity score: 78%, identity score for the first 350 residues: 86%.

PRIMARY STRUCTURE OF THE ALPHA3 SUBUNIT.

The alpha3 subunit mRNA is abundant in chicken ciliary ganglion neurons and rare in the avian CNS (Boyd et al., 1988). It has been detected in the rat PC12 pheochromocytoma cell-line and at a few, well-defined sites in rat cerebrum (Boulter et al., 1986). In figure 2 below, we compare the derived protein sequences of the alpha3 subunits in those two species.

```
               .         .         .         .         .         .         .         .
c a3  SEPEHRLYAALFKNYNQFVRPVKNASDPVIIQFEVSMSQLVKVDEVNQIMETNLWLKHIWNDYKLRWNPVDYGGAEFIRV
r a3  SEAEHRLFQYLFEDYNEIIRPVANVSHPVIIQFEVSMSQLVKVDEVNQIMETNLWLKQIWNDYKLKWKPSDYQGVEFMRV
        *    ***  **  ***   * * *                              *       * * *  * *  *

c a3  PSGQIWKPDIVLYNNAVGDFQVDDKTKALLKYTGDVTWIPPAIFKSSCKIDVTYFPFDYQNCTMKFGSWSYDKAKIDLVL
r a3  PAEKIWKPDIVLYNNADGDFQVDDKTKALLKYTGEVTWIPPAIFKSSCKIDVTYFPFDYQNCTMKFGSWSYDKAKIDLVL
       ***            *                 *

c a3  IGSTMNLKDYWESGEWAIIKAPGYKHDIKYNCCEEIYTDITYSLYIRRLPLFYTINMIIPCLLISFLTVLVFYLPSDCGE
r a3  IGSSMNLKDYWESGEWAIIKAPGYKHEIKYNCCEEIYQDITYSLYIRRLPLFYTINLIIPCLLISFLTVLVFYLPSDCGE
         *                      *          *            ------*------------------

c a3  KVTLCISVLLSLTVFLLVITETIPSTSLVIPLIGEYLLFTMIFVTLSIVITVFVLNVHYRTPKTHTMPVWVRTIFLNLLP
r a3  KVTLCISVLLSLTVFLLVITETIPSTSLVIPLIGEYLLFTMIFVTLSIVITVFVLNVHYRTPTTHTMPTWVKAVFNLLLP
       --------------------              -----------------------    *     *  ***
```

```
c a3   RIMFMTRPTSDEENNQKPKPFYTSEFSNLNCFNSSEIKCCKDGFVCQDMACSCCQYQRMKFSDFSGNLTRSSSSESVDPL
r a3   RVMFMTRPTSGEGDTPKTRTFYGAELSNLNCFSRADSKSCKEGYPCQDGTCGYCHHRRVKISNFSANLTRSSSSESVNAV
        *        * **** ***  ** *      ***** *  * **   ** ** *** * * *  *           ***

c a3   FSFSVLSPEMRDAIESVKYIAENMKMQNEAKEIQDDWKYVAMVIDRIFLWVFILVCILGTAGLFLQPLMTGDDM   474
r a3   LSLSALSPEIKEAIQSVKYIAENMKAQNVAKEIQDDWKYVAMVIDRIFLWVFILVCILGTAGLFLQPLMARDDT   474
       * * *    ***  *          *  *                 --------------------   **  *
```

Figure 2. A comparison between the mature avian (Nef et al., 1988 and S.C., unpublished) and rat (Boulter et al., 1986) alpha3 subunits. Notations are as in fig. 1. Overall identity score: 83%, identity score for the first 350 residues: 87%.

STRUCTURAL FEATURES OF THE NEURONAL RECEPTOR SUBUNITS.

Analysis of the nucleic acid and protein sequences of the avian neuronal receptor subunits reveals important differences, although it appears that the neuronal subunits are more closely related to one another genetically and structurally than to any of the endplate subunits. All neuronal subunit genes have the same structure consisting of six protein-coding exons, with a very large fifth exon encoding two thirds to three fourths of the protein and fusing together sequences encoded by many exons in the endplate receptor subunits. Moreover, neuronal subunit protein sequences, be they alphas or non-alphas, yield better identity scores when compared to one another than when compared to any of the endplate receptor subunits (Nef et al., 1988). In table 1, we list some structural features of the cDNAs and derived protein sequences for endplate alpha and neuronal subunits.

	%AT	AA	MW	N-gly
alpha	51.1	437	50180	141
alpha2	42.0	505	58090	31, 81
alpha3	57.3	474	54800	24, 141
alpha4	47.7	599	68440	29, 79
non-alpha	37.3	473	54030	26, 143

Table 1. Base composition of the cDNAs and properties of the avian alpha and neuronal AChR subunits. % AT, base composition of the sequences encoding the mature subunits. AA, number of residues in the mature subunits. MW, molecular weight in daltons of the mature, unglycosylated subunits. N-gly, potential asparagine-linked glycosylation sites (extracellular domain only).

The very large differences in the base compositions of related subunits from the same organism is surprising indeed (e.g. alpha3 and non-alpha, which differ by 20% AT). Closer analysis (not shown) demonstrates that the main cause of these large differences comes from systematic third-base bias. We suggest that the evolutionary advantage deriving from such marked biases resides in avoiding recombinations leading to gene loss or gene conversion. It may also reflect the varied composition of the DNA at different loci in the genome. In the corresponding proteins, extracellular asparagine-linked glycosylation sites, and intracellular cAMP dependent protein kinase and protein kinase C sites also differ in locations and number. These sites are prime targets for site-directed mutagenesis of the subunits since reconstitution systems are now available (in the Xenopus oocyte or in cell lines) allowing detailed electrophysiological analysis of cloned neurotransmitter receptors.

Acknowledgements
Work in our laboratory is funded by the State of Geneva, the Swiss National Fund and the Sandoz Foundation.

References

Ballivet M et al. (1989) Neuron, in press

Boulter J, Evans K, Goldman D, Martin G, Treco D, Heinemann S, Patrick J (1986) Isolation of a cDNA clone coding for a possible neural nicotinic acetylcholine receptor alpha-subunit. Nature 319:368-374

Boulter J, Connolly J, Deneris E, Goldman D, Heinemann S, Patrick J (1987) Functional expression of two neuronal nicotinic acetylcholine receptors from cDNA clones identifies a gene family. Proc Natl Acad Sci USA 84:7763-7767

Boyd RT, Jacob MH, Couturier S, Ballivet M, Berg DK (1988) Expression and regulation of neuronal acetylcholine receptor mRNA in chick ciliary ganglia. Neuron 1:495-502

Deneris ES, Connolly J, Boulter J, Wada E, Wada K, Swanson LW, Patrick J, Heinemann S (1988) Primary structure and expression of β_2: a novel subunit of neuronal nicotinic acetylcholine receptos. Neuron 1:45-54

Goldman D, Deneris E, Luyten W, Kochhar A, Patrick J, Heinemann S (1987) Members of a nicotinic acetylcholine receptor gene family are expressed in different regions of the mammalian central nervous system. Cell 48:965-973

Kao PN, Karlin A (1986) Acetylcholine receptor binding site contains a disulfide crosslink between adjacent half-cystinyl residues. J Biol Chem 261:8085-8088

Nef P (1986) PhD Thesis, University of Geneva

Nef P, Mauron A, Stalder R, Alliod C, Ballivet M (1984) Structure linkage, and sequence of the two genes encoding the delta and gamma subunits of the nicotinic acetylcholine receptor. Proc Natl Acad Sci USA 81:7975-7979

Nef P, Oneyser C, Alliod C, Couturier S, Ballivet M (1988) Genes expressed in the brain define three distinct neuronal nicotinic acetylcholine receptors. EMBO J 7:595-601

Takai T, Noda M, Mishina M, Shimizu S, Furutani Y, Kayano T, Ikeda T, Kubo T, Takahashi H, Takahashi T, Kuno M, Numa S (1985) Cloning, sequencing and expression of cDNA for a novel subunit of acetylcholine receptor from calf muscle. Nature 315:761-764

Whiting P, Esch F, Shimasaki S, Lindstrom J (1987a) Neuronal nicotinic acetylcholine receptor β-subunit is coded for by the cDNA clone α_4. FEBS Lett 219:459-463

Whiting PJ, Lindstrom JM (1987b) Affinity labelling of neuronal acetylcholine receptors localizes acetylcholine binding sites to their β-subunits. FEBS Lett 213:55-60

Witzemann V, Barg B, Nishikawa Y, Sakmann B, Numa S (1987) Differential regulation of muscle acetylcholine receptor γ-and ϵ-subunit mRNAs. FEBS Lett 223:104-112

STRUCTURE OF MUSCLE AND NEURONAL NICOTINIC ACETYLCHOLINE RECEPTORS

Ralf Schoepfer, Paul Whiting, Mike Luther, Kent Keyser,[1] *Harvey Karten,*[1] *and Jon Lindstrom*
The Salk Institute for Biological Studies
P.O. Box 85800
San Diego, CA 92138, USA

INTRODUCTION

Nicotinic acetylcholine receptors (AChRs) of muscle, brain, and ganglia belong to the ligand-gated cation channel family, whose most well-studied member is the AChR of the electric organs of *Torpedo californica* and related species. The strychnine-binding glycine receptor of the spinal cord and the brain γ-aminobutyric $acid_A$ ($GABA_A$) receptor are ligand-gated anion channels which share sequence homologies with AChRs, which suggest that they are part of a ligand-gated receptor superfamily with a common evolutionary origin (Lindstrom et al., 1987b; Barnard et al., 1987). The neuronal α-bungarotoxin- (α-Bgt) binding protein, whose function and exact structure is still elusive, is probably a member of this superfamily, and surely more members will be discovered. Muscarinic AChRs (Peralta et al., 1988), on the other hand, belong to a completely different superfamily of receptors, with rhodopsin (Nathans & Hogness, 1983), adrenergic (Dixon et al., 1986), and serotonin receptors as members, and mediate their action through coupling proteins (G-proteins) (Kerlavage, 1987; Julius et al., 1988).

Monoclonal antibodies (mAbs) and cDNA clones initially derived from studying *Torpedo* AChRs have been useful tools in elucidating the structure of human muscle and neuronal AChRs. Here we will combine the results of these two complimentary approaches we have taken to elucidate the structure of AChRs.

---

1 Department of Neurosciences
The University of California
San Diego, CA 92093
USA

NATO ASI Series, Vol. H 32
Molecular Biology of Neuroreceptors and Ion Channels. Edited by A. Maelicke

HUMAN MUSCLE AChR

Discussing the structure of muscle-type AChRs first will serve as a basis for later comparisons with neuronal AChRs.

In searching for cell lines expressing neuronal AChRs, we studied the AChRs expressed by the human medulloblastoma cell line TE671 and found that this cell line, which was derived from a cerebellar neoplasm (McAllister et al., 1977), actually expresses muscle AChRs (Lindstrom et al., 1987a; Schoepfer et al., 1988a; Luther et al., in press). In collaboration with Mauricio Montal (Luther et al., in press) we confirmed earlier reports (Syapin et al., 1982) that TE671 produces a functional AChR that can be blocked by α-Bgt. The linear current/voltage relationship (10-100 mV) with a single-channel conductance (γ) of 44-45 pS and channel openings with time constants (τ) of 0.82 ms and 3.3 ms are consistent with those found in muscle AChR (Sakmann & Neher, 1983). Sine (in press) has made similar single-channel observations on AChRs from TE671 cells. Pharmacological properties of these human AChRs differ somewhat from muscle AChRs of other species. Sine (in press) demonstrated that their affinity for α-Bgt was appreciably lower than that of AChRs from mouse BC3H-1 cells or *Torpedo* electric organ, and that the two ACh binding sites in each AChR did not differ greatly in affinity for curare, as is the case in mice and *Torpedoes*. Antisera raised against the α-Bgt-binding protein purified from chicken brain do not react with the α-Bgt-binding protein of TE671 cells, even though they do react with the α-Bgt-binding protein of human brain (Luther et al., in press). Autoantibodies to muscle AChRs in the sera of patients with myasthenia gravis do not react with α-Bgt-binding proteins from human brains (Whiting et al., 1987a), but they do immunoprecipitate α-Bgt-labeled AChRs from TE671 cells (Lindstrom et al., 1987a) as do mAbs against human muscle AChRs (Luther et al., in press). TE671 AChRs react with mAbs specific for extrajunctional AChRs of human muscle. Figure 1 shows that TE671 AChRs purified by α-Bgt affinity chromatography consist of four polypeptides corresponding to the α, β, γ, and δ subunits of *Torpedo* AChRs by apparent molecular weight (42,000; 52,500; 55,000; and 62,000; respectively) and by subunit-specific antibody labeling on western blots (Luther et al., in press). After reduction of the TE671 AChR, the ACh binding site affinity label MBTA labels its α subunits, suggesting that the α subunits have disulfide-linked cysteines homologous to cysteines α192 and 193, near the ACh binding site on α subunits of *Torpedo* AChRs (Kao et al., 1984; Kao & Karlin, 1986). Northern blot analysis of TE671 poly-A^+ RNA revealed RNA species for α, β, γ, and δ subunits when probed at high stringency with cDNA clones (obtained from Dr. Jim Boulter) coding for the four mouse muscle AChR subunits.

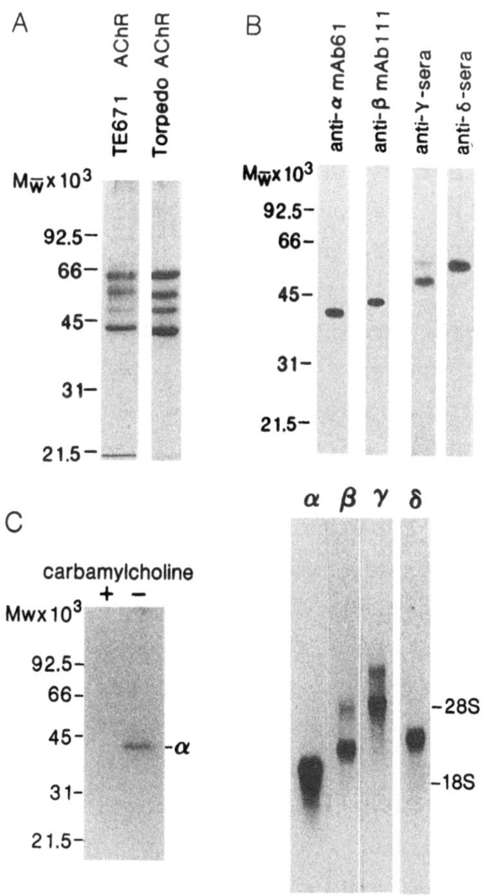

Figure 1. Subunits of AChRs from TE671. (A) AChRs affinity purified from TE671 and *Torpedo* electric organ have similar

subunits. Purification at 43% yield from 12 roller bottles typically yields 100 μg. AChRs (10 μg) were resolved into their subunits by electrophoresis on a 10% acrylamide gel in SDS under reducing conditions and stained with Coomassie blue. (B) Subunits from TE671 AChRs correspond to those of electric organ AChRs by western blotting. Purified TE671 AChR (50 ng/lane) was resolved into subunits by electrophoresis and then blotted onto DPT paper. Each lane was incubated with the indicated antibody, mAb 61 (Tzartos et al., 1981) and mAb 111 (Tzartos et al., 1986) at 10 nM and γ and δ anti-subunit sera (Lindstrom et al., 1979) at 1 nM. Bound antibodies were localized by autoradiography using a ^{125}I mouse anti-rat IgG mAb. (C) Affinity labeling with ^{3}H-MBTA identifies the α subunits of AChR from TE671 as forming the ACh binding site. (D) Poly A^{+} mRNAs for the four subunits of TE671 AChR are detected by high stringency hybridization with cDNAs for mouse muscle AChR α, β, γ, and δ subunits. The cDNA probes used were described by Heinemann et al. (1986). Reproduced from Luther et al. (in press).

cDNA clones coding for the α and δ subunits of AChR from TE671 cells displayed, respectively, 95% and 91% amino acid identity with the homologous mouse muscle subunits (Schoepfer et al., 1988a; Luther et al., in press). The deduced amino acid sequence of the α subunit was identical to the amino acid sequence deduced from the previously reported genomic sequence of human muscle AChR α subunits (Noda et al., 1983). In both subunits the characteristic features of the muscle AChR subunit family are found: a signal peptide; four hydrophobic putative transmembrane segments, M1-M4 of highly conserved sequences; an extended putative cytoplasmic loop between M3 and M4 of poorly conserved sequences; a highly conserved region including two cysteines linked by a disulfide bond at position *Torpedo* α128 and *Torpedo* α142 with a putative N-glycosylation site at α141; and cysteines 192-193 found on the α subunit, which are characteristic of binding subunits, but are absent in structural subunits of AChRs.

In summary, the neuromedulloblastoma cell line TE671 expresses a functional human muscle AChR and is a suitable source for analyzing AChR on the protein and gene level and therefore a useful tool for studying myasthenia gravis, an autoimmune disease mediated by autoantibodies against muscle AChR (reviewed by Lindstrom et al., 1988).

NEURONAL AChRs

mAbs have proven extremely useful in studying neuronal AChRs. The initial evidence that a mAb raised against the muscle-type AChRs of eel electric organ might be used to study neuronal AChRs was the observation that mAb 35 specifically labeled a nucleus in chicken brains which did not bind *α*-Bgt (Swanson et al. 1983). Subsequently mAb 35 was found to react with the functional AChRs of chicken ciliary ganglion neurons (Halvorsen & Berg, 1986) and bovine adrenal medulla cells (Higgins & Berg, 1987). Paul Whiting used mAb 35 to purify AChRs from chicken brains (Whiting & Lindstrom, 1986a; Whiting et al., 1987c), then used these chicken brain AChRs to raise a library of mAbs for subsequent structural characterization of these AChRs (Whiting et al., 1987c). These mAbs also proved useful for identifying AChRs in rat brains (Whiting & Lindstrom, 1987b; Swanson et al., 1987), and immunoaffinity purifying AChRs from rat brain (Whiting & Lindstrom, 1987b). Immunoaffinity purified AChRs from rat brain were then used as immunogen to raise another library of mAbs for studies of AChRs from the brains of rats, cattle, and humans (Whiting & Lindstrom, in press). These studies are summarized in Figure 2.

Immunoaffinity-purified neuronal AChRs were found to consist of only two kinds of subunits. The larger subunit was identified as the ACh-binding subunit by nicotine-blockable MBTA labeling after reduction, suggesting that it contains cysteines homologous to cysteines 192,193 of *Torpedo* AChR *α* subunits. (Whiting & Lindstrom, 1987a, in press; Whiting et al., 1987c). We now refer to the lower molecular weight subunit as the structural subunit. This nomenclature is more informative than referring to the subunits as *α* and *β* in order of increasing apparent molecular weight, is applicable to other members of the gene family, and avoids confusion with other nomenclature which refers to AChR subunit homologue cDNAs containing cysteines 192,193 as "*α*" and other AChR subunit homologues as "non-*α*" or "*β*2" (Nef et al., 1988; Boulter et al., 1987; Goldman et al., 1986; Deneris et al.,

1988). mAb 35 and mAb 210, which bind to the α subunits of muscle-type AChRs, bind to structural subunits of neuronal AChRs on western blots (Whiting & Lindstrom, 1986a, and unpublished). AChRs are thought to have evolved from a primordial AChR with only a single type of subunit (Raftery et al., 1980; Breer et al., 1985). Apparently, in the muscle AChR lineage, the ACh binding site and the main immunogenic region (MIR) recognized by mAb 35 and 210 remained on α subunits, and both the ACh binding site and the MIR were lost by all of the structural subunits. By contrast, in the neuronal lineage, apparently structural subunits lost the ACh binding site and retained a vestige of the MIR, while the ACh-binding subunit lost all vestiges of the MIR.

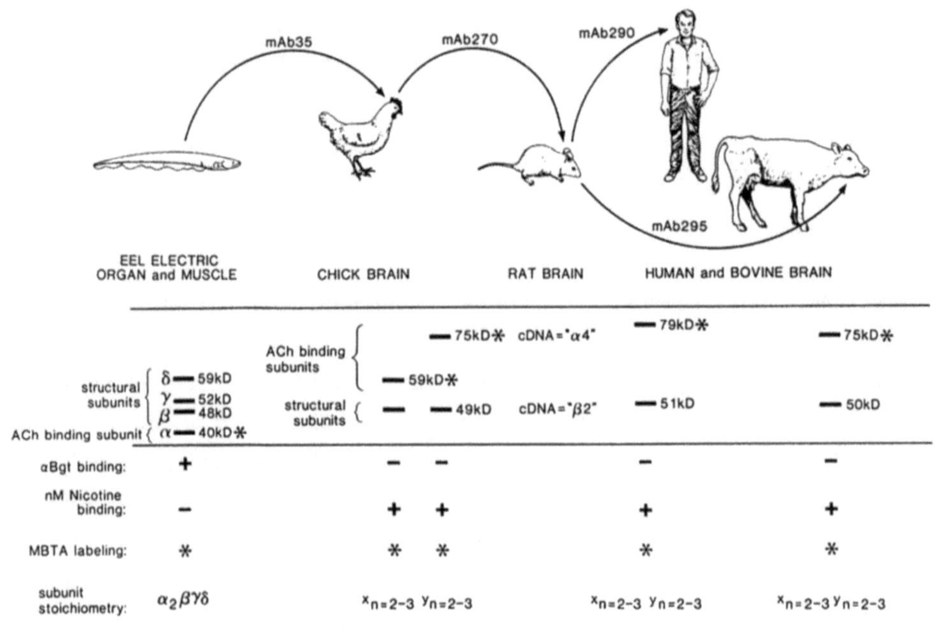

Figure 2. Summary of immunoaffinity purification of neuronal nicotinic AChRs. mAbs to AChRs from one species were used to purify AChRs from another, and then mAbs to these AChRs were used to purify AChRs from other species. ACh-binding subunits were identified by affinity labeling with MBTA and are indicated by *. The other subunits are referred to as "structural." Modified from Lindstrom et al. (1987b)

In chicken brain two AChR subtypes with almost identical pharmacology, but different ACh-binding subunits (apparent molecular weights of 59 kD and 75 kD) were isolated. As judged by peptide maps and N-terminal protein sequences, they share the same structural subunit (apparent molecular weight 49 kD) (Whiting et al., 1987c; Schoepfer et al., 1988b). mAb 270 binds to the structural subunits of chicken and rat neuronal AChRs. Only one subtype of AChR was immunoaffinity purified from brains of rats and cattle. It corresponded to the AChR from chicken brain with the 75 kD ACh-binding subunit by apparent molecular weight, mAb cross-reactivity, and N-terminal amino acid sequence. The observation that mAb 270 identifies the functional ganglionic-type AChR in PC12 cells (Whiting et al., 1987d) suggests that in rats as well as chickens, AChR subtypes with differing ACh-binding subunits may use the same type of structural subunit. Since immunoaffinity purification of AChRs from rat brain with mAb 270 to the structural subunit (Whiting & Lindstrom, 1987b), or other mAbs to the structural subunit, or a mAb to the ACh-binding subunit (Whiting & Lindstrom, unpublished), all yield preparations with the same subunit pattern on SDS-PAGE, the AChR subtype observed must be much more prevalent in brain than any other subtypes which use the same structural subunit.

Immunoisolated brain AChRs had nanomolar affinity for nicotine and cytisine, but no measurable affinity for α-Bgt (Whiting & Lindstrom, 1986b, in press; Whiting et al., 1987b). These mAbs precipitate >90% of the high affinity nicotine binding sites in the brains of chickens (both subtypes together, each in approximately equal amounts), rats, cattle, and humans.

Several lines of evidence demonstrate that the AChRs identified by these mAbs function as ACh-gated cation channels. Antiserum to AChRs purified from chicken brains block the electrophysiological function of AChRs in chicken ciliary ganglia (Stollberg et al., 1986). mAb 270 binds to the functional AChRs in rat PC12 cells (Whiting et al., 1987d). mAb 35 binds to the functional AChRs in chicken ciliary ganglion cells and bovine adrenal cells (Halvorsen & Berg, 1986; Higgens & Berg, 1987). cDNAs coding for the structural subunits and the ACh-binding subunits of AChRs from rat and chicken brains, when coexpressed in *Xenopus* oocytes, produce functional AChRs (Boulter et al., 1987; Deneris et al., 1988; Ballivet et al., in press). Synaptosomes from rat brains immunoaffinity purified using mAbs to neuronal AChRs release ACh and dopamine in response to ACh or nicotine, demonstrating that *in vivo,* some of these AChRs act as presynaptic neuromodifiers (Irons et al., 1988).

^{125}I-mAb 270 has been used to localize AChRs throughout the rat brain (Swanson et al., 1987), binding to basically the same areas which bind nicotine with high affinity (Clarke et al., 1985). mAbs have also been used to histologically localize AChRs in the brains of frogs (Sargent et al., in press), chicken brains (Swanson et al., 1983, 1987), finch brains (Watson et al., 1988), and chicken retina (Keyser et al., in press). There is immunochemical evidence that many of these AChRs are in pre- or extrasynaptic locations in goldfish (Henley et al., 1986), frogs (Sargent et al., in press), chickens (Swanson et al., 1987), and rats (Swanson et al., 1987). There is corresponding evidence that AChRs identified histologically by high affinity binding of ^{3}H-nicotine are frequently in presynaptic locations (Clarke et al., 1986; Prusky et al., 1988; Prusky & Cynader, 1988) The expression of AChRs on central processes of retinal ganglion cells and throughout the visual system is quite striking. AChRs have been implicated in visual synaptic plasticity (Bear & Singer, 1986; Prusky et al., 1988) and shown to affect process outgrowth by retinal ganglion cells (Lipton et al., 1988; McCobb et al., 1988). Thus, although neuronal AChRs, like muscle AChRs, are ACh-gated ion channels, they are found not only in postsynaptic locations, as in ganglia where they directly mediate signal transmission, but also in presynaptic locations where they modulate transmission, and in pre- and extrasynaptic locations where they may affect neuronal growth and synaptic plasticity.

N-terminal sequences obtained on both subunits of rat brain AChR and subunits of the chick brain AChR with the 75 kD ACh-binding subunit were used to identify corresponding cDNA clones of the AChR gene family which were isolated by us and others (Whiting et al., 1987b; Schoepfer et al., 1988b; Whiting et al., unpublished; Nef et al., 1988; Goldman et al., 1987; Deneris et al., 1988) (Fig. 3). The ACh-binding subunit is encoded by the gene termed *α*4, the structural subunit gene is termed non-*α* by Nef et al. (1988) and *β*2 by Deneris et al. (1988). Genes coding for ACh-binding subunits of the predominant brain subtype (*α*4) or the ganglionic type (*α*3), in combination with structural subunits, give rise to functional AChRs when coinjected into *Xenopus* oocytes (Boulter et al., 1987; Ballivet et al., in press), confirming our previous observations (Whiting & Lindstrom, 1986a, 1987b; Whiting et al., 1987c, in press) that brain AChRs consist of two types of subunits (Fig. 2).

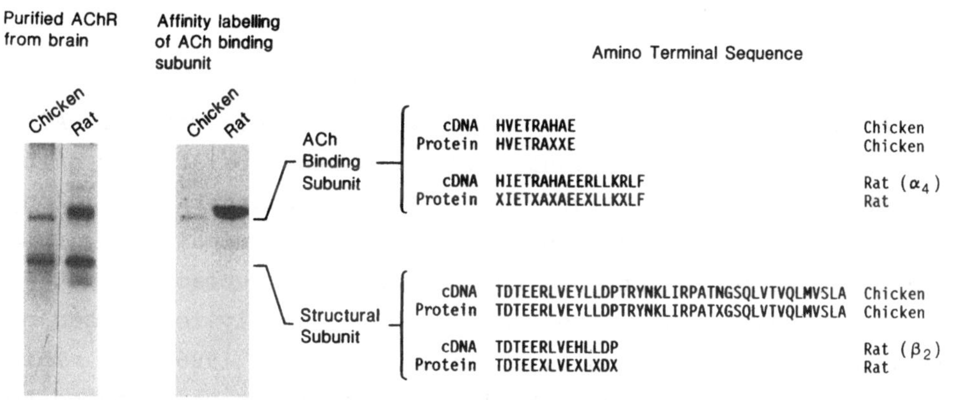

Figure 3. Data summary of gel electrophoresis, N-terminal amino acid sequences, and cDNA sequences of subunits of AChRs purified from the brains of chickens and rats. Acrylamide gels reproduced from Whiting and Lindstrom (1987a,b) and Whiting et al. (1987c). N-terminal amino acid sequences are from Whiting et al. (1987c) and Schoepfer et al. (1988b). cDNA sequences of subunits of AChRs from chicken brain are from Schoepfer et al. (1988b) and Whiting et al. (unpublished). cDNA sequences of the corresponding rat subunits are from Goldman et al. (1987) and Deneris et al. (1988).

Figure 4. Comparison of the amino acid sequences of AChR subunits deduced from cDNAs. The sequence of chicken brain AChR structural subunit is from Schoepfer et al. (1988b). Mouse muscle AChR ACh-binding subunit and rat PC12 ACh-binding subunit are from Boulter et al. (1987). Rat brain AChR ACh-binding subunit is from Goldman et al. (1987). Reproduced from Schoepfer et al. (1988b).

Alignment of subunit cDNA sequences of various AChR subtypes reveals highly conserved segments separated by quite variable stretches (Fig. 4). Most prominent is a long sequence unique to each subunit located C-terminal of M3 in a region known to exist on the cytoplasmic surface of native *Torpedo* AChRs in a loose conformation easily recognized by mAbs to denatured subunits (Ratnam et al., 1986a,b). This highly variable putative cytoplasmic loop might be expected to interact specifically with the axonal transport system, the outer surface of vesicles, other cytoskeletal proteins, and phosphorylation mechanisms potentially affecting activation, desensitization, and turnover. The poor conservation of this segment between homologous chick and rat genes might reflect the fast rate of evolution in this neuronal lineage and is in contrast to the well-conserved muscle genes. Alternatively, the variability in this region could indicate lack of selective pressure due to the unimportance of this region. The putative cytoplasmic loop of the chicken brain 75 kD ACh-binding subunit expressed in bacteria bound to mAb 289 and gave rise to subunit-specific antisera which bound to native AChR (Whiting & Lindstrom, unpublished).

The retina is an especially interesting region for studies of neuronal AChR subtypes because its architecture is well characterized. AChR localization in brain by binding of ^{125}I-mAbs, ^{3}H-nicotine, or ^{35}S-cRNA *in situ* hybridization provides low resolution. Cell-level resolution of neuronal AChRs in chick retina has been obtained using mAbs localized with fluorescein and peroxidase (Keyser et al., in press). In Figure 5, AChRs are localized in retina using both mAbs and *in situ* hybridization. There are AChRs on dendrites, cell bodies, and central processes of retinal ganglion cells and displaced ganglion cells. Future studies involving many methodologies will be required to localize the various AChR subtype mRNAs and proteins, and to determine where these AChRs function and what their functional roles are. Ultrastructural studies will also be required. mAbs have been used to localize AChRs by electron microscopy in frog brains (Sargent et al., in press) and ciliary ganglia (Jacob et al., 1984, 1986), but many more studies will be required. Developmental studies will also be important. Preliminary studies of developmental regulation of AChR expression in chicken brain (Whiting et al., 1987c) and retina (Whiting et al., unpublished) reveal developmental changes in AChR subtype expression. Between embryonic days 8 and 20 there is good correlation between the mRNA levels for both subunits of brain AChR and the amount of immunoisolated, high affinity nicotine binding sites.

In conclusion, the primary structure of the predominant AChR subtype in avian and mammalian brains has been identified, and this AChR has been shown to be the high affinity nicotine-binding component in brain. mAbs and cDNAs are continuing to provide probes for localizing and characterizing these AChRs and other members of the nicotinic

AChR family. Vastly more remains to be done to fully characterize the structure of these proteins, determine their functional roles, and study their development and evolution.

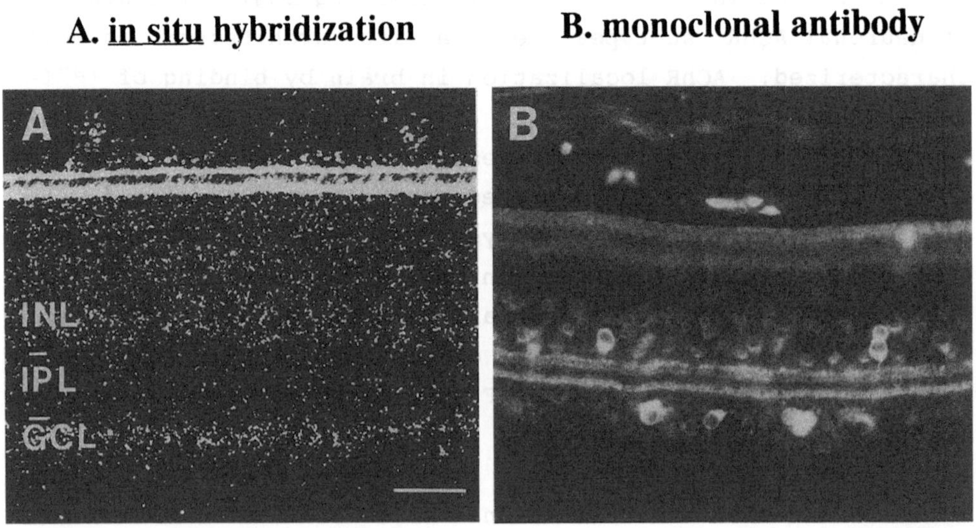

Figure 5. Localization of the structural subunit in chicken retina. (A) mRNA for the structural subunit is localized using an ^{35}S-labeled anti-sense RNA probe corresponding to the putative cytoplasmic loop (amino acids 306 to 410) of the structural subunit (Schoepfer et al. 1988). Cell bodies in the ganglion cell layer (GCL) are labeled, as are cells in the inner nuclear layer (INL), whereas dendritic processes in the inner plexiform layer (IPL) are not heavily labeled. The pigment layer produces an intense artifactual birefringent double layer at the top of this darkfield micrograph. A control sense cRNA probe gave uniform low background labeling. The scale bar is 40 μ. (B) Structural subunit protein is localized by indirect immunofluorescence using mAb 270. The subunit protein is localized in the cell bodies of ganglion cells and displaced ganglion cells, as well as in the dendritic and axonal processes of these cells, and some others, located in the inner and outer plexiform layers. The pigment layer at the top appears dark in this fluorescence micrograph. The sections in A and B are from different parts of the retina and differ in sizes. Immunofluorescence localization of the structural subunit in chick retinal ganglion cells has been studied in detail by Keyser et al. (in press).

ACKNOWLEDGMENTS

Support for the laboratory of J.L. is provided by grants from the National Institutes of Health (NS11323), the Muscular Dystrophy Association, the U.S. Army (DAMD17-86-C-6148), the Alexander Onassis Public Benefit Foundation, the Council for Tobacco Research-USA, Inc., and the California chapter of the Myasthenia Gravis Foundation. R.S. was supported by a "Boehringer Ingelheim Fonds" fellowship. M.L. was supported by a fellowship from the Muscular Dystrophy Association.

REFERENCES

Ballivet M, Nef P, Couturier S, Rungger D, Bader CR, Bertrand D, Cooper E (in press) Electrophysiology of a chick neuronal nicotinic acetylcholine receptor expressed in *Xenopus* oocytes after cDNA injection. Neuron

Barnard E, Darlison M, Seeburg P (1987) Molecular biology of the $GABA_A$ receptor: The receptor/channel superfamily. Trends in Neurosci 10:502-509

Bear M, Singer W (1986) Modulation of visual cortical plasticity by acetylcholine and noradrenalin. Nature 320:172-176

Boulter J, Connolly J, Deneris E, Goldman D, Heinemann S, Patrick J (1987) Functional expression of two neuronal nicotinic acetylcholine receptors from cDNA clones identifies a gene family. Proc Natl Acad Sci USA 84:7763-7767

Boulter J, Evans K, Goldman D, Martin G, Treco D, Heinemann S, Patrick J (1986) Isolation of a cDNA clone coding for a possible neural nicotinic acetylcholine receptor α subunit. Nature 319:368-374

Breer H, Kleene R, Hinz G (1985) Molecular forms and subunit structure of the acetylcholine receptor in the central nervous system of insects. J Neurosci 5:3386-3392

Clarke P, Hamill G, Nadi N, Jacobowitz D, Pert A (1986) ^{3}H-nicotine and ^{125}I-α-bungarotoxin-labeled nicotinic receptors in the interpeduncular nucleus of rats. II. Effects of habenular deafferentation. J Comp Neurol 251:407-413

Clarke P, Schwartz R, Paul S, Pert C, Pert A (1985) Nicotinic binding in rat brain: Autoradiographic comparison of [^{3}H]acetylcholine, [^{3}H]nicotine, and [^{125}I]-α-bungarotoxin. J Neurosci 5:1307-1315

Deneris E, Connolly J, Boulter J, Wada E, Wada K, Swanson L, Patrick J, Heinemann S (1988) Primary structure and expression of β2: A novel subunit of neuronal nicotinic acetylcholine receptors. Neuron 1:45-54

Dixon RAF, Kobilka BK, Strader DJ, Benovic JL, Dohlman HG,

Frielle T, Bolanowski MA, Bennett CD, Rands E, Diehl RE, Mumford RA, Slater EE, Sigal IS, Caron MG, Lefkowitz RJ, Strader CD (1986) Cloning of the gene and cDNA for mammalian β-adrenergic receptor and homology with rhodopsin. Nature 321:75-79

Goldman D, Deneris E, Luyten W, Kochhar A, Patrick J, Heinemann S (1987) Members of a nicotinic acetylcholine receptor gene family are expressed in different regions of the mammalian central nervous system. Cell 48:965-973

Halvorsen S, Berg D (1986) Identification of a nicotinic acetylcholine receptor on neurons using an α neurotoxin that blocks receptor function. J Neurosci 6:3405-3412

Heinemann S, Boulter J, Connolly J, Goldman D, Evans K, Treco D, Ballivet M, Patrick J (1986) Molecular biology of muscle and neural acetylcholine receptors. In: Maelicke A (ed) Nicotinic Acetylcholine Receptor Structure and Function. Springer-Verlag, Heidelberg, pp 360-387

Henley J, Lindstrom J, Oswald R (1986) Acetylcholine receptor synthesis in retina and transport to the optic tectum in goldfish. Science 232:1627-1629

Higgins L, Berg D (1987) Immunological identification of a nicotinic acetylcholine receptor on bovine chromaffin cells. J Neurosci 7:1792-1798

Irons J, Lunt G, Wonnacott S, Whiting P, Lindstrom J (1988) Immunoaffinity isolation of hippocampal nerve terminals bearing [^{3}H]-nicotine receptors. Neurosci Soc Abs 440.3

Jacob M, Berg D, Lindstrom J (1984) A shared antigenic determinant between the *Electrophorus* acetylcholine receptor and a synaptic component on chick ciliary ganglion neurons. Proc Natl Acad Sci USA 81:3223-3227

Jacob MH, Lindstrom JM, Berg DK (1986) Surface and intracellular distribution of a putative neuronal nicotinic acetylcholine receptor. J Cell Biol 103:205-214

Julius D, MacDermott AB, Axel R, Jessell TM (1988) Molecular characterization of a functional cDNA encoding the serotonin 1c receptor. Science 241:558-564

Kao P, Dwork A, Kaldany R, Silver M, Wideman J, Stein S, Karlin A (1984) Identification of the α subunit half cystine specifically labeled by an affinity reagent for the acetylcholine receptor binding site. J Biol Chem 259:11662-11665

Kao P, Karlin A (1986) Acetylcholine receptor binding site contains a disulfide crosslink between adjacent half-cystinyl residues. J Biol Chem 261:8085-8088

Kerlavage AR, Fraser CM, Venter JC (1987) Muscarinic cholinergic receptor structure: Molecular biological support for subtypes. TIPS 8:426-431

Keyser KT, Hughes TE, Whiting PJ, Lindstrom JM, Karten HJ (in press) Cholinoceptive neurons in the retina of the chick: An immunohistochemical study of the nicotinic acetylcholine receptors. Vis Neurosci

Lindstrom J, Criado M, Ratnam M, Whiting P, Ralston S, Rivier J, Sarin V, Sargent P (1987a) Using monoclonal antibodies to determine the structures of acetylcholine receptors from electric organs, muscles, and neurons. Ann

NY Acad Sci 505:208-225
Lindstrom J, Schoepfer R, Whiting P (1987b) Molecular studies of the neuronal nicotinic acetylcholine receptor family. Molec Neurobiol 1:218-337
Lindstrom J, Shelton GD, Fujii Y (1988) Myasthenia gravis. Adv Immunol 42:233-284
Lindstrom J, Walter B, Einarson B (1979) Immunochemical similarities between subunits of acetylcholine receptors from *Torpedo, Electrophorus,* and mammalian muscle. Biochem 18:4470-4480
Lipton S, Frosch M, Phillips M, Tauck D, Aizenman E (1988) Nicotinic antagonists enhance process outgrowth by rat retinal ganglion cells in culture. Science 239:1293-1296
Luther M, Schoepfer R, Whiting P, Blatt Y, Montal MS, Montal M, Lindstrom J (in press) A muscle-like acetylcholine receptor is expressed by the human medulloblastoma cell line TE671. J Neurosci
McAllister RM, Isaacs H, Rongey R, Peer M, Au W, Soukup SW, Gardner MB (1977) Establishment of a human medulloblastoma cell line. Int J Cancer 20:206-212
McCobb DP, Cohan CS, Connor JA, Kater SB (1988) Interactive effects of serotonin and acetylcholine on neurite elongation. Neuron 1:377-385
Nathans J, Hogness DS (1983) Isolation, sequence analysis and intron-exon arrangement of the gene encoding bovine rhodopsin. Cell 34:807-816
Nef P, Oneyser C, Alliod C, Couturier S, Ballivet M (1988) Genes expressed in the brain define three distinct neuronal nicotinic acetylcholine receptors. EMBO J 7:595-601
Noda M, Furutani Y, Takahashi H, Toyosato M, Tanabe T, Shimizu S, Kikyotani S, Kayano T, Hirose T, Inayama S, Numa S (1983) Cloning and sequence analysis of calf cDNA and human genomic DNA encoding α subunit precursor of muscle acetylcholine receptor. Nature 305:818-823
Peralta EG, Winslow JW, Ashkenazi A, Smith DH, Ramachandran J, Capon DJ (1988) Structural basis of muscarinic acetylcholine receptor subtype diversity. Trends in Pharmacol, Feb Suppl: 6-11
Prusky G, Cynader M (1988) [^{3}H]nicotine binding sites are associated with mammalian optic nerve terminals. Vis Neurosci 1:245-248
Prusky G, Shaw C, Cynader M (1988) The distribution and ontogenesis of ^{3}H-nicotine binding sites in cat visual cortex. Dev Brain Res 39:161-176
Raftery M, Hunkapillar M, Strader C, Hood L (1980) Acetylcholine receptor: Complex of homologous subunits. Science 208:1454-1457
Ratnam M, LeNguyen D, Rivier J, Sargent PB, Lindstrom J (1986a) Transmembrane topography of nicotinic acetylcholine receptor: Immunochemical tests contradict theoretical predictions based on hydrophobicity profiles. Biochem 25:2633-2643
Ratnam M, Sargent PB, Sarin V, Fox JL, LeNguyen D, Rivier J, Criado M, Lindstrom J (1986b) Location of antigenic determinants on primary sequences of subunits of nicotinic acetylcholine receptor by peptide mapping. Biochem 25:2621-2632

Sakmann B, Neher, E (1983) Single Channel Recording. Plenum Press, New York
Sargent P, Pike S, Nadel D, Lindstrom J (in press) Nicotinic acetylcholine receptor-like molecules in the retina, retinotectal pathway, and optic tectum of the frog. J Neurosci
Schoepfer R, Luther M, Lindstrom J (1988a) The human medulloblastoma cell line TE671 expresses a muscle-like acetylcholine receptor: Cloning of the *α* subunit cDNA. FEBS Lett 226:235-240
Schoepfer R, Whiting P, Esch F, Blacher R, Shimasaki S, Lindstrom J (1988b) cDNA clones coding for the structural subunit of a chicken brain nicotinic acetylcholine receptor. Neuron 1:241-248
Sine SM (in press) Functional properties of human skeletal muscle acetylcholine receptors expressed by the TE671 cell line. J Biol Chem
Stollberg J, Whiting PJ, Lindstrom J, Berg D (1986) Functional blockade of neuronal acetylcholine receptors by antisera to a putative receptor from brain. Brain Res 376:179-182
Swanson LW, Lindstrom J, Tzartos S, Schmued LC, O'Leary DDM, Cowan WM (1983) Immunohistochemical localization of monoclonal antibodies to the nicotinic acetylcholine receptor in the midbrain of the chick. Proc Natl Acad Sci USA 80:4532-4536
Swanson LW, Simmons DM, Whiting PJ, Lindstrom JM (1987) Immunohistochemical localization of neuronal nicotinic receptors in the rodent central nervous system. J Neurosci 7:3334-3342
Syapin PJ, Salvaterra PM, Engelhardt JK (1982) Neuronal-like features of TE671 cells: Presence of a functional nicotinic acetylcholine receptor. Brain Res 231:365-377
Tzartos S, Langeberg L, Hochschwender S, Swanson LW, Lindstrom J (1986) Characteristics of monoclonal antibodies to denatured *Torpedo* and to native calf acetylcholine receptors: Species, subunit, and region specificity. J Neuroimmunol 10:235-253
Tzartos SJ, Rand DE, Einarson BL, Lindstrom JM (1981) Mapping of surface structures of electrophorus acetylcholine receptor using monoclonal antibodies. J Biol Chem 256:8635-8645
Watson JT, Adkins-Regan E, Whiting P, Lindstrom JM, Podleski TR (1988) Autoradiographic localization of nicotinic acetylcholine receptors in the brain of the zebra finch (peophila guttata). J Comp Neurol 274:255-264
Whiting PJ, Lindstrom JM (1986a) Purification and characterization of a nicotinic acetylcholine receptor from chick brain. Biochem 25:2082-2093
Whiting P, Lindstrom J (1986b) Pharmacological properties of immunoisolated neuronal nicotinic receptors. J Neurosci 6:3061-3069
Whiting P, Lindstrom J (1987a) Affinity labeling of neuronal acetylcholine receptors localizes acetylcholine binding sites to their *β* subunits. FEBS Lett 213(1):55-60
Whiting P, Lindstrom J (1987b) Purification and characterization of a nicotinic acetylcholine receptor from rat brain. Proc Natl Acad Sci USA 84:595-599
Whiting PJ, Lindstrom JM (in press) Characterization of bovine

and human neuronal nicotinic acetylcholine receptors using monoclonal antibodies. J Neurosci

Whiting P, Cooper J, Lindstrom JM (1987a) Antibodies in sera from patients with myasthenia gravis do not bind to nicotinic acetylcholine receptors from human brain. J Neuroimmunol 16:205-213

Whiting P, Esch F, Shimasaki S, Lindstrom J (1987b) Neuronal nicotinic acetylcholine receptor β subunit is coded for by the cDNA clone α4. FEBS Lett 219(2):459-463

Whiting P, Liu R, Morley BJ, Lindstrom J (1987c) Structurally different neuronal nicotinic acetylcholine receptor subtypes purified and characterized using monoclonal antibodies. J Neurosci 7:4005-4016

Whiting PJ, Schoepfer RD, Swanson LW, Simmons DM, Lindstrom J (1987d) Functional acetylcholine receptor in PC12 cells reacts with a monoclonal antibody to brain nicotinic receptors. Nature 327:515-518

NICOTINIC ACETYLCHOLINE RECEPTORS IN THE NERVOUS SYSTEM OF INSECTS

H. Breer, W. Hanke*, D. Benke*, E. Tareilus, J. Krieger
Universität Stuttgart-Hohenheim
Institut für Zoophysiologie
7000 STUTTGART 70, West Germany

Acetylcholine (ACh) appears to be the predominant excitatory neurotransmitter in the nervous system of insects (Florey, 1963) and specific receptors for ACh have been detected using electrophysiological approaches and binding studies. In contrast to vertebrates, the receptors for acetylcholine in insects are confined to the central nervous system, probably due to the different innervation pattern in arthropods. Pharmacological research has revealed that the receptors for ACh in insects as in vertebrates can be distinguished in nicotinic and muscarinic receptor types. However, a different receptor type predominates in each animal group: in vertebrate mostly muscarinic receptors, in insect nervous tissue very high concentrations of nicotinic receptors (Breer and Sattelle, 1987). The insect ganglia thus provides a rich source for nicotinic receptors of nerve cells. The elucidation of functional and molecular properties of the acetylcholine receptors from arthropods, which are far apart from vertebrates on the evolutionary scale, may allow some interesting comparisons that are of particular interest for understanding the phylogenentic evolution of receptor molecules; furthermore a combination of data on the receptor structure with physiological and pharmacological findings on receptor function from different species, may provide some valuable insights in molecular mechanisms of receptor proteins. Furthermore, as many of the today`s insecticides act on cholinergic synapses the receptors for ACh may represent sensitive targets for future insecticides.

* Universität Osnabrück
Fachbereich Biologie
4500 OSNABRÜCK, West Germany

NATO ASI Series, Vol. H 32
Molecular Biology of Neuroreceptors and
Ion Channels. Edited by A. Maelicke

Localisation of acetylcholine receptors

Electrophysiological studies on the pharmacology of cholinergic synaptic transmission in insects have revealed that receptors for ACh are widely distributed on the synapse-free cell bodies of neurones as well as throughout the complex, synapse-rich neuropil. On both sites the action of acetylcholine, activating specific cation channels, was blocked by nanomolar α-bungarotoxin (Sattelle et al., 1983), indicating that synaptic as well as extrasynaptic membranes in insect ganglia apparently do contain nicotinic receptor-types for acetylcholine.

Functional properties of the receptors

Voltage clamp experiments on insect nerve cell bodies have revealed that nicotinic acetylcholine receptors gate cation channels permeable to sodium, potassium and possibly calcium (David and Sattelle, 1984). Patch clamp single channel analysis on dissociated nerve cells (Sattelle et al., 1986; Beadle and Lee,1986) as well as on membrane patches from nerve cells in an outside-out-configuration (Tareilus and Hanke, 1988) revealed an ACh-activated cation channel with a conductance of approximately 40pS. When purified nicotinic acetylcholine receptors from locust ganglia were reconstituted in planar lipid bilayer membranes, channel conductance of a somewhat larger amplitude (75pS) was detected. The reconstituted receptor probably derived predominantly from membranes of synaptic origin, whereas the cell soma receptors studied by patch clamp techniques are not in synaptic regions. The differences observed for single channel conductance of the cell body receptor and the reconstituted neuronal acetylcholine receptor are quite similar to those reported for nicotinic receptors in membranes of vertebrate muscle cells when synaptic and extrasynaptic regions are compared (Sakmann et al., 1983).

dependent manner. The capability of the extract to change the binding properties of the membrane was sensitive to trypsin-treatment, suggesting that protein molecules are primarily responsible. The endogenous modulator was purified with a series of procedures including gel filtration and ion exchange chromatography. The activity was recovered in a few fractions corresponding to a molecular size of about 80,000, however after polyacrylamide gelelectrophoresis only one polypeptide band of about 20,000 was detected, suggesting that the native molecule is dissociated by SDS-treatment into several subunits. This view was substantiated by experiments using a monoclonal antibody which inactivated and precipitated the modulator activity; the

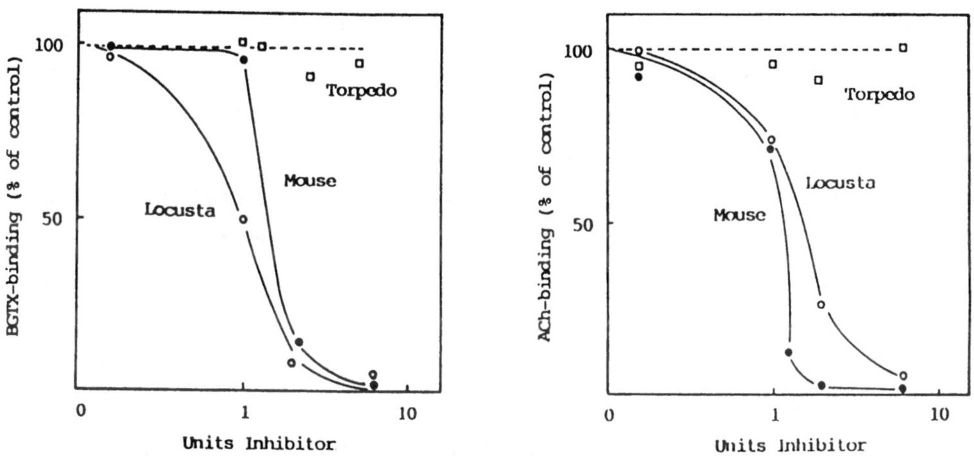

Fig. 1 Inhibitory effect of an endogenous factor on the interaction of α-bungarotoxin and acetylcholine with receptors in membrane preparations from the electric tissue of Torpedo and from nervous tissue of mouse and locust.

Ligand binding

Alternative to the most direct, electrophysiological approaches to study ligand-receptor interaction, binding studies have been performed using radiolabelled ligands in order to quantify specific receptors and to determine kinetic parameters as well as the pharmacological profile of ligand-receptor interactions. Based on high affinity, specific and saturable binding, nicotinic receptors have been probed by labelled snake α-neurotoxins, noteably α-bungarotoxin. Putative nicotinic receptors were found in the nervous tissue of various insect species at concentrations several orders of magnitude higher than in vertebrate brain tissue (Breer and Sattelle, 1987). The observation that the relative levels of nicotinic and muscarinic receptors in the vertebrate and insect nervous tissue is reversed reflects a major contrast between cholinergic binding sites in arthropods and vertebrates although its functional implications are still unknown. When tritiated acetylcholine was used as ligand under appropriate conditions (block of muscarinic receptors and ACh-esterase) two populations of binding sites with different affinity were detected; a small population with a high affinity (5 nM) and a population with an affinity of about 300 nM. The sum of binding sites for ACh excactly matched the number of binding sites for α-bungarotoxin. The differences in affinity of the putative ACh-receptors may be due to desensitisation of some of the receptors; it is wellknown that desensitisation coincides with an increase in agonist affinity (Anholt et al, 1983).

Modulation of ligand binding sites

Pretreatment of neuronal membrane preparations from insects was found to increase the number of binding sites for α-toxins but decrease the number of ACh-binding sites. These changes in binding activities are supposed due to a wash-out of an endogenous modulator, as readding of the extract reconstituted the original binding properties in a dose-antigen was identified in Western blots as a polypeptide with a molecular mass of 20,000. Further experiments exploring the regulatory role of the endogenous modulator indicate that phosphorylation of the protein via cAMP-dependent kinases inactivates the molecules, whereas dephosphorylation increases the modulatory activity.

Biochemical characterisation of the insect nicotinic ACh-receptor

Size determination of the solubilized native receptor protein using various approaches including gradient centrifugation, polyacrylamide gelelectrophoresis and radiation-inactivation has shown that the toxin binding site, the putative nicotinic acetylcholine receptor, in the nervous tissue of different insect species apparently represents a macromolecular protein complex with a sedimentation coefficient of about 10S and a M_r-value between 236-300,000 (Breer et al., 1985; Sattelle and Breer, 1985). In the presence of SDS, however, the receptor protein obviously dissociated into several subunits; the major protein migrated as a band corresponding to M_r=65,000 together with a minor conponent of about 55,000. The minor band has been considered as a proteolysis product, but may in fact represent a second receptor subunit. This observation is of considerable interest as recent results of biochemical, immunological und molecular approaches suggest that the nicotinic acetylcholine receptor of the vertebrate brain is apparently composed of two different polypeptides and thus also quite different from the muscle receptor (Whiting and Lindstrom, 1986; Boulter et al., 1987).

In immunological approaches (ELISA, Western blots) it was found that some of the monoclonal antibodies raised against the nicotinic acetylcholine receptor from *Torpedo* significantly crossreacted with neuronal membrane preparations from locust as well as with purified receptor proteins (Fels et al., 1983; Breer et al., 1985). These results point to some molecular similarities in the structure of the ACh-receptor proteins from these two distinct sources. Antisera raised against purified receptor proteins from locust were used for a topochemical localisation of receptors in the nervous system. The antibodies were found to stain selectively distinct areas in the central neuropil of ganglia from different insect species including regions which are known to contain functional nicotinic acetylcholine receptors (Breer and Sattelle, 1987).

Reconstitution of nicotinic acetylcholine receptors

The reconstitution of purified receptor proteins in artificial lipid membranes is considered as the appropriate approach to ultimately demonstrate that the isolated polypeptides represent functional receptors, i.e. contain the ligand binding site as well as the ion channel. After incorporation of the purified insect proteins into planar lipid bilayer it was found that addition of cholinergic agonists induced the fluctuation of ion channels with a conductance of about 75pS and an open time of a few msec, thus demonstrating that the purified toxin binding protein from locust nervous tissue represents a functional nicotinic receptor for acetylcholine (Hanke and Breer, 1986).

After a more detailed analysis some characteristic features of this neuronal acetylcholine receptor channel have emerged. The channel was only permeable for cations with a mean life time of about 4 msec. Two closed states were found: a mean closed time of 25 msec within the bursts and 300 msec between bursts. The channel was activiated by various cholinergic agonists; suberyldicholine appeared to be the most potent agonist followed by acetylcholine and

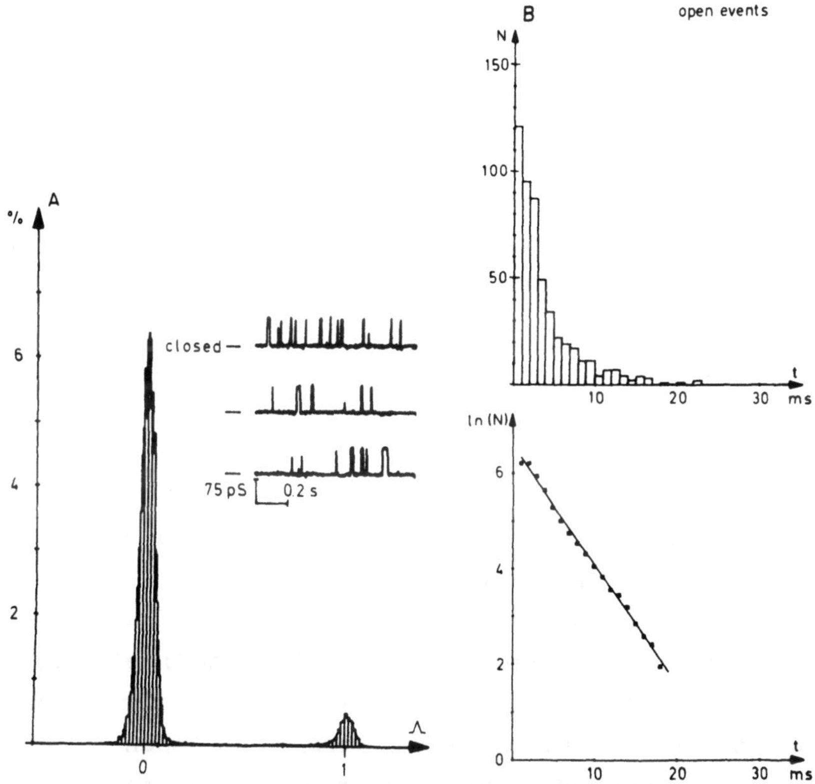

Fig. 2 Evaluation of single channel properties of the neuronal acetylcholine receptor reconstituted in planar lipid bilayer. The amplitude histogram demonstrates a well defined closed state (0) and a well defined open state (1). The lifetime distribution of the open state is depicted in a linear (top) and halflogarithmic plot. The lifetime is single exponential distributed; mean life time of about 4 msec.

carbamylcholine. The mean lifetime of the open state was only slightly dependent on the agonist concentration, the closed-state lifetime, however, decreased significantly with increasing agonist concentration. Furthermore, the dependence of channel activity on agonist concentrations showed no cooperativity suggesting that the neuronal ACh-receptor from insects may be activated by one agonist molecule only; this characteristic feature is an interesting difference to the peripheral vertebrate receptor which is supposed to be activated by two ACh-molecules. Pharmacological studies revealed that various antagonists, like d-tubocurarine or hexamethonium, did not affect the conductance of the channel but rather reduced the probability of the channel being in the open state. Hexamethonium appeared to be quite ineffective in blocking the insect receptor, whereas micromolar concentrations of hexamethonium almost completely blocked the reconstituted Torpedo receptor (Hanke et al., 1981). This finding confirms results from electrophysiological and binding studies, which document a low affinity of hexamethonium for insect acetylcholine receptors (Breer, 1981; David and Sattelle, 1984). These results can be considered as further indication for the pharmacological integrity of the purified and reconstituted receptor protein.

At low agonist concentrations, the open- and closed-state lifetimes were single exponentially distributed; at higher concentrations (5 μM) multiple gating events were observed, i.e. series of not statistically independent open events. Application of carbamylcholine concentrations as high as 50 μM induced a characteristical bursting behavior of the reconstituted channel. At higher agonist concentrations typical desensitization effects were observed (Hanke and Breer, 1987). Thus, all functional properties of the reconstituted protein support the view that it represents a functional ACh-receptor channel. This first reconstituted neuronal ACh-receptor exhibits properties (multiple gating, burst behavior, desensitization) very similar to the receptor from electric tissue (Boheim et al., 1981) and to the nicotinic acetylcholine receptors in muscle cells analysed <u>in situ</u> (Sakmann et al., 1983).

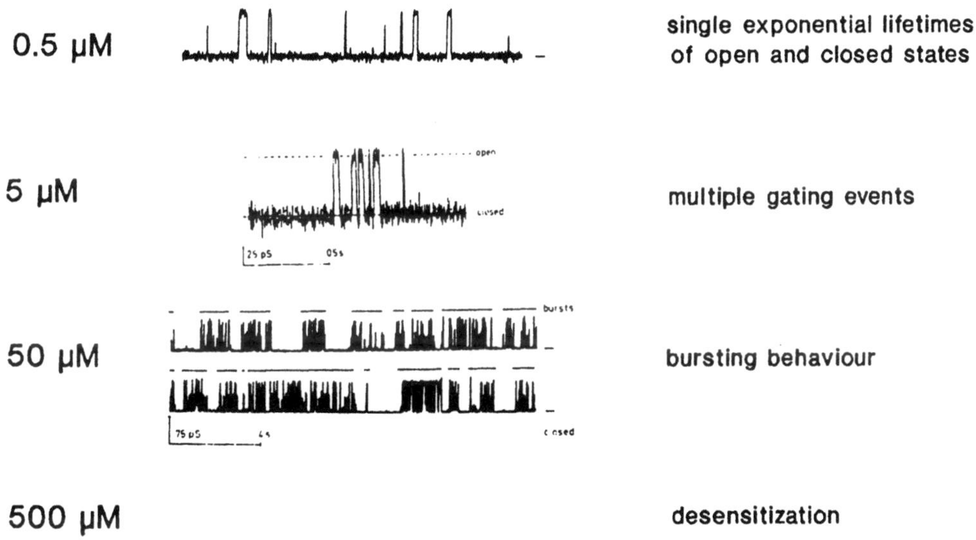

Fig. 3 Effect of different agonist concentrations on the kinetics of the acetylcholine-receptor channel

Very recently procedures have been developed to reconstitute polypeptides in lipid bilayer after separation by SDS-polyacrylamide gelelectrophoresis followed by electroelution. After incorporation of the isolated receptor polypeptide into lipid membranes under appropriate conditions it was found that cholinergic agonists induced increased cation permeability of the membrane and

fluctuations of single channels with similar properties as channels formed by the native protein.
Thus, a single population of receptor polypeptides formed functional channels which can be activated by cholinergic agonists.

Table 1

CHANNEL PROPERTIES	LOCUSTA	TORPEDO	REFERENCE
Maximum conductance pS	100	95*; 90**	* Montal et al., 1984 ** Tank et al.,1983
Sodium concentration for half-max. conductance mM	50	395*; 129**	* Montal et al. 1984
Open state lifetime at low Carb (< 1uM) ms	3	5	
App. cooperativity of agonists	1	2	Labarca et al.,1984
Potency of agonists	Sub > ACh > Carb	Sub > ACh > Carb	
Sensitivity to d-tubo-curarine	++	++	
Sensitivity to hexa-methonium§	+	++	Hanke et al., 1981
Selectivity	$Na^+ \sim K^+ \gg Cl^-$	$Na^+ \sim K^+ \gg Cl^-$	Boheim et al.,1981

Expression of receptor-specific mRNA

In order to recognize mRNA encoding receptor-specific polypeptides, isolated poly A^+ -RNA was expressed in a cell-free translation system as well as in Xenopus oocytes. RNA isolated from nervous tissue of young locusts was translated in a rabbit reticulocyte lysate in the presence of labelled aminoacids. In immunoprecipitation experiments followed by electrophoretic analysis it was found that about 0.1% of the newly synthesized polypeptides could be separated indicating that antigenic sites already exist in non-processed polypeptides and that the poly A^+-fractions obviously contained receptor-specific mRNA (Benke and Breer, 1987). Furthermore, the electrophoretic analysis revealed that the precipitated polypeptides were significantly smaller than the native receptor subunits. Expression experiments in a reticulocyte system supplemented with microsomes gave a polypeptide band of larger molecular size (M_r=60,000) indicating that some posttranslational modifications have been performed in this system. In order to estimate if intact receptor polypeptides could be synthesized, isolated mRNA was expressed in oocytes. When poly A^+-RNA isolated from the nervous tissue of young locusts was injected into oocytes, after 1 day specific binding activity for α-bungarotoxin was detected (Breer and Benke, 1986). Immunoprecipitation experiments revealed that receptor polypeptides synthesized in ovo display the same or a similar size as the native subunits, indicating that obviously most of the posttranslational modification of the insect receptor protein were performed in oocytes. Ion flux studies using ^{86}Rb isotops demonstrated that 24h after microinjection the oocytes contain functional ion channels that can be activated by cholinergic agonists. These results indicate that insect mRNA can direct synthesis and processing of specific membrane proteins and furthermore demonstrate that the isolated RNA-fraction does contain receptor-specific mRNA. This nucleic acid fraction was used as template to produce complementary DNA which is screened with specific antibodies and oligonucleotide probes for receptor-specific clones which can then be used to decipher the primary structure of the locust receptor polypeptides.

Conclusion

In many cases the potential of invertebrate nervous tissue for studying general aspects of cellular and molecular neurobiology has been demonstrated (Goodman, 1985). Concerning the neuronal acetylcholine receptor, insect material has provided the first evidence for a functional postsynaptic role of an α-bungarotoxin-sensitive nicotinic acetylcholine receptor at an identified central cholinergic synapse; furthermore an affinity purified toxin binding protein isolated from insect nervous tissue was the first neuronal nicotinic acetylcholine receptor functionally reconstituted in lipid membranes. As the receptor for acetylcholine is a key element in cholinergic transmission of signals in the nervous system of insects, it may be considered as a potential target for insect control agents; thus a more detailed knowledge of insect ACh-receptor molecules may contribute to design and develop more selective and safer insecticides.

References

Anholt, R., Lindstrom, J. and Montal, M. (1983) the molecular basis of neurotransmission:Structure and function of the nicotinic acetylcholine receptor. In: The enzymes of biological membranes. (ed. Martenosi, A.) Plenum Press, New York. pp. 335-401.

Benke, D., and Breer, H. (1987) Expression of neuronal acetylcholine receptor polypeptides in vitro. Cell. Molec. Neurobiol. 7, 391-401.

Boheim, G., Hanke, W.,Barrantes, F.J., Eibl, H., Sakmann, B., Fels, G. and Maelicke, A. (1981) Agonist activated ionic channels in acetylcholine receptor reconstituted into planar lipid bilayers. Proc. Natl. Acad. Sci. (USA) 78, 3586-3590.

Boulter, J., Evans, K., Goldmann, D., Martin, G., Treco, D., Heineman, S., and Patrick, J. (1987) Isolation of a cDNA clone coding for a possible neural nicotinic acetylcholine receptor α-subunit. Nature 319, 368-374.

Breer, H. (1981) Properties of putative nicotinic and muscarinic cholinergic receptors in the central nervous system of Locusta migratoria.. Neurochem. Int. 3, 43-52.

Breer, H., Kleene, R., Hinz, G. (1985) Molecular forms and subunit structure of the acetylcholine receptor in the central nervous system of insects. J. Neurosci. 5, 3386-3392.

Breer, H., and Benke, D. (1986) Messenger RNA from insect nervous tissue induces expression of neuronal acetylcholine receptors in Xenopus oocytes. Molec. Brain Res. 1, 111-117.

Breer, H., Sattelle, D.B. (1987) Molecular properties and functions in insect acetylcholine receptors. J. Insect Physiol. 33, 771-790.

David, J.A., Sattelle, D.B. (1984) Actions of cholinergic pharmacological agents on the cell body membrane of the fast coxal depressor motoneurone of the cockroach (Periplaneta americana). J. Exp. Biol. 108, 119-136.

Fels, G., Breer, H., and Maelicke, A. (1983) Are there nicotinic acetylcholine receptors in invertebrate ganglionic tissue? In: Hucho, F. and Ovchinnow, A. (eds.). Toxins as Tools in Neurochemistry. De Gruyter Berlin, 127-140.

Florey, E., Acetylcholine in invertebrate nervous system. (1963) Can. J. Biochem. Physiol. 41, 2619-2626.

Goodman C. S. (1985) Flies and worms: molecular genetic approaches to neurobiology. TINS 8: 229-231.

Hanke, W., Eibl, H., and Boheim, G. (1981) A new method for membrane reconstitution: fusion of protein containing vesicles with planar lipid bilayers below phase transition temperature. Biophys. Struct. Mech. 7, 131-137.

Hanke, W., and Breer, H. (1986) Channel properties of a neuronal acetylcholine receptor protein purified from central nervous system of insect reconstituted in planar lipid bilayers. Nature (Lond) 321, 171-174.

Hanke, W., and H. Breer (1987) Characterization of the channel properties of a neuronal acetylcholine receptor reconstituted into planar lipid bilayer. J. Gen. Physiol. 90, 855-879.

Sakmann, B., Bormann, J. Hamill, O.P. (1983) Ion transport by single receptor channels. In: Molecular Neurobiology (Ed. by Watson J.D. and McKay, R.). Cold Spring Harbor Symposia on Quantitative Biology XLVII, 247-257.

Sattelle, D.B., David, J.A. (1983) Voltage-dependent block by histrionicotoxin of the acetylcholine-induced current in a insect motoneurone cell body. Neurosci. Lett. 43, 37-41.

Sattelle, D.B., and Breer, H. (1985) Purification by affinity chromatography of nicotinic acetylcholine receptor from the CNS of the cockroach Periplaneta americana. Comp. Biochem. Physiol. 82C, 349-352.

Sattelle, D.B., Sun, Y.A., Wu, C.F. (1986) Neuronal acetylcholine receptor: patch clamp recording of single channel properties from dissociated insect neurones. IRCS Med. Sci. 14, 65-66.

Tareilus, E., Hanke, W. (1988) Single channel analyis of the neuronal acetylcholine receptor from insects. Hoppe-Seyler's J. Physiol. Chem. (in press).

Whiting, P.J., and Lindstrom, J.L. (1986) Purification and characterization of a nicotinic ACh-receptor from chick brain. Biochemistry 25, 2082-2093.

LIGAND-GATED ION CHANNELS OF <u>DROSOPHILA</u>

Eckart D. Gundelfinger*, Irm Hermans-Borgmeyer°, Patrick Schloß, Erich Sawruk, Conny Udri, Martin Vingron$, Heinrich Betz, and Bertram Schmitt.
ZMBH, Universität Heidelberg, Im Neuenheimer Feld 282, D-6900 Heidelberg, FRG, and $EMBL, Postfach 102209, D-6900 Heidelberg, FRG.

INTRODUCTION

Protein subunits of ligand-gated ion channels including excitatory nicotinic acetylcholine receptors (AChR) as well as inhibitory γ-amino butyric acid (GABA) and glycine receptors (GlyR) are encoded by a large super-family of related genes (Schofield et al., 1987; Grenningloh et al., 1987a, b). Using DNA probes of vertebrate receptor subunits we have started to isolate genomic and cDNAs encoding ion channel proteins of the fruitfly <u>Drosophila melanogaster</u>. Two of these genes/cDNAs and the deduced proteins will be described: the <u>ard</u> gene coding for an AChR-like <u>Drosophila</u> protein (ARD protein) and the <u>grd</u> gene encoding a potential ligand-gated chloride channel polypeptide.

NEURONAL NICOTINIC ACETYLCHOLINE RECEPTOR

Acetylcholine is a major excitatory neurotransmitter in the nervous system of insects. At least part of its action appears to be mediated by nicotinic AChR(s) since (i) nicotine is an efficient insecticide and (ii) high levels of binding sites for α-bungarotoxin (α-Btx), a potent antagonist of AChR at the vertebrate neuromuscular junction, have been detected in the

---

Present addresses: *Zentrum für Molekulare Neurobiologie, Universitätskrankenhaus Eppendorf, Martinistraße 52, D-2000 Hamburg 20, FRG; °Neurobiology Laboratory, The Salk Institute, San Diego CA 92138, USA.

NATO ASI Series, Vol. H 32
Molecular Biology of Neuroreceptors and Ion Channels. Edited by A. Maelicke

neuropil region of Drosophila central nervous system (Schmidt-Nielsen et al., 1977). By molecular cloning, primary structures of two AChR-like subunits from Drosophila have been elucidated recently: the α-like subunit (ALS) which resembles ligand binding subunits of vertebrate receptors (Bossy et al., 1988), and the ARD protein, a potential non-ligand binding AChR subunit (Hermans-Borgmeyer et al., 1986; Wadsworth et al., 1988). This first chapter deals with structural and functional features of the ARD protein and the corresponding gene.

1. The ard gene

A cDNA probe from the γ-subunit of the Torpedo electroplax AChR (Claudio et al., 1983) was used to select genomic and cDNA clones encoding the ARD protein (Hermans-Borgmeyer et al., 1986; Gundelfinger et al., 1986; Sawruk et al., 1988). In situ hybridization of genomic DNA to polytene salivary gland chromosomes revealed that the ard gene maps to region 64B of the left arm of the 3rd chromosome from D. melanogaster. The ard gene extends over more than 7 kb and has at least six introns, five of which interrupt the protein coding region and one is located upstream of the translation start site.

Compared to genes encoding α-, γ-, and δ-subunits of vertebrate muscle AChRs, the ard gene contains fewer introns (Figure 1A). Almost all of the intron positions occurring in all of these genes are exactly conserved. Only the intron separating the first two membrane spanning segments, M1 and M2, of AChR proteins is placed 8 to 14 nucleotides downstream in the ard gene relative to the different muscle receptor genes (Figure 1B). Neither vertebrate neuronal AChR genes (Nef et al., 1988; Wada et al., 1988) nor the Drosophila gene encoding ALS (Bossy et al., 1988) contain an intron at this position. Since the region between M1 and M2 is highly conserved (Figure 1B), independent acquisition of introns during evolution rather than intron sliding may explain the differential placement of this intron (for a more detailed discussion see Sawruk et al., 1988).

A

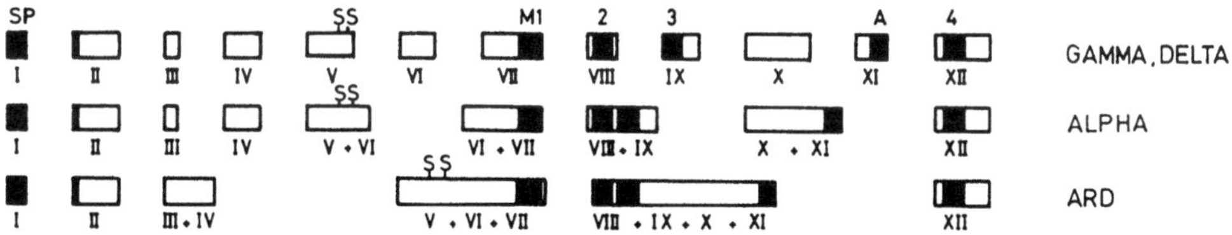

B

Exon I/II

ARD	-6	PheSerLeuVal - - - SerAla	-1
human α	-9	LeuCysSerAlaGlyLeuValLeuGly	-1
human γ	-7	ValCysLeuGlyAla - - GlnGly	-1
chick γ	-7	ValLeuProGlyIle - - SerCys	-1
chick δ	-7	ValLeuSerGlyGly - - LeuCys	-1

Exon VII/VIII

ARD	237	ProAlaGluAlaGly - GluLysValThr	245
human α	236	ProThrAspSerGly - GluLysMetThr	244
human γ	244	ProAlaLysAlaGlyGlyGlnLysCysThr	253
chick γ	245	ProAlaLysAlaGlyGlyGlnLysCysThr	254
chick δ	252	ProAlaAspSerGly - GluLysMetThr	260

Exon II/III

ARD	41	IleAsnValAsnGluLys	46
human α	41	IleAsnValAspGluVal	46
human γ	41	IleSerLeuAsnGluArg	46
chick γ	41	IleSerLeuAsnGluArg	46
chick δ	43	IleSerLeuLysGluVal	48

Exon XI/XII

ARD	441	TyrIleGlnThrArgGluAspTrp	448
human α	391	SerAsnAsnAlaAlaAlaGluTrp	399
human γ	436	PheAspAsnGlyAsnGluGluTrp	443
chick γ	435	PheSerSerGluAsnGluGluTrp	442
chick δ	433	TyrAsnGluGluLysAspAsnTrp	440

Exon IV/V

ARD	93	PheAsnAsnAlaAspGly	98
human α	93	TyrAsnAsnAlaAspGly	98
human γ	93	GluAsnAsnValAspGly	98
chick γ	93	GluAsnAsnIleAspGly	98
chick δ	95	GluAsnAsnAsnAspGly	100

Figure 1. Comparison of AChR genes. (A) Schematic presentation of exon-intron organization of the protein coding regions of the α-, γ-, and δ-subunit genes of the vertebrate muscle AChR (Noda et al., 1983; Nef et al., 1984; Shibahara et al., 1985) and the <u>ard</u> gene (Sawruk et al., 1988; Wadsworth et al.,1988). (B) Location of splice positions (▾) on the aligned amino acid sequences. Only introns occurring in the <u>ard</u> gene are considered. Numbers give the positions of the amino acid residues within the mature proteins. Negative numbers indicate cleavable signal peptides.

2. Mature and immature transcripts of the <u>ard</u> gene during development

Both Northern analysis and <u>in situ</u> hybridization with cDNA probes encoding the ARD protein to frozen sections of different developmental stages of <u>Drosophila</u> were performed to study the

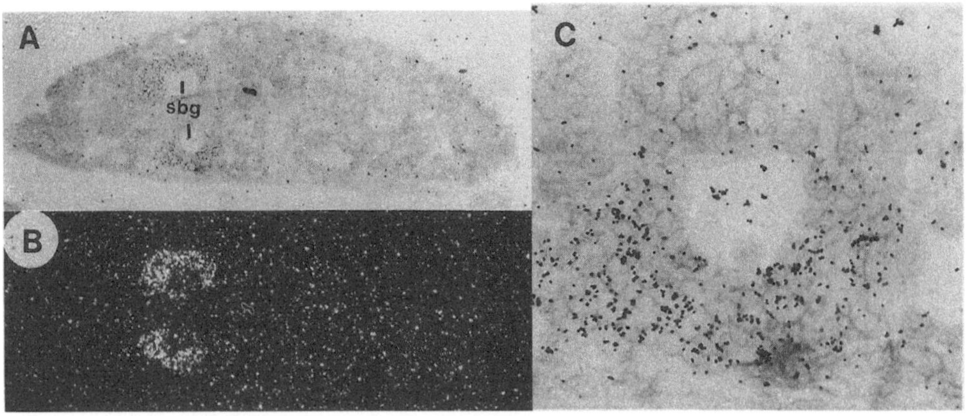

Figure 2. Localization of ard transcripts in a late embryo by in situ hybridization with ard cDNA. (A) Phase contrast and (B) darkfield photomicrographs (50-fold magnification) of a horizontal section of a stage 17 (16-22 hrs) embryo. Suboesophageal ganglia (sbg) are heavily labelled. (C) 250-fold magnification of one of the ganglia shows that not all cell bodies are labelled.

spatial and temporal expression pattern of the ard gene. Significant hybridization was found only over the cell body layers of the central nervous system at all stages examined (Figure 2 and not shown). Transcripts were detected first in the ventral cord and oesophageal ganglia of 10 hrs-old embryos. During late embryogenesis a strong increase of ard transcripts was observed (Figure 2). At larval stages, lower RNA levels are present while in late pupae and newly eclosed flies transcripts accumulate again (Wadsworth et al., 1988; Hermans-Borgmeyer et al., submitted). These findings are consistent with the assumption that the ARD protein is part of a neuronal receptor which is synthesised predominantly during terminal differentiation of the larval and the adult nervous systems.

As shown in Figure 3, a major transcript of about 3.2 kb is found in the late embryo while in fly heads three RNA bands of 3.2 kb, 4.2 kb, and ≈7.5 kb can be detected on Northern blots hybridized with a radiolabelled ARD cDNA fragment from the 6th exon. Since several ARD cDNA clones contained intron sequences of the ard genes, we probed Northern blots of embryonal and adult head poly(A)RNA with radiolabelled fragments derived from

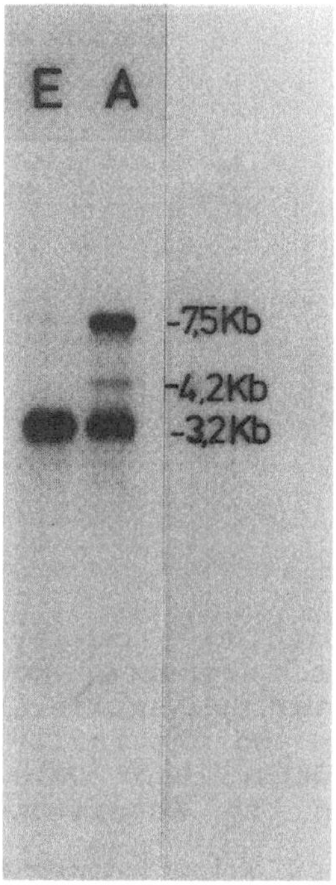

Figure 3. Northern blot analysis of embryonal (E) and adult (A) ard transcripts using probe an ARD cDNA probe.

introns 2, 3, 4, and 5. The results indicate that the 7.5 kb transcripts contain all four introns while the 4.2 kb band hybridized to probes from introns 3 and 4 only. Thus during late stages of Drosophila development, high amounts of partially or unspliced ard RNAs accumulate. In situ hybridization using a fragment derived from the 3rd intron indicated a co-distribution of intron-containing and mature ard transcripts (Hermans-Borgmeyer et al., submitted).
A similar developmentally regulated accumulation of immature transcripts has also been reported for the $su(w^a)$ gene of Drosophila (Chou et al., 1987). Whether such partially spliced RNAs are functionally relevant or whether they are just by-products of RNA processing remains to be clarified. In any case, these findings suggest regulation of ARD protein biosynthesis at the level of primary transcript processing.

3. The ARD protein, a component of an α-Btx binding complex

A Scatchard analysis of ^{125}I-α-Btx binding to membranes of Drosophila heads revealed the existence of two different high-affinity binding sites for the toxin. Site 1 ($K_d \approx 0.15$ nM) represents 20 to 30% of the total number of toxin binding complexes, whereas site 2 ($K_d \approx 4$ nM) represents 70 to 80%. Both toxin binding sites exhibit a typical nicotinic cholinergic pharmacology (Schloß et al., 1988).

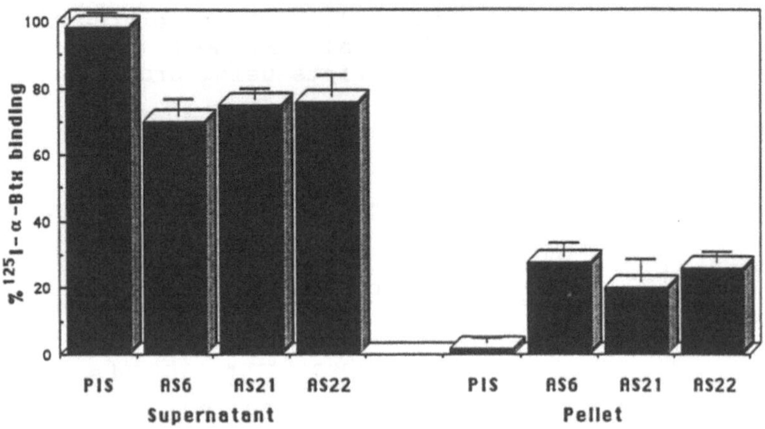

Figure 4. Immunoprecipitation of ^{125}I-α-Btx binding sites by ARD antisera. ^{125}I-α-Btx-labelled detergent extracts of Drosophila head membranes were incubated with various antisera (see text). IgGs were precipitated by Pansorbin and the radioactivity in supernatant and immun-pellet was determined. Data are expressed as % binding sites present in the detergent extract.

We have used rabbit antisera, obtained against bacterially expressed fusion proteins of the N-terminal region (AS 21, AS 22) and the cytoplasmic domain (AS 6) of the ARD protein, to immunoprecipitate ^{125}I-α-Btx binding sites solubilized from head membranes (Schloß et al., 1988). Both groups of antisera precipitated about 20 to 30% of the toxin binding complexes while pre-immune sera (PIS) left all binding sites in the supernatant (Figure 4). Scatchard analyses of toxin binding to AS 6 and AS 21 supernatants revealed that they contain exclusively high-affinity binding site 2. Therefore, we concluded that the ARD protein is associated with ^{125}I-α-Btx binding site 1 and most probably represents a subunit of a neuronal nicotinic AChR.

The existence of two physically separable α-Btx binding sites suggests heterogeneity of insect AChRs. Since the ARD protein is lacking structural features of toxin binding α-subunits, it is unlikely that it constitutes a homo-oligomeric receptor as was proposed for a locust neuronal AChR isolated by Breer et al. (1985). Whether the recently described α-like subunit ALS

(Bossy et al., 1988) is part of any of the α-Btx sites described here is unknown. However, since the temporal expression patterns of ALS and ARD transcripts are different, the corresponding proteins are unlikely to constitute subunits of the same AChR complex.

4. Implications for the evolution of AChRs

The sequences of nine paralogous subunits of vertebrate AChRs have been established. Five of these are expressed in muscle where they constitute two developmentally regulated isoforms of extrasynaptic and synaptic AChRs. Four different subunits - three α-like and one of non-α type - were identified in neural tissues. Additionally, the primary structures of two invertebrate potentially neuronal AChR subunits, ALS and ARD protein of Drosophila, have now been elucidated. How did this heterogeneity evolve from an ancesteral gene, presumed to encode the subunits of a homo-oligomeric receptor progenitor? Table I summarizes homologies and evolutionary distances of 320 to 337 amino acid long segments of the various AChR subunits spanning the N-terminus of mature proteins to a proline residue common to all subunits, 23 amino acids beyond the C-terminus of the third transmembrane region. These segments were chosen since they could be aligned almost unambigously while the cytoplasmic regions are too variable for proper alignment.

Neuronal receptors, including the invertebrate ones show significantly higher homology among each other (50 to 82%) than to the non-α subunits of the muscle receptors (39 to 50%). The muscle α-subunit also seems to be more closely related to the neuronal than to other muscular subunits. The ARD protein and the ß2 neuronal subunit both share a higher homology with neuronal α-subunits than with each other; they rather may be paralogous than homologous proteins. Preliminary constructions of phylogenetic trees suggest that vertebrate AChR proteins may have evolved from the "ALS branch" of invertebrate receptor polypeptides. However, more invertebrate sequences are required to precisely reconstruct the evolution of AChRs.

Table I. Homology and evolutionary distance matrices for ligand-gated ion channel proteins[1)]

	% IDENTITY													
	a	b	g	e	d	a2	a3	a4	b2	ALS	ARD	48K	Ga	Gb
AChR														
Muscle a (6)[2)]		44	40	36	42	56	56	57	50	51	45	20	19	17
b (3)	97		49	45	47	47	43	46	50	43	46	19	17	18
g (6)	112	84		59	55	42	41	43	46	41	41	18	17	15
e	126	92	58		50	42	42	41	44	39	40	16	15	15
d (5)	108	90	71	81		45	44	44	48	40	46	17	17	17
Neural a2 (2)	67	92	110	111	103		70	82	58	60	53	20	18	18
a3	63	99	111	106	100	40		70	57	57	51	20	17	18
a4 (2)	63	93	105	111	103	23	39		59	58	52	20	18	19
b2 (2)	81	84	107	114	98	65	64	63		55	50	20	19	19
D.m. ALS	78	103	114	121	113	60	63	63	73		53	19	19	18
D.m. ARD	94	92	112	117	101	79	81	81	84	77		21	17	18
GlyR 48K						n.d.[3)]							38	41
GABAR a						n.d.						119		39
GABAR b						n.d.						108	118	
	EVOLUTIONARY DISTANCE (mutations / 100 residues)													

Distance values were calculated according to Feng et al. (1985) using unitary matrix scores obtained from a multiple alignment of 30 different AChR proteins. [1)]For reasons of proper alignment only amino acids 1 to 320-337 were considered (see text). [2)]If more than one homologous subunit was included, the number of different ones is given in parenthesis. a=alpha, b=beta, g=gamma, d=delta, e=epsilon. [3)]Not determined, since values obtained with the unitary matrix of proteins with less than 30% homology are not reliable (Feng et al., 1985). Sequences were taken from the following references: Nef et al., 1984, 1988; Kubo et al., 1985; Takai et al.,1985; Heinemann et al., 1986; Hermans-Borgmeyer et al., 1986; Grenningloh et al., 1987a; Schofield et al., 1987; Baldwin et al., 1988; Bossy et al., 1988; Deneris et al., 1988.

Although homology studies place the muscle non-α subunits as a separate group, an analysis of structural features of AChR proteins and genes suggests a common phylogenetic descent for all muscle AChR polypeptides. They lack a potential N-glycosylation site close to the amino terminus which is present in all neuronal receptors. Another N-glycosylation site (Asn 141) is shared by all muscle subunits while absent in some neuronal ones. Furthermore, the intron separating hydrophobic regions M1 and M2 is placed in homologous positions in all muscle receptor genes investigated but is either absent or placed in a different position in neuronal genes (see above). These data suggest that the vertebrate muscle AChR subunits may have diverged from the neuronal receptors once as a common original branch. The appearent distance may then be due to a much higher mutation rate of non-α vs ligand binding α-subunits. Indeed, evolution rates can be estimated to be between 4 to 5 exchanges per 100 amino acids in 100 million years for α-subunits and about 11 exchanges for the non-α polypeptides.

LIGAND-GATED CHLORIDE CHANNELS

In vertebrates glycine and GABA receptor subunits are about 40% identical (Table I). The regions of highest homology are between the two conserved cysteines in the extracellular domain and in the first three membrane spanning regions. The potential channel forming transmembrane segment, M2, is characterized by 8 consecutive amino acids which are identical in the various GABA and glycine receptor polypeptides (Grenningloh et al., 1988b). We have used cDNAs encoding the conserved regions of the rat GlyR 48K subunit to screen a Drosophila genomic library (obtained from V. Pirotta) for homologous clones. Of several hybridization positives, one clone (GRD C3) was identified as encoding two exons of a neurotransmitter receptor-like protein. The deduced amino acid sequence of the putative second transmembrane region indicates that this polypeptide is a potential ligand-gated chloride channel subunit, because 7 consecutive amino acids are identical to the M2 consensus sequence of vertebrate GABA and glycine receptors (Fig. 5).

GABA acts as a neurotransmitter in both muscle and central nervous system of insects (Gerschenfeld, 1973) whereas nothing is known about glycine action in this group of

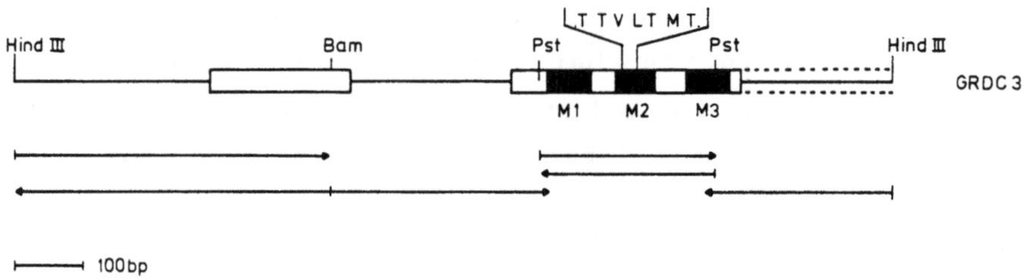

Figure 5. Physical map and sequencing strategy of the genomic DNA clone GRD C3. The M2 consensus sequence conserved in all ligand-gated chloride channels is given. Bam = Bam HI; Pst = Pst I. Dotted lines: 3'end of the exon encoding M1 to M3 region was not defined, since homology to vertebrate subunits ceases.

organisms. The protein encoded by GRD C3 therefore may represent a subunit of an invertebrate GABA receptor. Cloning and expression of the corresponding cDNA should help to identify the nature of the encoded protein.

PERSPECTIVES

In vertebrates, a high degree of heterogeneity of AChRs (reviewed by Lindstrom et al., 1987; see above), GABA receptors (Levitan et al., 1988), and GlyRs (Becker et al., 1988) has been described. A correlation of pharmacologically characterized ligand binding sites with receptor proteins identified by means of gene technology appears to be difficult. In particular the possiblility to obtain "functional" receptors/ion channels from arteficially combined subunits expressed in frog oocytes complicates the investigation of these proteins. As shown here, invertebrate ligand-gated ion channels are structurally and pharmacologically related to vertebrate ones. Since fewer and simpler receptor complexes may exist in invertebrates (e.g. Breer et al., 1985), their analysis should, however, be correspondingly simpler. The lower complexity of invertebrate nervous systems may facilitate the correlation of receptor subunits by in situ hybridization and immunocytochemistry. Finally, the possibility to create receptor mutants in Drosophila should help to elucidate the role of these receptors both in the developing and mature nervous system.

Acknowledgements: We thank Dr. C. Garner for comments on the manuscript and S. Raabe for help during its preparation. E. S. is the recipient of a Boehringer Ingelheim Fonds predoctoral fellowship. This work was supported by grants of the Deutsche Forschungsgemeinschaft (Gu230/1-1 and SFB317), the Bundesministerium für Forschung und Technologie (BCT 365/1), and the Fonds der Chemischen Industrie.

REFERENCES

Baldwin TJ, Yoshihara CM, Blackmer K, Kintner CR, Burden SJ (1988) Regulation of Acetylcholine Receptor Transcript Expression during Development in Xenopus laevis. J Cell Biol 106:469-478

Becker CM, Hoch W, Betz H (1988) Glycine receptor heterogeneity in rat spinal cord during postnatal development. EMBO J 12:(in press)

Bossy B, Ballivet M, Spierer P (1988) Conservation of neural nicotinic acetylcholine receptors from Drosophila to vertebrate central nervous system. EMBO J 7:611-618

Breer H, Kleene R, Hinz G (1985) Molecular forms and subunit structure of the acetylcholine receptor in the central nervous system of insects. J Neurosci 5:3386-3392

Chou TB, Zachar Z, Bingham PM (1987) Developmental expression of a regulatory gene is programmed at the level of splicing. EMBO J 6:4095-4104

Claudio T, Ballivet M, Patrick J, Heinemann S (1983) Nucleotide and deduced amino acid sequences of Torpedo californica acetylcholine receptor γ-subunit. Proc Natl Acad Sci USA 80: 1111-1115

Deneris ES, Connolly J, Boulter J, Wada E, Wada K, Swanson W, Patrick J, Heinemann S (1988) Primary Structure and Expression of β2: A Novel Subunit of Neuronal Nicotinic Acetylcholine Receptors. Neuron 1:45-54

Feng DF, Johnson MS, Doolittle RF (1985) Aligning Amino Acid Sequences: Comparison of Commonly Used Methods. J Mol Evol 21: 112-125

Gerschenfeld HM (1973) Chemical Transmission in Invertebrate Central Nervous Systems and Neuromuscular Junctions. Physiol Reviews 53: 1-119

Grenningloh G, Rienitz A, Schmitt B, Methfessel C, Zensen M, Beyreuther K, Gundelfinger ED, Betz H (1987a) The strychnine-binding subunit of the glycine receptor shows homology with nicotinic acetylcholine receptors. Nature 328:215-220

Grenningloh G, Gundelfinger ED, Schmitt B, Betz H, Darlison MG, Barnard EA, Schofield PR, Seeburg PH (1987b) Glycine vs GABA receptors. Nature 330:25-26

Gundelfinger ED, Hermans-Borgmeyer I, Zopf D, Sawruk E, Betz H (1986) Characterization of the mRNA and the gene of a putative neuronal nicotinic acetylcholine receptor from Drosophila. In: Maelicke A (ed) Nicotinic Acetylcholine Receptor, vol H3. NATO ASI Series, Springer Verlag Berlin, Heidelberg, pp 437-446

Heinemann S, Boulter J, Connolly J, Goldman D, Evans K, Treco D, Ballivet M, Patrick J (1986) Molecular Biology of Muscle and Neural Acetylcholine Receptors. In: Maelicke A (ed) Nicotinic Acetylcholine Receptor, vol H3. NATO ASI Series, Springer Verlag Berlin, Heidelberg, pp 359-387

Hermans-Borgmeyer I, Zopf D, Ryseck RP, Hovemann B, Betz H, Gundelfinger ED (1986) Primary structure of a developmentally regulated nicotinic acetylcholine receptor protein from Drosophila. EMBO J 5:1503-1508

Kubo T, Noda M, Takai T, Tanabe T, Kayano T, Shimizu S, Tanaka K, Takahashi H, Hirose T, Inayama S, Kikuno R, Miyata T Numa S (1985) Primary structure of δ subunit precursor of calf muscle acetylcholine receptor deduced from cDNA sequence. Eur J Biochem 149:5-13

Levitan ES, Schofield PR, Burt DR, Rhee LM, Wisden W, Köhler M, Fujita N, Rodriguez HF, Stephenson A, Darlison MG, Barnard EA, Seeburg PH (1988) Structural and functional basis for $GABA_A$ receptor heterogeneity. Nature 335:76-79

Lindstrom J, Schoepfer R, Whiting P (1987) Molecular Studies of the Neuronal Nicotinic Acetylcholine Receptor Family. Molec Neurobiol 1:281-337

Nef P, Mauron A, Stalder R, Alliod C, Ballivet M (1984) Structure, linkage, and sequence of the two genes encoding the δ and γ-subunits of the nicotinic acetylcholine receptor. Proc Natl Acad Sci USA 81:7975-7979

Nef P, Oneyser C, Alliod C, Couturier S, Ballivet M (1988) Genes expressed in the brain define three distinct neuronal nicotinic acetylcholine receptors. EMBO J 7:595-601

Noda M, Furutani Y, Takahashi H, Toyosato M, Tanabe T, Shimizu S, Kikyotani S, Kayano T, Hirose T, Inayama S, Numa S (1983) Cloning and sequence analysis of calf cDNA and human genomic DNA encoding α-subunit precursor of muscle acetylcholine receptor. Nature 305:819-823

Sawruk E, Hermans-Borgmeyer I, Betz H, Gundelfinger ED (1988) Characterization of an invertebrate nicotinic acetylcholine receptor gene: the ard gene of Drosophila melanogaster. FEBS Lett 235:40-46

Schloß P, Hermans-Borgmeyer I, Betz H, Gundelfinger ED (1988) Neuronal acetylcholine receptors in Drosophila: The ARD protein is a component of a high-affinity α-bungarotoxin binding complex. EMBO J 7:2889-2894

Schmidt-Nielsen BK, Gepner JI, Teng NNH, Hall LM (1977) Characterization of an α-Bungarotoxin binding component from Drosophila melanogaster. J Neurochem 29:1013-1029

Schofield PR, Darlison MG, Fujita N, Burt DR, Stephenson FA, Rodriguez H, Rhee LM, Ramachandran J, Reale V, Glencorse TA, Seeburg PH, Barnard EA (1987) Sequence and functional expression of the $GABA_A$ receptor shows a ligand-gated receptor super-family. Nature 328:221-227

Shibahara S, Kubo T, Perski HJ, Takahashi H, Noda M, Numa S (1985) Cloning and sequence analysis of human genomic DNA encoding γ-subunit precursor of muscle acetylcholine receptor. Eur J Biochem 146:15-22

Takai T, Noda M, Mishina M, Shimizu S, Furutani T, Kayano T, Ikeda T, Kubo T, Takahashi H, Takahashi T, Kuno M, Numa S (1985) Cloning squencing and expression of cDNA for a novel subunit of acetylcholine receptor from calf muscle. Nature 315:761-764

Wada K, Ballivet M, Boulter J, Connolly J, Wada E, Deneris ES, Swanson LW, Heinemann S, Patrick J (1988) Functional Expression of a New Pharmacological Subtype of Brain Nicotinic Acetylcholine Receptor. Science 240:330-334

Wadsworth SC, Rosenthal LS, Kammermeyer LK, Potter MB, Nelson DJ (1988) Expression of a Drosophila acetylcholine receptor-related gene in the central nervous system. Molec Cell Biol 8:778-785

Chandler [illegible]

[illegible]

THE STRUCTURE AND EXPRESSION OF THE $GABA_A$ RECEPTOR AS DEDUCED BY MOLECULAR GENETIC STUDIES

Mark G. Darlison, Eric A. Barnard, Alan N. Bateson, Thora A. Glencorse, Robert J. Harvey, Andrew A. Hicks, Stephen P. Hunt, Brian J. Morris, Mario Vallejo, Erno Vreugdenhil and William Wisden

MRC Molecular Neurobiology Unit
Medical Research Council Centre
Hills Road
Cambridge, CB2 2QH
United Kingdom

INTRODUCTION

It is well established that the inhibitory neurotransmitter γ-aminobutyric acid (GABA) mediates many of its effects by binding to the $GABA_A$ receptor, which is present on the majority of mammalian brain neurons (Enna, 1983), resulting in the opening of an integral chloride channel. This receptor has considerable pharmaceutical importance as it contains binding sites for anti-convulsant (barbiturates), anxiogenic (β-carbolines) and convulsant (picrotoxin) drugs (reviewed in Turner and Whittle, 1983; Olsen and Venter, 1986). Although it is known that anxiolytic agents such as benzodiazepines also bind to the $GABA_A$ receptor, there exists a population of benzodiazepine-insensitive receptors (Study and Barker, 1981; Unnerstall et al., 1981; de Blas et al., 1988). Our understanding of the molecular structure and function of this receptor complex has been greatly increased by the application

NATO ASI Series, Vol. H 32
Molecular Biology of Neuroreceptors and Ion Channels. Edited by A. Maelicke

of recombinant DNA methodology and the subsequent cloning and expression of cDNA clones that encode its constituent subunits. This chapter will review these advances and describe some recent results arising from the use of the cloned sequences.

PROTEIN BIOCHEMISTRY AND MOLECULAR CLONING

When the $GABA_A$ receptor is purified by benzodiazepine affinity chromatography from bovine (Sigel et al., 1983; Sigel and Barnard, 1984; Schoch et al., 1985), chick (Mamalaki et al., 1987), rat (Taguchi and Kuriyama, 1984) or pig (Kirkness and Turner, 1986) brain and analysed by sodium dodecyl sulphate (SDS)-polyacrylamide gel electrophoresis, two bands are detected. These have been designated as α (M_r 53000) and β (M_r 57000). In the rat, immunohistochemical distributions of monoclonal antibodies that recognise epitopes present either in the α band or in the β band are identical (Richards et al., 1987), supporting the notion that these two components are present in the same oligomeric complex. In addition, photo-affinity labelling studies on the purified bovine receptor have shown that the benzodiazepine flunitrazepam reacts with the α band (Mamalaki et al., 1987) whereas the GABA agonist muscimol reacts with the β band (Casalotti et al., 1986). As with other membrane-bound receptors, the $GABA_A$ receptor is glycosylated as evidenced by a mobility shift of the α and β bands upon treatment of the complex with endoglycosidase F: indeed, upon deglycosylation and SDS-polyacrylamide gel electrophoresis the α band is no longer detected and three bands of lower molecular weight (one major and two minor) are seen (Mamalaki et al., 1987).

The $GABA_A$ receptor complex as purified from adult bovine cerebral cortex and the isolated α band were treated with either cyanogen bromide or trypsin and the resultant peptides separated by high pressure liquid chromatography. Several of these were subjected to gas-phase micro-sequencing. 'Best-guess' oligonucleotide probes (see Wood, 1987) were then used to screen calf cerebral cortex and adult bovine brain cDNA libraries. In this way two clones were isolated that, based on the encoded polypeptide sequences, specified an α and a β subunit (Schofield et al., 1987). Subsequently, screening of a bovine brain cDNA library with a 60-base oligonucleotide probe (corresponding to nucleotides 554 to 593 of the α-subunit cDNA sequence, encoding part of the proposed extra-cellular amino-terminal domain) resulted in the isolation of two additional cDNA clones (Levitan et al., 1988). Because of the strong homology of the polypeptides encoded by these cDNAs with that encoded by the original α-subunit cDNA clone and because the new clones (as well as the α-subunit cDNA) encoded a peptide sequence that was determined by chemical sequencing of the gel-purified α band, the encoded polypeptides were designated α2 and α3. The originally isolated α-subunit cDNA was consequently deemed to encode an α1 polypeptide. It is interesting to note that the deduced α2 polypeptide also con- tained a peptide sequence (not present in the α1, α3 or β polypeptides) that was also determined by chemical sequencing of the receptor.

Hydropathy profiles (Kyte and Doolittle, 1982) of the four putative $GABA_A$ receptor subunits were highly reminiscent of those of the electric organ and muscle acetylcholine receptor (AChR) subunits (Schofield et al., 1987) and of the strychnine-binding subunit (M_r 48000) of the rat glycine receptor (Barnard et al., 1987; Grenningloh et al., 1987) suggesting that the $GABA_A$ receptor subunits possess four hydrophobic membrane-spanning domains (designated M1 to M4). Since, in common with other neurotransmitter receptor subunits, $GABA_A$ receptor polypeptides possess a signal sequence, the amino-termini and carboxy-termini are presumed to be extracellular and there exists a long cytoplasmic loop between the third and fourth proposed transmembrane helices. The $GABA_A$ receptor subunits exhibit amino-acid sequence homology with other neurotransmitter receptor subunits (Table

1). As may be seen, the $GABA_A$ receptor α subunits exhibit 70 to 80% amino-acid identity with each other but each of these displays only approximately 35% identity to the $GABA_A$ receptor β subunit. It is noteworthy that the $GABA_A$ receptor and glycine receptor subunits (all of which are presumed to form part of the chloride channel in each receptor) are more homologous to one another than any is to the AChR receptor which possesses an integral sodium channel. Other structural features of the $GABA_A$ receptor and some features common to all neurotransmitter receptor subunits have been described in detail elsewhere (Barnard et al., 1987; Grenningloh et al.,

Table 1. Amino-acid sequence homology of $GABA_A$ receptor and other neurotransmitter receptor subunits

	GABA α1	GABA α2	GABA α3	GABA β	GLYR 48K	AChR α
GABA α1	-	79	74	35	34	19
GABA α2		-	73	34	35	17
GABA α3			-	34	34	18
GABA β				-	38	15
GLYR 48K					-	15
AChR α						-

Sequences, excluding the proposed signal peptides, were aligned with the aid of the computer program DIAGON (Staden, 1982). The values correspond to the percentage amino-acid identity between pairs of sequences. Note that for comparison purposes the insertion of gaps in the sequences in order to achieve maximal alignment are counted as one regardless of their length. The deduced amino-acid sequences, derived from the corresponding cDNA clones, of the bovine $GABA_A$ receptor α1 (GABA α1), α2 (GABA α2), α3 (GABA α3) and β (GABA β) subunits (Schofield et al., 1987; Levitan et al., 1988), the rat spinal cord M_r 48000 subunit (GLYR 48K) (Grenningloh et al., 1987) and the bovine muscle acetylcholine receptor α subunit (AChR α) (Noda et al., 1983), are compared. Although it is the bovine muscle AChR α subunit that is compared here, similar values are obtained if one instead compares *Torpedo* or neuronal AChR subunit sequences.

1987; Schofield *et al.*, 1987). Taken together these observations point to the existence of a superfamily of neurotransmitter receptor subunit genes that encode the electric organ, muscle and neuronal acetylcholine receptors, the $GABA_A$ receptor, the glycine receptor and perhaps some members of the glutamate receptor series (kainate, quisqualate and N-methyl-D-aspartate).

EXPRESSION IN XENOPUS OOCYTES

In order to confirm that the isolated clones did indeed encode subunits of the $GABA_A$ receptor they were subcloned into suitable plasmid vectors containing bacteriophage promoters and RNA was synthesised and capped *in vitro* using, as appropriate, either SP6 or T7 RNA polymerase. These RNAs were then injected singly or in combination into *Xenopus* oocytes and the responses to GABA, in the absence or presence of drugs that act on the $GABA_A$ receptor, recorded. Significant GABA-evoked currents were only detected when an α RNA was co-injected with the β RNA (Schofield *et al.*, 1987; Levitan *et al.*, 1988). The reversal potential for the GABA-induced currents in oocytes injected with either α1 + β, α2 + β or α3 + β was approximately -20mV (Schofield *et al.*, 1987; Levitan *et al.*, 1988), close to the chloride equilibrium potential (Kusano *et al.*, 1982), indicating that the ligand was indeed gating a chloride ion channel. Furthermore, these induced receptors exhibited the same pharmacological responses detected when *Xenopus* oocytes are injected with poly$(A)^+$ RNA isolated from either chick brain (Smart *et al.*, 1983; Sigel and Bauer, 1988) or rat brain (Houamed *et al.*, 1984). For example they exhibit sensitivity to bicuculline and picrotoxin, potentiation by pentobarbital and also direct activation by pentobarbital. However, the large potentiation by benzodiazepines such as diazepam and chlorazepate was not observed with any of the α + β combinations. A small potentiation has been observed in receptors induced by α1 + β RNAs in undefolliculated oocytes (Schofield *et al.*, 1987; N. Fujita unpublished data); however, this response is variable and has not been observed in receptors induced either by α2 + β or by α3 + β RNAs (Levitan *et al.*, 1988). An additional feature of these three types of channel is that they have Hill slopes of less than or equal to 1.0 rather than 1.4 to 2.0 as estimated for $GABA_A$ receptors expressed from brain poly$(A)^+$ RNA (Smart *et al.*, 1987; Sigel and Bauer, 1988). The reason for these differences is at present unclear. It is conceivable that some specific post-translational processing event, that is lacking or is very inefficient in the *Xenopus* oocyte, is required for correct assembly of the complex and for normal benzodiazepine potentiation. Alternatively, additional receptor subunits may be missing. A third possibility is that they represent benzodiazepine-insensitive $GABA_A$ receptors (Study and Barker, 1981; Unnerstall *et al.*, 1981; de Blas *et al.*, 1988). In this context it is interesting to note that the group of Peter Seeburg report the isolation of several additional clones encoding novel subunits, one of which they term γ (see chapter by P. Seeburg, this Volume). When a plasmid containing this clone is transfected into mammalian cell-lines along with plasmids containing clones encoding an α and a β subunit, a strong potentiation of the GABA response by benzodiazepines is observed. The relationship of the polypeptides encoded by these new cDNAs to, for example, the BZI/BZII classification of $GABA_A$ receptors (Braestrup and Nielsen, 1981; Lippa *et al.*, 1982) remains to be established. To this end, permanently transformed mammalian cell-lines will be invaluable in studying the binding of the various agonists and antagonists to these expressed receptors.

It is clear, however, that the different α subunits confer different apparent GABA sensitivities on the β subunit when these are co-expressed in the *Xenopus* oocyte (Levitan *et al.*, 1988). Although desensitisation of these receptors is rapid and variable, the α2 + β receptor is consistently the most sensitive (half-maximal dose for GABA = 1.3 ± 0.2 μM), the α1 + β receptor has intermediate sensitivity (12 ± 1 μM) and the α3 + β receptor is the least sensitive (42 ± 8 μM) to GABA. These results suggest that the different α subunits might define subtypes of the $GABA_A$ receptor.

EXPRESSION IN BOVINE BRAIN

In order to unequivocally distinguish between the three mRNAs encoding the different α subunits, 45-base oligonucleotide probes corresponding to part of the cytoplasmic region between the third and fourth putative transmembrane domains (a segment that is highly diverged at both the DNA and amino-acid levels) were synthesised. These were used to examine the presence and abundance of α-subunit-specific transcripts both by Northern blotting and by in situ hybridisation. Preliminary Northern blot data (Levitan et al., 1988) indicated that the three α subunit genes were expressed to different extents in the three different brain regions examined (cortex, hippocampus and cerebellum). The α1 mRNA was found to be the most abundant transcript in all three regions and was the predominant species in the bovine cerebellum. In contrast, the α2 transcript was enriched in the hippocampus compared to cortex and cerebellum. The α3 mRNA was predominant in cortex and interestingly, at least in the cerebellum, the level appeared to decline during development.

A more extensive study of the expression of the α-subunit genes has now been performed (Figure 1). These Northern blots show the presence of α transcripts in many areas of bovine brain, specifically: colliculi, midbrain, olfactory bulb, striatum, retina and hypothalamus. These areas have previously been shown by ligand-binding and immunocytochemical studies to contain significant amounts of $GABA_A$ receptors (Schoch et al., 1985; Richards et al., 1986, 1987; de Blas et al., 1988). Although not all lanes on these blots contain equal amounts of poly$(A)^+$ RNA, as assessed by hybridisation with a mouse β-actin genomic fragment (Figure 1D), it is

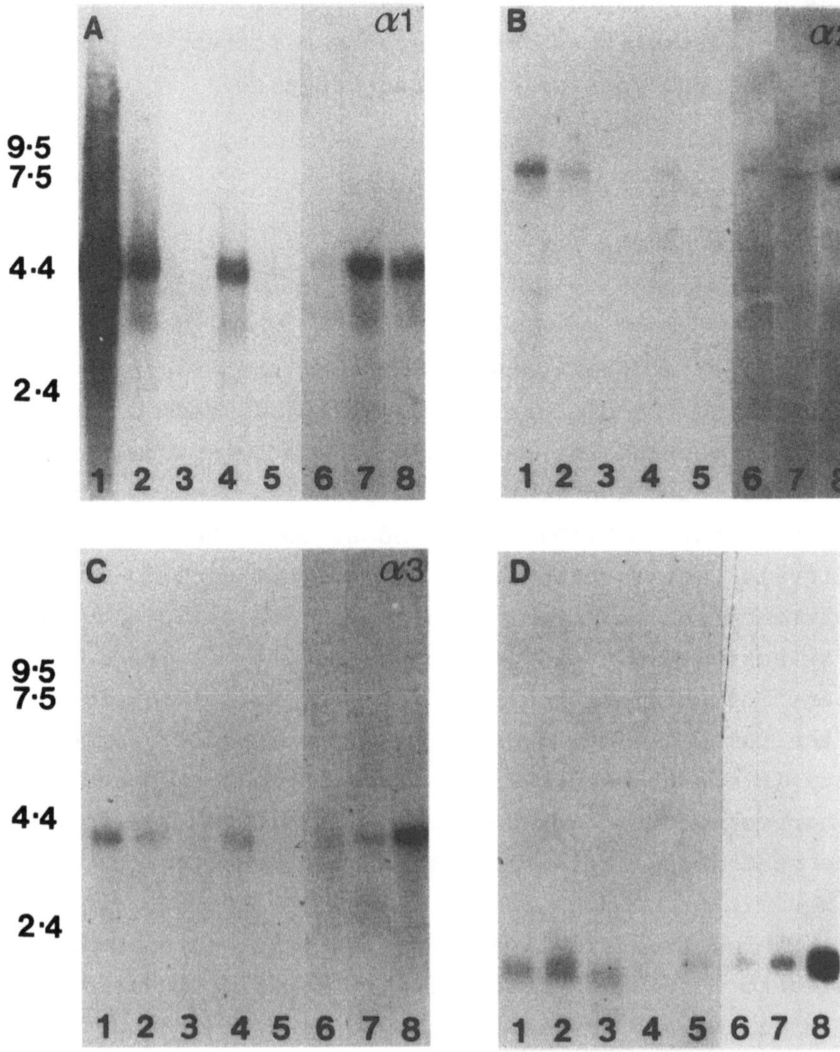

Figure 1. Northern blot analysis of $GABA_A$ receptor α-subunit transcripts. Approximately 20 µg per lane of bovine poly(A)$^+$ RNA was electrophoresed in a formaldehyde-agarose gel and transferred to a nylon membrane. This was hybridised with 45-base oligonucleotide probes (Levitan et al., 1988) specific for the α1-subunit (A), the α2-subunit (B) and the α3-subunit (C) transcripts and a mouse β-actin genomic probe (D). Lanes contained poly(A)$^+$ RNA from the following brain regions: 1, colliculi; 2, midbrain; 3, brainstem; 4, olfactory bulb; 5, pituitary; 6, striatum; 7, retina and 8, hypothalamus. Exposure to X-ray film, with an intensifying screen at -70°C, was either overnight (D), 3 days (A), 4 days (C) or 5 days (B). Size markers are indicated (in kilobases).

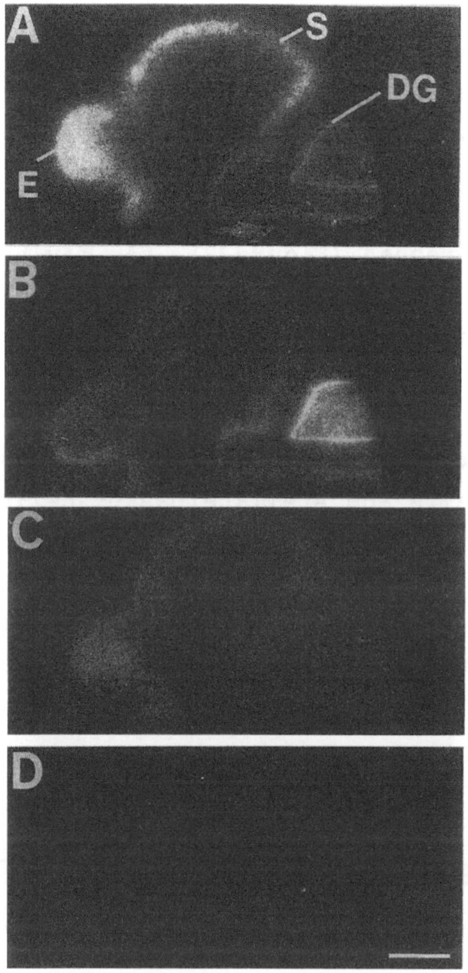

Figure 2. Localisation of α1-, α2- and α3-subunit transcripts in adult bovine brain. In situ hybridisation (shown in dark field), using 45-base oligonucleotide probes (Wisden et al., 1988) specific for the α1-subunit (A), the α2-subunit (B) and the α3-subunit (C) transcripts and a 45-base sense (control) oligonucleotide (D). The indicated hippocampal regions are: DG, dentate gyrus; E, entorhinal cortex and S, subiculum.

clear that the α1 transcript is the predominant one in most brain regions. A striking example of this is in the colliculi (Figure 1A, lane 1).

A detailed analysis of the neuronal localisation of the α-subunit transcripts has also been performed using in situ hybridisation (Wisden et al., 1988). In frontal cortex, the α1 and α2 transcripts are most abundant in layers II to IV. In contrast, the α3 transcript predominates in layers V and VI. In the hippocampal complex, the α-subunit genes are also differentially expressed (Figure 2). As may be seen, α1 predominates in the entorhinal cortex and the subiculum, whereas the α2 transcript is abundant in the dentate gyrus and hilar region. The α3 mRNA is expressed at much lower levels throughout the hippocampus. In the cerebellum, α1 is the predominant transcript and has been localised to granule cells, Purkinje cells and stellate/basket cells. Taking the Northern and in situ data together these results again suggest that these α subunits define receptor subtypes in the brain. It should be pointed out, however, that we cannot at present rule out the possibility that more than one type of α subunit can occur in the same receptor complex.

CLONING OF DNAs ENCODING THE CHICKEN $GABA_A$ RECEPTOR

The chicken is an excellent model for the study of development. Indeed, work in our laboratory on the cloning of muscle and neuronal acetylcholine receptor subunit DNAs has concentrated on this animal (see Barnard et al., 1986; Darlison et al., 1988). Because much of the $GABA_A$ receptor electrophysiology has been carried out on chick brain poly(A)$^+$ RNA-injected oocytes (Smart et al., 1983; Sigel and Bauer, 1988), we have used bovine cDNAs to cross-hybridise to the chicken. We now have a family of cDNA and genomic clones that are being analysed by their ability to hybridise to 51- to 64-base consensus sequence oligonucleotide probes, that correspond either to part of the proposed extracellular domain, the M2 transmembrane helix or the M4 transmembrane helix, and by DNA sequence analysis. Interestingly, we have

```
Chicken    D Y T I T M Y F Q Q S W R D K R L A Y T T L P L N L T L
           * * * | * * * * * * * * | * * * * | * | | | * * * * * *
Bovine     D Y T L T M Y F Q Q S W K D K R L S Y S G I P L N L T L
           56                                                    83

Chicken    D N R V A D Q L W L P D T Y F L N D K K S F L H G V T V
           * * * * * * * * * | * * * * * * * * * * * * | * * * * *
Bovine     D N R V A D Q L W V P D T Y F L N D K K S F V H G V T V
           84                                                   111

Chicken    K N R M I R L H P D G T V L Y G L R
           * * * * * * * * * * * * * * * * * *
Bovine     K N R M I R L H P D G T V L Y G L R
          112                               129
```

Figure 3. Amino-acid sequence of part of a chicken $GABA_A$ receptor β subunit. The predicted amino-acid sequence (shown in single-letter code), derived from the DNA sequence of an exon of a chicken $GABA_A$ receptor β-subunit gene, is aligned with the corresponding segment of the bovine β subunit (Schofield et al., 1987). Numbers below the bovine sequence correspond to those of the mature polypeptide. Asterisks and vertical bars between the two sequences indicate identical and conservatively-substituted (Dayhoff et al., 1978) amino-acid residues respectively.

isolated two genomic clones that by restriction mapping appear to be different, yet by DNA sequence analysis both encode polypeptides with strong amino-acid homology to the bovine β subunit. The DNA sequence of an exon of one of these is presented in Figure 3. It is interesting to note that the positions of the introns, that are found at either end of this sequence, are not conserved in this gene when compared to muscle and neuronal acetylcholine receptor genes (Nef et al., 1988). Our results lead us to conclude that in the chicken there appears to exist at least two β-subunit genes in addition to several α-subunit genes.

INVERTEBRATE GABA RECEPTORS

The presence of GABA receptors has been shown in a wide variety of invertebrate species, including nematodes, annelids, insects, crustaceans, arthropods and molluscs (reviewed in Tremblay *et al.*, 1988) thus indicating the early evolutionary presence of GABA as a neurotransmitter. As discussed above, the structural similarities of the different ligand-gated ion channels characterised to date, suggests a common ancestral origin. In this context the pharmacological and electrophysiological characteristics of several molluscan GABA receptors is highly interesting. In molluscs such as *Aplysia* GABA is capable of inducing three types of ionic current (Na^+, K^+ and Cl^-) rather than just the classical chloride current seen in vertebrates (Yarowsky and Carpenter, 1976). Of these, only the chloride current is blocked by bicuculline and picrotoxin. In *Helix*, excitatory effects of GABA (for example, depolarisation of some neurons in the subesophogeal ganglion) and acetylcholine, which can both be blocked by bicuculline, have been reported (Walker *et al.*, 1971). In addition, in *Aplysia*, the fast depolarising response to GABA is depressed by curare, a classical acetylcholine receptor antagonist (Yarowsky and Carpenter, 1978). Finally it should be noted that no high-affinity benzodiazepine binding sites have been detected in *Aplysia* (Nielsen *et al.*, 1978) although the GABA-evoked chloride conductance *is* augmented by benzodiazepines (due to an increased affinity between the receptor and GABA). Taken together these results suggest that, at least in molluscs, the GABA receptor may contain different integral ion channels (having different ionic selectivities) and that the GABA and acetylcholine receptors *may* have binding sites in common.

In addition, some molluscs have important advantages in the study of the molecular biology of neurotransmitter receptors. Firstly, they possess very simple nervous systems which are comprised of only 20000 to 25000 neurons. Secondly, the cell bodies of many of these neurons are extremely large

```
                 _____________________________________________________
Lymnaea      S Y V K A I D I Y L V M C F V F V F A A L L E Y A A
             | * | * * * * * * * |   * * * * * *   * * * * * *
Bovine       P Y V K A I D I Y L M G C F V F V F L A L L E Y A F
            276                                                301

             _
Lymnaea      V N Y T Y W G A R A K R K A K R L R E R A T S V R K
             * * *   | | *     |   | * |         |   | |     |
Bovine       V N Y I F F G K G P Q K K G A G K Q D Q S A N E K N
            302                                                327
```

Figure 4. Amino-acid sequence of part of an invertebrate $GABA_A$ receptor β subunit. The predicted amino-acid sequence (shown in single-letter code), derived from the DNA sequence of a *Lymnaea stagnalis* β-subunit genomic fragment, is aligned with the corresponding segment of the bovine β subunit (Schofield *et al*., 1987). Numbers below the bovine sequence correspond to those of the mature polypeptide. Asterisks and vertical bars between the two sequences indicate identical and conservatively-substituted (Dayhoff *et al*., 1978) amino-acid residues respectively. The third proposed transmembrane segment, M3, is denoted by a solid line above the *Lymnaea* sequence.

(up to 1mm in diameter) and they are easily identifiable by their size, shape and colour. Because of these properties, electrophysiological recording from individual neurons is possible (Ono and Salvaterra, 1981). The large size of these neurons should allow the easy identification of cells expressing GABA receptors by means of *in situ* hybridisation (McAllister *et al*., 1983; Van Minnen *et al*., 1988). This feature also facilitates the introduction (by injection) of modified cloned genes (see DesGroseillers *et al*., 1987) and subsequent electrophysiological recording either *in situ* or *in vitro*.

In view of the complex pharmacology of GABA receptors from molluscs and the advantages of these animals, we have begun to isolate DNA clones encoding this receptor from the fresh-water snail *Lymnaea stagnalis* by cross-hybridisation using bovine cDNAs. Genomic clones encoding putative α- and β-subunit genes have been isolated and partially characterised by DNA sequence analysis. An alignment of part of the predicted amino-acid sequence from a Lymnaean β-subunit gene with the bovine β subunit is shown in Figure 4. This sequence, which extends from just before the third proposed transmembrane domain into the proposed intracellular loop between M3 and M4, exhibits 48% amino-acid identity. Clearly, however, there is much stronger homology (83%) over the M3 region alone. A subfragment encoding the M2 and M3 region has now been used to isolate clones from a *Lymnaea stagnalis* central neryous system cDNA library. Expression of these either by injection of RNA into *Xenopus* oocytes or by injection of the cloned DNAs into molluscan neurons should elucidate the ionic selectivity and pharmacology of the receptors that these clones encode. In addition to providing information on the structure/function relationship of neurotransmitter receptors, these experiments may yield important insights into the evolution of this superfamily of ligand-gated ion channels.

CONCLUSIONS AND FUTURE PROSPECTS

Within a period of only five years or so, a combination of conventional protein biochemistry and molecular biology has resulted in a rapid increase in our understanding of the structure and function of the $GABA_A$ receptor. From these studies we can deduce that the tertiary and perhaps even the quaternary structure of this receptor is similar to that of the acetylcholine receptor and the glycine receptor and that the genes encoding these neurotransmitter receptors probably arose by duplication from a common ancestral gene. In addition, by reconstitution of subunits of the $GABA_A$ receptor in *Xenopus* oocytes it can be shown that this protein does indeed possess an integral chloride channel. Finally, molecular cloning has revealed the existence of hitherto unknown components of this receptor that, at least for the α subunits, appear to define receptor subtypes. However, much remains to be determined, for example the absolute composition and stoichiometry of the receptor (including the correlation of subunit composition with the benzodiazepine BZI/BZII pharmacology), the mechanism of ligand-gating, the molecular mechanisms of ligand/receptor interactions, the mechanism of receptor desensitisation and the regulation of expression of the $GABA_A$ receptor genes in individual neurons. Both molecular genetic (including in vitro mutagenesis) and protein biochemical studies will be required to address these important questions.

REFERENCES

Barnard, E.A., Beeson, D.M.W., Cockcroft, V.B., Darlison, M.G., Hicks, A.A., Lai, F.A., Moss, S.J. and Squire, M.D. (1986) in Nicotinic Acetylcholine Receptor: Structure and Function (Maelicke, A. ed.), NATO ASI Series H, Vol. 3, pp 389-415, Springer-Verlag, Berlin.

Barnard, E.A., Darlison, M.G. and Seeburg, P. (1987) Trends Neurosci. 10, 502-509.

Braestrup, C. and Nielsen, M.J. (1981) J. Neurochem. 37, 333-341.

Casalotti, S.O., Stephenson, F.A. and Barnard, E.A. (1986) J. Biol. Chem. 261, 15013-15016.

Darlison, M.G., Hicks, A.A., Cockcroft, V.B., Squire, M.D. and Barnard, E.A. (1988) in Cellular and Molecular Basis of Synaptic Transmission (Zimmermann, H. ed.), NATO ASI Series H, Vol. 21, pp 475-492, Springer-Verlag, Berlin.

Dayhoff, M.O., Schwartz, R.M. and Orcutt, B.C. (1978) in Atlas of Protein Sequence and Structure (Dayhoff, M.O. ed.), Vol. 5, Suppl. 3, pp 345-352, National Biomedical Research Foundation, Washington, DC.

de Blas, A.L., Vitorica, J. and Friedrich, P. (1988) J. Neurosci. 8, 602-614.

DesGroseillers, L., Cowan, D., Miles, M., Sweet, A. and Scheller, R.H. (1987) Mol. Cell Biol. 7, 2762-2771.

Enna, S.J. (1983) in The GABA Receptors (Enna, S.J. ed.), pp 1-23, Humana Press, Clifton, New Jersey.

Grenningloh, G., Rienitz, A., Schmitt, B., Methfessel, C., Zensen, M., Beyreuther, K., Gundelfinger, E.D. and Betz, H. (1987) Nature 328, 215-220.

Houamed, K.M., Bilbe, G., Smart, T.G., Constanti, A., Brown, D.A., Barnard, E.A. and Richards, B.M. (1984) Nature

310, 318-321.
Kirkness, E.F. and Turner, A.J. (1986) Biochem. J. 233, 265-270.
Kusano, K., Miledi, R. and Stinnakre, J. (1982) J. Physiol. 328, 143-170.
Kyte, J. and Doolittle, R.F. (1982) J. Mol. Biol. 157, 105-132.
Levitan, E.S., Schofield, P.R., Burt, D.R., Rhee, L.M., Wisden, W., Köhler, M., Fujita, N., Rodriguez, H., Stephenson, A., Darlison, M.G., Barnard, E.A. and Seeburg, P.H. (1988) Nature 335, 76-79.
Lippa, A.S., Meyerson, L.R. and Beer, B. (1982) Life Sci. 31, 1409-1417.
Mamalaki, C., Stephenson, F.A. and Barnard, E.A. (1987) EMBO J. 6, 561-565.
McAllister, L.B., Scheller, R.H., Kandel, E.R. and Axel, R. (1983) Science 222, 800-808.
Nef, P., Oneyser, C., Alliod, C., Couturier, S. and Ballivet, M. (1988) EMBO J. 7, 595-601.
Nielsen, M., Braestrup, C. and Squires, R.F. (1978) Brain Res. 141, 342-346.
Noda, M., Furutani, Y., Takahashi, H., Toyosato, M., Tanabe, T., Shimizu, S., Kikyotani, S., Kayano, T., Hirose, T., Inayama, S. and Numa, S. (1983) Nature 305, 818-823.
Olsen, R.W. and Venter, J.C., eds (1986) in Benzodiazepine/ GABA Receptors and Chloride Channels: Structural and Functional Properties, Alan R. Liss, New York.
Ono, J.K. and Salvaterra, P.M. (1981) J. Neurosci. 1, 259-270.
Richards, J.G., Möhler, H. and Haefely, W. (1986) in Neurology and Neurobiology (Panula, P., Päivärinta, H. and Soinila, S. eds), Vol. 16: Neurohistochemistry, Modern Methods and Applications, pp 629-677, Alan R. Liss, New York.
Richards, J.G., Schoch, P., Häring, P., Takacs, B. and Möhler, H. (1987) J. Neurosci. 7, 1866-1886.
Schoch, P., Richards, J.G., Häring, P., Takacs, B., Stähli, C., Staehelin, T., Haefely, W. and Möhler, H. (1985) Nature 314, 168-171.
Schofield, P.R., Darlison, M.G., Fujita, N., Burt, D.R., Stephenson, F.A., Rodriguez, H., Rhee, L.M., Ramachandran, J., Reale, V., Glencorse, T.A., Seeburg, P.H. and Barnard, E.A. (1987) Nature 328, 221-227.
Sigel, E. and Barnard, E.A. (1984) J. Biol. Chem. 259, 7219-7223.
Sigel, E. and Bauer, R. (1988) J. Neurosci. 8, 289-298.
Sigel, E., Stephenson, F.A., Mamalaki, C. and Barnard, E.A. (1983) J. Biol. Chem. 258, 6965-6971.
Smart, T.G., Constanti, A., Bilbe, G., Brown, D.A. and Barnard, E.A. (1983) Neurosci. Lett. 40, 55-59.
Smart, T.G., Houamed, K.M., Van Renterghem, C. and Constanti, A. (1987) Biochem. Soc. Trans. 15, 117-122.
Staden, R. (1982) Nucleic Acids Res. 10, 2951-2961.
Study, R.E. and Barker, J.L. (1981) Proc. Natl Acad. Sci. USA 78, 7180-7184.
Taguchi, J. and Kuriyama, K. (1984) Brain Res. 323, 219-226.
Tremblay, J.P., Robitaille, R. and Atwood, H.L. (1988) in

GABA and Benzodiazepine Receptors (Squires, R.F. ed.), Vol. I, pp 23-57, CRC Press, Boca Raton, Florida.
Turner, A.J. and Whittle, S.R. (1983) Biochem. J. 209, 29-41.
Unnerstall, J.R., Kuhar, M.J., Niehoff, D.L. and Palacios, J.M. (1981) J. Pharm. Exp. Ther. 218, 797-804.
Van Minnen, J., Van De Haar, Ch., Raap, A.K. and Vreugdenhil, E. (1988) Cell Tissue Res. 251, 477-484.
Walker, R.J., Crossman, A.R., Woodruff, G.N. and Kerkut, G.A. (1971) Brain Res. 33, 75-82.
Wisden, W., Morris, B.J., Darlison, M.G., Hunt, S.P. and Barnard, E.A. (1988) Neuron, in press.
Wood, W.I. (1987) in Guide to Molecular Cloning Techniques (Berger, S.L. and Kimmel, A.R. eds), Methods in Enzymology Vol. 152, pp 443-447, Academic Press, Orlando.
Yarowsky, P. and Carpenter, D.O. (1976) Fed. Proc. 35, 543.
Yarowsky, P.J. and Carpenter, D.O. (1978) Brain Res. 144, 75-94.

GABA$_A$ receptor: subunit expression in Xenopus oocytes and localization of receptor mRNAs in rat brain

P. Malherbe, J.G. Richards, J.M. Séquier, S. Mathews,
A. Draguhn* and H. Möhler
Pharmaceutical Research Department
F. Hoffmann-La Roche & Co., Ltd.,
CH-4002 Basle, Switzerland

Introduction

γ-Aminobutyric acid (GABA), the major inhibitory neurotransmitter in the brain, exerts its main synaptic action by interacting with a GABA$_A$ receptor complex located in the subsynaptic membrane of target neurons. The receptor complex consists of a GABA-gated chloride ionophore which can be modulated through several drug binding sites on the receptor, of which the benzodiazepine receptor (Möhler and Okada, 1977; Squires and Braestrup, 1977) is of major pharmacological and therapeutical interest. Depending on the type of benzodiazepine receptor ligand, GABAergic synaptic transmission can be modulated in opposite directions. Receptor agonists enhance synaptic transmission while inverse agonists reduce it (Fig. 1).

Thereby complex CNS states such as the level of anxiety, the extent of muscle tension, the state of vigilance and the likelihood of convulsions can be chemically regulated in

* Max-Planck-Institut für biophysikalische Chemie, D-3400 Göttingen, FRG

NATO ASI Series, Vol. H 32
Molecular Biology of Neuroreceptors and Ion Channels. Edited by A. Maelicke

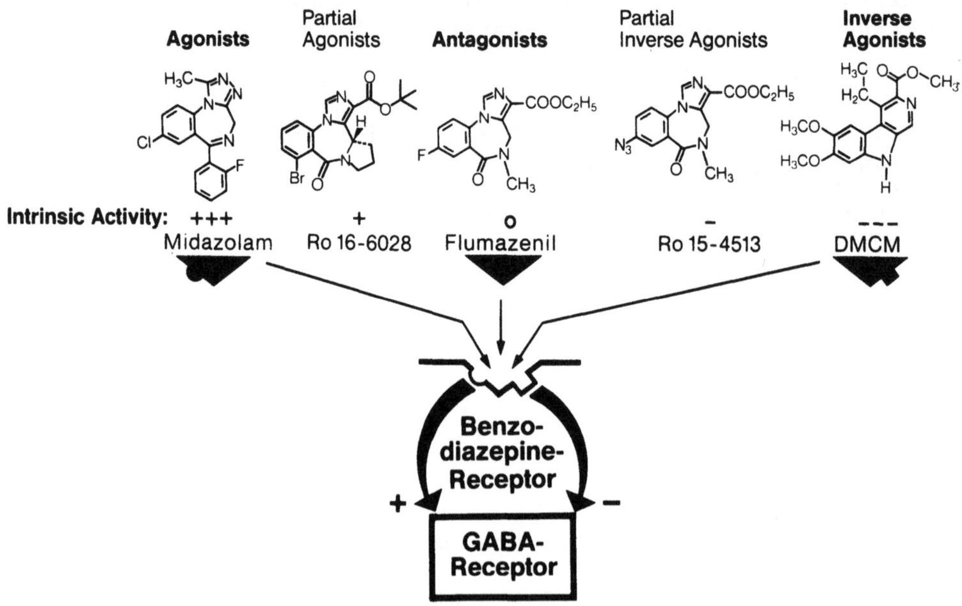

Fig. 1. Schematic representation of the spectrum of intrinsic activities of ligands of the benzodiazepine receptor. Intrinsic activities are schematically indicated by + as positive, - as negative, o indicates a lack of intrinsic activity.

opposite directions (Table 1). Both types of effects can be competitively antagonized by a third type of ligand, the benzodiazepine antagonists e.g. flumazenil, which per se lack major intrinsic activity (Haefely et al., 1985; Haefely and Polc, 1986; Hunkeler et al., 1981).

In the present chapter, we describe the structure and functional expression of the rat $GABA_A$ receptor α- and ß-subunit cDNAs. Furthermore, the sites of gene expression of both subunits are demonstrated in rat brain by *in situ* hybridization histochemistry studies. Finally, structural information on the heterogeneity of the $GABA_A$ receptor in rat brain is presented.

Table 1. Modulation of CNS function by benzodiazepine receptor ligands with opposite intrinsic activities.

Agonists		Inverse agonist
-	Anxiety	+
-	Muscle Tension	+
-	Convulsions	+
-	Vigilance	+

Agonists are ligands which enhance GABAergic synaptic transmission; inverse agonists reduce GABAergic transmission.
- Indicates decrease of symptoms; + increase of symptoms.

Structure of the $GABA_A$ receptor

Biochemical experiments such as photoaffinity labelling with photoreactive benzodiazepines (Möhler et al., 1980), affinity purification (Schoch et al., 1984; Sigel and Barnard, 1983; Stauber et al., 1987) and immunoprecipitation studies (Häring et al., 1985; Schoch et al., 1985) suggested that the $GABA_A$ receptor is a hetero-oligomer consisting of α- and ß-subunits with molecular sizes of 50kD and 55kD respectively (Möhler et al., 1986; Stephenson, 1988). Molecular size determination of the intact receptor (Chang and Barnard, 1982) suggested that it consists of four subunits (α2ß2; Fig. 2) although a pentameric structure could not be excluded. The binding sites for benzodiazepines were located by photoaffinity labelling on the α-subunit (Möhler et al., 1980) as well as in certain brain regions on the ß-subunit (Sieghart and Karobath, 1980), the high affinity GABA site was located on the ß-subunit (Casalotti et al., 1986) (Fig. 2).

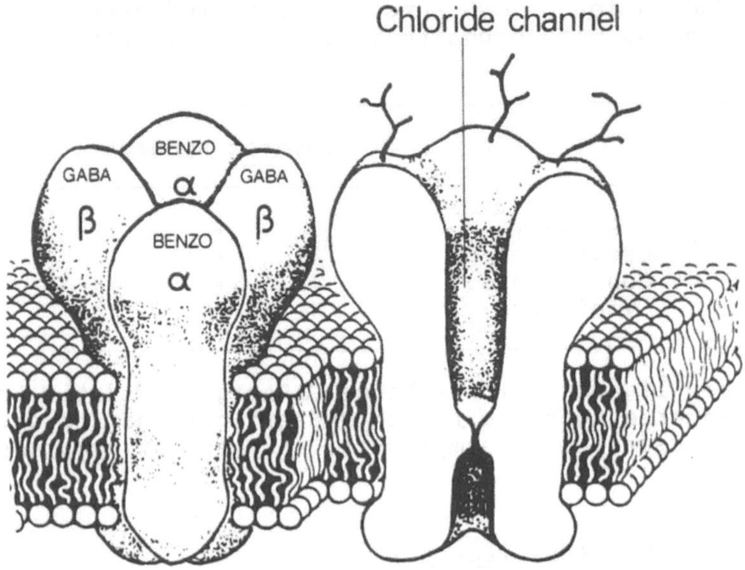

Fig. 2. Working hypothesis for the structure of the $GABA_A$ receptor complex (Möhler et al., 1986).

In order to obtain more detailed information on the general architecture and topology of the receptor protein, cDNAs coding for the α- and ß-subunit were isolated from rat brain. This species was chosen since most of the electrophysiological (Bormann et al., 1987; Hamill et al., 1983) and histochemical (Palacios et al., 1981; Richards and Möhler, 1984; Richards et al., 1986, 1987, 1988; Schoch et al., 1985; Young and Kuhar, 1980) investigations on the $GABA_A$ receptor have been performed in rat brain. After partial amino acid sequences had been obtained from the purified rat brain α- and ß-subunits (collaboration with K. Beyreuther and G. Multhaup, Center of Molecular Biology, Heidelberg, FRG), the α- and ß-subunit cDNAs were isolated from a rat brain cDNA library.

The predicted amino acid sequences of the α- and ß-subunit cDNA clones contained the partial amino acid sequences determined in the native protein. Furthermore, the deduced amino acid sequences of the rat α-and ß-subunit cDNA clones corresponded largely to those of bovine brain reported by Schofield et al. (1987).

The secondary structure prediction of both rat clones is in line with putative membrane proteins. Apart from a hydrophobic N-terminal sequence (putative signal peptide), four

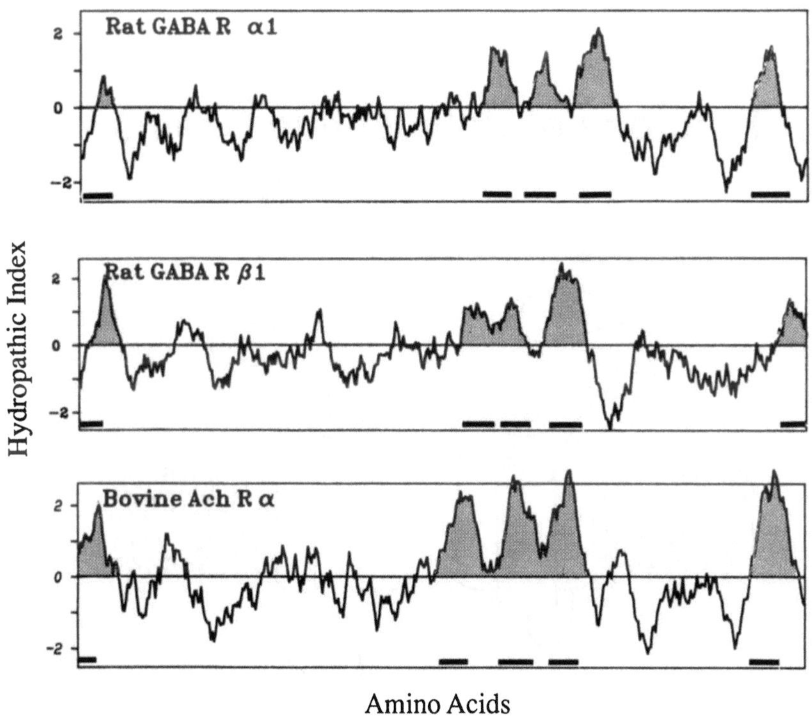

Fig. 3. Comparison of hydropathy plots of the rat $GABA_A$ receptor α- and ß-subunits with that of the bovine nicotinic acetylcholine receptor (AchR) α-subunit. The putative signal peptide sequence and the membrane-spanning domains are indicated by solid lines.

hydrophobic domains are apparent which may form transmembrane domains (Fig. 3). The number and distribution of the hydrophobic domains are very similar in the α- and ß-subunits of the rat and correspond to those of the nicotinic acetylcholine receptor (nAchR) (Noda et al., 1983; Guy and Hucho, 1987) (Fig. 3).

Based on these homologies, the following model for the topology of the amino acid chain is proposed (Fig. 4). In both α- and ß-subunits, a large N-terminal domain is located extracellularly, followed by a segment in which the membrane is spanned four times (M1 - M4). The C-terminal domain is

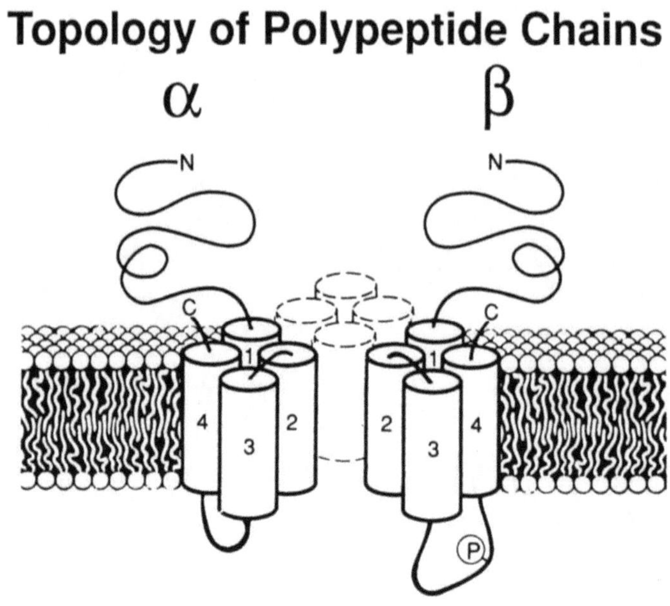

Fig. 4. Topology of the polypeptide chains according to the primary amino acid sequence deduced from cDNAs of the α- and ß-subunit and their hydropathy profiles.

located extracellularly. The N-terminal domains of both subunits contain 2 cysteine-residues which may form a loop in analogy to the cysteine-connecting loop of the nAchR (Kao and Karlin, 1986). The α- and ß-subunits of the rat $GABA_A$ receptor differ mainly in their cytoplasmic loop connecting transmembrane domains M3 and M4 where the ß-subunit but not the α-subunit contains a consensus site for phosphorylation by protein kinase A which may be involved in receptor sensitivity regulation (Huganir et al., 1986). The first transmembrane domain in both subunits of rat brain contains a proline residue which is located at an identical position in the bovine α- and ß-subunits (Schofield et al., 1987), the strychnine-binding subunit of the glycine receptor (Grenningloh et al., 1987) and the nAchR (Noda et al., 1983). This proline residue introduces a kink in the M1 region, which may represent the structural determinant for channel gating. In analogy to the finding with the AChR (Giraudat et al., 1986; Imoto et al., 1986; Oberthühr et al., 1986), one can assume that the transmembrane region M2 forms the channel lumen of the $GABA_A$ receptor (Fig. 4). A high proportion of serine and threonine residues, located in the M2 transmembrane region, could form a hydrophilic lining of the ion channel. The presence of positive charges at both ends of the transmembrane regions may determine the ion selectivity, in analogy to the distribution of negative charges at either end of the channel region of the acetylcholine receptor.

Heterologous expression in Xenopus oocytes

To identify the functional relevance of the α- and ß-cDNA clones isolated from rat brain, synthetic cRNA was prepared in an in vitro SP6 transcription system. A mixture of the α- and ß-cRNAs was microinjected into Xenopus oocytes and tested by measuring inward currents induced by superfusion of GABA under voltage clamp at -70mV.

GABA reproducibly induced currents in oocytes (Fig. 5) having a reversal potential of -20mV, which corresponds to the oocyte chloride equilibrium potential in ringer solution (-22mV) (Dascal et al., 1984). The amplitude of the currents was dependent on the concentration of GABA and could be blocked by picrotoxin. However, when log (current amplitude) was plotted against log [GABA] the slope was close to one and thus lower than the Hill coefficient reported for native receptor (Sakmann et al., 1983).

Some of the drug modulatory sites of the $GABA_A$ receptor

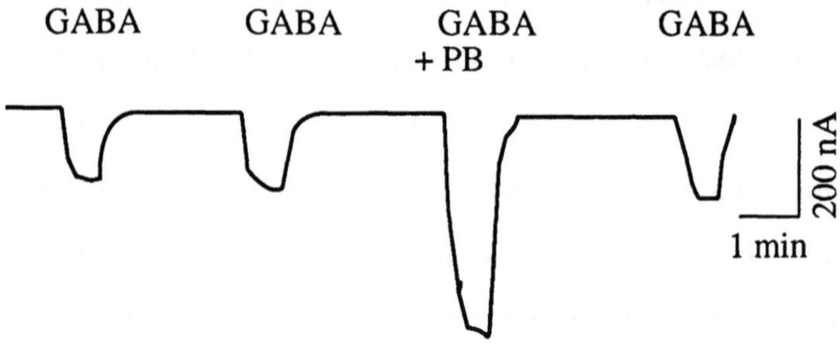

Fig. 5. Expression of cloned $GABA_A$ receptor in Xenopus oocytes. GABA-induced current recorded from an individual oocyte injected with 50ng of a mixture of α- and ß-subunit cRNAs. Voltage-clamp, -70mV, downward deflections indicate inward currents. Pentobarbital (PB) enhanced the GABA (10μM)-induced current.

were operative. For instance, the GABA (10μM)-induced peak current was increased 3.5-fold by pentobarbital (25μM). However, potentiation of GABA-induced currents with benzodiazepines, which has been reported for neurones (Bormann, 1988) as well as for poly-A^+-RNA injected oocytes (Sigel and Bauer, 1988), was not obtained reproducibly.

Sites of $GABA_A$ receptor gene expression

To identify the sites of $GABA_A$ receptor action in brain, the receptor protein was previously visualized radioautographically using radioligands of the BZR such as $[^3H]$ flumazenil (Richards et al., 1986), and immunohistochemically using monoclonal antibodies (Richards et al., 1986, 1987; Schoch et al., 1985). Because of the complexity of synaptic connections, the protein distribution often does not allow a precise localization of the cell bodies which generate the receptors. However, these neurons can be mapped by _in situ_ hybridization histochemistry revealing the distribution of those cells which respond to GABA and to the drugs acting at the $GABA_A$ receptor. On rat brain sections hybridizing RNA was detected by _in situ_ hybridization using a ^{35}S-labelled antisense α-subunit cRNA probe. In general, areas with high contents of hybridizing RNA e.g. the cerebral cortex, cerebellum, olfactory bulb and limbic system also contain the high receptor densities (Fig. 6a). In the control experiments with a ^{35}S-labelled sense α-subunit cRNA probe, only background hybridization was found. On a cellular level, a striking correspondence between the somatic localization of

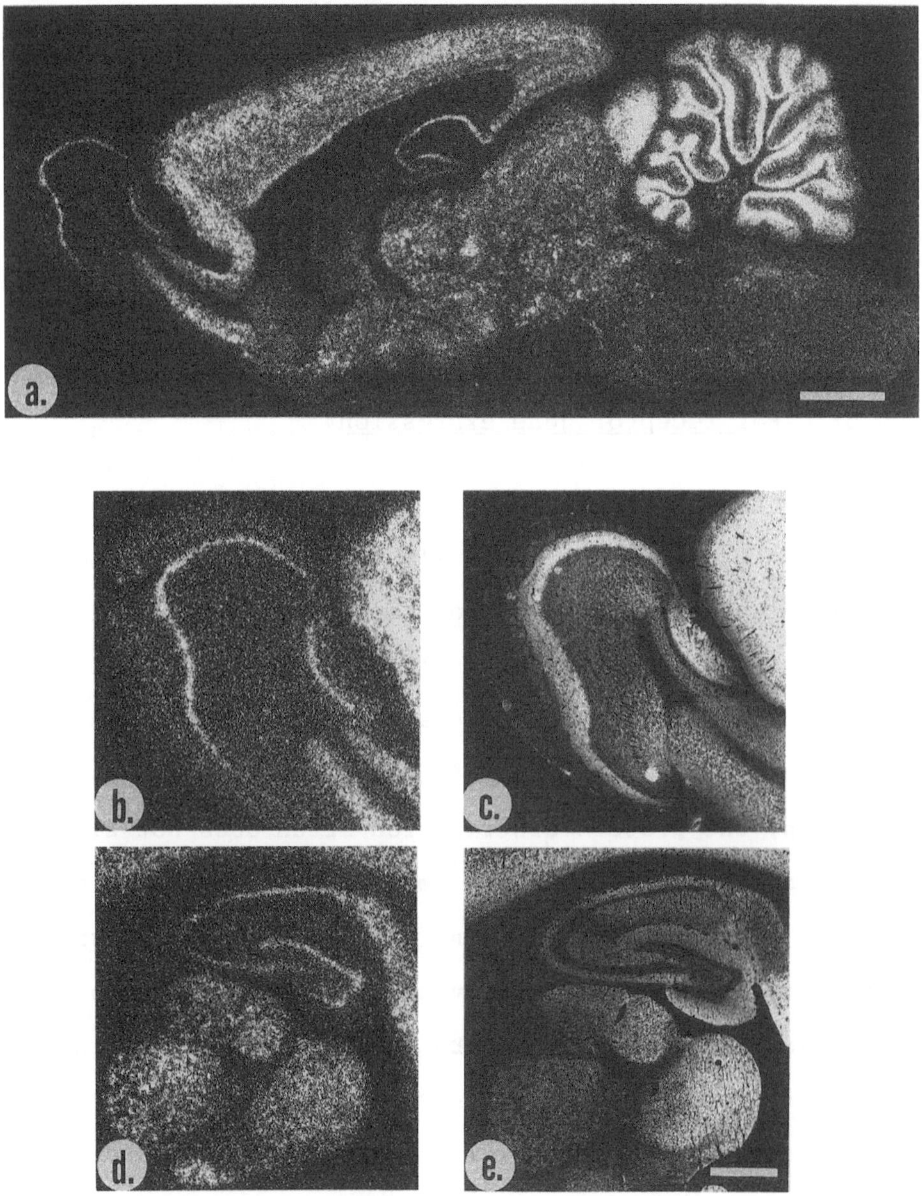

Fig. 6. In situ hybridization of the ^{35}S-labelled RNA antisense strand of the α-subunit in a parasagittal section of rat brain (a). White areas correspond to brain regions with an intense signal. (Bar = 2mm). Details of adjacent sections of the olfactory bulb (b,c) and hippocampus-thalamus (d,e), illustrating the regional distribution of hybridizing RNA (b and d) and the receptor protein revealed by bd-17 immunohistochemistry (c and e). (Bar = 1 mm). Reproduced, with permission, from Séquier et al. (1988).

hybridizing RNA and the dendritic receptor protein localization was found (Séquier et al., 1988). For instance, mitral cells of the olfactory bulb which receive a dense dendritic GABAergic input in the external plexiform layer showed a strong hybridization signal (Fig. 6b,c). Similarly, the pyramidal cells of hippocampus, which contain high densities of $GABA_A$ receptors on their basal and apical dendrites, contain a high content of hybridizing RNA in their cell bodies (Fig. 6d,e). Likewise, the granule cells of the dentate gyrus show hybridizing RNA in their cell bodies and high receptor density in their dendrites (Fig. 6d,e). Hybridizing RNA has been identified in additional cells which are known to receive GABAergic input such as the Purkinje, stellate and basket cells in the cerebellum and cells in the substantia nigra.

Since the $GABA_A$ receptor is a hetero-oligomer of α- and ß-subunits, mRNA encoding the ß-subunit was expected to show a distribution similar to that of the α-subunit. The hybridization signal obtained with the ß-subunit antisense ^{35}S- labelled cRNA corresponded in many areas to that of the α-subunit probe e.g. cerebral cortex, cerebellum, inferior colliculus, olfactory bulb (mitral cells), hippocampal formation (pyramidal cells) and granular cells of the dentate gyrus (all areas of high receptor content). However, there were also areas in which the α- and ß-signal intensities did not correspond. For instance, while the α-hybridizing RNA is present in the substantia nigra, the ß-hybridizing RNA is almost totally absent from this region

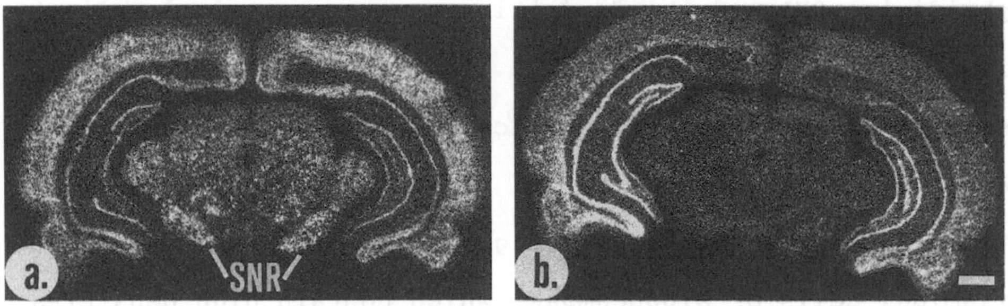

Fig. 7. Comparison between the regional hybridization of the ^{35}S-labelled RNA anti-sense strands of the α-subunit (a) and the ß-subunit (b) in adjacent sections of the midbrain. SNR, substantia nigra reticular part. (Bar = 2 mm). Reproduced, with permission, from Séquier et al. (1988).

(Fig. 7a,b). Conversely, the α-hybridizing RNA is much weaker than ß-hybridizing RNA in bed nucleus. This regional heterogeneity in the hybridization pattern of α- and ß-subunit RNAs may reflect regional differences in RNA stability or subunit composition.

$GABA_A$ receptor heterogeneity

A structural basis for the heterogeneity of the $GABA_A$ receptor became apparent through the isolation of two cDNA clones encoding proteins closely related to the α-subunit. Among the additional α-cDNA clones which have been isolated from rat brain, only two clones, termed α2 and α3, have been sequenced (the original α-cDNA clone is now termed α1). The amino acid sequences of all three α-subunits are highly homologous, in particular in the transmembrane regions M1, M2 and M3 (Fig. 8). The primary structures differ between the α-subunit subtypes mainly in their cytoplasmic region

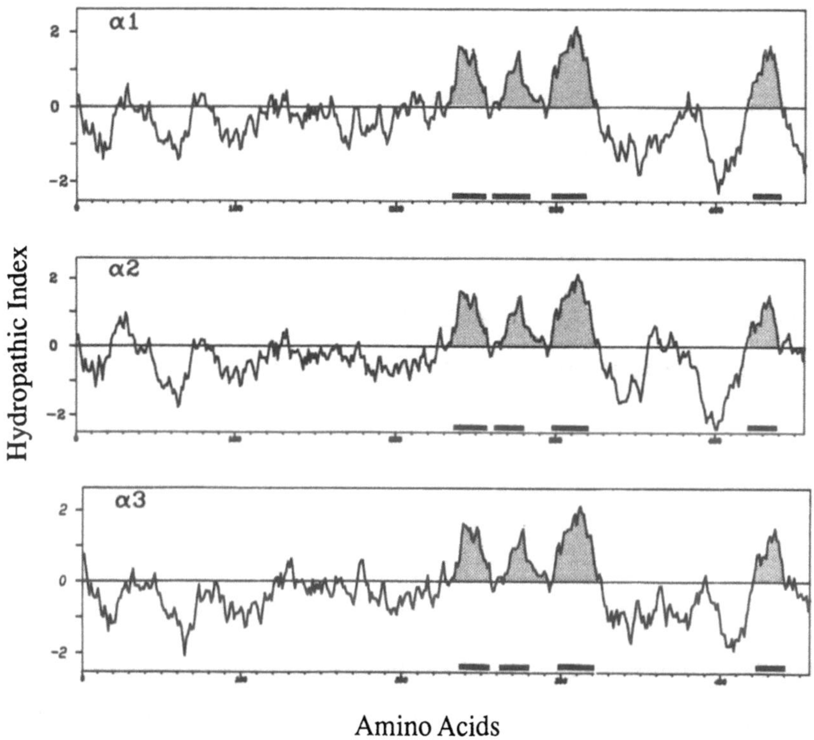

Fig. 8. Comparison of hydropathy plots of three rat $GABA_A$ receptor α-subunit subtypes. The membrane-spanning domains are indicated by solid lines.

linking M3 and M4, and in their extreme N-terminal sequences. Furthermore, they contain different numbers of potential N-linked glycosylation sites on their presumed extracellular N-terminal domain (2 sites in α1, 3 in α2 and 4 in α3). The sizes of the mRNA as detected in Northern blot analysis were different for the α-subunit subtypes (4.5 kb, 4.2 kb for α1, 3.8 kb for α2 and 3.0 kb for α3). Preliminary findings from in situ hybridization histochemistry revealed that the α-subunit subtypes differ in their distribution in the rat brain. Cellular and regional differences in the subunit repertoire may represent the structural basis for a functional diversity of $GABA_A$ receptors in rat brain.

Outlook

Receptor-structure: Heterologous expression of rat α- and ß-subunit cDNAs in *Xenopus* Oocytes fails to reconstitute a fully functional $GABA_A$ receptor. It remains to be established whether this deficit is due to insufficient post-translational processing in *Xenopus* Oocytes or due to the lack of a yet unknown structural entity of the native $GABA_A$ receptor.

Heterogeneity: Isolation of further α- and ß-subunit subtypes and their differential expression in different brain regions will generate a variety of subunit repertoires for $GABA_A$ receptor assembly. It will be of major interest to determine the physiological significance of the structural receptor heterogeneity.

Gene-structure: Isolation of the genes coding for the $GABA_A$ receptor subunits will help to identify possible control elements for the regulation of gene expression in vivo. Such information may provide insight into the control of $GABA_A$ receptor expression in normal and pathological brain.

References

Bormann J, Hamill OP, Sakmann B (1987) Mechanism of anion permeation through channels gated by glycine and γ-aminobutyric acid in mouse cultured spinal neurones. J Physiol 385: 243-286

Bormann J (1988) Electrophysiology of $GABA_A$ and $GABA_B$ receptor subtypes. TINS 11: 112-116

Casalotti SO, Stephenson FA, Barnard EA (1986) Separate subunits for agonist and benzodiazepine binding in the γ-aminobutyric $acid_A$ receptor oligomer. J Biol Chem 261: 15013-15016

Chang L-R, Barnard EA (1982) The benzodiazepine GABA-receptor complex: molecular size in brain synaptic membranes and in solution. J Neurochem 39: 1507-1518
Dascal N, Landau EM, Lass Y (1984) Xenopus oocyte resting potential, muscarinic responses and the role of calcium and guanosine 3'-5'-cyclic monophosphate. J Physiol 352: 551-574
Giraudat J, Dennis M, Heidmann T, Chang JY, Changeux JP (1986) Structure of the high-affinity binding site for noncompetitive blockers of the acetylcholine receptor: serine-262 of the δ subunit is labeled by [^{3}H]chlorpromazine. Proc Natl Acad Sci USA 83: 2719-2723
Grenningloh G, Rienitz A, Schmitt B, Methfessel C, Zensen M, Beyreuther K, Gundelfinger ED, Betz H (1987) The strychnine-binding subunit of the glycine receptor shows homology with nicotinic acetylcholine receptor. Nature 328: 215-220
Guy HR, Hucho F (1987) The ion channel of the nicotinic acetylcholine receptor. Trends Neurosci 10: 318-321
Haefely W, Kyburz E, Gerecke M, Möhler H (1985) Recent advances in the molecular pharmacology of benzodiazepine receptors and in the structure activity relationship of their agonists and antagonists. Advances in Drug Research 14: 165-322
Haefely W, Polc P (1986) Physiology of GABA enhancement by benzodiazepines and barbiturates. In: Olsen RW, Venter JC (eds) Benzodiazepine/GABA Receptors and Chloride Channels: Structural and Functional Properties. Alan R. Liss, New York, p 97
Hamill OP, Bormann J, Sakmann B (1983) Activation of multiple conductance state chloride channels in spinal neurones by glycine and GABA. Nature 305: 805-808
Häring P, Stähli C, Schoch P, Takacs B, Staehelin T, Möhler H (1985) Monoclonal antibodies reveal structural homogeneity of γ-aminobutyric acid/benzodiazepine receptors in different brain areas. Proc Natl Acad Sci USA 82: 4837-4841
Huganir RL, Delcour AH, Greengard P, Hess GP (1986) Phosphorylation of the nicotinic acetylcholine receptor regulates its rate of desensitization. Nature 321: 774-776
Hunkeler W, Möhler H, Pieri L, Polc P. Bonetti EP, Cumin R, Schaffner R, Haefely W (1981) Selective antagonists of benzodiazepines. Nature 290: 514-516
Imoto K, Methfessel C, Sakmann B, Mishina M, Mori J, Konno T, Fukuda K, Kurasaki M, Bujo H, Fujita J, Numa S (1986) Location of a δ-subunit region determining ion transport through the acetylcholine receptor channel. Nature 324: 670-674
Kao PN, Karlin A (1986) Acetylcholine receptor binding site contains a disulfide cross-link between adjacent half-cystinyl residues. J Biol Chem 261: 8085-8088
Möhler H, Okada T (1977) Demonstration of benzodiazepine receptors in the central nervous system. Science 198: 849-851
Möhler H, Battersby MK, Richards JG (1980) Benzodiazepine receptor protein identified and visualized in brain tissue by a photoaffinity label. Proc Natl Acad Sci USA 77: 1666-1670

Möhler H, Schoch P, Richards JG, Häring P, Takacs B, Stähli C (1986) Monoclonal antibodies as probes for studying the structure and location of the GABA receptor/benzodiazepine receptor/chloride channel complex. In: Olsen RW, Venter JC (eds) Benzodiazepine/GABA Receptors and Chloride Channels: Structural and Functional Properties. Alan R Liss, New York, p 285

Noda M, Furutani Y, Takahashi H, Toyosato M, Tanabe T, Shimizu S, Kikyotani S, Kayano T, Hirose T, Inayami S, Numa S (1983) Cloning and sequence analysis of calf cDNA and human genomic DNA encoding α-subunit precursor of muscle acetylcholine receptor. Nature 305: 818-823

Oberthür W, Muhn P, Baumann H, Lottspeich F, Wittmann-Liebold B, Hucho F (1986) The reaction site of a non-competitive antagonist in the δ-subunit of the nicotinic acetylcholine receptor. EMBO J 5: 1815-1819

Palacios JM, Wamsley JK, Kuhar MJ (1981) High-affinity GABA receptors-autoradiographic localization. Brain Res 222: 285-307

Richards JG, Möhler H (1984) Benzodiazepine receptors. Neuropharmacology 23: 233-242

Richards JG, Möhler H, Haefely W (1986) Mapping benzodiazepine receptors in the CNS by radiohistochemistry and immunohistochemistry. In: Panula P, Päivärinta H, Soinita S (eds) Neurology and Neurobiology Vol. 16, Neurohistochemistry, Modern Methods and Applications. Alan R. Liss, New York, p 629

Richards JG, Schoch P, Häring P, Takacs B, Möhler H (1987) Resolving $GABA_A$/benzodiazepine receptors: cellular and subcellular localization in the CNS with monoclonal antibodies. J Neurosci 7: 1866-1886

Richards JG, Glinz R, Schoch P, Möhler H (1988) New Trends in mapping benzodiazepine receptors. In: Biggio G, Costa E (eds) Advances in Biochemical Psychopharmacology, Vol. 45. Raven Press, New York, p 27

Sakmann B, Hamill OP, Bormann J (1983) Patch-clamp measurements of elementary chloride currents activated by the putative inhibitory transmitters GABA and glycine in mammalian spinal neurons. J Neur Transm, Suppl 16: 83-95

Schoch P, Häring P, Takacs B, Stähli C, Möhler H (1984) A GABA/benzodiazepine receptor from bovine brain: purification, reconstitution and immunological characterization. J Receptor Res 4: 189-200

Schoch P, Richards JG, Häring P, Takacs B, Stähli C, Staehelin T, Haefely W, Möhler H (1985) Co-localization of $GABA_A$ receptors and benzodiazepine receptors in the brain shown by monoclonal antibodies. Nature 314: 168-171

Schofield PR, Darlison MG, Fujita N, Burt DR, Stephenson FA, Rodriguez H, Rhee LM, Ramachandran VR, Glencorse TA, Seeburg PH, Barnard EA (1987) Sequence and functional expression of the $GABA_A$ receptor shows a ligand-gated receptor super-family. Nature 328: 221-227

Séquier JM, Richards JG, Malherbe P, Price GW, Mathews S, Möhler H (1988) Mapping of brain areas containing RNA homologous to cDNAs encoding the α- and ß-subunits of the rat $GABA_A$ receptor. Proc Natl Acad Sci USA (in press)

Sieghart W, Karobath M (1980) Molecular heterogeneity of benzodiazepine receptors. Nature 286: 285-287

Sigel E, Barnard EA (1983) A γ-aminobutyric acid/benzodiazepine receptor complex of bovine cerebral cortex: purification and partial characterization. J Biol Chem 258: 6965-6971

Sigel E, Bower R (1988) Allosteric modulation of benzodiazepine receptor ligands of the $GABA_A$-receptor channel expressed in Xenopus oocytes. J Neurosci 8: 289-295

Squires RF, Braestrup C (1977) Benzodiazepine receptors in rat brain. Nature 266: 732-734

Stauber GB, Ransom RW, Dilber AI, Olsen RW (1987) The γ-aminobutyric acid/benzodiazepine receptor protein from rat brain: large scale purification and preparation of antibodies. Europ J Biochem 167: 125-133

Stephenson FA (1988) Understanding the $GABA_A$ receptor: a chemically gated ion channel. Biochem J 249: 21-32

Young WSIII, Kuhar MJ (1980) Radiohistochemical localization of benzodiazepine receptors in rat brain. J Pharm Exp Ther 212: 337-346

MOLECULAR HETEROGENEITY OF $GABA_A$-BENZODIAZEPINE RECEPTORS

W. Sieghart and K. Fuchs
Department of Biochemical Psychiatry
Psychiatrische Universitätsklinik
Währinger Gürtel 18-20
A-1090 Vienna
Austria

Since the introduction of benzodiazepines into clinical use and the establishment of their therapeutic action as anxiolytics, anticonvulsants, muscle relaxants and sedative hypnotics (Randall, 1961) a wealth of information has been accumulated on the molecular mechanism of action of these compounds. Thus, electrophysiological investigations have indicated that benzodiazepines exert most of their actions by enhancing the postsynaptic effects of the putative neurotransmitter GABA (Haefely et al., 1981). Biochemical and pharmacological investigations have established the existence of a high affinity binding site for benzodiazepines in brain membranes (Braestrup and Squires, 1977; Möhler and Okada, 1977) and demonstrated an excellent correlation between the clinical potency of a series of benzodiazepines and their ability to displace ^{3}H-diazepam or ^{3}H-flunitrazepam from their high affinity binding sites. Thus, it was concluded that these sites are the physiological receptors by which benzodiazepines exert their pharmacologically and clinically relevant actions. The close association of these receptors with a $GABA_A$-receptor, a chloride ion channel and several different drug binding sites (Olsen, 1982) indicate that benzodiazepine receptors are allosteric modulatory sites on the $GABA_A$ receptor gated chloride-ion channel.

In addition to these "central" benzodiazepine receptors, two other types of benzodiazepine binding sites have been identified which are not associated with GABA receptors. The "peripheral" binding sites exhibit a high affinity for some benzodi-

NATO ASI Series, Vol. H 32
Molecular Biology of Neuroreceptors and Ion Channels. Edited by A. Maelicke

azepines and are present in liver, kidney and other peripheral tissues (Braestrup and Squires, 1977) as well as in brain (Schoemaker et al., 1981). The "micromolar" benzodiazepine binding sites so far have been found in brain tissue only and exhibit a rather low affinity for benzodiazepines (Bowling and De Lorenzo, 1982). These two types of binding sites differ from "central" benzodiazepine receptors in their binding properties and kinetic and pharmacological characteristics, and the physiological significance of these sites is presently not known.

Originally, it was assumed that there is only one type of the "central" high affinity benzodiazepine receptor in the brain. Recently, however, several compounds with distinct chemical structures have been identified which provide some evidence for the existence of at least two benzodiazepine receptor subtypes. Thus, the triazolopyridazine Cl 218872 (Klepner et al., 1979), some esters of ß-carboline-3-carboxylate (Braestrup and Nielsen, 1983), some new benzodiazepines like quazepam, SCH 15725 and cinolazepam (Sieghart and Schuster 1984) and the imidazopyridine zolpidem (Arbilla et al., 1985) have been shown to have a several times higher affinity for benzodiazepine receptors in cerebellum than for those in hippocampus and other brain regions. These and other results led to the hypothesis that in cerebellum a benzodiazepine receptor subtype (type I receptor, BZ_1 receptor) is enriched which exhibits a high affinity for Cl 218872 or the various selective benzodiazepines. In addition, it was concluded that at least one other benzodiazepine receptor, a type II receptor (BZ_2 receptor) has to exist which exhibits a low affinity for the above mentioned compounds and is enriched in hippocampal membranes (Klepner et al., 1979). Other experiments suggested that both the BZ_1 and BZ_2 receptors are coupled to GABA- and anion recognition sites (Stapleton et al., 1982).

These data from reversible binding experiments are supported and supplemented by results from irreversible binding studies. ^{3}H-flunitrazepam on irradiation with UV-light has been shown to irreversibly bind to a membrane protein with apparent molecular weight 51,000 (P_{51}) in all brain regions investigated (Sieghart and Karobath, 1980). Whereas in cerebellum only P_{51} was significantly labeled, in hippocampus and other brain regions this compound irreversibly bound to several other proteins with apparent molecular weights 51,000 (P_{51}), 53,000 (P_{53}), 55,000 (P_{55}) and 59,000 (P_{59}) (Sieghart and Karobath, 1980; Lippa et al., 1982). All these proteins specifically and irreversibly labeled by ^3H-flunitrazepam seem to be associated with the "central", GABA-receptor associated type of benzodiazepine receptors, since irreversible binding to these proteins was abolished by diazepam, was stimulated by GABA and was unaffected by the "peripheral" benzodiazepine receptor ligand Ro 5-4864 (Sieghart and Karobath, 1980). Since in cerebellum BZ_1 receptors were most highly enriched and only one protein (P_{51}) was significantly labeled by ^3H-flunitrazepam, and since labeling of P_{51} by ^3H-flunitrazepam was inhibited with higher potency by the BZ_1-selective ligands Cl 218872 or SCH 15725 than that of the other photolabeled proteins, P_{51} seems to be associated with BZ_1 receptors (Sieghart et al., 1983). The different and distinct regional distribution of the other photolabeled proteins and the differential interaction of the BZ_1 receptor selective compounds with ^3H-flunitrazepam binding to the individual proteins (Sieghart et al., 1983) seem to indicate that proteins P_{53}, P_{55}, and P_{59} are associated with separate and different $GABA_A$-benzodiazepine receptor subtypes. This conclusion was supported by several experiments indicating a difference in the molecular structure of the $GABA_A$-benzodiazepine receptors associated with proteins P_{51} or P_{55} (Sieghart et al., 1987; Sieghart, 1988).

Recently, the $GABA_A$-benzodiazepine receptor complex has been highly purified by affinity chromatography (Sigel et al., 1983). The purified receptor preparation seemed to contain only two major protein bands with apparent molecular weight 53,000 (α-subunit) and 57,000 (ß-subunit) as revealed by SDS-polyacrylamide gel electrophoresis. Molecular mass determination suggested the complex to be a heterotetramer with an $\alpha_2\beta_2$ stoichiometry (Barnard et al., 1984). SDS-polyacrylamide gel electrophoretic techniques with higher resolution however revealed that the α- and β-subunits each consisted of several different proteins with quite similar molecular weight (Fuchs and Sieghart, unpublished results). Using monoclonal antibodies specific for α- or β-subunits, it was possible to demonstrate that all the various proteins irreversibly photolabeled by ^3H-flunitrazepam are different α-subunits of the $GABA_A$-benzodiazepine receptor complex (Fuchs et al., 1988). In addition, the existence of several different β-subunits was demonstrated. These subunits are specifically and irreversibly photolabeled by the GABA agonist ^3H-muscimol and are recognized by a β-subunit-selective antibody (Fuchs and Sieghart, submitted for publication). Interestingly, the apparent molecular weights of the various α- and β-subunits are closely similar and thus, α- and β-subunits cannot be distinguished from each other by their molecular weight.

Immunological and immunohistochemical evidence seems to indicate that α-subunits might be associated with β-subunits in all brain regions investigated (Häring et al., 1985; Schoch et al., 1985). Assuming a molecular structure of $\alpha_2\beta_2$ for the $GABA_A$-benzodiazepine receptor complex, a large variety of structurally different receptor subtypes could possibly arise by a combination of various α- and β-subunits. Thus, an α-subunit could be associated with the same or a different α-subunit or with the same or different β-subunits. Since the α-subunits seem to contain the benzodiazepine - and the β-subunits the GABA/muscimol binding sites, different combinations of α- and β-subunits could lead to a different way of modulation by benzodiazepines of the GABA-induced chloride ion flux but also could possibly infer allosterically changed binding properties to the α-subunit associated benzodiazepine receptor site. The demonstration of a different regional distribution of the various α-subunits labeled by ^3H-flunitrazepam (Sieghart and Karobath, 1980) and the availability of compounds preferentially interacting with one of these subunits (Sieghart et al., 1983) indicate that the development of highly specific benzodiazepine receptor ligands could offer the possibility to selectively modulate specific GABA neurons in various brain regions.

Acknowledgement: This work was supported by the Fonds zur Förderung der wissenschaftlichen Forschung in Österreich.

References

Arbilla S, Depoortere H, George P, Langer SZ (1985) Pharmacological profile of the imidazopyridine zolpidem at benzodiazepine receptors and electrocorticogram in rats. Naunyn-Schmiedeberg's Arch Pharmacol 330:248-251

Barnard EA, Stephenson FA, Sigel E, Mamalaki C, Bilbe G, Constanti A, Smart TE, Brown DA (1984) Structure and properties of the brain GABA/benzodiazepine receptor complex. In: Kito S, Segawa T, Kuriyama K, Yamamura HI, Olsen RW (eds) Neurotransmitter receptors: mechanisms of action and regulation. Plenum Press, New York, p 235

Bowling AC, De Lorenzo RJ (1982) Micromolar affinity benzodiazepine receptors identification and characterization in central nervous system. Science 216:1247-1250

Braestrup C, Squires R (1977) Specific benzodiazepine receptors in rat brain characterized by high affinity ^{3}H-diazepam binding. Proc Natl Acad Sci USA 74:3804-3809

Braestrup C, Nielsen MJ (1983) Benzodiazepine receptors. In: Iversen L, Iversen SD, Snyder, SH (eds) Handbook of psychopharmacology, vol. 17. Plenum Publishing Corp., p 285

Fuchs K, Möhler H, Sieghart W (1988) Various proteins from rat brain, specifically and irreversibly labeled by ^{3}H-flunittrazepam are distinct α-subunits of the GABA-benzodiazepine receptor complex. Neurosci Letters 90:314-319

Haefely W, Pieri L, Polc P, Schaffner R (1981) General pharmacology and neuropharmacology of benzodiazepine derivatives. In: Hoffmeister E, Stille G (eds) Handbook of Experimental Pharmacology, Vol.55/II, Springer Verlag, Berlin, p 134

Häring P, Stähli C, Schoch P, Takacs B, Staehelin T, Möhler H (1985) Monoclonal antibodies reveal structural homogeneity of γ-aminobutyric acid/benzodiazepine receptors in different brain areas. Proc Natl Acad Sci USA 82:4837-4841

Klepner CA, Lippa AS, Benson DI, Sano MC, Beer B (1979) Resolution of two biochemically and pharmacologically distinct benzodiazepine receptors. Pharmacol Biochem Behav 11:457-462

Lippa AS, Jackson D, Wennogle LP, Beer B, Meyerson LR (1982) Non benzodiazepine agonists for benzodiazepine receptors. In: Usdin E, Skolnick Ph, Tallman JF, Greenblatt D, Paul SM (eds) Pharmacology of Benzodiazepines, Macmillan Press Ltd., London and Basingstoke, p 431

Möhler H, Okada T (1977) Benzodiazepine receptors - demonstration in the central nervous system. Science 198:849-851

Olsen RW (1982) Drug interactions at the GABA receptor ionophore complex. Ann Rev Pharmacol Toxicol 22:245-277

Randall LO (1961) Pharmacology of chlordiazepoxide (Librium). Dis Nerv System 22, Sect 2, Suppl 7:7-15

Schoch P, Richards JG, Häring P, Takacs B, Stähli C, Staehelin T, Haefely W, Möhler H (1985) Co-localization of $GABA_A$ receptors and benzodiazepine receptors in the brain shown by monoclonal antibodies. Nature 314:168-171

Schoemaker H, Bliss, M, Yamamura HI (1981) Specific high affinity binding of ^{3}H-Ro 5-4864 to benzodiazepine binding sites in rat cerebral cortex. Europ J Pharmacol 71:173-175

Sieghart W, Karobath M, (1980) Molecular heterogeneity of benzodiazepine receptors. Nature 286:285-287

Sieghart W, Mayer A, Drexler G (1983) Properties of ^{3}H-flunitrazepam binding to different benzodiazepine binding proteins. Europ J Pharmacol 88:291-299

Sieghart W, Schuster A (1984) Affinity of various ligands for benzodiazepine receptors in rat cerebellum and hippocampus. Biochem Pharm 33:4033-4038

Sieghart W, Eichinger A, Zezula J (1987) Comparison of tryptic peptides of benzodiazepine binding proteins photolabeled with ^{3}H-flunitrazepam or ^{3}H-Ro 15-4513. J Neurochem 48:1109-1114

Sieghart W (1988) Comparison of two different benzodiazepine binding proteins by peptide mapping after limited proteolysis. Brain Res 450:387-391

Sigel E, Stephenson FA, Mamalaki C, Barnard E (1983) A γ-aminobutyric acid/benzodiazepine receptor complex of bovine cerebral cortex. Purification and partial characterization. J Biol Chem. 258:6965-6971

Stapleton SR, Prestwich SA, Horton RW (1982) Regional heterogeneity of benzodiazepine binding sites in rat brain. Europ J Pharmacol 84:221

THE POSTSYNAPTIC GLYCINE RECEPTOR - A MEMBER OF THE NEUROTRANSMITTER-GATED CHANNEL PROTEIN FAMILY

Dieter Langosch, Gabriele Grenningloh, Volker Schmieden, Peter Prior, Maria-Luisa Malosio, Bertram Schmitt, and Heinrich Betz.

ZMBH, Zentrum für Molekulare Biologie, 6900 Heidelberg, F.R.G.

Introduction

Postsynaptic inhibition in the central nervous system is mediated primarily by two pharmacologically distinguishable integral membrane proteins, the receptors for the amino acid neurotransmitters glycine and γ-aminobutyric acid. Whereas the γ-aminobutyric acid receptor ($GABA_AR$) is distributed throughout higher brain regions, the glycine receptor (GlyR) predominates in spinal cord and brain stem (Barnard et al.,1987; Betz and Becker, 1988). Both proteins contain an intrinsic chloride channel which opens after binding of the respective agonists to the receptor. Thereby permeation of small anions through the postsynaptic membrane is permitted to antagonize postsynaptic depolarization. Chloride flux through the GlyR is blocked after binding of the plant alkaloid strychnine, resulting in overexcitation of the motor system, i.e. muscular convulsions (Barnard et al., 1987; Betz and Becker, 1988). In our laboratory, immobilized 2-aminostrychnine, a biologically active derivative of this antagonist, has been exploited to affinity-purifiy the GlyR from detergent extracts of mammalian spinal cord membranes. Biochemical and immunological analysis of the purified GlyR showed that it contains two glycosylated integral membrane polypeptides of M_r 48 K and 58 K and a non-glycosylated, cytoplasmically localized component of M_r 93 K (Pfeiffer et al., 1982; Graham et al.,1985; Becker et al. 1986). The

NATO ASI Series, Vol. H 32
Molecular Biology of Neuroreceptors and
Ion Channels. Edited by A. Maelicke

membrane spanning subunits apparently are homologous to each other and thought to form the chloride channel of the receptor whereas the 93 K subunit has been proposed to be involved in the postsynaptic anchoring of the GlyR (Schmitt et al., 1987; Pfeiffer et al.,1984; Langosch et al. 1988). The strychnine binding site has been assigned to the 48 K subunit on the basis of photo-affinity labelling experiments (Graham et al., 1983). Lately, variants of the 48 K polypeptide have been shown to result from alternative splicing of the corresponding pre-mRNA (G. G. and M.-L. M., unpublished results).

The glycine receptor is homologous to other ligand-gated ion channel proteins

Recently, the primary structure of the GlyR 48 K subunit has been established by cDNA cloning (Grenningloh et al., 1987a). Within this sequence hydropathy analysis revealed four hydrophobic segments (M1 to M4) proposed to form transmembrane α-helices. This transmembrane topology is reminiscent of the nicotinic acetycholine receptor (nAchR) polypeptides and of subunits of the $GABA_AR$ (Schofield et al., 1987). Furthermore, significant amino acid sequence homology exists between all these diffferent receptor proteins. An alignment of the sequence of the GlyR 48 K subunit with those of the $GABA_AR$ polypeptides is shown in Fig.1.
Particularly well conserved are the structures of the respective membrane spanning domains. Quite remarkably, uncharged polar amino-acid residues are abundant in segments M1 and particularly in M2 suggesting that these domains are likely to form the hydrophilic inner lining of the chloride channel. The highest degree of amino acid conservation between inhibitory receptor polypeptide domains is displayed by segment M2 (Grenningloh et al., 1987b). Interestingly, transmembrane segment M2 of nAchR subunits has been shown to be involved in cation translocation and channel blocker

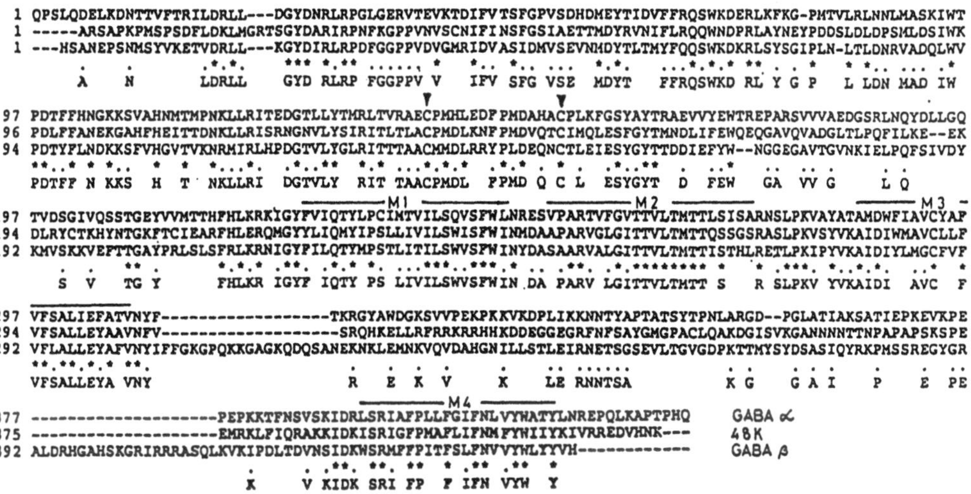

Fig. 1. Alignment of glycine receptor 48 K subunit and $GABA_A$ receptor α and β subunit sequences. Amino acid residues conserved in all three or in two of the polypeptides are indicated by an asterisk or a dot, respectively.

binding (Giraudat et al.,1986; Hucho et al., 1986; Imoto et al. 1986) and may thus be a common major determinant of ligand-gated ion channel function. Further, a proline residue in M1 is also found in all subunits of the nAchR and the $GABA_A$R. The kink introduced by this residue into the proposed helical structure of M1, therefore, may have significant implications for ion channel function.

At the interphases between the membrane spanning domains and the hydrophilic parts of the GlyR 48 K as well as of $GABA_A$R subunits clusters of positively charged residues are found. These charges may explain the existence of two sequentially occupied anion binding sites in the chloride channel of both GlyR and $GABA_A$R identified by patch clamp analysis (Bormann et al., 1987). A corresponding charge distribution is found at the presumptive inner and outer mouths of the ion channel of the nAchR (Stroud and Finer-Moore, 1985) and therefore can be assumed to provide the ion selectivity filter of the various ion channel proteins.

The extracellular region of the GlyR 48 K subunit contains two cysteine residues which are precisely conserved in polypeptides of the $GABA_AR$ and the nAchR. In the latter, these residues have been proposed to form a disulfide bridge essential for ligand binding and receptor tertiary structure (Stroud and Finer-Moore, 1985). Similar folding patterns may therefore be shared by different ligand-gated ion channel proteins.

Analysis of the quarternary structure of the GlyR by various crosslinking procedures showed that the membrane-spanning 48 K and 58 K subunits, but not the peripheral 93 K polypeptide, can be linked to a series of adducts up to M_R 260 K. This indicates a pentameric core possibly composed of three copies of the 48 K and two copies of the 58 K subunits, respectively (Langosch et al. 1988). A similar structural design, i.e. a pentameric functional complex, has been established for the nAchR (Stroud et al., 1985). In contrast, the $GABA_AR$ was proposed to be tetrameric; an analysis based, however, on gel scanning data (Mamalaki et al., 1987). Hence, a common quasi-symmetrical pentameric quarternary structure shared by all ligand-gated ion channel proteins is suggested.

Structural homologies reflect the existence of a gene family

In summary, our data show that the GlyR is homologous to other ion channel proteins at different levels of protein structure - notably amino acid sequence, transmembrane topology, and quarternary structure. A common evolutionary relationship between the members of the neurotransmitter-gated ion channel protein family is primarily indicated by the conservation of primary structure. Presumably, all subunits of these proteins are encoded by a large super-family of genes which evolved from a primitive channel ancestor. Specific functional properties such as ligand specificity and ion selectivity would then have developed by divergent evolution to suit the discriminating needs of complex neuronal circuits.

REFERENCES

Barnard, E.A., Darlison, M.G., & Seeburg, P.H. (1987) Trends Neurosci. **10**, 502-509.

Betz, H. (1987) Trends Neurosci. **10**, 113-117.

Becker, C.M., Hermans-Borgmeyer, I., Schmitt, B., & Betz, H. (1986) J. Neurosci. **6**, 1358-1364.

Betz H. (1987) Trends Neurosci. 10, 502-509.

Betz H. & Becker C.-M. (1988) Neurochem. Int., in press.

Bormann J., Hamill O.P., Sakmann B. (1987) J. Physiol. **385**, 243-286.

Giraudat, J., Dennis, M., Heidmann, T., Haumont, P.T., Lederer, F., & Changeux, J.P. (1987) Biochemistry **26**, 2410-2418.

Graham, D., Pfeiffer, F., & Betz, H. (1983) Eur. J. Biochem. **131**, 519-525.

Graham, D., Pfeiffer, F., & Betz, H. (1985) Biochemistry **24**, 990-994.

Grenningloh, G., Rienitz, A., Schmitt, B., Methfessel, C., Zensen, M., Beyreuther, K., Gundelfinger, E.D., & Betz, H. (1987) Nature (London) **328**, 215-220.

Grenningloh, G., Gundelfinger, E.D., Schmitt, B., Betz, H., Darlison, M.G., Barnard, E.A., Schofield, P., & Seeburg, P.H. (1987) Nature (London) **330**, 25-26.

Hucho, F.L., Oberthür, W., & Lottspeich, F. (1986) FEBS Lett. **205**, 137-142.

Imoto, K., Methfessel, C., Sakmann, B., Mishina, M., Mori, Y., Konno, T., Fukuda, K., Kurasaki, M., Bujo, H., Fujita, Y., & Numa, S. (1986) Nature (London) **324**, 670-674.

Langosch D., Thomas L., & Betz H. (1988) Proc. Natl. Acad. Sci. USA, in press.

Mamalaki, C., Stephenson, F.A., & Barnard, E.A. (1987) EMBO J. **6**, 561-565.

Pfeiffer, F., Graham D., & Betz H. (1982) J. Biol. Chem. **257**, 9389-9393.

Pfeiffer, F., Simler, R., Grenningloh, G., & Betz, H. (1984) Proc. Natl. Acad. Sci. USA **81**, 7224-7227.

Schmitt, B., Knaus, P., Becker, C.-M., & Betz, H. (1987) Biochemistry **26**, 805-811.

Schofield, P.R., Darlison, M.G., Fujita, N., Burt, D.R., Stephenson, F.A., Rodriguez, H., Rhee, L.M., Ramachandran, J., Reale, V., Glencorse, T.A., Seeburg, P.H., & Barnard, E.A. (1987) Nature (London) **328**, 221-227.

Stroud, R.M. & Finer-Moore, J.A. (1985) Rev. Cell. Biol. 1, 317-351.

Immunological identification of a neonatal glycine receptor isotype and its expression in primary cultures of spinal cord

Werner Hoch, Heinrich Betz, and Cord-Michael Becker
Zentrum für Molekulare Biologie, Universität Heidelberg,
Im Neuenheimer Feld 282,
D-6900 Heidelberg, Federal Republic of Germany

Development of chemosensitivity is a major focus of interest in neurobiology. During formation of the mammalian neuromuscular junction, the embryonic subtype of the nicotinic acetylcholine receptor is replaced by an adult receptor type. This process coincides with a redistribution of receptors from extrasynaptic to subsynaptic areas of the muscle cell membrane (for reviews, see Schuetze and Role, 1987; Brehm et al., 1988). It is not known whether, in the CNS, synaptogenesis is also accompanied by changes in receptor subtype expression and distribution. So far, studies have been hampered by low abundance of these proteins and lack of suitable methodology.

The inhibitory glycine receptor (GlyR) appears well suited for a developmental analysis. The GlyR is a ligand-gated chloride channel of the postsynaptic membrane (reviewed by Betz and Becker, 1988) which has been purified by affinity chromatography using a derivative of its specific antagonist strychnine (Pfeiffer et al., 1982). The 48 kD and 58 kD GlyR subunits form the transmembrane 'core' of the receptor (Langosch et al., 1988), whereas the 93 kD polypeptide is a peripheral membrane protein facing the cytoplasm (Triller et al., 1985; Schmitt et al., 1987).

Immunomethods employing monoclonal antibodies (mAbs) raised against GlyR (Pfeiffer et al.,.1984) have led to the detection of a GlyR isotype characterized by low affinity for strychnine which predominates in spinal cord of neonatal rats (Becker et al., 1988; see also Becker and Hoch, this volume). As shown here, differences in antigenic epitopes of our mAbs permit an immunological identification and a separation of neonatal and adult GlyR subtypes. Employing these methods, we show that primary cultures of embryonic mouse spinal cord almost exclusively express the neonatal GlyR isotype.

NATO ASI Series, Vol. H 32
Molecular Biology of Neuroreceptors and
Ion Channels. Edited by A. Maelicke

1. Immunomethods distinguish between GlyR isotypes

Expression of GlyR was analyzed by a comparison of [^{3}H]strychnine binding activity and antigenic epitopes. In adult tissue, no evidence for receptor heterogeneity was found as antibody and ligand binding correlated well throughout the CNS (Becker et al., submitted). Out of four mAbs directed against GlyR, only mAb 2b efficiently precipitated [^{3}H]strychnine binding sites from detergent extracts of adult spinal cord (table I). As expected, epitopes for the other antibodies, e.g. mAbs 4a and 7a, coprecipitated under these conditions (Becker et al., submitted).

A marked difference between ligand and antibody binding became apparent with neonatal spinal cord preparations. Here, the concentration of GlyR-like mAb 4a-antigen also was high (Becker et al., 1988; Becker and Hoch, this volume), but only low numbers of [^{3}H]strychnine binding sites were detectable (table II). Further studies revealed that mab 4a recognized two populations of GlyRs in neonatal spinal cord. One group of receptors resembled that found in the adult animal, i.e. carried the mab 2b-epitope and bound [^{3}H]strychnine with high affinity. This is evident from the efficient removal of [^{3}H]strychnine binding activity from membrane extracts by chromatography on a mAb 2b-column (table II). At birth, however, mab 4a-antigen also comprises a population of "neonatal" GlyRs which still bind strychnine, but at an affinity too low to be detected in radioligand filtration assays. These

Table I. Characteristics of mAbs used in this study (see also Pfeiffer et al., 1984; Schmitt et al., 1987)

monoclonal antibody	subunit recognized on Western blots	immunoprecipitation of [^{3}H]strychnine binding sites
1a	48 kD	-
2b	48 kD	+
4a	48 kD, 49 kD (58 kD)*	-
7a	93 kD	-

*) This subunit is visualized in purified preparations only.

neonatal GlyRs lack the mAb 2b-epitope. This is demonstrated by the following experiment (table III).

As a first step, detergent extracts from either neonatal or adult spinal cord were subjected to mAb 2b-affinity chromatography. In a second step, the flow through fractions of mAb 2b-columns were subsequently applied to 2-aminostrychnine agarose-columns. In agreement with [^{3}H]strychnine binding data (table II), most of the GlyR from adult extracts could be recovered from the mAb 2b-column (table III). This indicates that both high affinity binding site for strychnine and mAb 2b-epitope characterize the adult type GlyR. With neonatal samples, however, high amounts of GlyR-like mAb 4a-antigen passed the antibody column but were recovered from the 2-aminostrychnine column. Apparently, receptors of the low affinity type which in

Table II. Retention of high affinity [^{3}H]strychnine binding sites on a mAb 2b-column

	neonatal extract		adult extract	
	[cpm]	[%]	[cpm]	[%]
detergent extract	947±48	100	2481±43	100
flow through of mAb 2b-column	233± 7	25	188±49	8

Table III. Distribution of GlyR epitopes in eluates from sequentially loaded mAb 2b and 2-aminostrychnine agarose columns

	neonatal extract		adult extract	
	mAb 4a	mAb 7a	mAb 4a	mAb 7a
• step 1:				
mAb 2b-column	33	10	195	210
• step 2:				
2-aminostrychnine agarose-column	131	40	54	53

Immunoreactivity is assessed in arbitrary units.

filtration binding assays eluded detectability were retained on the highly substituted agarose matrix.

The 93 kD protein recognized by mAb 7a is expressed at comparatively low amounts around birth (Becker et al., 1988). In detergent extracts from neonatal spinal cord, the protein associates with both GlyR isotypes to the same extent. This was deduced from a similar distribution of mAb 7a and mAb 4a-epitopes (table III; see also Becker et al., 1988).

Besides reduced affinity for strychnine and altered antigenicity, neonatal GlyR shows a molecular weight of its ligand binding subunit (49 kD) slightly higher than the adult isotype (Becker et al., this volume). Table IV summarizes the properties of the two isotypes.

2. Primary cultures of mouse spinal cord predominantly express the neonatal GlyR isotype.

Electrophysiological investigations (Ransom et al., 1977; Bormann et al., 1987) have established that dissociated cell cultures of rodent spinal cord possess functional GlyR. Primary cultures appear suitable for a study of GlyR expression since they offer several advantages over an exclusive use of in vivo studies:

I.) Metabolic labelling of cell cultures allows studies on biosynthesis, subunit assembly, and degradation of GlyR.

Table IV. Properties of GlyR isotypes from rat spinal cord

GlyR isotype:		neonatal	adult
affinity for strychnine		low	high
glycine binding		++	++
mol. wt of ligand binding subunit		49 kD	48 kD
association with 93 kD protein		++	++
binding of mAbs		4a only	1a, 2b, and 4a
distribution:	neonatal rat	70-80%	20-30%
	adult rat	≤ 5%	≥ 95%
	neuronal culture	≥ 90%	≤ 10%

II.) In vitro systems are more amenable to experimental manipulations and therefore should facilitate an identification of factors involved in regulation of GlyR isotype expression during development.
III.) Differences in GlyR expression between in vivo and in vitro conditions could give hints about the influence of proper neuronal connections on receptor subtype pattern.

Primary cultures were prepared from mouse and rat spinal cord at embryonic day 13. The expression of antigens recognized by mAbs 2b, 4a, and 7a was determined after 1 and 4 weeks in culture (Figure 1). The mAbs used cross-react with mouse GlyR (Becker et al., 1986), and

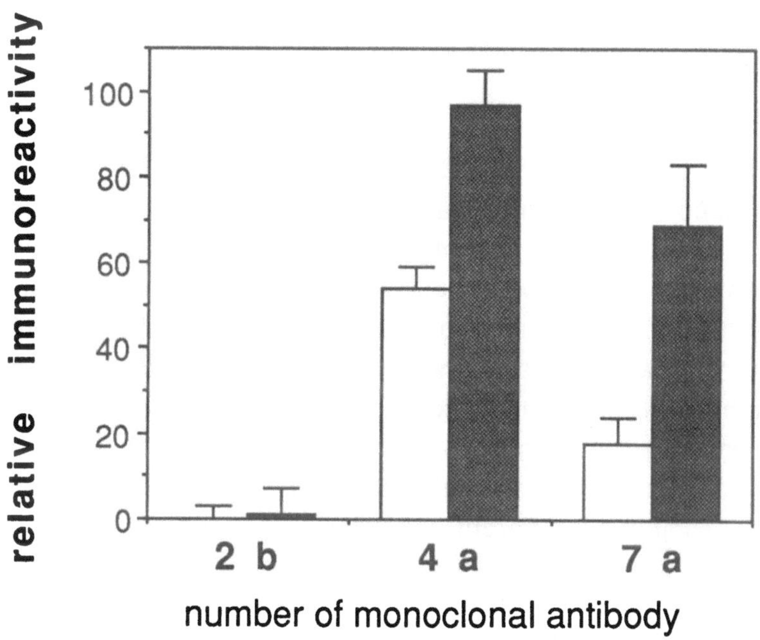

Figure 1: Binding of GlyR-mAbs to membrane fractions from 1 week (open bars) and 4 weeks old (shaded bars) primary cultures of mouse spinal cord. Immunoreactivity of GlyR-antigen per mg protein was normalized using spinal cord membranes from adult rats as a standard [% of standard value]. Binding of mAbs 4a and 7a was determined by dot receptor assay (Becker et al., submitted), binding of mAb 2b was monitored by a centrifugation assay using intact membranes (Hoch et al., in preparation).

no difference of GlyR expression in cell culture was observed between both species. The content of mAb 4a-antigen in membrane fractions from cell cultures increased about twofold from 1 to 4 weeks in culture. After 4 weeks in culture, these membranes contained about as much mAb 4a-antigen per mg of protein as membrane fractions from spinal cord of adult rats. In contrast, expression of mAb 2b-epitope was barely detectable throughout the culture period. Thus, only small amounts of adult GlyR isotype are expressed in primary cultures. This conclusion was confirmed by a determination of high affinity [^{3}H]strychnine binding to these cultures. Based on the concentration of mAb 4a-antigen, less than 10% of the [^{3}H]strychnine binding activity expected for adult GlyR could be detected in cultured cells. This implies that GlyR in cultures as detected by mAb 4a predominantly represented the neonatal isotype. Apparently, the concentration of this GlyR isotype increased during later times in culture albeit it rapidly decreases during the corresponding period in vivo (Becker et al., 1988; Becker and Hoch, this volume).

Expression in cell cultures of 93 kD protein as monitored by mAb 7a showed a steep increase from rather low levels after the first week to much higher concentrations after 4 weeks. However, it reached only 70% of the level seen in vivo with adult GlyR. In rat (Becker et al., 1988; Becker and Hoch, this volume) and mouse (C.-M. Becker, unpublished) spinal cord, increases in expression of adult GlyR isotype and 93 kD protein exhibit very similar kinetics. In culture, however, both processes are dissociated since induction of adult GlyRs was prevented whereas 93 kD protein accumulated. This demonstrates that these proteins are subject to different regulatory mechanisms.

3. Conclusions

Here, we employed immunological techniques to characterize two subtypes of the inhibitory GlyR expressed during postnatal development. These GlyR isotypes could be separated using the selective antibody, mAb 2b, which recognizes only the adult receptor species. Both subtypes contain distinct ligand binding subunits and show pronounced differences in pharmacological properties. Heterogeneity of nicotinic acetylcholine receptors during muscle development is well documented and found to result from an exchange of a non-ligand binding subunit upon muscle innervation (Mishina et al., 1986). Interestingly, cDNAs coding for variants of the strychnine binding subunit of the GlyR have recently been identified (G. Grenningloh and C.-M. Becker, unpublished).

Our observation that GlyR expressed in dissociated cultures of embryonic spinal cord mostly comprises the neonatal isotype may warrant a reconsideration of previous electrophysiological studies. It remains to be shown whether GlyR channel properties reported for cultured neurons (Bormann et al., 1987) represent a particular receptor isoform or whether the adult GlyR isotype encountered in vivo shares the same single channel characteristics. Certainly, dissociated cell cultures may provide a useful tool for identification of the mechanisms governing expression of particular GlyR subtypes and factors responsible for their selective regulation during development.

Acknowledgements

We thank I. Wolters and C. Schroeder for expert technical assistance. This work was supported by the Deutsche Forschungsgemeinschaft (Be 718/8-1 and SFB 317) and the Fonds der Chemischen Industrie.

References

Becker, C.-M., Hermans-Borgmeyer, I., Schmitt, B., and Betz, H. (1986) J. Neurosci. 6: 1358-1364.

Becker, C.-M., Hoch, W., and Betz, H. (1988) EMBO J. 7: 3717-3726.

Betz, H., and Becker, C.-M. (1988) Neurochem. Int. 13: 137-146.

Bormann, J., Hamill, O.P., and Sakmann, B. (1987) J. Physiol. 385: 243-286.

Brehm, P., and Henderson, L. (1988) Dev. Biol. 129: 1-11.

Graham, D., Pfeiffer, F., Simler, R., and Betz, H. (1985) Biochemistry 24: 990-994.

Grenningloh, G., Rienitz, A., Schmitt, B., Methfessel, C., Zenssen, M., Beyreuther, K., Gundelfinger, E.D., and Betz, H. (1987) Nature 328: 215-220.

Langosch, D., Thomas, L., and Betz, H. (1988) Proc. Natl. Acad. Sci. USA 85: 7394-7398.

Mishina, M., Takai, T., Imoto, K., Noda, M., Takahashi, T., Numa, S., Methfessel, C., and Sakmann, B. (1986) Nature 321: 406-411.

Pfeiffer, F., Graham, D., and Betz, H. (1982) J. Biol. Chem. 257: 9389-9393.

Pfeiffer, F., Simler, R., Grenningloh, G., and Betz, H. (1984) Proc. Natl. Acad. Sci. USA 81: 7224-7227.

Ransom, B.R., Bullock, P.N., and Nelson, P.G. (1977) J. Neurophysiol. 40: 1163-1177

Schmitt, B., Knaus, P., Becker, C.-M., and Betz, H. (1987) Biochemistry 26: 805-811.

Schuetze, S.M., and Role, L.W. (1987) Annu. Rev. Neurosci. 10: 403-457.

Triller, A., Cluzeaud, F., Pfeiffer, F., Betz, H., and Korn, H. (1985) J. Cell Biol. 101: 683-688.

Glycine receptor regulation in rodent spinal cord

Cord-Michael Becker and Werner Hoch
Zentrum für Molekulare Biologie, Universität Heidelberg,
Im Neuenheimer Feld 282,
D-6900 Heidelberg, Federal Republic of Germany

The inhibitory action of glycine on spinal cord neurons is effectively antagonized by the convulsant alkaloid strychnine. Conversely, glycine-displaceable [^{3}H]strychnine binding has been widely used as a reliable assay for the postsynaptic glycine receptor [GlyR] (Young and Snyder, 1973, 1974). A single class of high-affinity [^{3}H]strychnine binding sites is found with affinity constants of 3-10 nM (for review, see Betz and Becker, 1988). Affinity-purified GlyR contains three polypeptides of 48 kD, 58 kD, and 93 kD molecular weight (Pfeiffer et al., 1982; Graham et al., 1985; Becker et al., 1986; see also Langosch et al., this volume). The primary structure of the ligand-binding 48 kD subunit has been determined by cDNA cloning (Grenningloh et al., 1987). Both the 48 kD and 58 kD polypeptides are integral membrane proteins, whereas the 93 kD polypeptide is a peripheral membrane component (Schmitt et al., 1987) found at the cytoplasmic face of the postsynaptic membrane (Triller et al., 1987).

Little is known about the regulation of glycine and other neurotransmitter receptors in the CNS. To elucidate the mechanisms underlying GlyR expression, we have studied two examples of low [^{3}H]strychnine binding activity in rodent spinal cord. A physiological state of low [^{3}H]strychnine binding activity occurs in spinal cord of newborn rats (Benavides et al., 1981). Here, we show that this results from the predominance at birth of a GlyR isoform characterized by low strychnine binding affinity. This isoform, therefore, remains undetected by standard assays of ligand binding (Becker et al., 1988). In contrast, the pathological reduction of [^{3}H]strychnine binding in the mutant mouse *spastic* (White and Heller, 1982) results from a reduced number of GlyRs. The mutation appears to affect the regulation, rather than the structure, of inhibitory GlyRs (Becker et al., 1986).

NATO ASI Series, Vol. H 32
Molecular Biology of Neuroreceptors and Ion Channels. Edited by A. Maelicke

I. Glycine receptor heterogeneity during postnatal development of the rat

In rat spinal cord, high affinity [^{3}H]strychnine binding activity was low at birth accounting for about 20-30% of the adult level. A continuous increase to adult values was found during the first two postnatal weeks (Figure 1a). Scatchard analysis yielded B_{max} values of 229 and 643 fmol/µg protein for neonatal and adult spinal cord, respectively (Figure 1b). Apparent receptor affinity did not significantly change during development as the respective K_D values were determined to be 6.8 nM and 4.0 nM. These observations are consistent with earlier reports for embryonic chick (Zukin et al., 1975) and postnatal rat spinal cord (Benavides et al., 1981).

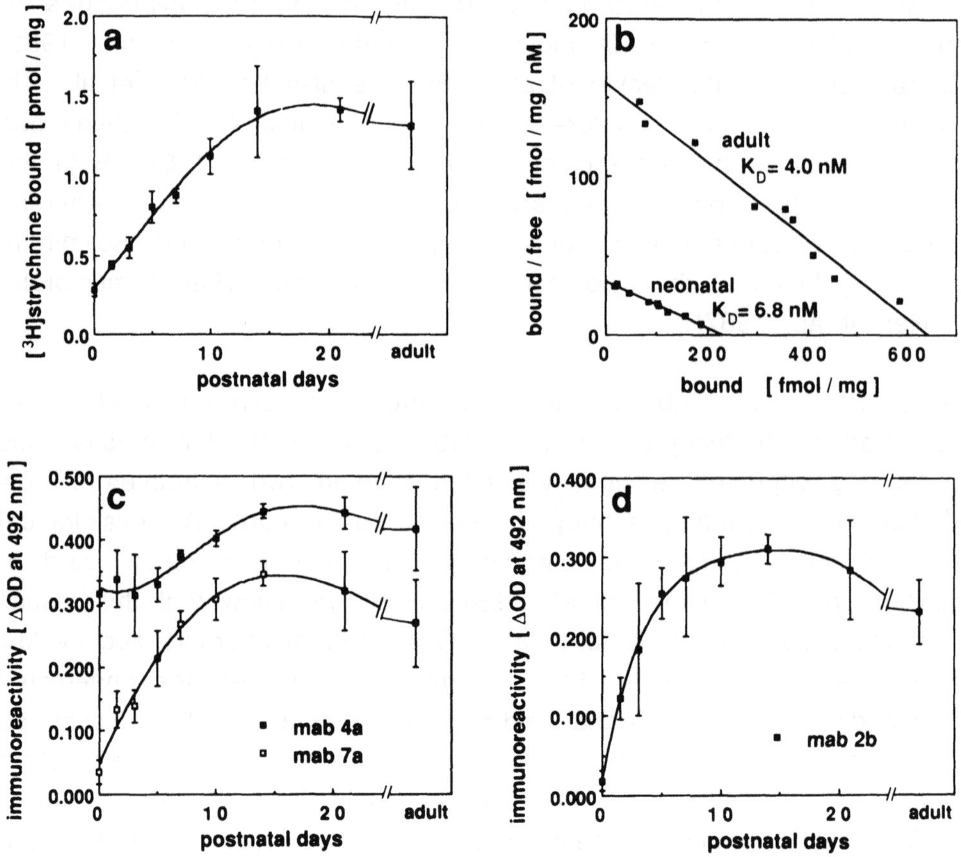

Figure 1: Postnatal development of [^{3}H]strychnine binding activity and GlyR-antigen content in rat spinal cord (from Becker et al., 1988).

The ligand-binding 48 kD polypeptide of GlyR is recognized by monoclonal antibodies (mAbs) 1a, 2b, and 4a originally raised against receptor from adult rat spinal cord (Pfeiffer et al., 1984). Postnatal development of GlyR antigenic epitopes in rat spinal cord as determined by sensitive immunoassays employing mAbs 2b (Figure 1d) and 1a (not shown) correlated well with [^{3}H]strychnine binding activity. In contrast, reactivity with mAb 4a was high in neonatal samples (Figure 1c), a finding inconsistent with other "markers" of the ligand binding subunit. Thus, some GlyR-like antigen was present at close to adult levels already at birth even though [^{3}H]strychnine binding was low.

Content of mAb 7-antigen showed a postnatal increase which quantitatively matched the appearance of high affinity [^{3}H]strychnine binding (Figure 1c). However, both changes might reflect independent processes since the antagonist binding site and the mAb 7a-epitope are located on distinct proteins, i.e. the 48 kD and 93 kD GlyR polypeptides, respectively.

Western blot analysis confirmed these quantitative changes in GlyR-antigens (Figure 2). The 93 kD polypeptide of GlyR recognized by mAb 7a was intensely stained in adult spinal cord, but barely visible at

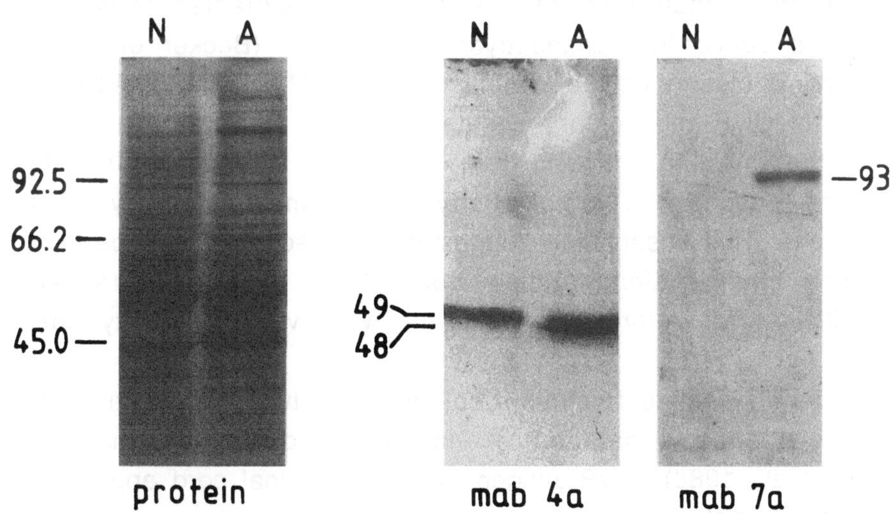

Figure 2: Western blot of membranes from neonatal (N) and adult (A) spinal cord.

birth. In membranes from adult spinal cord, mAb 4a detected the 48 kD subunit as previously reported (Schmitt et al., 1987; Becker et al., submitted). At birth, however, a band of 49 kD immunoreactive with mAb 4a was abundant, indicating a heterogeneity of GlyR subunit molecular weight during early development. In most neonatal preparations, the 48 kD component was also detectable, and could be resolved from the 49 kD band (Becker et al., 1988) which is thought to represent low amounts of the adult type GlyR-antigen. The neonatal GlyR is a macromolecular complex indistinguishable from adult receptor in its apparent sedimentation coefficient (7.7 S) upon sucrose density centrifugation (Becker et al., 1988). After separation from the 93 kD associated protein (Schmitt et al., 1987; Becker et al., submitted), the transmembrane 'core' of the adult GlyR possesses a molecular weight of ≈250 kD (Pfeiffer et al., 1982; Langosch et al., this volume). Therefore, the GlyR-antigen present in neonatal spinal cord represents oligomeric complexes rather than unassembled monomers of the ligand binding subunit.

Since the neonatal GlyR-like antigen detected by mAb 4a did not correlate with [^{3}H]strychnine binding activity, we attempted to separate strychnine binding and non-binding forms of GlyR by affinity chromatography on 2-aminostrychnine columns (Pfeiffer et al., 1982). However, more than 95% of mAb 4a-antigen extracted from neonatal spinal cord bound to 2-aminostrychnine agarose (Becker et al., 1988). Furthermore, elution of the column by 200 mM glycine yielded similar amounts of GlyR-antigen for neonatal and adult samples. Thus, virtually all of the mAb 4a-antigen from neonates bound to immobilized strychnine despite the low [^{3}H]strychnine binding activity of these preparations. This observation suggested that neonatal GlyR possesses a strychnine binding affinity below detectability by filtration assay, but still sufficient for retention on the highly-derivatized affinity column.

Elution from 2-aminostrychnine columns by two consecutive glycine gradients allowed separation of neonatal and adult forms of GlyR (Becker et al., 1988). GlyR-antigen from adult spinal cord appeared as a single peak only after application of high glycine concentrations (about 11-13 mM). This GlyR form was detected by both mAbs 2b and 4a (Figure 3A). A different elution pattern was obtained with extracts from neonatal spinal cord (Figure 3B). Two GlyR peaks were observed; one of these was reactive with both mAbs 2b and 4a, eluted at high glycine concentrations, and corresponded to that obtained with adult material. However, the other peak was recognized by mAb 4a only and appeared specific for neonatal tissue. It eluted from the affinity column already at low concentrations of glycine indicating a low affinity for the 2-aminostrychnine agarose matrix. Thus, two populations of GlyR could be distinguished by glycine gradient elution in neonatal spinal cord extracts.

The 93 kD polypeptide which binds to the transmembrane 'core' of the GlyR (Schmitt et al., 1987; Becker et al., submitted) accumulated postnatally as shown above. This protein was found to associate with both isoforms of the GlyR during differential elution from 2-aminostrychnine agarose (Becker et al., 1988). Thus, a possible function of the 93 kD protein in the structural organization of the glycinergic synapse may relate to both GlyR isoforms.

The neonatal form of low strychnine binding affinity accounted for about 70% of the GlyR in spinal cord of newborn rats. Immunoassays employing mAb 2b and Scatchard analysis of high-affinity

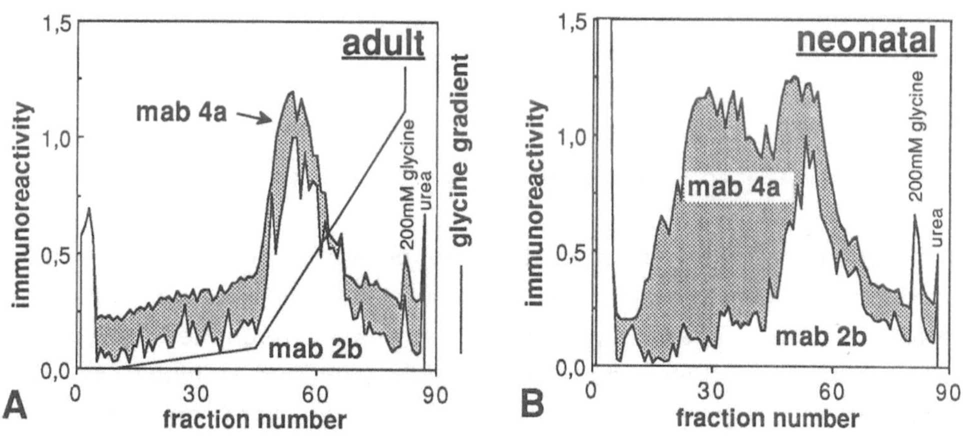

Figure 3: Elution of GlyR from adult **(A)** and neonatal **(B)** spinal cord by two consecutive glycine gradients. The gradients ranged from 0 to 4mM (fractions 9-44) and from 4 to 30mM (fractions 45-80).

[^{3}H]strychnine binding are indicative of the adult form of the GlyR. Both showed that of the total GlyR, i. e. mAb 4a-antigen, present at birth about 30% corresponded to the adult isotype. The neonatal isoform of the GlyR awaits to be described in more detail. It binds glycine and, with low affinity, strychnine as well as mAb 4a, but not mAb 2b (see Hoch et al., this volume). Its ligand binding subunit differs in molecular weight (49 kD) from that of the adult receptor (48 kD). Preliminary data from cDNA cloning suggest differences in primary structure of neonatal and adult ligand binding subunits (G. Grenningloh and C.-M. Becker, unpublished). A developmental heterogeneity in molecular weight of the ligand-binding subunit has also been reported for the $GABA_A$ receptor (Eichinger and Sieghardt, 1986).

Although the significance of the developmental heterogeneity of GlyRs is not known, the neonatal isotype probably represents a functional GlyR channel involved in neuronal inhibition. Besides a prevalence in neonatal spinal cord, this isotype almost exclusively comprises the GlyR in primary cell cultures of mouse spinal cord (Hoch et al., this volume) which are highly sensitive to glycine application. Furthermore, the postnatal switch from low to high affinity GlyRs in spinal cord coincides with an increase in sensitivity to strychnine intoxication of rodents as reported by an old pharmacological study (Falck, 1884, 1885). From a relative "immunity" at birth, the lethal dose for mice of this convulsant decreases postnatally approaching adult values around day 15 when, as shown here, the high affinity GlyR isoform is fully expressed.

2. Glycine receptors are reduced in the mutant mouse *spastic*

Homozygotes of the mutant mouse *spastic* suffer from disturbed motor functions. Heller and Hallet (1982) showed that the electromyographical pattern of the spastic phenotype strongly resembled strychnine intoxication, and White and Heller (1982) identified a highly selective reduction of [^{3}H]strychnine binding in spinal cord membranes of homozygotic mutants. Scatchard analysis (Figure 4) showed that this reduction in the mutant of ligand binding resulted from a 70 to 80% loss of [^{3}H]strychnine binding sites with no major alteration in receptor affinity detectable (White, 1985; Becker et al., 1986). After affinity purification on 2-aminostrychnine agarose, GlyR from spastic mutants (S) aged 6 to 8 weeks showed the same composition of 48 kD, 58 kD, and 93 kD polypeptides on SDS-polyacrylamide gels (Figure 5) as GlyR from unaffected control mice (C) or from rat (R). Photoaffinity labelling with [^{3}H]strychnine identified the 48 kD protein as the antagonist binding subunit in the mutant as well (Becker et al., 1986).

The antigenic epitopes of mAbs raised against GlyR from adult rat were shared by the receptor of normal mice as evident from analysis by enzyme linked immunosorbent assay (ELISA). Figure 6 shows the same ELISA profile of mAb binding for GlyR isolated from spastic and control mouse and from rat. This indicates that adult homozygotes (6-8 weeks) of the mutant mouse *spastic* express the adult GlyR isotype in spinal cord, although at a drastically reduced concentration. Note that the number of [^{3}H]strychnine binding sites applied to SDS-PAGE (Figure 5) and ELISA (Figure 6) was the same for each of these three preparations (for details see Becker et al., 1986). No evidence was obtained for

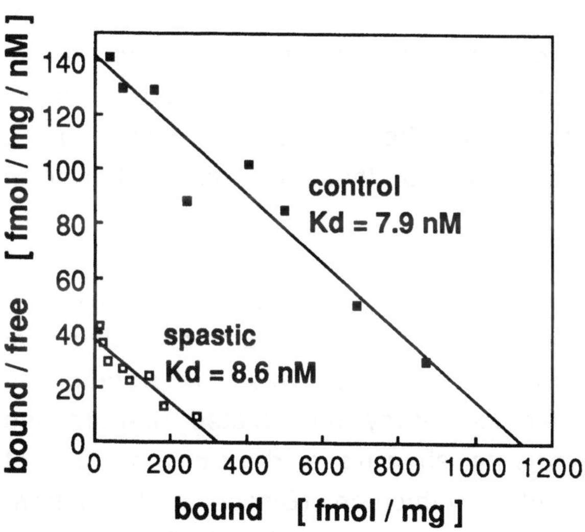

Figure 4 : Scatchard analysis of [^{3}H]strychnine binding to spinal cord membranes from control mice and spastic mutants (Figures 4 and 5 are from Becker et al., 1986).

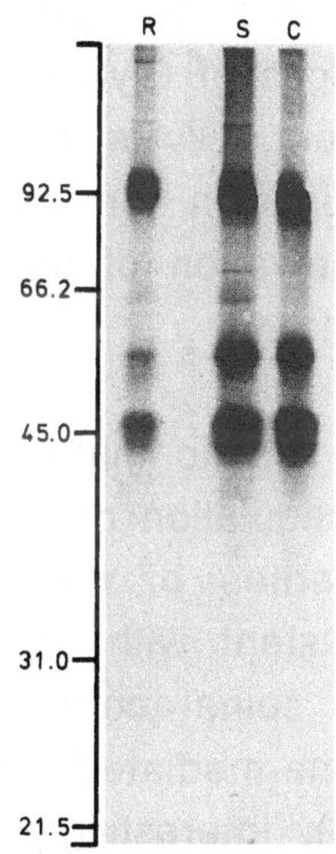

Figure 5: SDS-PAGE of GlyR isolated from spinal cord of rat (R), *spastic* (S) and control mouse (C).

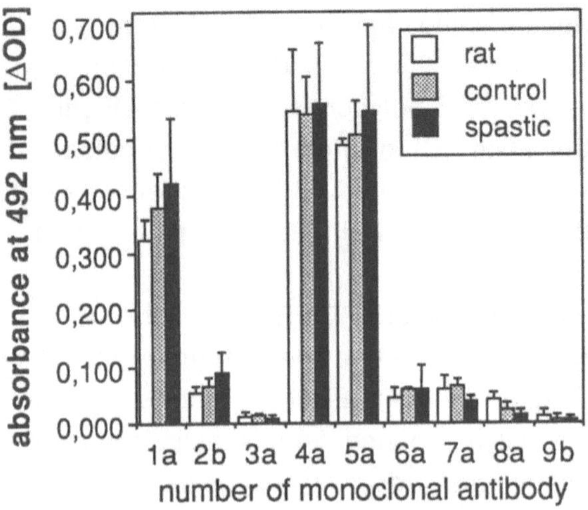

Figure 6: ELISA pattern of antibody binding to GlyR from spinal cord of rat, control and spastic mouse. The same number of [^{3}H]strychnine binding sites was used for each preparation (from Becker et al., 1986).

defective or non-ligand binding forms of GlyR by immunoassay or Western blot of spinal cord membranes from the spastic mouse (not shown). These data indicate that the spastic GlyR deficiency is not caused by major deletions or alterations in subunit composition, antigenic epitopes, or ligand binding site(s). Nevertheless, minor changes of GlyR primary structure in the mutant can not be ruled out at present. GlyRs in spinal cord of the spastic mouse showed a patchy distribution on neurons and within the neuropil arguing for a conservation of the synaptic localization of these receptors (Becker et al., 1986).

The spastic mutation apparently defines a gene which interferes with the regulation of GlyR receptor number leaving structural and functional properties of the receptor protein unaffected. This interpretation is consistent with an electrophysiological study on isolated preparations from spinal cord of spastic mutants showing a drastic reduction of glycine-mediated, chloride-dependent inhibition (Biscoe and Duchen, 1986). Interestingly, the developmental isotypes of the GlyR appear to be differently affected by the spastic mutation. Preliminary data indicate that, in the early postnatal period, spastic mutants express close to normal levels of GlyR-antigen as detected by immunoassays and Western blots using mAb 4a. It is only at later stages when the adult GlyR prevails in spinal cord, that GlyR levels of mutants were found drastically below those of age-matched control animals (C.-M. Becker, unpublished).

3. Conclusions

A combination of immunological and radioligand binding techniques has permitted the identification of a GlyR isotype of low strychnine-binding affinity prevailing in spinal cord of neonatal rats. In muscle, the nicotinic acetylcholine receptor undergoes both a developmental change of subtypes and redistribution during synaptogenesis (reviewed by Schuetze and Role, 1987). It remains to be established whether the developmental switch in GlyR isotypes reported here is accompanied by a similar redistribution of receptors from extrasynaptic areas of the neuronal membrane to subsynaptic sites. Adult *spastic* mice express the correct GlyR isotype, although at dramatically reduced levels. Further research should elucidate whether the primary defect of the mutation lies within the regulatory region of GlyR structural genes, or whether GlyR levels are only affected by indirect mechanisms.

Acknowledgements
We thank I. Wolters and C. Schroeder for expert technical assistance and Dr. Heinrich Betz for helpful discussions and continuous support. This work was supported by the Deutsche Forschungsgemeinschaft (Be 718/8-1 and SFB 317) and the Fonds der Chemischen Industrie. C.-M. B. was supported in part by a fellowship from the Boehringer Ingelheim Fonds.

References
Becker, C.-M., Hermans-Borgmeyer, I., Schmitt, B., and Betz, H. (1986) J. Neurosci. 6: 1358-1364.
Becker, C.-M., Hoch, W., and Betz, H. (1988) EMBO J. 7: 3717-3726.
Benavides, J., Lopez-Lahoya, J., Valdivieso, F., and Ugarte, M. (1981) J. Neurochem. 37: 315-320.
Betz, H., and Becker, C.-M. (1988) Neurochem. Int. 13: 137-146.
Biscoe, T.J., and Duchen, M.R. (1986) J. Physiol. 379: 275-292.

Eichinger, A., and Sieghart, W. (1986) J. Neurochem. 46: 173-180.
Falck, F.A. (1884) Archiv f. d. ges. Physiol. 34: 530-575.
Falck, F.A. (1885) Archiv f. d. ges. Physiol. 36: 285-308.
Graham, D., Pfeiffer, F., Simler, R., and Betz, H. (1985) Biochemistry 24: 990-994.
Grenningloh, G., Rienitz, A., Schmitt, B., Methfessel, C., Zenssen, M., Beyreuther, K., Gundelfinger, E.D., and Betz, H. (1987) Nature 328: 215-220.
Heller, A.H., and Hallett, M. (1982) Brain Res. 234: 299-308.
Pfeiffer, F., Graham, D., and Betz, H. (1982) J. Biol. Chem. 257: 9389-9393.
Pfeiffer, F., Simler, R., Grenningloh, G., and Betz, H. (1984) Proc. Natl. Acad. Sci. USA 81: 7224-7227.
Schmitt, B., Knaus, P., Becker, C.-M., and Betz, H. (1987) Biochemistry 26: 805-811.
Schuetze, S.M., and Role, L.W. (1987) Annu. Rev. Neurosci. 10: 403-457.
Triller, A., Cluzeaud, F., and Korn, H. (1987) J. Cell Biol.104: 947-956.
White, W.F. (1985) Brain Res. 329: 1-6.
White, W.F., and Heller, A.H. (1982) Nature 298: 655-657.
Young, A. B., and Snyder, S.H. (1973) Proc. Natl. Acad. Sci. USA 70: 2832-2836.
Young, A. B., and Snyder, S.H. (1974) Mol. Pharmacol. 10: 790-809.
Zukin, S. R., Young, A.B., and Snyder, S.H. (1975) Brain Res. 83: 525-531.

EXPRESSION-CLONING AND ANALYSIS OF THE SEROTONIN 1c RECEPTOR FROM MOUSE BRAIN

Hermann Lübbert
Preclinical Research 386/226
Sandoz Ltd.
CH-4002 Basel
Switzerland

SUMMARY

We developed a strategy for the cloning of neurotransmitter-receptor and ion-channel genes that is based on electrophysiological assays of mRNA-injected Xenopus oocytes. The procedure circumvents the purification of these membrane proteins, which is hindered by their low abundance and their hydrophobic nature. It involves methods for RNA fractionation by gel electrophoresis, directional cDNA-cloning in a single-stranded vector, and screening of the cDNA library by voltage-clamp measurements of currents induced by serotonin in mRNA-injected oocytes. Clones are tested by hybrid-depletion and therefore even incomplete cDNA-clones or clones coding for an individual subunit of a multisubunit protein can be identified.

The method was used for the cloning of a serotonin receptor cDNA from a mouse choroid plexus papilloma. The receptor expressed in oocytes injected with hybrid-selected RNA is fully functional and displays 5-HT1c pharmacology, indicating that it is composed of a single subunit encoded by a 5-kilobase RNA.

NATO ASI Series, Vol. H 32
Molecular Biology of Neuroreceptors and Ion Channels. Edited by A. Maelicke

Two independent cDNA clones isolated from the choroid plexus tumor library had a 94 bp deletion compared to clones from a total brain cDNA library. By analysis of mouse genomic clones isolated by hybridization-screening with the cDNA clones, we found that the 5-HT1c receptor gene contains several introns, unlike the unspliced genes coding for the muscarinic and adrenergic receptors analyzed so far. The deletion that we observed in choroid plexus tumor cDNA appears to be the result of a splicing artifact, possibly due to a loss of splicing specificity in these transformed cells. Evidence for the formation of different receptor types by differential splicing could not be found.

INTRODUCTION

Serotonin (5-HT) is one of the major modulators of brain and body function. It can act as a neurotransmitter (Aghajanian and Wang, 1978), a hormone (Kravitz et al., 1980) or a mitogen (Nemecek et al., 1986). Furthermore, it has an effect on growth cone mobility in the developing nervous system (Haydon et al., 1984). In the central nervous system, the cells that use serotonin as their transmitter are present in nine "Raphe Nuclei", all located in brain stem or pons. They innervate every major part of the brain and project down the spinal cord. Serotonergic nerve endings are also present in the smooth muscle cell layers lining the blood vessels (Steinbusch, 1981).

The cells in the raphe nuclei have an autonomous electrical activity which is selfregulated via autoreceptors (Aghajanian and Wang, 1978). They fire spontaneously at a rate of 0.5-5 spikes per second. With this constant pattern of electrical activity the raphe neurons modulate the basic activity of their target cells, which has diverse psycological and physiological effects.

Although the firing rate of the raphe nuclei can vary under certain circumstances (it decreases greatly during REM-sleep or after LSD-uptake) (Aghajanian and Wang, 1978) the pattern of constant firing is quite rigid. Heterogeneity is introduced into the system by the presence of a variety of serotonin receptor types with different physiological effects on the target cells.

At least eight serotonin receptor types can be distinguished by pharmacological methods (Peroutka, 1986; Heuring and Peroutka, 1987; Richardson et al., 1985). Following the original classification into 5-HT1 and 5-HT2 receptors, four subtypes of the 5-HT1 binding sites, denoted 1a, 1b, 1c, and 1d, have been characterized by pharmacological criteria in the central nervous system. 5-HT2 receptors are found in brain, in smooth muscle and in blood platelets (Leysen et al., 1983; Van Nueten et al., 1981).

5-HT2 and 1c receptors stimulate phosphoinositol turnover (Conn and Sanders-Bush, 1985; de Chaffoy de Courcelles et al. 1985; Conn et al. 1986) while 5-HT1a, 1b, and 1d receptors mediate adenylate cyclase activity (Devivo and Maayani, 1985; Shenker et al. 1985; Weiss et al. 1986; Hoyer et al. 1988). G-proteins can also directly couple 5-HT1a receptors to a potassium-channel (Andrade et al. 1986). In addition to the G-protein-coupled 5-HT1 and 2 receptors, there are at least three 5-HT3 (M) receptor types. Apparently, those are receptor-channel complexes (Neijt et al., 1988) which are located primarily in the peripheral nervous system (Richardson et al., 1985).

EFFECT OF SEROTONIN ON THE MEMBRANE-CONDUCTANCE OF XENOPUS OOCYTES INJECTED WITH BRAIN mRNA

Xenopus oocytes have been used to study neurotransmitter receptors and ion channels expressed on the oocyte surface after injection of mRNA isolated from electrically excitable tissues. Using electrophysiological techniques, functional serotonin receptors can be examined in oocytes injected with RNA from rat, mouse or human brain (Gundersen et al., 1984). The typical response to serotonin of an oocyte injected with mouse brain RNA is shown in figure 1. Typically, the response consists of two main components: an initial, large transient phase followed by a delayed, long-lasting component, which in some oocytes is superimposed with current fluctuations. The amplitude of the initial transient peak reflects the amount of mRNA injected into the oocyte or the serotonin concentration used (Lübbert et al., 1987a).

We and others have characterized the pathways that are involved in the mediation of this serotonin response in oocytes. The response has a reversal potential between -20 and -24 mV, as expected for a chloride current (Gundersen et al., 1983). It can be suppressed by injection of EGTA and mimicked by injection of IP3 or calcium, but not of cAMP (Miledi and Parker, 1984; Dascal et al., 1985; Oron et al.,

1985; Dascal et al., 1986). Involvement of a G-protein is supported by the finding that pertussis toxin inhibited the response to serotonin by about 50% (Dascal et al., 1986).

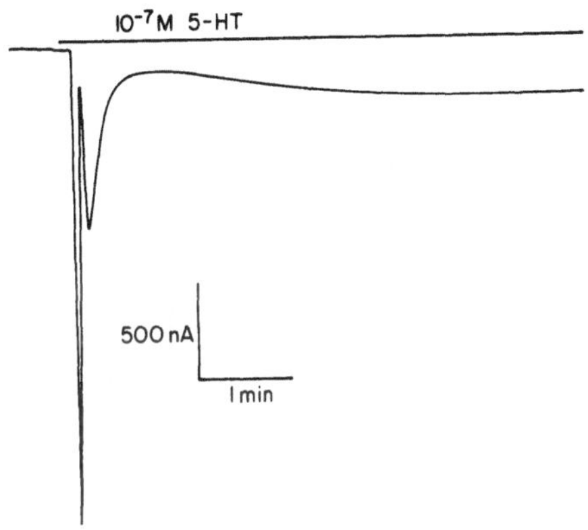

Figure 1: Serotonin-evoked currents in Xenopus oocytes. Currents from oocytes injected with 50 ng brain poly(A)-RNA were recorded with a two-electrode voltage clamp at a holding potential of -70 mV. Bars indicate the duration of agonist application.

The serotonin response in injected oocytes is similar to the oocyte's endogenous cholinergic muscarinic response, which was shown in previous studies to be mediated by IP3 synthesis leading to Ca release from intracellular stores (Oron et al., 1985). In oocytes injected with total brain RNA, the responses to acetylcholine (ACh) and 5-HT exhibited self- and cross-desensitization, i.e., application of either ACh or 5-HT inhibited the subsequent response to either one of the two transmitters (Dascal et al., 1986). This observation implies that the underlying second messenger systems are identical. Interestingly, this desensitization was not observed in oocytes injected with RNA isolated from choroid plexus, suggesting that one or several components involved in this desensitization process are encoded by the brain RNA injected into the oocyte (Lübbert et al., 1987a). Since injection of the isolated serotonin receptor RNA is sufficient for the generation of the serotonin response in oocytes (Lübbert et al., 1987a), it can be concluded that all other components involved in the pathway (i.e., the G-protein, the Cl-channels, and the enzymes involved in PI-hydrolysis and Ca-release) are already present in the oocyte before the injection of exogenous RNA. A function for this naturally occurring pathway can be predicted since, after expression of serotonin receptors in mature Xenopus eggs, serotonin could initiate a series of events similar to those normally initiated by sperm. Thus, the serotonin receptor may replace a hypothetical sperm receptor in this pathway (Kline et al., 1988).

Pharmacological Characterization of the Receptor

drug	apparent K_i in oocytes	$5\text{-}HT_{1A}^{*}$	$5\text{-}HT_{1B}^{*}$	$5\text{-}HT_{1C}^{*}$	$5\text{-}HT_{1D}^{*}$	$5\text{-}HT_{1C}^{+}$
serotonin	50±10	2.2	7.6	11	2.1	46
8-OH-DPAT	>10000	1.0	24000	7200	1600	—
spiperone	4000±1000	130	47000	4800	—	6200
mesulergine	250±100	630	12600	2.5	10000	—
mianserin	40±15	800	10000	110	—	12
ketanserin	35±20	>10000	>10000	—	>1000	130
ritanserin	30±20	—	—	—	—	—
cyproheptadine	300±100	110	840	2600	—	—
yohimbine	>7000	—	—	—	10-1000	—

Conclusion: The serotonin receptor functional in Xenopus oocytes resembles the $5\text{-}HT_{1C}$ type

Table 1: Affinity of serotonergic ligands to the receptor in the oocytes compared with those for the individual 5-HT1 receptor types determined by membrane binding (*) or by the determination of antagonist effects on the serotonin-stimulated PI-turnover in rat choroid plexus (+).

To characterize the serotonin receptor type we determined its affinity to specific ligands. The pharmacological profile of the receptor is shown in table 1. It is most consistent with that of the 5-HT1c receptor. This conclusion is supported by the observation that oocytes injected with RNA isolated from choroid plexus, which is the richest source for 5-HT1c receptors, respond 16 or 8 times more strongly to serotonin than do oocytes injected with RNA isolated from cortex or substantia nigra, respectively. Additional electrophysiologically active 5-HT receptor types could not be detected in mRNA-injected oocytes (Lübbert et al., 1987a).

USING THE OOCYTES FOR EXPRESSION-CLONING

To clone the 5-HT1c receptor cDNA we designed the cloning strategy outlined in figure 2 which is based on electrophysiological assays of Xenopus oocytes injected with mRNA (Lübbert et al., 1987b). In general, there are three steps: RNA-fractionation by gel-electrophoresis, construction of a directional cDNA library in a single-stranded vector, and library-screening by hybrid depletion. The use of hybrid-depletion has several advantages over the expression of RNA synthesized in vitro as has later been done by other groups (Masu et al., 1987; Julius et al., 1988): First, it is possible to find incomplete cDNA clones (which, under certain circumstances, may be considered a disadvantage) and second, individual subunits of a multisubunit protein can be identified (provided that they are functionally significant). In the first step, poly-A RNA was separated on an agarose-gel containing the reversible denaturing agent methylmercury-hydroxyde. RNA was then eluted from gel slices by electrophoresis into a step gradient comprised of various concentrations of salt and sucrose (experimental details can be obtained upon request). To test which RNA-fraction contains the serotonin receptor RNA, a portion of each fraction was injected into oocytes. Oocytes injected with RNA from one particular fraction (about 4.8 to 5.2 kb in size) were responsive to 5-HT. The remaining RNA from this fraction was then used for the construction of a cDNA library. For this we developed a method that should facilitate the subsequent screening. The vector used (pUC119) allowed the formation of single-stranded DNA. The direction of the cDNA insert was chosen such that the single-stranded DNA contained anti-sense inserts which were readily available for hybridization with their corresponding RNAs.

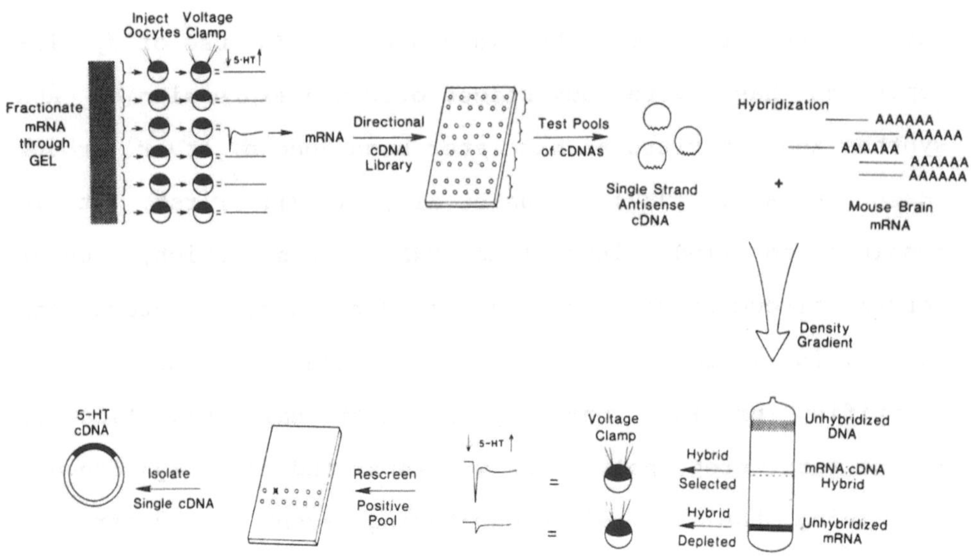

Figure 2: Outline of the procedure used to isolate a 5-HT1c cDNA-clone. This figure was reprinted with permission from T.P. Snutch, TINS 11:250-256 (1988).

Pools of single-stranded cDNAs were screened by hybridization to choroid plexus tumor RNA and subsequent separation of RNA, DNA:RNA-hybrids, and DNA by density gradient centrifugation. Upon injection into oocytes of the unhybridized mRNA, a cDNA pool that contained a 5-HT1c receptor cDNA resulted in the specific removal of 5-HT receptor activity. By re-screening the positive pool we isolated a single cDNA-clone. The identity of this clone was confirmed by various hybrid-depletion and hybrid-selection experiments.

ANALYSES OF GENOMIC AND cDNA CLONES

The originally isolated cDNA clone contains a 1.9 kb insert which did not contain an open reading frame of a significant length. With the help of this clone we isolated additional cDNA and genomic clones. Analyses of these clones showed that the receptor has structural similarity with other receptors that are coupled to G-proteins. They all have seven hydrophobic regions that are putatively membrane-spanning (Kerlavage et al., 1986). The hydrophobicity profile (Kyte and Doolittle, 1982) of the serotonin 1c receptor is shown in figure 3. The seven regions that are putatively membrane-spanning (positive hydrophobicity indexes) are easily visible. An additional hydrophobic region is present at the N-terminus of the protein.

Unlike most other G-protein-coupled receptors the gene coding for the 5-HT1c receptor is spliced (own unpublished results). In choroid plexus tumors, we found a high level of uncorrectly spliced serotonin receptor RNA. To find evidence for the formation of different serotonin receptor types by differential splicing, we hybridized RNA blots and brain slices with probes that were specific for individual exons. Since the patterns of hybridizations were identical with all the probes used, differential splicing does not seem to be responsible for the diversity of serotonin receptors.

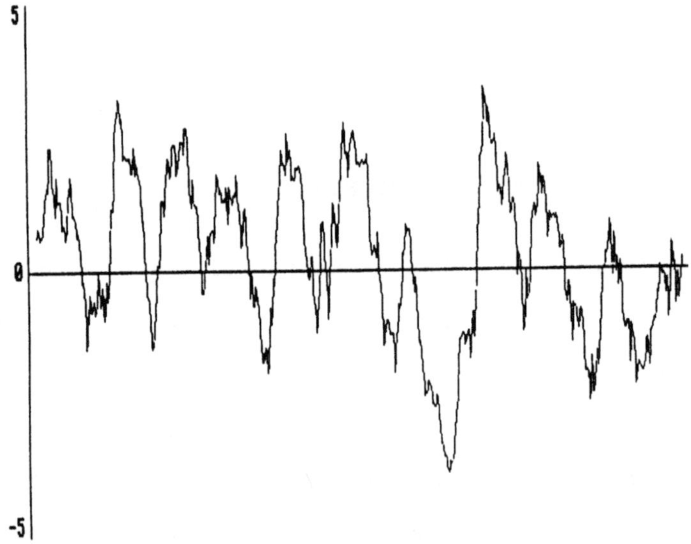

Figure 3: Hydrophobicity profile of the 5-HT1c receptor according to Kyte and Doolittle (1982).

By comparing the RNA-levels (examined by RNA-blot and in situ hybridization analyses) and receptor levels (examined by receptor autoradiography) we found that the protein turnover rates vary in different tissues. The levels of receptor protein are significantly higher in choroid plexus than in any other brain region, while the RNA in hippocampus, amygdala, subthalamic nucleus, or olfactory nucleus is nearly as abundant as in choroid plexus.

While the present study was in progress, Julius et al. (1988)

also isolated a cDNA clone of the 5-HT 1c receptor. They expressed their clone in NIH 3T3 cells and found serotonin-stimulated calcium release in these cells by using fluorescence-activated cell sorting.

The serotonin 1c receptor is an example for a mediator of signals that activate phophoinositol hydrolysis. Cloning, sequencing and functional expression of this receptor provides a tool for the examination of signals that are transferred by this pathway.

REFERENCES

Aghajanian, G.K., and Wang, R.Y. (1978) Physiology and pharmacology of central serotonergic neurons. in: Psychopharmacology: a generation of progress, eds.: Lipton, M.A., DiMascio, A., and Killman, K.F., Raven Press, New York.

Andrade, R., Malenka, R.C., and Nicoll, R.A. (1986) A G-protein couples serotonin and GABA-b receptors to the same channels in hippocampus. Science 234: 1261-1265.

Conn, P.J., and Sanders-Bush, E. (1985) Serotonin-stimulated phosphoinositide turnover: mediation by the S2 binding site in rat cerebral cortex but not in subcortical regions. J. Pharmacol. and Exp. Therapeutics 234:195-203.

Conn, P.J., Sanders-Bush, E., Hoffman, B.J., and Hartig, P. (1986) A unique serotonin receptor in choroid plexus is linked to phosphatidylinositol turnover. Proc. Natl. Acad. Sci. USA 83: 4086-4088.

Dascal, N., Gillo, B., and Lass, Y. (1985) Role of calcium mobilization in mediation of acetylcholine-evoked chloride currents in Xenopus laevis oocytes. J. Physiol. 366: 299-313.

Dascal, N., Ifune, C., Hopkins, R., Snutch, T.P., Lübbert, H., Davidson, N., Simon, A.I., and Lester, H.A. (1986) Involvement of a GTP-binding protein in mediation of serotonin and acetylcholine responses in Xenopus oocytes injected with rat brain messenger RNA. Mol. Brain Res. 1: 201-209.

De Chaffoy de Courcelles, D., Leysen, J.E., De Clerck, F., Van Belle, H., and Janssen, P.A.J. (1985) Evidence that phospholipid turnover is the signal transducing system coupled to serotonin-S2 receptor sites. J. Biol. Chem. 260: 7603-7608.

Devivo, M., and Maayani, S. (1985) Inhibition of forskolin-stimulated adenylate cyclase activity by 5-HT receptor

agonists. Eur. J. Pharmacol. 119: 231-234.
Gundersen, C.B., Miledi, R., and Parker, I. (1983) Serotonin receptors induced by exogenous messenger RNA in Xenopus oocytes. Proc. R. Soc. Lond. B 219: 103-109.
Gundersen, C.B., Miledi, R., and Parker, I. (1984) Messenger RNA from human brain induces drug- and voltage-operated channels in Xenopus oocytes. Nature 308: 421-424.
Haydon, P.G., McCobb, D.P., and Kater, S.B. (1984) Serotonin selectively inhibits growth cone motility and synaptogenesis of specific identified neurons. Science 226: 561-564.
Heuring, R.E., and Peroutka, S.J. (1987) Characterization of a novel 5-hydoxytryptamine binding site. J. Neurosci. 7: 894-903.
Hoyer, D., Schoeffter, P., and Waeber, C. (1988) 5-HT1d receptors are negatively coupled to adenylate cyclase in calf substantia nigra. Naunyn Schmiedeberg's Arch. Pharmacol. 337(suppl.): R104.
Julius, D., MacDermott, A.B., Axel, R., and Jessel, T.M. (1988) Molecular characterization of a functional cDNA encoding the serotonin 1c receptor. Science 241: 558-564.
Kerlavage, A.R., Fraser, C.M., Chung, F.-Z., and Venter, C. (1986) Molecular structure and evolution of adrenergic and cholinergic receptors. Proteins: Structure, Function, and Genetics 1: 287-301.
Kline, D., Simoncini, L., Mandel, G., Maue, R.A., Kado, R.T., and Jaffe, L.A. (1988) Fertilization events induced by neurotransmitters after injection of mRNA in Xenopus eggs. Science 241: 464-467.
Kravitz, E.A. et al. (1980) J. Exp. Biol. 89: 159.
Kyte, J., and Doolittle, R.F. (1982) A simple method for displaying the hydropathic character of a protein. J. Mol. Biol. 157: 105-132.
Leysen, J.E., Gommeren, W., and De Clerck, F. (1983) Demonstration of S2-receptor binding sites on cat blood platelets using ketanserin. Eur. J. Pharmacol. 88: 125-130.
Lübbert, H., Snutch, T.P., Dascal, N., Lester, H.A., and Davidson, N. (1987a) Rat brain 5-HT1c receptors are encoded by a 5-6 kbase mRNA size class and are functionally expressed in injected Xenopus oocytes. J. Neurosci. 7: 1159-1165.
Lübbert, H., Hoffman, B.J., Snutch, T.P., van Dyke, T., Levine, A.J., Hartig, P.R., Lester, H.A., and Davidson, N. (1987b) cDNA cloning of a serotonin 5-HT1c receptor by electrophysiological assays of mRNA-injected Xenopus oocytes. Proc. Natl. Acad. Sci. USA 84: 4332-4336.
Masu, Y., Nakayama, K., Tamaki, H., Harada, Y., Kuno, M., and Nakanishi, S. (1987) cDNA cloning of bovine substance-K receptor through oocyte expression system. Nature 329: 836-838.
Miledi, R., and Parker, I. (1984) Chloride current induced by injection of calcium into Xenopus oocytes. J. Physiol. 357: 173-183.
Neijt, H.C., Plomp, J.J., and Vijverberg, H.P.M. (1988) Kinetics of the membrane currents mediated by serotonin

5-HT3 receptors in cultured mouse neuroblastoma cells. J. Physiol., submitted.
Nemecek, G.M., Coughlin, S.R., Handley, D.A., and Moskowitz, M.A. (1986) Stimulation of aortic smooth muscle cell mitogenesis by serotonin. Proc. Natl. Acad. Sci. USA 83: 674-678.
Oron, Y., Dascal, N., Nadler, E., and Lupu, M. (1985) Inositol 1,4,5-trisphosphate mimics muscarinic response in Xenopus oocytes. Nature 313: 141-143.
Peroutka, S.J. (1986) Pharmacological differentiation and characterization of 5-HT1a, 5-HT1b, and 5-HT1c binding sites in rat frontal cortex. J. Neurochem 47: 529-540.
Richardson, B.P., Engel, G., Donatsch, P., and Stadler, P.A. (1985) Identification of serotonin M-receptor subtypes and their specific blockade by a new class of drugs. Nature 316: 123-131.
Schenker, A., Maayani, S., Weinstein, H., and Green, J.P. (1985) Two 5-HT receptors linked to adenylate cyclase in guinea pig hippocampus are discriminated by 5-carboxamidotryptamine and spiperone. Eur. J. Pharmacol. 109: 427-429.
Steinbusch, H.W.M. (1981) Distribution of serotonin-immunoreactivity in the central nervous system of the rat - cell bodies and terminals. Neurosci. 6: 557-618.
Van Nueten, J.M., Janssen, P.A.J., Van Beek, J., Xhonneux, R., Verbeuren, T.J., and Vanhoutte, P.M. (1981) Vascular effects of ketanserin (R 41 468), a novel antagonist of 5-HT2 serotonergic receptors. J. Pharmacol. and Exp. Therapeutics 218: 217-230.
Weiss,S., Sebben, M., Kemp, D.E., and Bockaert, J. (1986) Serotonin 5-HT1 receptors mediate inhibition of cyclic AMP production in neurons. Eur. J. Pharmacol. 120: 227-230.

CLONED SEROTONIN 5HT-2 RECEPTOR: STRUCTURE AND SECOND MESSENGER COUPLING

Dolan B. Pritchett, Alfred Bach[#], Omar Taleb,
Roberto Dal Toso, and Peter H. Seeburg

Laboratory of Molecular Neuroendocrinology
University of Heidelberg,
ZMBH, INF 282, D-6900 Heidelberg, F.R.G.

and [#](present address)
BASF, AG., Abt. Biotechnology,
Carl-Bosch-Str. 38, D-6700 Ludwigshafen, F.R.G.

Serotonin (5-hydroxytryptamine, 5HT) was first discovered as a serum factor and later shown also to be a neurotransmitter in the central and peripheral nervous systems (reviewed by Leysen, 1988). It interacts with a family of pharmacologically distinct receptors on the surface of neurons and other cells in the body (reviewed by Peroutka, 1988). These receptors have been grouped into three major classes (5HT-1, 5HT-2 and 5HT-3) based on interactions with various agonists and antagonists (Bradley et al., 1986). The first class contains at least four subtypes (5HT-1A to 5HT-1D). The primary structure of a 5HT-1C receptor was recently described (Julius et al., 1988) and characterizes a new member of the G-protein coupled receptor family containing seven transmembrane regions (Lefkowitz et al., 1986). This receptor is located primarily in the choroid plexus, and its functional role there is unknown.

In contrast to the 5HT-1 class, 5HT-2 sites seem to be homogeneous (Leysen et al., 1981). The highest level of 5HT-2 binding is seen in layer IV of the cerebral cortex (Pazos et al., 1985). A number of central (Glennon et al., 1984) and peripheral effects of 5HT have been attributed to this subtype. Thus, 5HT-elicited seizures can be prevented by 5HT-2 receptor specific

NATO ASI Series, Vol. H 32
Molecular Biology of Neuroreceptors and
Ion Channels. Edited by A. Maelicke

antagonists (Leysen et al., 1978). This, and its involvement in tracheal smooth muscle contraction, bronchoconstriction (Leysen et al., 1984) and control of aldosterone production (Matsuoka et al., 1985) make it desirable to describe the structure of this receptor. Because both the 5HT-1C and 5HT-2 subtypes appear to exert their cellular effects through phosphoinositol turnover, they may be structurally closely related in analogy to the M1 and M4 muscarinic acetylcholine receptor subtypes (Peralta et al., 1988). Using oligonucleotides encoding 5HT-1C sequences, we have previously isolated a cloned cDNA encoding the complete 5HT-2 receptor (Pritchett et al., 1988b). This report describes a model for the structure based on analysis of the primary sequence as well as studies of the second messenger coupling of this receptor.

MATERIALS AND METHODS

Materials

Pharmacological agents were obtained from Janssen Pharmaceutica or Research Biochemicals International. Other chemicals were from Sigma or Merck.

Cell Culture and Transfection

Human embryonic kidney 293 cells (ATCC) were grown in MEM supplemented with 10% fetal calf serum. For transfection, these cells were plated at 1 - 3 x 10^6 cells/10 cm plate and were used for transfection at 50% confluency. SacI-EcoRI restriction fragment of the cloned cDNA (approximately 1600bp and containing the complete coding region) was cloned into the HpaI restriction site of an expression vector plasmid (Peralta et al., 1987; Pritchett et al., 1988). DNA of this construct was prepared and purified by equilibrium centrifugation banding on a cesium chloride gradient. The purified DNA was used to transfect exponentially growing 293 cells, using the $CaPO_4$ method as modified by Chen and Okayama (1986). The precipitate was washed off after 16-20 hrs and fresh growth medium was added. The cells were allowed to grow up to 48 hrs from the time the DNA precipitate was added.

Phosphoinositol Turnover Assay

Assays on transfected cells were performed as described by Peralta et al. (1988). Briefly, 2 x 10^{-6} uCi/ml of ^{3}H myo-inositol (NEN) were added to the growth media 16 hr after transfection. After 32 hr, the cells were washed three times in serum free media and incubated in this medium for an additional 45 min. Media were removed, and serum free medium containing 10 mM LiCl was added for 15 min. This medium was again removed, and the cells were suspended in serum free medium containing 10 mM LiCl at 2 x 10^{6} cell/ml. After addition of the agonist or antagonsit to 0.5 ml cell suspensions, the incubation was teminated by the addition o 1ml of 15% trichloric acid. After three ether extractions and neutralization with 6 mM Borax, the samples were chromatographed on Dowex AG1x8 columns (Biorad) in the formate form. Elution of IP_1, IP_2 and IP_3 was as follows: 10 ml H_2O (wash), 10 ml of 5 mM Borax/60 mM Na-formate (wash), 10 ml of 0.1 M formic acid/0.2 M NH_4-formate (IP_1), 10 ml of 0.1 M formic acid/0.4 M NH_4-formate (IP_2) and 10 ml of 0.1 M formic acid/1.0 M NH_4 formate (IP_3). The radioactivity in 2 ml of each eluate was determined after the addition of 10 ml of scintillation fluid.

Oocyte Injection and Current Recordings

After surgical removal, Xenopus laevis oocytes were dissected and collagenase treated for 8 hr to obtain single defolliculated oocytes. The cells at maturation stage IV were selected and pressure injected with RNA (25 ng in 100 nl) using pipettes with a tip of 10-15 um diameter. Injected oocytes were incubated at 20°C in Barth's medium containing penicillin and streptomycin (100 units/ml) . Voltage clamp experiments on these oocytes were performed 3 days after injection. The cells were held in a recording chamber (volume appr. 0.4 ml), using a two electrode voltage clamp amplifier (Takahashi, 1987) and perfused at 4 ml/min with frog Ringer solution (115 mM NaCl, 5 mM KCl, 1.8 mM $CaCl_2$, 10 mM HEPES, pH 7.2). The current injection and voltage recording microelectrodes were filled with 3 M KCl. 5HT and spiperone were applied by manual switching of the perfusion lines.

RESULTS

From many previous experiments it has been predicted (Hoyer, 1988) that the 5HT-2 receptor would be a member of the second messenger coupled receptor gene family. It has further been proposed that these receptors have seven alpha helices that span the lipid bilayer. To test this hypothesis for the recently isolated serotonin 5HT-2 receptor, we analysed the deduced amino-acid sequence for hydrophobic sequences long enough to span the lipid bilayer. Figure 1 shows the results of computer analysis by the method of Kyte and Doolittle (1982). The deduced polypeptide does indeed contain seven putative transmembrane spanning regions, typical of G protein-coupled receptors. The amino acids that form these hydrophobic regions also contain many of the residues that are highly conserved between the members of this gene family, suggesting that this receptor does indeed have a structure similar to the other receptors. This proposed structure does have one theoretical flaw and that is the extracellular location of the amino terminus and the lack of signal sequence. The location of the amino-terminus is likely to be extracellular since the receptor is known to be glycosylated and the only glycosylation sites are on the amino-terminal portion. A similar lack of a signal sequence has been proposed for the 5HT-1C receptor, even though it has a long hydrophobic amino-terminus (Julius et al.). Overall, the amino acid sequence of the 5HT-2 receptor is 51% identical to that of the 5HT-1C receptor and the highest identity is seen in the transmembrane regions (one of these, the 5th, is identical in both receptors), and lowest sequence homologies occur in the N-terminal 60 amino acid residues, the intracellulary located loop between transmembrane regions V and VI as well as in the C-terminus. Such a pattern of sequence identity is also observed between different muscarinic acetylcholine receptors (Peralta et al., 1987; Bonner et al., 1987). Based on the evidence of the hydrophobicity plot and the sequence similarity with other members of this gene family we propose that the topology of the serotonin 5HT-2 receptor is as shown in figure 2.

Effector Coupling to PI Hydrolysis and Ca Release

To determine the second messenger system affected by the receptor encoded by the cloned cDNA, we expressed this cDNA transiently in a

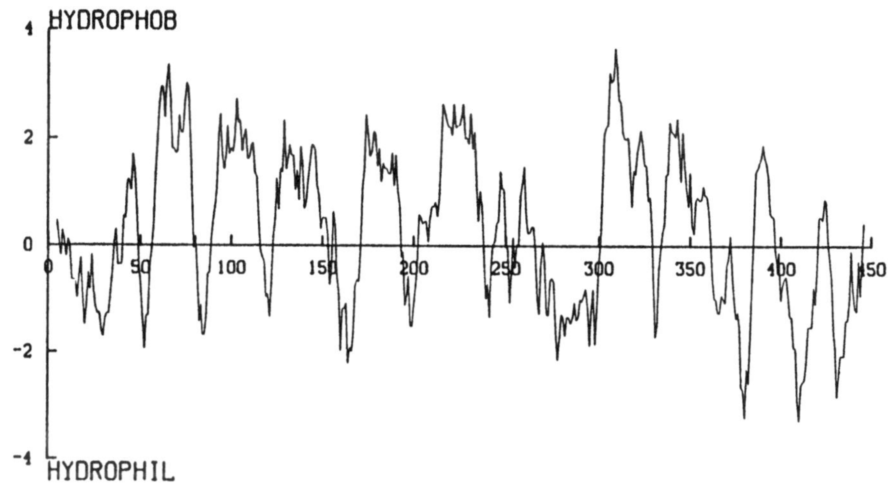

Figure 1. Hydropathy profile computed according to Kyte and Doolittle. The window size was set at 7. Similar results were obtained using window sizes of 9 and 11. When the hydropathy index of Hopp and Wood was used, the same segments were indicated as possible membrane spanning segments.

mammalian cell line by transfection with a recombinant plasmid DNA containing the entire coding region of the cDNA under the control of a cytomegalovirus promoter (Gorman et al, 1983; Peralta et al., 1987). Membranes prepared from these cells were analyzed previously (Pritchett et al., 1988b) by binding of a variety of 5HT receptor ligands. Results suggested that our cloned cDNA encodes a 5HT-2 receptor. 5HT-induced phosphoinositide (PI) hydrolysis appears to occur as a result of 5HT-2 receptor activation in cortex (Kendall and Nahorski, 1985), thoracic aorta (Roth et al., 1984), and platelets (De Chaffoy et al., 1985). The same second messenger system is stimulated by the 5HT-1C receptor and may modulate a number of cellular processes, including Ca^{2+} release from intracellular stores and protein kinase C activation. To determine if the cloned receptor is coupled to this pathway, we analysed serotonin-triggered changes in the levels of inositol phosphates. Following treatment of cells expressing 5HT-2 receptors with a saturating dose of serotonin (10^{-5} M), kinetic analysis of the accumulation of inositol mono-, bis-, and trisphosphate was monitored using an assay described by Peralta et al. (1988) for the analysis of muscarinic

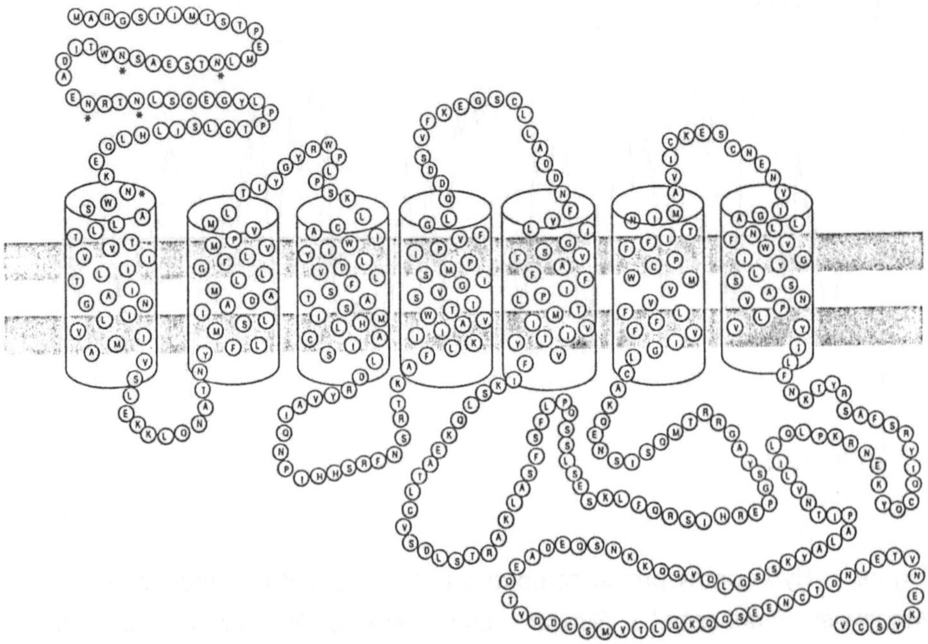

Figure 2. A model for the transmembrane topology of the rat 5HT-2 receptor. The amino acid sequence transverses the lipid bilayer seven times. The amino-terminus is extracellular. The carboxy terminus is intracellular. The cylinders represent alpha helices crossing the membrane, each composed of approximately twenty amino acids. Potential N-glycosylation sites are indicated by asterisks.

receptors. A substantial rise in the levels of mono- and bisphosphates were observed (Fig. 3). Only a modest rise in the amount of trisphosphates was seen but this reflects the rapid conversion of trisphosphates into bis-, and monophosphates as the shape of the curves suggest. All increases were completely blocked by 100 nM spiperone. These results demonstrate that the 5HT-2 receptor activates phospholipase C and substantially elevates free Ca^{2+} levels. This activation of PI turnover strongly suggests that the receptor is a 5HT-2 subtype, in excellent agreement with its pharmacological profile.

<u>5HT-2 receptor expression in Xenopus oocytes</u>

<u>In-vitro</u> generated RNA having the same nucleotide sequence as the cloned

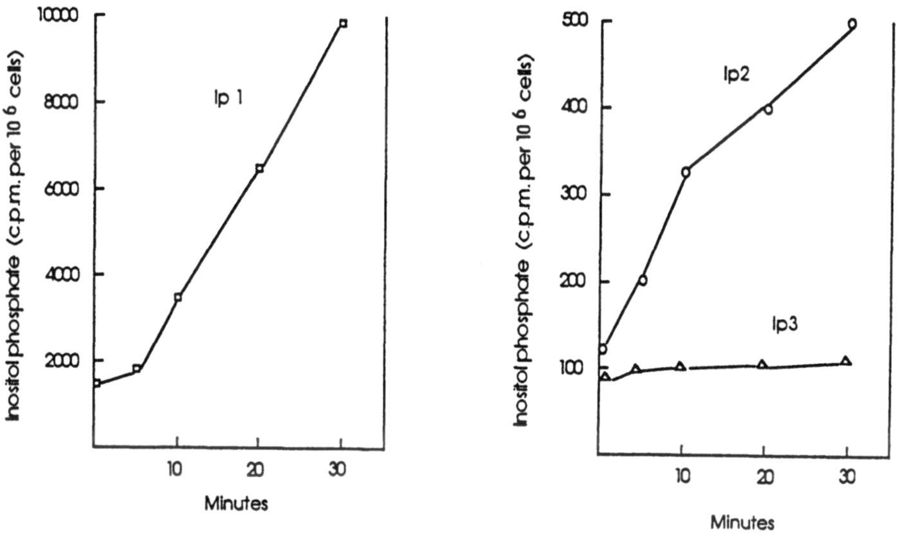

Figure 3. Time course of serotonin stimulated accumulation of inositol phosphates in cells transiently transfected with a 5HT-2 receptor cDNA expression plasmid. Untransfected cells showed no increase upon serotonin treatment.

cDNA and encoding the 5HT-2 receptor was injected into Xenopus oocytes. In these oocytes serotonin (10^{-5} M) evoked fast desensitizing inward currents of several hundred nA, (Fig. 4), when recorded under voltage clamp conditions at a holding potential of -50 mV. No response to serotonin was seen in uninjected or antisense 5HT-2 RNA injected oocytes. The current-voltage relationship for the RNA-induced 5HT response was linear and showed a reversal potential of -27 mV which corresponds to the equilibrium potential of chloride in Xenopus oocytes (Dascal et al., 1984). Superfusion with 10 nM spiperone greatly reduced 5HT-evoked currents (data not shown). Only small effects on 5HT-1C receptor activity were seen using high concetrations (10^{-6} M) of spiperone (Julius et al., 1988). These results demonstrate that the 5HT-2 receptor can activate a second messenger system in oocytes which controls Ca^{2+}-sensitive chloride channels by Ca^{2+} released from IP_3 sensitive intracellular stores (Takashi et al., 1987). For 5HT receptors, the activation of these channels has previously been only reported for the 5HT-1C subtype (Lübbert et al., 1987b; Julius et al., 1988).

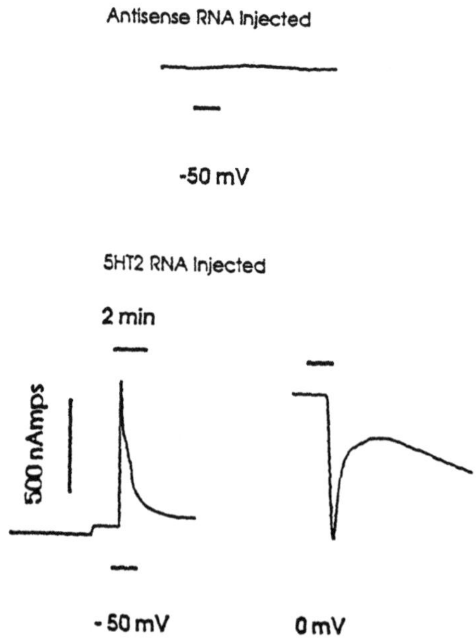

Figure 4. Current traces recorded from Xenopus oocytes injected with in vitro synthesized RNA encoding the 5HT-2 receptor or control RNA. Oocytes were clamped at the voltages indicated. A fast desensitizing inward current is seen in response to the application of 5HT, (10^{-5} M), indicated by the bar.

DISCUSSION

The structure suggested for the 5HT-2 receptor is similar to that of other second messenger coupled receptors. Thus, its design shows the presence of seven putative transmembrane regions detected from hydrophobicity analysis. These seven regions together form the ligand binding domain in receptors of this class. Of all known members of this class, the structure of the described receptor is closest to that of the 5HT-1C receptor (Julius et al., 1988). Although the amino acid sequences and binding profiles of the 5HT-2 and 5HT-1C receptors show that they are closely related, major differences in the hydrophobicity profiles and glycosylation sites of the respective amino-termini may cause some functional differences. Further experiments using chimeric receptors constructed from portions of the 5HT-2 and the 5HT-1C should aid in predicting which of the differences between the receptors are important in establishing their profiles. Obviously, residues important for ligand specificity must have extracellular access but many of these residues may lie deep in a pocket formed by the receptor itself. Therefore, the model shown in figure 2 does not rule out the possibility that residues in the transmembrane helices may be important in ligand specificity. The cell biological question of how the amino-terminus gets to the extracellular surface with no signal sequence should be further investigated.

Consistent with the observed activation of phospholipase C in mammalian cells, the 5HT-2 receptor activates a chloride channel in Xenopus oocytes (Nomura et al., 1987). This finding was expected on theoretical grounds but not anticipated from existing literature, where only 5HT-1C expression had been reported using injections of brain RNA (Lübbert et al., 1987a; Julius et al., 1988). It is known that receptors which stimulate inositol phosphate turnover in cells can stimulate a variety of channels in mammalian cells (Neher, 1988), as well as cause the phosphorylation of many structural proteins (de Chaffoy de Courcelles, 1984). Studies to determine which of these effects are important for the various known functions of the serotonin 5HT-2 receptor will be greatly aided by the availability of this clone. Experiments expressing both the 5HT-2 and the 5HT-1C clones in various cell types will show to what extent the coupling to the second messenger sytem is the same for both receptors. Preliminary indications from our work and indications from the experiments of others suggest that the receptors may differ in rates of desensitization (Lübbert, 1987). These differences may be due to differences in second messenger coupling.

REFERENCES

Bonner, T.I., Buckley, N.J., Young, A.C. and Brann, M.R. (1987) Science 237, 527-532.

Bradley, P.B., Engel, G., Feniuk, W., Fozard, J.R., Humphrey, P.P.A., Middlemiss, D.N., Mylecharane, E.J., Richardson, B.P. and Saxena, P.R. (1986) Neuropharmacology 25, 563-576.

Dascal, N., Landan, E.M., and Lass, Y. (1984) J. Physiol. 352, 551-574.

De Chaffoy de Courcelles, D., Leysen, J.E., De Clerck, F., Van Belle, H. and Janssen, P.A.J. (1985) J. Biol. Chem. 260, 7603-7608.

Glennon, R.A., Titeler, M. and McKenney, J.D. (1984) Life Sci. 35, 2505-2511.

Gorman, C., Padmanabram, R. and Howard, B. (1983) Science 221, 551-553.

Hoyer, D. (1988) Trends in Pharm. Sci. 9, 89-94.

Julius, D., MacDermott, A.B., Axel, R. and Jessell, T.M. (1988) Science 241, 558-564.

Kendall, D.A. and Nahorski, S.R. (1985) J. Pharm. Exp. Therapeutics 233, 473-479.

Kyte, J. and Doolittle, R.F. (1982) J. Molec. Biol. 157, 105-132.

Lefkowitz, R.J., Benovic, J.L., Kobilka, B. and Caron, M.G. (1986) Trends in Pharmac. Sci. 7, 444-448.

Leysen, J.E. (1981) J. Physiol. Paris 77, 351-362.

Leysen, J.E. (1988) In: Boulton, A.A., Baker, G.B. and Juorio, A.V. (eds.) Neuromethods, Neuropharmacology II; Humana Press, Clifton, in press.

Leysen, J.E., Awguters, F., Kennis, L., Laduron, P., Vandenberk, J. and Janssen, P. (1981) Life Sci. 28, 1015-1018.

Leysen, J.E., de Courcelles, D.C., De Clerck, F., Niemegeers, J.E. and Van Nueten, J.M. (1984) Neuropharmacology 23, 1493-1501.

Leysen, J.E., Niemegeers, C.J.E., Tollenaere, J.P. and Laduron, P.M. (1978) Nature 272, 163-166.

Lübbert, H., Hoffman, B.J., Snutch, T.P., Van Dyke, T., Levine, A.J., Hartig, P.R., Lester, H.A., and Davidson, N. (1987) Proc. Natl. Acad. Sci. USA 84, 4332-4336.

Lübbert, H., Snutch, T.P., Pascal, N., Lester, H.A. and Davidson, D. (1987) J. Neuroscience 4, 1159-1165.

Matsuoka, H., Ishii, M., Goto, A. and Sugimoto, T. (1985) Am. J. Physiol. 234, 234-238.

Nomura, Y., Kaneko, S., Kato, K., Yamagishi, S. and Sugiyama, H. (1987) Mol. Brain Res. 2, 113-123.

Neher, E. (1988) Neuroscience 26, 727-734.

Pazos, A., Cortes, R. and Palacios, J.M. (1985) Brain Res. 346, 231-249.

Peralta, E.G., Ashkenazi, A., Winslow, J.W., Smith, D.H., Ramachandran, J. and Capon, D.J. (1987) EMBO J. 6, 3223-3229.

Peralta, E.G., Ashkenazi, A., Winslow, J.W., Ramachandran, J. and Capon, D.J. (1988) Nature 334, 434-436.

Peroutka, S.J. (1988) Annu. Rev. Neuroscience 11, 45-60.

Pritchett, D.B., Sontheimer, H., Kettenmann, H., Gorman, C. Seeburg, P.H. and Schofield, P.R.(1988) Science, in press.

Pritchett, D.B., Bach, A.W.J. Taleb, O., Dal Toso, R. and Seeburg, P.H. (1988) EMBO J. 7, in press.

Roth, B.L., Nakaki, T., Chuang, D.M. and Costa, E. (1984) Neuropharmacology 23, 1223-1236.

Takahashi, T., Neher, E. and Sakmann, B. (1987) Proc. Natl. Acad. Sci. U.S.A. 84, 5063-5067.

AN APPARENT NEW L-GLUTAMATE RECEPTOR IN THE MAMMALIAN-PERIPHERAL NERVOUS SYSTEM

Pål Aas
Norwegian Defence Research Establishment
Division for Environmental Toxicology
P.O. Box 25
N-2007 Kjeller
Norway

The amino acids glutamate and aspartate function as important excitatory neurotransmitters in the mammalian central nervous system (CNS) [Curtis et al., 1960; Fonnum, 1984] and in invertebrate striated muscle [Usherwood, 1981, Crawford and McBurney, 1976]. The present report is the first demonstration of an excitatory effect of L-glutamate (L–glu) in the mammalian peripheral nervous system. The effect of L-glu cannot be explained in terms of the three well known receptors for excitatory amino acids in the CNS, N-methyl-D-aspartate (NMDA), quisqualate and kainate receptors [Foster and Fagg, 1984]. The results show that L-glu specifically enhance the electrical field stimulated (EFS) release of acetylcholine (ACh) from cholinergic nerves in the vagus nerve in rat bronchial smooth muscle. L-aspartate (L-asp), D-glu and L-α-amino adipic acid (L-α-AA) do not exert such an effect. The results are particularly important in view of the recent demonstration of N-acetyl-aspartyl-glutamate (NAAG) in mammalian motoneurons [Ory-Lavollee, 1987].

MATERIALS AND METHODS

Bronchial smooth muscle from the albino rat was isolated and mounted in an organ bath for measurement of contraction [Aas and Helle, 1982]. The Krebs buffer (37° C, pH=7.4) had the following composition (mM): NaCl, 118.4; KCl, 4.7; $CaCl_2$, 2.6; $MgSO_4$, 1.2; $NaHCO_3$, 24.9; KH_2PO_4, 1.2; Glucose, 11.1. The buffer was gassed with 95% O_2 and 5% CO_2. For EFS the bronchi were mounted between platinum electrodes and stimulated (20 Hz, 1.0 ms, supramaximal voltage) by a Grass Stimulator. The preparations were given a preload of 1.0 g and equilibrated for 60 min before start of the experiments. The contractions were recorded isometrically by Grass Force Displacement Transducers (FT-03C) and monitored on a Grass Polygraph (Model 7) fitted with amplifiers (7 P 1A).

NATO ASI Series, Vol. H 32
Molecular Biology of Neuroreceptors and
Ion Channels. Edited by A. Maelicke

RESULTS AND DISCUSSION

L-glu specifically enhanced the EFS stimulated contraction of rat bronchial smooth muscle (Fig. 1). This contraction has previously been demonstrated to be due to release of ACh [Aas and Helle, 1982]. L-glu potentiated the contraction in concentrations from 1.3 μM to 8.6 mM and the maximal increase in tonus was 104.3 ± 17.1 (n = 30) percent of control stimulation. This result is the first direct evidence for the presence of a modulatory role of L-glu in the peripheral nervous system in vertebrates. The concentration of L-glu inducing halfmaximal response (ED_{50}) was 3.5 ± 0.1 mM. Concentrations of L-glu higher than 22 mM, however, decreased the EFS induced contractions to zero. This might be due to a reversal of an elevated membrane conductance of cations induced by L-glu. L-glu has previously shown to enhance the influx of Na^{+} and Ca^{2+}, while conductance changes are virtually absent for Cl^{-} and unlikely for K^{+} [Sonnhof and Burle, 1981]. The effect of L-glu was not due to enhanced levels of ACh following inhibition of acetylcholinesterase (AChE) activity, since the AChE-activity was not significantly altered (less than 15% inhibition) by exposure to L-glu in concentrations up to 50 mM. The AChE-inhibition must exceed 60 percent if spontaneously released ACh should enhance the tonus of the muscle, as previously shown following exposure to AChE-inhibitors [Aas et al., 1986]. The AChE-activity was determined by a radiochemical method [Sterri and Fonnum, 1978].

The effect of L-glu on the EFS induced contraction was not inhibited by glutamate diethyl ester (L-GDEE) (1 mM) or by DL-2-amino-5-phosphonovalerate (DL-APV) (20 μM and 1.0 mM), which are considered to be selective antagonists of quisqualate [Hicks et al., 1978] and NMDA [Davies et al., 1981] receptors. Quisqualate (4 μM-2mM) enhanced the EFS induced contractions only negligible (less than 10 percent). The L-glu receptor agonist kainate, at high concentrations (> 5.2 mM), temporarily elevated EFS induced contractions. At 5.2 mM it was increased by 21 ± 4 percent (n = 4). The effect was, however, only seen at high concentrations of kainate and the effect was transient and lasted less than 30 seconds. The results provide little evidence for a kainate-like receptor on the cholinergic nerve terminals in the rat bronchi. NAAG, which is a transmitter candidate, is present in the CNS and it has been recently demonstrated also in the rat sciatic nerve [Orly-Lavollee et al., 1987].

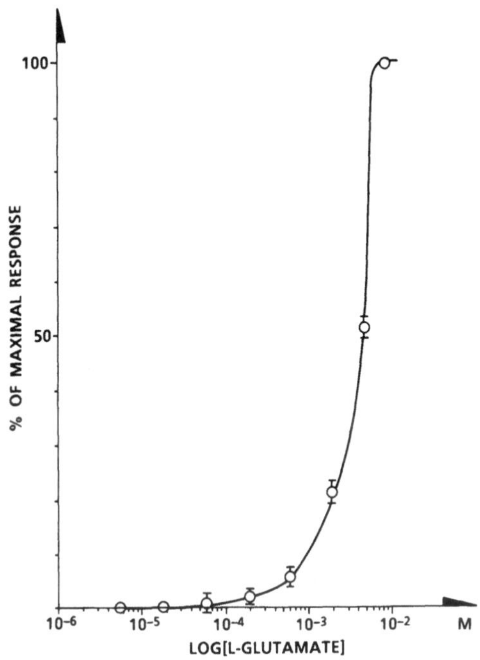

Figure 1 Concentration-response curve of L-glutamate. The responses are enhancement of EFS (20 Hz, 1.0 ms, supramaximal voltage) induced contraction of rat bronchial smooth muscle, due to release of ACh. The results are mean ± SEM of 30 experiments, expressed as percentage of the maximal response induced by L-glu.

NAAG (0.6 mM) had no effect on EFS induced contraction of rat bronchi. It is therefore evident that NAAG itself is not a candidate at the L-glu receptor shown in the present work.

The site of action of L-glu could either be presynaptic by facilitating the release of ACh or postsynaptic by interaction with the muscarinic cholinergic receptor. Tetrodotoxin (TTX) (3.0 µM), which inhibit the release of ACh by blocking sodium-ion channels [Kao, 1966], abolished the effect of L-glu on EFS induced contraction suggesting a presynaptic site of action of L-glu. Furthermore, complete suppression of ACh release by hemicholinium-3 (HC-3) (10 µM) [Birks and Maclntosh, 1957] also inhibited the effect of L-glu. Inhibition of the muscarinic receptors by atropine (1.1 µM) also abolished the EFS induced contraction and thereby the effect of L-glu. Since there was no alteration of the concentration-response curve following exogenously applied ACh by L-glu (8.6 mM), shown by the apparent affinity (ED_{50}) and the intrinsic activity (α) of ACh, it is clear that the effect of L-glu (> 22 mM) is a presynaptic facilitation of ACh release and not a postsynaptic interaction at the muscarinic receptor. The ED_{50} of ACh was 63 ± 35 µM (n=6) and in the presence of 8.6 mM L-glu it was not significantly altered (ED50 = 69 ± 35 µM, n=6). The intrinsic activity was increased non-significantly by 11.8 ± 8.1 percent.

The observed facilitation of EFS induced contraction by L-glu in mammalian respiratory smooth muscle was a specific effect by L-glu, since only an inhibition of the contraction was seen with the analogues D-glu, L-asp and L-α-AA. At 31 mM the reduction was 50%. L-asp and L-α-AA were shown to have the same effect as D-glu in concentrations from 2-80 mM and reduced the EFS enhanced contraction to the same extent as D-glu. Similar results were seen with the NMDA-receptor agonist ibotenic acid. The inhibition was, however, higher than at a comparable concentration of D-glu, L-asp and L-α-AA. These results provide therefore evidence for a specific presynaptic facilitation of ACh release from peripheral cholinergic nerves by L-glu, where L-glu stimulate a receptor-site not sensitive to neither D-glu, L-asp, L-α-AA, NMDA, quisqualate nor kainate.

REFERENCES

Aas P and Helle KB (1982) Neurotensin receptors in the rat bronchi, Regulatory Peptides 3, 405–413

Aas P, Veiteberg T and Fonnum F (1986) In vitro effects of soman on bronchial smooth muscle, Biochem Pharmacol 35, 11, 1793–1799

Birks RI and MacIntosh FC (1957) Acetylcholine metabolism at nerve endings, Br Med Bull 13 , 157–161

Curtis DR, Phillis JW and Watkins JC (1960) The chemical excitation of spinal neurons by certain acidic amino acids, J Physiol Lond 150, 656–682

Crawford AC and McBurney RN (1976) The post-synaptic action of some putative excitatory transmitter substances, Proc R Soc Lond B, 192, 481–489

Davies J, Francis AA, Jones AW and Watkins JC (1981) 2-amino-5-phosphonovalerate (2APV), a potent and selective antagonist of amino acid-induced and synaptic excitation, Neurosci Lett 21, 77–81

Fonnum F (1984) Glutamate: A neurotransmitter in the mammalian brain, J Neurochem 42, 1–11

Foster AC and Fagg GE (1984) Acidic amino acid binding sites in mammalian neuronal membranes: Their characteristics and relationship to synaptic receptors, Brain Res Review 7, 103–164

Hicks TP, Hall JG and McLennan H (1978) Ranking of excitatory amino acids by the antagonists glutamic acid diethylester and D-α-amino adipic acid, Can J Physiol Pharmacol 56, 901–906

Kao CY (1966) Tetrodotoxin, saxitoxin and their significance in the study of excitation phenomena, Pharmacol Rev 18, 997–1049

Ory-Lavollee L, Blakely RD and Coyle JT (1987) Neurochemical and immunocytochemical studies on the distribution of N-acetyl-aspartylglutamate and N-acetylaspartate in rat spinal cord and some peripheral nervous tissues J Neurochem 48, 895–899

Sonnhof U and Burle C (1981) An analysis of glutamate-induced ion fluxes across the membrane of spinal motoneurons of the frog, In: Glutamate as a Neurotransmitter (eds G Di Chiara and Gessal, GL) 195–204

Sterri SH and Fqnnum F (1978) Isolation of organic anions by extraction with liquid anion exchangers and its application to micromethods for acetylcholinesterase and 4–amino butyrate aminotransferase, Eur J Biochem 91, 215–222

Usherwood PNR (1981) Glutamate synapse and receptors in insect muscle, in: Glutamate as a neurotransmitter (eds G Di-Chiara and GL Gessal), 183–192

MOLECULAR PROPERTIES OF VOLTAGE-SENSITIVE SODIUM AND CALCIUM CHANNELS

William A. Catterall
Department of Pharmacology, SJ-30
University of Washington
Seattle, WA 98195 USA

INTRODUCTION

Voltage-sensitive sodium and calcium channels mediate the inward movement of those ions during the action potential in nerve and muscle cells. These membrane proteins have now been purified, their physiological activity has been restored by reconstitution *in vitro*, the primary structures of their principal subunits have been determined, and models for the relationship between their structure and function have been proposed. In this chapter, I review work on the molecular properties of voltage-sensitive sodium channels from brain and voltage-sensitive calcium channels from skeletal muscle.

SUBUNIT STRUCTURE

Sodium channels. Biochemical studies of voltage-sensitive sodium channels have depended upon use of a number of specific neurotoxins that act at five different receptor sites as molecular probes of channel structure and function (reviewed in Catterall, 1980; Catterall, 1988). Sodium channels can be solubilized from excitable membranes by treatment with nonionic detergents, detected in solubilized form by high affinity binding of tetrodotoxin or saxitoxin at neurotoxin receptor site 1, and purified by a combination of conventional methods including ion exchange chromatography, lectin-Sepharose chromatography, and sucrose gradient sedimentation. Highly purified, functional preparations with well-defined subunit composition have been obtained from mammalian brain (Hartshorne et al, 1982; Hartshorne & Catterall, 1984) and skeletal muscle (Casadei et al, 1986; Kraner et al, 1985) and eel electroplax (Agnew et al, 1980; Miller et al, 1983). The sodium channel from mammalian brain consists of a heterotrimeric complex of alpha (260

NATO ASI Series, Vol. H 32
Molecular Biology of Neuroreceptors and Ion Channels. Edited by A. Maelicke

kDa), beta1 (36 kDa), and beta2 (33 kDa) subunits. The beta2 subunit is attached to the alpha subunit by disulfide bonds. The sodium channel from mammalian skeletal muscle contains subunits of 260 kDa and 38 kDa analogous to the alpha and beta1 subunits of the brain sodium channel. The sodium channel purified from eel electroplax contains only a single 260 kDa subunit. These results imply that the large alpha subunit in each of these preparations contains the binding site for tetrodotoxin and saxitoxin and that this subunit is the main functional component of the sodium channel.

The overall arrangement of α, beta1 , and beta2 subunits of the rat brain sodium channel in the membrane has been inferred from biochemical experiments. The alpha and beta1 subunits are covalently labeled by scorpion toxins that act at neurotoxin receptor sites 3 and 4 from outside the cell (Beneski & Catterall, 1980; Darbon et al, 1983), and all three subunits are heavily glycosylated with up to 30% of their apparent mass due to carbohydrate (Miller et al, 1983; Roberts & Barchi, 1987; Grishin et al, 1984; Messner & Catterall, 1985), indicating that they are all exposed to the extracellular surface. The alpha subunit is phosphorylated by cAMP-dependent protein kinase in intact cells or synaptosomes (Costa & Catterall, 1984; Rossie & Catterall, 1987), indicating that it is exposed at the intracellular surface and therefore is a transmembrane protein. The beta1 and beta2 subunits are also intrinsic membrane proteins which have substantial hydrophobic domains labeled by hydrophobic photoaffinity probes and are preferentially extracted into hydrophobic detergent phases (Reber & Catterall, 1987). These biochemical properties of the three sodium channel subunits have led to the proposal of an overall model of their disposition in the membrane as illustrated in Figure 1.

Highly purified voltage-sensitive sodium channels with well-defined subunit composition from skeletal muscle, brain, or electroplax have been incorporated into phospholipid vesicles and isotopic flux methods have been used to show that these preparations mediate selective ion conductance activated by neurotoxins such as veratridine and batrachotoxin acting at neurotoxin receptor site 2 and blocked by tetrodotoxin and saxitoxin (Tamkun et al, 1984; Kraner et al, 1985; Recio-Pinto et al, 1987). These results show that the protein components of the isolated channels (alpha, beta1, and beta2 subunits from brain; alpha and beta subunits from skeletal muscle; or alpha subunits alone from electroplax) are sufficient to mediate selective, neurotoxin-activated ion flux with the flux rate, ion selectivity, and pharmacological properties expected of native sodium channels.

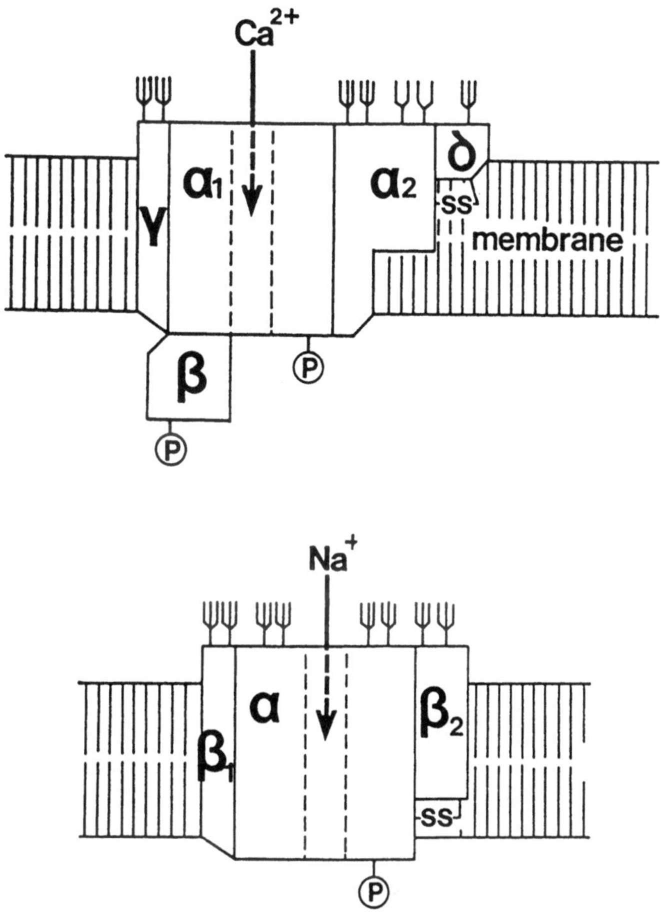

Figure 1. Subunit structures of sodium channels and calcium channels. The subunit structure and transmembrane organization of the rat brain sodium channel is illustrated as adapted from Catterall (1986a). The subunit structure of the rabbit skeletal muscle calcium channel is illustrated as adapted from Takahashi et al, 1987.

The properties of individual purified sodium channels incorporated into macroscopic planar phospholipid bilayers have been studied after activation of the channels by batrachotoxin. These experiments on sodium channels from brain, skeletal muscle, and electroplax showed that purified sodium channels retain the single channel conductance, ion selectivity, and voltage dependence of native sodium channels activated by batrachotoxin (Recio-Pinto et al, 1987; Hartshorne et al, 1985; Furman et al, 1986). The conductance increase upon channel opening corresponds to 25 pS in 500 mM Na^+ and the fraction of time that an individual channel is in the open state is strongly voltage dependent. The probability of opening is 50% at -91 mV and reaches 96% at voltages more positive than -60 mV. The steepness of voltage-dependent activation of the single purified sodium channels is consistent with a gating charge movement of 3.8 (see below for a discussion of gating charge). These results indicate that the voltage-sensing and ion-conducting structures of the sodium channel are intact in the purified proteins.

The roles of the different subunits in the functional properties of the purified sodium channels have also been studied by reconstitution methods. The beta1 and beta2 subunits of the rat brain sodium channel can be selectively dissociated from the alpha subunit by treatment with high ionic strength or disulfide reducing agents, respectively (Messner and Catterall, 1986; Messner et al, 1986). The beta2 subunit can be removed without effect on the neurotoxin binding, ion flux, single channel conductance, or voltage-dependence of the purified channels, while removal of beta1 subunits causes loss of all these functional activities. The dissociation of beta1 subunits and loss of functional activities can both be prevented by tetrodotoxin, demonstrating energy coupling between toxin binding and association of beta1 subunits in the functional state of the purified brain sodium channel. This functional state can also be stabilized in the absence of beta1 subunits by formation of intramolecular isopeptide bonds within the alpha subunit itself, suggesting that the role of beta1 subunit in the purified channel is, at least in part, to stabilize the structure of the alpha subunit (Tejedor et al, 1988).

Considered together, these results indicate that the alpha subunit is the principal structural and functional unit of the sodium channel and is expressed in different tissues in association with beta1 subunits which can help to stabilize its normal functional state and with beta2 subunits which may have no direct influence on channel function but may modify channel biosynthesis or assembly.

Calcium channels: Subunit composition and functional reconstitution. There are multiple classes of calcium channels in most excitable cells (Byerly & Hagiwara, 1988). Calcium channels which mediate long-lasting calcium currents (L-type, Nowycky et al, 1985) are inhibited by three distinct classes of organic calcium channel antagonists: dihydropyridines like nifedipine, phenylalkylamines like verapamil, and benzothiazepines like diltiazem (reviewed in Janis & Scriabine, 1983). These structurally different agents act at three separate, allosterically linked receptor sites, which are reminiscent of the multiple interacting neurotoxin receptor sites previously described for sodium channels (Catterall, 1980; Catterall, 1988). In addition, structurally related dihydropyridine calcium channel activators bind at the same dihydropyridine receptor site and favor the activated state of the calcium channel. These calcium channel modulators bind to the channel with high affinity and have been used as molecular probes to identify and isolate the protein components of dihydropyridine-sensitive calcium channels in the same way that neurotoxins had previously been used to identify and isolate sodium channels.

Although dihydropyridine-sensitive calcium channels are present in essentially all excitable cells, by far the highest concentration of these channels is found in the transverse tubules of skeletal muscle (Fosset et al, 1983) where they mediate very slowly activated, long-lasting calcium currents (Sanchez & Stefani, 1978; Almers et al, 1984). Experiments on identification and purification of these calcium channels have therefore focused on skeletal muscle. The skeletal muscle calcium channel can be solubilized in digitonin and purified by a series of conventional purification methods including ion exchange chromatography, lectin-Sepharose chromatography, and sucrose gradient sedimentation (Curtis & Catterall, 1984). The purified calcium channel is a complex of five polypeptide chains as illustrated in Figure 1 (Takahashi et al, 1987). The central component of the complex is the $alpha_1$ subunit with an apparent mass of 175 kDa as analyzed by SDS-PAGE. This polypeptide contains substantial hydrophobic domains, as assessed by labeling with a hydrophobic photoaffinity probe (Takahashi et al, 1987), indicating that it has a transmembrane orientation, but it is not heavily glycosylated (Takahashi et al, 1987; Leung et al, 1987). The $alpha_1$ subunit contains the receptor site for the dihydropyridine calcium channel modulators as assessed from photoaffinity labeling experiments (Takahashi et al, 1987; Sharp et al, 1987; Striessnig et al, 1987; Sieber et al, 1987). The central $alpha_1$ subunit is associated with a disulfide-linked complex of $alpha_2$ (143 kDa) and delta (27 kDa) subunits that are heavily glycosylated, but do not have substantial hydrophobic domains (Curtis & Catterall, 1984; Takahashi et al, 1987; Leung et al, 1987; Sharp et al, 1987; Striessnig et al, 1987; Sieber et al, 1987; Morton & Froehner, 1987; Barhanin et al, 1987), beta subunits (54 kDa) (Curtis & Catterall, 1984; Takahashi et al, 1987; Leung et al, 1987; Sieber et al, 1987; Leung et al, 1988) that are neither glycosylated nor hydrophobic (Takahashi et al, 1987), and γ subunits (30 kDa) (Curtis & Catterall, 1984; Takahashi et al, 1987; Leung et al, 1987; Sharp et al, 1987; Striessnig et al, 1987; Sieber et al, 1987; Morton & Froehner, 1987; Barhanin et al, 1987) that have both substantial hydrophobic domains and carbohydrate moieties (Takahashi et al, 1987). The $alpha_1$ and beta subunits are substrates for phosphorylation by cAMP-dependent protein kinase (Takahashi et al, 1987; Leung et al, 1987; Sieber et al, 1987; Curtis & Catterall, 1985), and are the sites of regulation of calcium channel ion conductance activity by this enzyme (see below). Although the disulfide-linked complex of $alpha_2$ and delta has a combined mass of 170 kDa, these polypeptides differ from the $alpha_1$ subunits in immunoreactivity, glycosylation, hydrophobicity, and phosphorylation by protein kinases, confirming that they are independent polypeptides. All these components of the calcium channel complex are specifically associated as assessed by co-immunoprecipitation with monospecific antibodies and by co-sedimentation in sucrose gradients (Curtis & Catterall, 1984; Takahashi et al, 1987; Leung et al, 1987; Morton & Froehner, 1987; Takahashi & Catterall, 1987). Since it contains the calcium channel antagonist receptor sites and extensive hydrophobic domains expected of the calcium channel protein, the alpha1 subunit is considered to be the principal functional component of the calcium channel in analogy with the alpha subunit of the sodium channel (Figure 1).

The functional properties of purified dihydropyridine-sensitive calcium channels from skeletal muscle have been restored by reconstitution methods similar to those used for the voltage-sensitive sodium channel. Incorporation of highly purified calcium channels, prepared in the presence of the calcium channel activator Bay K 8644, into phospholipid vesicles results in increased calcium conductance of the reconstituted vesicles that is inhibited by calcium channel antagonists (Curtis & Catterall, 1986). Although the ability to reconstitute calcium channel ion conductance activity correlated closely with the concentration of the α, β, and γ subunits of the purified channel preparation, less than 5% of the purified channels are reconstituted in functional form under these conditions suggesting that some aspect of the *in situ* environment may be important for calcium channel function. Dihydropyridine-sensitive, or L-type, calcium channels in cardiac and neural cells are regulated by intracellular cAMP and cAMP-dependent protein kinase. The probability of calcium channel activation is increased and calcium channels are shifted from an inactivated to an activatable state. Calcium currents mediated by calcium channels in skeletal muscle transverse tubules are also increased by intracellular cAMP and cAMP-dependent protein kinase (Arreola et al, 1987). The $alpha_1$ and beta subunits of purified skeletal muscle calcium channels are good substrates for phosphorylation by cAMP-dependent protein kinase *in vitro* (Takahashi et al, 1987; Curtis & Catterall, 1985). Phosphorylation by cAMP-dependent protein kinase increases the initial rate of $^{45}Ca^{2+}$ influx 8-fold in reconstituted phospholipid vesicles containing purified calcium channels and increases the number of purified calcium channels that are active (Catterall et al, 1988). Thus, appropriate phosphorylation appears to be required for functional activity of purified preparations of calcium channels. Individual purified calcium channels in planar lipid bilayers are also activated by cAMP-dependent protein kinase (Flockerzi et al, 1986; Hymel et al, 1988). Considered together, the results of these studies of purified and reconstituted calcium channels show that these channels are functional in the purified state and that their functional properties depend critically upon cAMP-dependent protein phosphorylation. These purified preparations will continue to provide valuable tools for analysis of the structure and function of voltage-sensitive ion channels.

PRIMARY STRUCTURE

Primary Structures of Sodium and Calcium Channels. The availability of purified preparations of voltage-sensitive sodium and calcium channels which retain normal functional properties provided the basis for isolation of cDNA clones encoding the primary structure of the principal subunits of these two ion channels. Noda et al (1984) screened expression cDNA libraries prepared from eel electroplax mRNA with both antibodies against the purified electroplax sodium channel alpha subunit and oligonucleotides designed to recognize segments of its mRNA sequence. Analysis of the resulting cDNA sequences showed that they encode a large polypeptide of 1832 amino acids containing four repeated domains of 300 to 400 amino acids with approximately 50% overall amino acid identity. Each domain contains multiple hydrophobic segments with the potential for formation of transmembrane alpha-helices. cDNA clones encoding the corresponding alpha subunit of the sodium channel from rat brain were isolated by screening with antibodies against the purified protein (Goldin et al, 1986) or cDNA probes encoding the electroplax protein (Noda et al, 1986) and shown to encode homologous proteins. Three different mRNA's encoding sodium channel alpha subunits have been identified in rat brain and fully sequenced to define the R_I, R_{II}, and R_{III} sodium channel subtypes (Noda et al, 1986, Kayano et al, 1988). Like the electroplax sodium channel, these three rat brain sodium channel alpha subunits contain 4 homologous domains with multiple hydrophobic segments.

cDNA clones encoding the $alpha_1$ subunit of the dihydropyridine-sensitive calcium channel have also been isolated by screening cDNA libraries with oligonucleotides designed to recognize segments of the corresponding mRNA sequence (Tanabe et al, 1987). This protein of 2005 amino acids has an amino acid sequence that is 29% identical to that of the alpha subunit of the R_{II} rat brain sodium channel and an additional 36% of the residues represent substitutions of chemically similar amino acids. The $alpha_1$ subunit contains four homologous domains with multiple hydrophobic segments like the sodium channel alpha subunit.

Proposed Transmembrane Organization of the Sodium Channel Alpha Subunit and Calcium Channel $Alpha_1$ Subunit. Analyses of the hydrophobicity, probable secondary structure, and regional conservation of the amino acid sequence of the alpha and $alpha_1$ subunits of these ion channels have led to a number of proposed models for their transmembrane folding (Noda et al, 1984; Noda et al, 1986; Guy & Seetharamulu, 1986; Greenblatt et al, 1985; Catterall, 1986a). Noda et al (1984) first noted that each of the four homologous domains of the electroplax sodium channel contains six regions of probable alpha-helical structure long enough to be membrane-spanning segments and designated them S1 through S6. The S1, S2, S3, S5, and S6 segments are predominantly hydrophobic with occasional negative charges, while the S4 segments are both hydrophobic and highly positively charged. Specific models of voltage-dependent gating have been developed which require the S4 segments to have a transmembrane orientation (Guy & Seetharamulu, 1986; Greenblatt et al, 1985; Catterall, 1986a; Catterall, 1986b and see below). Consistent with these proposals, current models of the transmembrane arrangement of the sodium channel alpha subunits from eel electroplax and rat brain place the S4 segments as one of six or eight transmembrane segments in each homologous domain (Noda et al, 1986; Guy & Seetharamulu, 1986; Greenblatt et al, 1985; Catterall, 1986a; Catterall, 1986b and see below).

The substantial amino acid sequence homology of the calcium channel $alpha_1$ subunit to the sodium channel alpha subunits led to a proposal of an analogous transmembrane structure (Tanabe et al, 1987). The conservation of amino acid sequence in the proposed transmembrane regions and the overall arrangement of hydrophilic and hydrophobic segments is even more striking than the amino acid sequence homology *per se*. The principal difference in overall structure is an extended hydrophilic domain at the carboxyl terminus of the calcium channel.

EXPRESSION OF FUNCTIONAL ION CHANNELS FROM mRNA

Sodium Channels. Injection of mRNA from excitable tissues into *Xenopus* oocytes results in expression of voltage-sensitive sodium, calcium, and potassium channels (Gundersen et al, 1983). Injection of a high molecular weight fraction of mRNA from rat brain is sufficient for functional expression of sodium channels suggesting that only mRNA encoding alpha subunits is required (Sumikawa et al, 1984; Hirono et al, 1985). Successful expression of sodium channels from alpha subunit mRNA isolated by hybrid selection (Goldin et al, 1986) and R_{II} mRNA made from cloned cDNA (Noda et al, 1986; Auld et al, 1988) also supports this conclusion. Assuming that *Xenopus* oocytes do not produce beta1 and/or beta2 subunits themselves, these results show conclusively that sodium channel alpha subunits of rat brain type II sodium channels are functionally autonomous when separately expressed in oocytes, even though they are expressed in association with beta1 and beta2 subunits in neurons (Catterall, 1986a) and a complex of alpha and beta1 subunits is the minimal unit capable of retaining sodium channel function under conditions of purification and reconstitution (Messner et al, 1986). The requirement for association of beta1 subunits with alpha subunits to maintain a functional state of the alpha subunit in purified form is likely to reflect a structural interaction between the two subunits which stabilizes the functional state of the purified alpha subunit (Tejedor et al, 1988).

The successful expression of the R_{II} and R_{III} sodium channel subtypes from mRNA encoding only the alpha subunit suggests that the corresponding protein components of the R_I and electroplax sodium channels and the dihydropyridine-sensitive calcium channel are the principal functional subunits of those proteins as well. However, the R_I mRNA is expressed poorly in *Xenopus* oocytes (Noda et al, 1986) and functional expression of cloned cDNA's encoding the electroplax sodium channel and the dihydropyridine-sensitive calcium channel has not been reported. Therefore, autonomous expression and function may not be a general characteristic of these homologous polypeptides.

Although high molecular weight mRNA from rat brain and mRNA encoding only the alpha subunit of R_{II} sodium channels can direct synthesis of functional sodium channels in *Xenopus* oocytes, the functional properties of these channels are not completely normal. For sodium channels expressed from high molecular weight mRNA or from mRNA transcribed from cloned DNA, the time course of inactivation is much slower and, for one clone of the R_{II} subtype, the voltage dependence of activation and inactivation is shifted to more positive potentials in comparison with sodium channels expressed in oocytes from total rat brain mRNA (Auld et al, 1988) or R_{II} sodium channels in brain neurons (Scheuer and Catterall, unpublished). Co-injection of a low molecular weight mRNA fraction from rat brain restores the time course of channel inactivation to normal. Evidently, proteins encoded by low molecular weight mRNA species can modulate the functional properties of R_{II} alpha subunits. These mRNA's might encode beta1 or beta2 subunits or other enzymes or factors that alter alpha subunit function by post translational modification.

STRUCTURAL MODELS OF ION CHANNEL FUNCTION

Transmembrane pore. Sodium channel alpha subunits are sufficient to form an active, voltage-gated ion channel. Nicotinic acetylcholine receptors and gap junction proteins form high conductance transmembrane pores in the center of a symmetric array of 5 homologous or 6 identical transmembrane subunits, respectively (Stroud & Finer-Moore, 1985; Unwin & Zampighy, 1980). Therefore, the presence of four homologous domains in the primary structure of the sodium channel alpha subunit led to the proposal that the transmembrane pore of the sodium channel is formed in the center of a pseudosymmetric square array of these four homologous transmembrane domains as illustrated in Figure 2A (Noda et al, 1984). All subsequent models of sodium channel structure have incorporated this basic architecture (Noda et al, 1986; Guy & Seetharamulu, 1986; Greenblatt et al, 1985; Catterall, 1986a; Catterall, 1986b). The strong conservation of the structure of the corresponding homologous domains in the amino acid sequence of the $alpha_1$ subunit of the calcium channel support the concept that they form transmembrane pores with a similar structure. Formation of a transmembrane pore in the center of a pseudosymmetric array of homologous transmembrane units may be a general motif among high conductance ion channels.

Voltage-dependent activation. The steep voltage dependence of activation of the voltage-sensitive ion channels is their most unique characteristic. It requires that they have charged amino acid residues or strongly oriented dipoles within the membrane electric field of the phospholipid bilayer. For example, the steepness of voltage-dependent activation of sodium channels requires the movement of 4 to 6 protein bound positive charges from the inner surface of the bilayer to the outer surface during activation (Armstrong, 1981). The movement of these gating charges or voltage sensors under the force of the electric field is believed to initiate a conformational change in the channel protein resulting in activation.

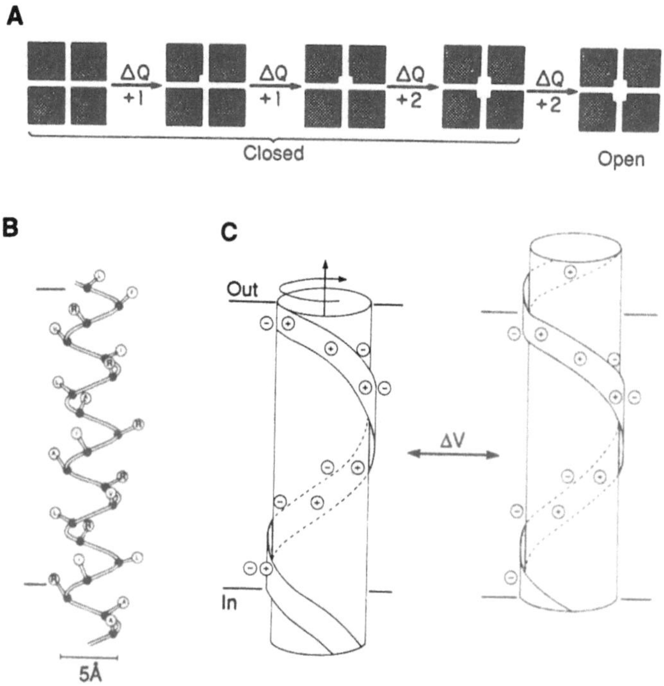

Figure 2. Sliding Helix model of voltage-dependent gating. **A.** The sodium channel alpha subunit is illustrated from a perspective perpendicular to the plane of the membrane as alpha array of four homologous domains surrounding a central transmembrane pore. Depolarization causes a sequential series of voltage-driven conformational changes in the individual domains. After each domain has changed conformation, an open ion channel is formed. Each conformational change is associated with an outward transfer to protein-bound positive gating charge (Δ Q) across the membrane. **B.** *Left.* A ball-and-stick, three-dimensional representation of the S4 helix of domain IV of the sodium channel. Darkened circles represent the alpha-carbon of each amino acid residue. Open circles show the direction of projection of the side chain of each residue away from the core of the helix. Nonpolar residues are illustrated in thin letters by their single letter code: F, Phe; A, Ala; I, Ile; L, Leu; V, Val; G, Gly; T, Thr. Positively charged amino acids are illustrated in bold letters. R, Arg. *Right.* Movement of the S4 helix in response to membrane depolarization. The proposed transmembrane S4 helix is illustrated as a cylinder with a spiral ribbon of positive charge. At the resting membrane potential (left), all positively charged residues are paired with fixed negative charges on other transmembrane segments of the channel and the transmembrane segment is held in that position by the negative internal membrane potential. Depolarization reduces the force holding the positive charges in their inward position. The S4 helix is then proposed to undergo a spiral motion through a rotation of approximately 60° and an outward displacement approximately 5Å. This movement leaves an unpaired negative charge on the inward surface of the membrane and reveals an unpaired positive charge on the outward surface to give a net charge transfer (Δ Q) of +1. Note that all intramembranous positive charges are paired in both conformations. In domains III and IV, there are enough arg residues to give a gating charge of +2.

The requirement for transmembrane movement of multiple charges during sodium channel activation has focused attention on the S4 segments of the voltage-sensitive ion channels which are both positively charged and hydrophobic. These unique structures consisting of repeated motifs of a positively charged amino acid residue, usually arginine, followed by two hydrophobic residues were first noted in the amino acid sequence of the electroplax sodium channel (Noda et al, 1984). Conservation of this amino acid sequence among different ion channels, first noted for sodium channels from electroplax and brain (Noda et al, 1986; Catterall, 1986b), is striking.

Several authors independently proposed that these S4 segments have a transmembrane orientation and are the gating charges or voltage sensors of the sodium channel (Noda et al, 1986; Guy & Seetharamulu, 1986; Greenblatt et al, 1985; Catterall, 1986b). Figure 2B illustrates the *'sliding helix'* or *'helical screw'* model of voltage-dependent gating (Catterall, 1986a; Catterall, 1986b; Guy & Seetharamulu, 1986). The S4 segments are proposed to adopt an alpha-helical conformation. In this conformation, the arginine residues form a spiral ribbon of positive charge around the core of the helix. They are stabilized in their transmembrane position by formation of ion pairs between the positive charges of the S4 segments and negatively charged amino acid residues from the surrounding transmembrane alpha helices of the S1, S2, and S3 segments (Figure 2B). This ion pairing arrangement is metastable, held in place by the force of the electric field drawing the positive charges into the cell and repelling the negative charges outward at a typical resting membrane potential of -80 mV. Upon depolarization, this force is relieved and the S4 segment is released to slide outward along a spiral path and form a new set of ion pairs (Figure 2B). This sliding motion of the S4 helix transfers the equivalent of a full positive charge across the membrane with only a 5Å outward movement and a 60° rotation. Sequential, voltage-driven movements of the S4 helices are proposed to initiate sequential conformational changes in the four domains of the sodium channel resulting in the transfer of at least four gating charges across the membrane and, finally, in activation of the ion channel (Figure 2A). The requirement for sequential conformational changes in the four domains before activation of the ion channel fits with the expected kinetics of sodium channel activation in which gating charge moves immediately upon depolarization while ionic current begins to flow only after a significant lag time (Armstrong, 1981).

<u>Inactivation.</u> The ion conductance activity of voltage-sensitive ion channels is transient because prolonged activation causes transition to an inactive state that cannot be activated by further depolarization. The mechanisms of inactivation of the

voltage-sensitive channels differ significantly. Inactivation of sodium channels is most thoroughly studied. It is intrinsically a weakly voltage-dependent process which derives its strong voltage-dependence by coupling to the voltage-dependent conformational changes leading to channel activation (Armstrong, 1981; Armstrong & Bezanilla, 1977; Aldrich et al, 1976). It can be slowed by neurotoxins acting at the extracellular surface of the channel (reviewed in Catterall, 1980; Catterall, 1988) and completely removed by proteolytic enzymes and group-specific reagents acting at the intracellular surface of the channel (Armstrong, 1975) and therefore must involve a transmembrane conformational change in the sodium channel protein.

Since segments of the channel protein involved in channel inactivation are accessible to macromolecular reagents, their functional activities might be inhibited by appropriate site-directed antibodies. Antibodies prepared against a synthetic peptide corresponding to the highly conserved proposed intracellular segment between homologous domains III and IV of the alpha subunit recognize sodium channels in brain, skeletal muscle, heart and locust nerve cord (Gordon et al, 1988). These antipeptide antibodies slow sodium channel inactivation without markedly altering the voltage dependence of sodium channel activation or inactivation when applied to the intracellular surface of sodium channels in skeletal muscle cells (Vassilev et al, 1988). Antibody effects appear more rapidly at a membrane potential of -110 mV than at -70 mV suggesting that the antibody binds preferentially to the "not inactivated" states of sodium channels. The extent of slowing of inactivation is greatest during strong depolarizations to positive membrane potentials indicating that pathways of inactivation that are most important at positive potentials are preferentially blocked. These results support an important role for this highly conserved intracellular segment in sodium channel inactivation and suggest that conformational changes associated with sodium channel inactivation alter the accessibility of this amino acid sequence to antibodies.

Phosphorylation sites. Purified and reconstituted calcium channels are regulated by phosphorylation of their $alpha_1$ and beta subunits (Flockerzi et al, 1986; Takahashi et al, 1987; Curtis & Catterall, 1985; Catterall et al, 1988). However, these channels have not been shown directly to be phosphorylated in intact cells and the sites of phosphorylation within their amino acid sequence are not known. In contrast, studies of sodium channels have defined their sites of phosphorylation *in situ*, but the functional significance of their phosphorylation is less clear. Sodium channel alpha subunits from brain are rapidly phosphorylated *in vitro* by cAMP-dependent protein kinase (Costa et al, 1982). In intact synaptosomes and neurons, activation of endogenous cAMP-dependent protein kinase results in phosphorylation of the same sites which can be phosphorylated *in vitro* (Costa & Catterall, 1984; Rossie & Catterall, 1987). Without stimulation, these sites contain approximately 2.5 mol phosphate per mol sodium channel and this is increased to 3.5 mol/mol upon stimulation. These major phosphorylation sites have been localized to the large intracellular segment of the alpha subunit between homologous domains I and II (see Figure 3) in experiments with sequence-directed antibodies (Rossie et al, 1987). The segment containing these phosphorylation sites is not present in the electroplax sodium channel, suggesting that functional effects of phosphorylation of these sites may be restricted to neuronal sodium channels.

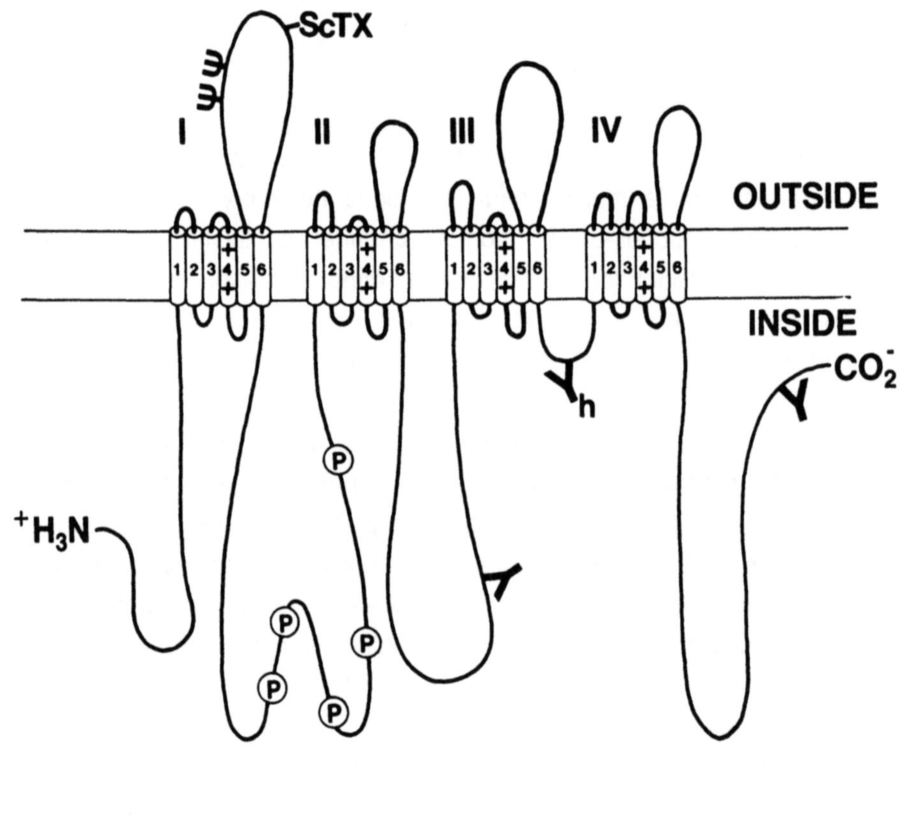

Figure 3. A functional map of the sodium channel alpha subunit. The transmembrane folding model of the alpha subunit is depicted with experimentally demonstrated sites of cAMP-dependent phosphorylation (**P**), interaction of site-directed antibodies that define transmembrane orientation (**Y**), covalent attachment of alpha-scorpion toxins (ScTx), glycosylation (ψψ), and modulation of channel inactivation (h).

Receptor sites. Because the neurotoxins and drugs that alter voltage-sensitive ion channel function have very specific effects, location of their sites of action will be a major contribution to understanding the functional architecture of these proteins. The site of covalent attachment of alpha-scorpion toxin to the alpha subunit of the rat brain sodium channel (Beneski & Catterall, 1980) has been investigated with site-directed antibodies (Tejedor & Catterall, 1988). Sodium channel alpha subunits having covalently bound alpha-scorpion toxin were subjected to cleavage with a series of proteases and with CNBr, and the resulting labeled fragments were analyzed by immunoprecipitation with a panel of antibodies against 20-residue segments of the protein. A fragment of 14 kDa contained the bound scorpion toxin. Immunoprecipitation with site-directed antibodies showed that this fragment is derived from the extracellular segment between transmembrane segments S5 and S6 of domain I as illustrated in Figure 3. Neurotoxin receptor site 3, at which alpha-scorpion toxins and sea anemone toxins specifically affect sodium channel inactivation, is therefore located near the largest extracellular segment of domain I in the three-dimensional structure of the sodium channel.

TRANSMEMBRANE TOPOLOGY OF THE VOLTAGE-SENSITIVE ION CHANNEL

The models for transmembrane folding of the principal subunits of the sodium, and calcium channels were derived from computer assisted analysis of hydrophobicity and secondary structure together with the assumptions that all four homologous domains should adopt similar transmembrane arrangements and that the N-terminal should be intracellular since no signal sequence was detected. Several pieces of experimental evidence now support the indicated arrangement of the major hydrophilic segments of the sodium channel as illustrated in Figure 3. Electron microscopic localization of bound antibodies directed against synthetic peptides corresponding to 20 amino acid sequences of the C-terminal hydrophilic segment and the hydrophilic segment between domains II and III of the electroplax sodium channel shows that they are both located on the intracellular surface of the membrane (Gordon et al, 1987; Gordon et al, 1988). Cyclic AMP-dependent phosphorylation of five sites in the hydrophilic segment between homologous domains I and II of the rat brain sodium channel in intact neurons defines this segment as intracellular (Rossie et al, 1987). Modulation of sodium channel inactivation by intracellular perfusion of antibodies against the hydrophilic segment between homologous domains III and IV places this segment on the intracellular surface of the membrane as well (Vassilev et al, 1988). These results provide four points of reference on the intracellular surface of the channel which agree with the overall orientation illustrated in Figure 3. Since the four homologous domains are considered by nearly all authors to adopt similar transmembrane orientations (Noda et al, 1984; Guy & Seetharamulu, 1986; Greenblatt et al, 1985; Catterall, 1986a; Catterall, 1986b), the N-terminal hydrophilic segment is also likely to be intracellular.

Each homologous domain of the sodium channel has a major hydrophilic segment that is proposed to be extracellular between the putative transmembrane segments S5 and S6 (Figure 3). This segment of domain 1 is covalently labeled by alpha-scorpion toxin and is highly glycosylated (Tejedor & Catterall, 1988). Since this segment is labeled by alpha-scorpion toxins which act from outside the cell and is glycosylated, it must be extracellular. Although direct evidence is not available at present, the strong homology among the four domains argues that segments between S5/S6 are likely to be at least partially extracellular in all four domains as proposed in nearly all of the transmembrane folding models (Noda et al, 1984; Guy & Seetharamulu, 1986; Greenblatt et al, 1985; Catterall, 1986a; Catterall, 1986b; and Figure 3). The results outlined here all support the location of the major hydrophilic domains as proposed in the general transmembrane folding schemes developed on the basis of computer analysis of amino acid sequence. However, clear definition of the topology of the voltage-sensitive ion channels requires determination of the transmembrane orientation of each of the segments within the homologous domains. Direct experimental evidence on this problem is still scanty.

CONCLUSIONS

Experiments on the molecular properties of voltage-sensitive sodium and calcium channels reviewed here show that these ion channels can be isolated in functional form and studied in detail in the purified state. Both of these channels consist of a homologous principal subunit of approximately 2000 amino acids that is expressed in association with multiple polypeptides of uncertain function. The principal alpha subunit of the sodium channel can form a functional ion channel by itself, but the expression and functional activity of this subunit are modulated by co-expression of low molecular weight brain mRNA. Development of a functional map of these principal ion channel subunits is underway and assignments of channel segments involved in voltage-dependent activation, inactivation, scorpion toxin binding, and cAMP-dependent phosphorylation have been proposed. Continuation of this approach is expected to lead to a detailed understanding of the molecular basis of electrical excitability.

REFERENCES

Agnew WS, Moore AC, Levinson SR, Raftery MA (1980) Identification of a large molecular weight peptide associated with a tetrodotoxin binding protein from the electroplax of *Electrophorus electricus*. Biochem Biophys Res Commun 92:860-866

Aldrich RW, Corey DP, Stevens CF (1976) A reinterpretation of mammalian sodium channel gating based on signal channel recording. Nature 306:436-441

Almers W, McCleskey EW, and Palade PT (1984) A nonselective cation conductance in frog muscle membrane blocked by micromolar external Ca^{++}. J Physiol 353:565-583

Armstrong CM (1975) Ionic pores, gates, and gating currents. Q Rev Biophys 7:179-210

Armstrong CM (1981) Sodium channels and gating currents. Physiol Rev 61:644-682

Armstrong CM, and Bezanilla F (1977) Inactivation of the sodium channel II. Gating current experiments. J Gen Physiol 70:567-590

Arreola J, Calvo J, García MC, Sánchez, JA (1987) Modulation of calcium channels of twitch skeletal muscle of the frog by adrenaline and cyclic adenosine monophosphate. J Physiol 393:307-330

Auld V, Goldin AL, Krafte DS, Marshall J, Dunn JM, Catterall WA, Lester HA, Davidson N, Dunn RJ (1988) A rat brain Na channel alpha subunit with novel gating properties. Neuron 1:449-461

Barhanin J, Coppola T, Schmid A, Borsotto M, Lazdunski M (1987) The calcium channel antagonist receptor from rabbit skeletal muscle. Reconstitution after purification and subunit characterization. Eur J Biochem 164:525-531

Beneski DA, Catterall WA (1980) Covalent labelling of protein components of the sodium channel with a photoactivable derivative of scorpion toxin. Proc Natl Acad Sci USA 77:639-643

Byerly H, Hagiwara S (1988) Calcium channel diversity. In: Calcium and Channel Modulation, edited by A D Grinnell, D Armstrong, M Jackson, New York, Plenum Press

Casadei JM, Gordon RD, Barchi RL (1986) Immunoaffinity isolation of Na^{+} channels from rabbit skeletal muscle. Analysis of subunits. J Biol Chem 261:4318-4323

Catterall WA (1980) Neurotoxins that act on voltage-sensitive sodium channels. Ann Rev Pharmacol Toxicol 20:15-43

Catterall WA (1986a) Molecular properties of voltage-sensitive sodium channels. Annu Rev Biochem 55:953-985

Catterall WA (1986b) Voltage-dependent gating of sodium channels: Correlating structure and function. Trends Neurosci 9:7-10

Catterall WA (1988) Molecular pharmacology of voltage-sensitive sodium channels. ISI Atlas of Science: Pharmacology 190-195

Catterall WA, Seagar MJ, Takahashi M, Nunoki K (1989) Molecular properties of voltage-sensitive calcium channels. In: Calcium Binding Proteins, edited by H Hidaka, New York, Plenum Press

Costa MRC, and Catterall WA (1984) Cyclic-AMP-dependent phosphorylation of the alpha subunit of the sodium channel in synaptic nerve ending particles. J Biol Chem 259:8210-8218

Costa MRC, Casnellie JE, Catterall WA (1982) Selective phosphorylation of the alpha subunit of the sodium channel by cAMP dependent protein kinase. J Biol Chem 257:7918-7921

Curtis BM, Catterall WA (1984) Purification of the calcium antagonist receptor of the voltage-sensitive calcium channel from skeletal muscle transverse tubules. Biochemistry 23:2113-2118

Curtis BM, Catterall WA (1985) Phosphorylation of the calcium antagonist receptor of the voltage-sensitive calcium channel by cAMP-dependent protein kinase. Proc Natl Acad Sci USA 82:2528-2532

Curtis BM, Catterall WA (1986) Reconstitution of the voltage-sensitive calcium channel purified from skeletal muscle transverse tubules. Biochemistry 25:3077-3083

Darbon H, Jover E, Couraud F, Rochat H (1983) Photoaffinity labeling of alpha- and β-scorpion toxin receptors associated with rat brain sodium channel. Biochem Biophys Res Commun 115:415-422

Flockerzi V, Oeken H-J, Hofmann F, Pelzer D, Cavalié A, Trautwein W (1986) Purified dihydropyridine-binding site from skeletal muscle t-tubules is a functional calcium channel. Nature 323:66-68

Fosset M, Jaimovich E, Delpont E, Lazdunski M (1983) [^{3}H]nitrendipine receptors in skeletal muscle. J Biol Chem 258:6086-6092

Furman ER, Tanaka JC, Mueller P, Barchi RL (1986) Voltage-dependent activation in purified reconstituted sodium channels from rabbit T-tubular membranes. Proc Natl Acad Sci USA 83:488-492

Goldin AL, Snutch T, Lubbert H, Dowsett A, Marshall J, Auld V, Downey W, Fritz LC, Lester HA, Dunn R, Catterall WA, Davidson N (1986) Messenger RNA coding for only the alpha subunit of the rat brain Na channel is sufficient for expression of functional channels in *Xenopus* oocytes. Proc Natl Acad Sci USA 83:7503-7507

Gordon D, Merrick D, Wollner DA, Catterall WA (1988) Biochemical properties of sodium channels in a wide range of excitable tissues studied with sited-directed antibodies. Biochemistry 27:7032-7038

Gordon RD, Fieles WE, Schotland DL, Hogue-Angeletti R, Barchi RL (1988) Topographical localization of the C-terminal region of the voltage-dependent sodium channel from *Electrophorus electricus* using antibodies raised against a synthetic peptide. Proc Natl Acad Sci USA 84:308-312

Gordon RD, Li Y, Fieles WE, Schotland DL, Barchi RL (to be published) Topological localization of a segment of the eel voltage-dependent sodium channel primary sequence (AA927-938) that discriminates between models of 3° structure. J Neurosci 8:3742-3749

Greenblatt RE, Blatt Y, Montal, M (1985) The structure of the voltage-sensitive sodium channel: Inferences derived from computer-aided analysis of the *Electrophorus electricus* channel primary structure. FEBS Lett 193:125-134

Grishin EV, Kovalenko VA, Pashkov VN, Shamotienko OG (1984) Isolation and characterization of sodium channel components. Biologicheskie Membrany 8:858-867

Gundersen CB, Miledi FRS, Parker I (1983) Voltage-operated channels induced by foreign messenger RNA in *Xenopus* oocytes. Proc R Soc Lond B220:131-140

Guy HR, Seetharamulu P (1986) Molecular model of the action potential sodium channel. *Proc Natl Acad Sci USA* 83:508-512

Hartshorne RP, Catterall WA (1984) The sodium channel from rat brain: Purification and subunit composition. J Biol Chem 259:1667-1675

Hartshorne RP, Keller BU, Talvenheimo JA, Catterall WA, Montal M (1985) Functional reconstitution of the purified brain sodium channel in planar lipid bilayers. Proc Natl Acad Sci USA 82:240-244

Hartshorne RP, Messner DJ, Coppersmith JC, Catterall WA (1982) The saxitoxin receptor of the sodium channel from rat brain. Evidence for two nonidentical beta subunits. J Biol Chem 257:13888-13891

Hirono C, Yamagishi S, Ohara B, Hisanaga Y, Nakayama T, Sugiyama, H (1985) Characterization of mRNA responsible for induction of functional sodium channels in *Xenopus* oocytes. Brain Res 359:57-64
Hymel L, Striessnig J, Glossmann H, Schindler H (1988) Purified skeletal muscle 1,4-dihydropyridine receptor forms phosphorylation-dependent oligomeric calcium channels in planar bilayers. Proc Natl Acad Sci USA 85:4290-4294
Janis R, Scriabine A (1983) Sites of action of Ca^{2+} channel inhibitors. Biochem Pharmacol 32:3499-3507
Kayano T, Noda M, Flockerzi V, Takahashi H, Numa S (1988) Primary structure of rat brain sodium channel III deduced from the cDNA sequence. FEBS Lett 228:187-194
Kraner SD, Tanaka JC, Barchi RL (1985) Purification and functional reconstitution of the voltage-sensitive sodium channel from rabbit T-tubular membranes. J Biol Chem 260:6341-6347
Leung AT, Imagawa T, Campbell KP (1987) Structural characterization of the 1,4-dihydropyridine receptor of the voltage-dependent Ca^{2+} channel from rabbit skeletal muscle. J Biol Chem 262:7943-7946
Leung AT, Imagawa T, Block B, Franzini-Armstrong C, Campbell KP (1988) Biochemical and ultrastructural characterization of the 1,4-dihydropyridine receptor from rabbit skeletal muscle. Evidence for a 52,000 Da subunit. J Biol Chem 263:994-1001
Messner DJ, Catterall WA (1985) The sodium channel from rat brain. Separation and characterization of subunits. J Biol Chem 260:10597-10604
Messner DJ, Catterall WA (1986) The sodium channel from rat brain. Role of the beta1 and beta2 subunits in saxitoxin binding. J Biol Chem 261:211-215
Messner DJ, Feller DJ, Scheuer T, Catterall WA (1986) Functional properties of rat brain sodium channels lacking the beta1 or beta2 subunit. J Biol Chem 261:14882-14890
Miller JA, Agnew, WS, Levinson SR (1983) Principal glycopeptide of the tetrodotoxin/saxitoxin binding protein from *Electrophorus electricus*: Isolation and partial chemical and physical characterization. Biochemistry 22:462-470
Morton ME, Froehner SC (1987) Monoclonal antibody identifies a 200-kDa subunit of the dihydropyridine-sensitive calcium channel. J Biol Chem 262:11904-11907
Noda M, Ikeda T, Kayano T, Suzuki H, Takeshima H, Kurasaki M, Takahashi H, Numa S (1986) Existence of distinct sodium channel messenger RNAs in rat brain. Nature 320:188-192
Noda M, Ikeda T, Suzuki T, Takeshima H, Takahashi H, Kuno M, Numa S (1986) Expression of functional sodium channels from cloned cDNA. Nature 322:826-828
Noda M, Shimizu S, Tanabe T, Takai T, Kayano T, Ikeda T, Takahashi H, Nakayama H, Kanaoka Y, Minamino N, Kangawa K, Matsuo H, Raftery M, Hirose T, Inayama S, Hayashida H, Miyata T, Numa S (1984) Primary structure of *Electrophorus electricus* sodium channel deduced from cDNA sequence. Nature 312:121-127
Nowycky M C, Fox AP, Tsien RW (1985) Three types of neuronal calcium channel with different calcium agonist sensitivity. Nature 316:440-443
Reber BFX, Catterall WA (1987) Hydrophobic properties of the beta1 and beta2 subunits of the rat brain sodium channel. J Biol Chem 262:11369-11374
Recio-Pinto E, Duch DS, Levinson SR, Urban BW (1987) Purified and unpurified sodium channels from eel electroplax in planar lipid bilayers. J Gen Physiol 90:375-395
Roberts RH, Barchi RL (1987) The voltage-sensitive sodium channel from rabbit skeletal muscle. Chemical characterization of subunits. J Biol Chem 262:2298-2303

Rossie S, Catterall WA (1987) Cyclic AMP-dependent phosphorylation of voltage-sensitive sodium channels in primary cultures of rat brain neurons. J Biol Chem 262:12735-12744

Rossie S, Gordon D, Catterall WA (1987) Identification of an intracellular domain of a sodium channel having multiple cyclic AMP-dependent phosphorylation sites. J Biol Chem 262:17530-17535

Sanchez JA, Stefani E (1978) Inward calcium current in twitch muscle fibers of the frog. J Physiol 283:197-209

Sharp AH, Imagawa T, Leung AT, Campbell KP (1987) Identification and characterization of the dihydropyridine-binding subunit of the skeletal muscle dihydropyridine receptor. J Biol Chem 262:12309-12315

Sieber M, Nastainczyk W, Zubor V, Wernet W, Hofmann F (1987) The 165-KDa peptide of the purified skeletal muscle dihydropyridine receptor contains the known regulatory sites of the calcium channel. Eur J Biochem 167:117-122

Striessnig J, Knaus H-G, Grabner M, Moosburger K, Seitz W, Lietz H, Glossmann H (1987) Photoaffinity labelling of the phenylalkylamine receptor of the skeletal muscle transverse-tubule calcium channel. FEBS Lett 212:247-253

Stroud RM, Finer-Moore J (1985) Acetylcholine receptor structure, function and evolution. Annu Rev Cell Biol 1:317-351

Sumikawa K, Parker I, Miledi R (1984) Partial purification and functional expression of brain mRNAs coding for neurotransmitter receptors and voltage-operated channels. Proc Natl Acad Sci USA 81:7994-7998

Takahashi M, Catterall, WA (1987) Dihydropyridine-sensitive calcium channels in cardiac and skeletal muscle membranes: Studies with antibodies against the alpha subunits. Biochemistry 26:5518-5526

Takahashi M, Seagar MJ, Jones JF, Reber BFX, Catterall, WA (1987) Subunit structure of dihydropyridine-sensitive calcium channels from skeletal muscle. Proc Natl Acad Sci USA 84:5478-5482

Tamkun MM, Talvenheimo JA, Catterall WA (1984) The sodium channel from rat brain: Reconstitution of neurotoxin-activated ion flux and scorpion toxin binding from purified components. J Biol Chem 259:1676-1688

Tanabe T, Takeshima H, Mikami A, Flockerzi V, Takahashi H, Kangawa K, Kojima M, Matsuo H, Hirose T, Numa S (1987) Primary structure of the receptor for calcium channel blockers from skeletal muscle. Nature 328:313-318

Tejedor FJ, and Catterall WA (1988) A site of covalent attachment of alpha-scorpion toxin derivatives in domain I of the sodium channel alpha subunit Proc Natl Acad Sci USA

Tejedor FJ, McHugh E, Catterall WA (1988) Stabilization of a sodium channel state with high affinity for saxitoxin by intramolecular crosslinking. Evidence for allosteric effects of saxitoxin binding. Biochemistry 27:2389-2397

Unwin PNT, Zampighy G (1980) Structure of the junction between communicating cells. Nature 283:545-549

Vassilev P, Scheuer T, Catterall, WA (1988) Identification of an intracellular segment involved in sodium channel inactivation. Science 24l:1658-1661.

GENETIC AND MOLECULAR ANALYSIS OF A GENE ENCODING SODIUM CHANNELS IN DROSOPHILA

Kate Loughney, Michael Stern, Robert Kreber and Barry Ganetzky
Laboratory of Genetics
University of Wisconsin
Madison, Wisconsin 53706
USA

Among the metazoan organisms used for studies of ion channels, *Drosophila* is unique in lending itself to detailed genetic analysis. Mutations that perturb membrane excitability can be readily isolated and provide useful material for studies of ion channel structure, function and regulation (reviewed by Ganetzky and Wu 1986). After a mutation has been identified, the relevant gene can be localized cytologically and cloned by recombinant DNA techniques. Because these mutations are isolated originally by virtue of their phenotypic effects on behavior and electrophysiology the product encoded by these genes is generally unknown until the gene has been cloned and sequenced. On the other hand, it is possible to isolate mutations affecting the structure or function of an ion channel even when the protein has not previously been purified biochemically or when its role in membrane excitability has not previously been recognized.

The cloning based on genetic methodology of the *Drosophila Sh* locus, a structural gene for A-type potassium channels, in the absence of any a priori knowledge about the biochemistry of potassium channels proteins is a case in point (reviewed by Miller 1988). As this example illustrates, by use of genetic techniques in *Drosophila* it may be possible to obtain information about ion channels that would be much more difficult to acquire in any other way.

NATO ASI Series, Vol. H 32
Molecular Biology of Neuroreceptors and Ion Channels. Edited by A. Maelicke

Even for sodium channels where the structural gene has already been cloned and sequenced from several organisms (Noda et al. 1984; 1986), many questions about the *in vivo* function, expression and regulation of this protein remain unanswered (Catterall 1986; Barchi 1988). The identification of mutations in genes encoding sodium channels or affecting their expression should prove useful in further detailed studies of sodium channels. Here we review the properties of mutations at the *para* locus in *Drosophila* and describe the results of sequence analysis that demonstrate *para* is a structural gene for sodium channels. In addition, we suggest a role of the *nap* locus in regulating the expression or function of sodium channels encoded by *para*.

ISOLATION AND CHARACTERIZATION OF *para* MUTATIONS

The $para^{ts}$ allele is a recessive X-linked mutation isolated because of its behavioral phenotype: when adult flies are shifted from 22°C to 29°C, they become paralyzed within seconds (Suzuki et al. 1971). The paralysis is reversible; within seconds after the flies are returned to 22°C they recover complete mobility. Mutant larvae are similarly affected although they require a higher temperature (37°C) to become paralyzed. This behavioral defect in $para^{ts}$ is correlated with a characteristic electrophysiological defect as well: there is a specific block in the propagation of action potentials in larval nerves at the restrictive temperature (Wu and Ganetzky 1980; Ganetzky et al. 1986). Similar results have been observed in electrophysiological studies of adults (Siddiqi and Benzer 1976; Benshalom and Dagan 1981).

A number of additional mutant *para* alleles have now been isolated that fall into several different categories (Ganetzky and Wu 1986): recessive temperature-sensitive alleles similar to $para^{ts}$ but usually requiring a higher temperature to cause paralysis; a dominant temperature-sensitive allele; null alleles, some of which are associated with chromosome rearrangements broken within the *para* locus, which appear to have lost all function of this gene and cause an unconditional recessive lethal phenotype; and temperature-sensitive alleles associated with the insertion of transposable elements. The latter class of mutant alleles has been especially important for the cloning of *para* (see below).

The defect in nerve conduction in *para*^{ts} and its synergistic interaction in combination with other mutants thought to affect sodium channels (see below) led to the proposal that *para* mutations in some way alter the structure or function of sodium channels (Wu and Ganetzky 1980). Subsequent studies supported this interpretation. For example, cultures of *para*^{ts} neurons from dissociated larval ganglia are more resistant than normal to the cytotoxic effects of veratridine (Suzuki and Wu 1984), which causes persistant activation of sodium channels. When the cells are cultured at 35°C vs. 22°C, resistance of *para*^{ts} neurons compared to normal neurons is further increased. More direct evidence for an effect on sodium channels is provided by recent experiments using the whole cell clamp technique to examine sodium currents in cultured embryonic neurons (O'Dowd and Aldrich 1987). In cultures of normal embryonic neurons, sodium currents are detected in about 65% of the cells examined. Neurons from three different temperature-sensitive *para* mutants were examined in the same way and each caused a reduction in the percentage of cultured cells that expressed sodium currents. The magnitude of this reduction varied in an allele-dependent manner ranging from 22-95%. It is not yet known whether the kinetics and voltage-dependence of sodium channels are normal in those remaining *para* neurons that do express sodium currents.

CLONING AND MOLECULAR ANALYSIS OF THE *para* LOCUS

The various results described above all supported the idea that the synthesis or function of sodium channels is altered by *para* mutations but the nature of this alteration was unclear. Therefore, to identify the protein encoded by the *para* locus and to determine its relationship with sodium channels we cloned the gene (Loughney, Kreber and Ganetzky in preparation). Our cloning strategy took advantage of the technique called transposon tagging (Kidwell 1986). Genetic crosses were carried out to mobilize the transposable element known as the P element (Engels 1983). New *para* mutations, caused by the insertion of the element close to or within the *para* locus, were recovered from these crosses. A P element probe was used to isolate genomic fragments from the mutant strain containing the inserted P element and the unique flanking DNA sequences. This flanking DNA was used in turn to isolate DNA from a wild-type genomic library. By successive chromosome walking steps, we isolated and mapped over 150 kb from the *para* locus (Loughney, Kreber and Ganetzky in preparation).

To determine what portion of the cloned region actually corresponded to the *para* gene, we localized by Southern blot analysis and in situ hybridization the lesions associated with various mutant *para* alleles on the molecular map of the region. These mutations, which included two associated with the breakpoints of chromosomal inversions and five associated with small insertions, were found to be distributed over approximately 50 kb of genomic DNA and provided a minimal estimate of the extent of the *para* locus.

DNA fragments from this 50 kb interval of genomic DNA were used to screen cDNA libraries to determine which portions of the gene were transcribed into mRNA and to recover the corresponding cDNA fragments. Several incomplete cDNAs have been isolated that hybridize back to the chromosome walk and together span those regions of the walk where the various mutant *para* alleles have been located. Sequence analysis of these cDNAs indicates that they are all derived from the same transcription unit. Because some cDNAs hybridize to portions of the cloned genomic DNA that lie outside the region defined by the mutations, the estimated size of the *para* transcription unit from the analysis of these cDNAs is at least 90 kb.

SEQUENCE ANALYSIS OF *para* cDNAS

Nucleotide sequence analysis of the various *para* cDNAs revealed that each contained an open reading frame capable of encoding an amino acid sequence closely related to different portions of the major subunit of vertebrate sodium channels.

Analysis of cDNAs encoding sodium channels in *Electrophorus* electroplax and rat brain has previously revealed that the protein comprises four repeated homology units each of which contains at least six presumptive membrane spanning regions of characteristic amino acid sequence (Noda et al. 1984; 1986). The deduced amino acid sequence of one *para* cDNA containing the initiating AUG codon can be readily aligned with the amino terminal half of the rat sequence, including all of the first two homology units. A second cDNA from the other end of the *para* locus can be translated into an amino acid sequence that aligns with the carboxy terminal third of the rat sequence, including about half of the third homology unit through the fourth homology unit.

The *para* gene product is thus very similar to the vertebrate sodium channel in terms of overall structure and amino acid sequence. As an example of the sequence comparisons, within the first homology unit the amino acid identity is 59% between the *para* and rat sequences. In the same region, the identity is 45% between rat and *DSC1*, a different *Drosophila* sodium channel gene cloned by its homology to the vertebrate sodium channel (Salkoff et al 1987). The identity between *para* and *DSC1* in this region is 47% . The amino acid sequence identity is highest in the presumptive membrane spanning regions and in the regions immediately flanking them. The cytoplasmic regions linking the first two homology domains shows very little conservation, in agreement with previous reports (Noda et al. 1986). It is of interest that the two *Drosophila* genes are no more closely related to one another than either is to the rat sequence. Apparently, the two *Drosophila* genes represent a very ancient gene duplication event. The complete nucleotide and amino acid sequence of the *para* cDNA clones will be published elsewhere (Loughney, Kreber and Ganetzky in preparation).

The results of the sequence analysis of *para* cDNAs together with the electrophysiological and pharmacological studies of *para* mutants leaves little doubt that *para* is a sodium channel structural gene. Although *DSC1* and other DNA sequences have been cloned from *Drosophila* by homology to the vertebrate sodium channel (Salkoff et al. 1987; Okamoto et al. 1987), so far *para* is the only one of these for which mutations exist. Therefore, the *para* locus provides a unique opportunity for genetic analysis of sodium channel function and regulation in the *Drosophila* nervous system.

INTERPRETATION OF *para* MUTANT PHENOTYPES

The location of the molecular lesions associated with various *para* alleles provides a basis for interpreting some of the phenotypes associated with these mutants. For example, a mutation associated with a physical break, such as an inversion endpoint, within the transcribed portion of the gene would not be expected to produce any functional *para* product. These mutations are lethal when homozygous (Ganetzky 1984) indicating that the $para^{+}$ function is essential for viability and that its role cannot be supplanted by other presumptive sodium channel genes in the *Drosophila* genome.

Surprisingly, however, we have isolated many *para* mutants in which the transcribed region is interrupted by the insertion of a transposable element up to several kb in length and these mutants are not lethal but show the phenotype of temperature-sensitive paralysis. These results suggest that the DNA insertions in these mutants map within introns rather than exons of *para* and do not alter the amino acid sequence of the *para* gene product. For some *para* mutants this interpretation has now been confirmed by the analysis of cDNAs. Why then do they cause a mutant phenotype? In other *Drosophila* genes that have been studied, P element insertions are known to interefere with the transcription of the gene in which they are inserted or with the correct processing of the primary transcript (cf. Tsubota et al. 1979). Therefore, we propose that the temperature-sensitive phenotype of the *para* insertional mutations is the result of an underproduction of sodium channel polypeptides rather than the production of functionally aberrant channels. In this view, neuronal membranes in these *para* mutants are thought to possess a sub-normal number of sodium channels. Experimental evidence in other organisms indicates that the fraction of available sodium channels required to maintain active conduction increases as temperature is raised (Colquhoun and Ritchie 1972). In temperature-sensitive *para* mutants the number of sodium channels is presumably sufficient to conduct enough sodium current to propagate action potentials at low temperature but not to meet the demands of elevated temperatures. A similar interpretation may apply to those temperature-sensitive *para* alleles not evidently associated with insertions of transposable elements.

When $para^{ts}$ or certain other temperature-sensitive *para* mutations are made heterozygous with a *para* null mutation, a lethal phenotype results (Ganetzky 1984). Together with the previous proposal for temperature-sensitivity this result suggests that a critical level of $para^{+}$ activity is essential for viability, presumably because if sodium channel levels drop too low conduction fails even at low temperatures.

An interesting consequence of this interpretation of *para* mutant phenotypes is that it implies that in *Drosophila* neurons, in contrast to those in rat (see Schmidt and Catterall 1986), there may be no regulatory mechanism to monitor the number of functional sodium channels that are incorporated into the cell membrane. Instead, the number of functional sodium channels in mature neurons may simply reflect the levels of transcription and translation of genes such as *para*.

POSSIBLE REGULATION OF *para* BY *nap*

The above hypothesis also provides a basis for interpreting the phenotype of nap^{ts}, another mutation suggested to affect sodium channels (Wu et al. 1978). The phenotype of nap^{ts} is similar to that of *para* mutants in many respects. It was isolated as a recessive mutation on the second chromosome that causes the rapid paralysis of adults and larvae when shifted from 22°C to 37°C. Paralysis is reversible and mutants regain mobility within seconds after being returned to 22°C. The similarity between $para^{ts}$ and nap^{ts} in behavioral phenotype is paralleled by a similar electrophysiological defect. At restrictive temperatures, nap^{ts} causes a specific block in the propagation of action potentials in larval nerves.

Even at 22°C nerve membrane properties in *nap*^{ts} are altered: the nerve refractory period is abnormally long, sensitivity of conduction to blockage by tetrodotoxin is greater than normal and sensitivity of cultured neurons to veratridine-induced cytotoxicity is less than normal (Wu and Ganetzky 1980; Suzuki and Wu 1984). Binding studies have indicated a reduced number of saxitoxin binding sites in membrane preparations from heads of *nap*^{ts} flies (Kauvar 1982; Jackson et al. 1984). Finally, in double mutant combinations at 22°C, *nap*^{ts} has been shown to suppress the behavioral and some of the physiological defects associated with mutations such as *Sh* and *eag*, which increase membrane excitability by reducing or eliminating the function of particular classes of potassium channels (Jan et al. 1977; Salkoff and Wyman 1981; Tanouye et al. 1981; Wu and Haugland 1985; Wu et al. 1983; Ganetzky and Wu 1983; Baumann et al. 1987; Kamb et al. 1987; Tempel et al. 1987). The apparent basis of suppression is the reduction in membrane excitability caused by *nap*^{ts} even at nominally permissive temperatures, which counterbalances the enhanced membrane excitability otherwise caused by *Sh* (Ganetzky and Wu 1982a,b). This collection of results all supported the idea that *nap*^{ts} affected sodium channels in some way.

The similarity of behavioral and electrophysiological phenotypes between *nap*^{ts} and *para*^{ts} reflects an important functional overlap between these genes. This overlap is revealed by the synergistic interaction between *nap*^{ts} and *para*^{ts} resulting in the unconditional lethality of the double mutant (Ganetzky 1984). The basis of the lethal phenotype may be the failure of doubly mutant neurons to propagate action potentials at all temperatures (Burg and Wu 1984).

What is the relationship between the *nap* and *para* genes? Results of genetic experiments described below suggest that the *nap* gene product somehow regulates the expression or function of the *para* gene product such that the number or activity of sodium channels encoded by *para*$^{+}$ is reduced in a homozygous *nap*^{ts} background. According to this interpretation, the temperature-sensitive phenotype of *nap*^{ts}, like that of some *para* mutants, results from the underproduction of functional sodium channels so that active conduction is impaired at high temperature (cf. Jackson et al. 1984). The only difference would be that in one case the mutation resides within the sodium channel structural gene itself and in the other case it resides within a gene necessary for the appropriate expression or function of the channel. Similarly, the lethality of the *para*^{ts}; *nap*^{ts} double mutant is explainable if *nap*^{ts} further decreases the level of *para* expression beyond the reduction already caused by the *para*^{ts} mutation. In this case, the number of functional sodium channels may be less than the critical level required for conduction at low temperature and therefore for viability. Thus, the *para*^{ts}; *nap*^{ts} genotype may be the functional equivalent of making *para*^{ts} heterozygous with a *para* null mutation in a *nap*$^{+}$ background which, as described earlier, also results in lethality.

This interpretation of the *nap* phenotype makes an easily testable genetic prediction: if the defects of *nap*^{ts} are caused by an underproduction of functional sodium channels encoded by *para*$^{+}$, then it should be possible to restore *nap*^{ts} to a normal phenotype by somehow increasing the level of *para*$^{+}$ expression. This prediction was tested by constructing homozygous *nap*^{ts} flies carrying three, rather than the usual two, copies of *para*$^{+}$ by use of a small chromosomal duplication that contained *para*$^{+}$. In agreement with the prediction, these flies are not paralyzed at elevated temperatures (Stern, Kreber and Ganetzky in preparation). With the availability of molecular probes for *para*, it will now be possible to carry out more direct tests of these ideas by assessing the expression of *para*$^{+}$ and various mutant *para* alleles in wild-type and *nap*^{ts} backgrounds.

EFFECT OF *para* GENE DOSAGE ON MEMBRANE EXCITABILITY

The rescue of *nap*^{ts} by the presence of an extra dose of *para*$^{+}$ is also consistent with the idea that the number of sodium channels in axonal membranes of *Drosophila* neurons, and therefore the electrical properties of these cells, can be altered just by varying the gene dosage of the *para* locus. In addition to the use of duplications to provide an extra copy of *para*$^+$ we can use small chromosomal deficiencies that delete one copy of the gene to test the relationship between membrane excitability and dosage of the *para* locus. Further experiments described below employing such deficiencies and duplications indicate that a decrease in *para*$^+$ gene dosage leads to underproduction of sodium channels and a decrease in membrane excitability whereas an increase in *para*$^+$ dosage leads to increased production of sodium channels and an enhancement of membrane excitability (Stern, Kreber and Ganetzky in preparation). These opposing effects of deficiencies and duplications of *para*$^+$ are most readily revealed by their phenotypic interactions with other mutations that alter membrane excitability.

For example, whereas a duplication of *para*$^+$ rescues *nap*^{ts}, a deficiency of *para*$^+$ is lethal in a homozygous *nap*^{ts} background. In this case, we believe that by reducing the copy number of *para*$^+$ from two to one, the expression of *para*$^+$ is diminished to the point where it is insufficient to provide the level of sodium channels required for survival. In effect, this situation is analogous to that of the *para*^{ts}; *nap*^{ts} double mutant already described.

As another example, the leg-shaking behavior of *Sh,* which is associated with increased membrane excitability owing to a defect in potassium channels, is suppressed in flies heterozygous for a deficiency of *para*$^+$ (Stern, Kreber and Ganetzky in preparation). Again, the likely basis of this suppression is that reduction in *para*$^+$ gene dosage leads to an underproduction of sodium channels and a consequent decrease in membrane excitability, which counterbalances the effect of *Sh*. Conversely, a duplication of *para*$^+$ enhances *Sh* so that the flies resemble *eag Sh* double mutants in appearance (Stern, Kreber and Ganetzky in preparation). In *eag Sh* double mutants the function of two different classes of potassium channels is disrupted causing extreme neuronal hyperexcitability (Ganetzky and Wu 1983). This hyperexcitability manifests itself in *eag Sh* adults as flies with downheld wings and an indentation of the cuticle of the dorsal thorax, which most likely result from hypercontraction of flight muscles. The same wing and thoracic phenotype is seen in *Sh* flies that carry an extra copy of *para*$^+$. We believe that in these flies it is the increased membrane excitability resulting from the production of excess sodium channels that further enhances the defect caused by *Sh*. Once again, the cloning of the *para* locus will now make direct molecular tests of predictions based on these ideas feasible.

RELATIONSHIP OF *para* TO OTHER SODIUM CHANNEL GENES IN *DROSOPHILA*

In *Drosophila*, at least one other gene (*DSC1*) besides *para* has been cloned that appears to encode sodium channels (Salkoff et al. 1987; Okamoto et al. 1987). What functional overlap, if any, exists among these different sodium channel genes? Although no mutations in *DSC1* have yet been isolated, it has been localized cytologically to region 60E of the polytene chromosomes, a region for which deficiencies and duplications have previously been generated for other reasons. It is thus possible to manipulate the gene dosage of *DSC1* to determine whether phenotypic effects analogous to those described above for *para*$^+$ can be produced.

We asked first whether the mutant phenotype of temperature-sensitive *para* alleles could be rescued by the addition of extra doses of *DSC1*. No such rescue was observed (Stern, Kreber and Ganetzky in preparation). We also asked whether the phenotypic expression of *nap*^{ts} or *Sh* were affected by altering dosage of *DSC1* as they were by changes in *para* gene dosage. An extra dose of *DSC1* also did not rescue the *nap*^{ts} phenotype nor enhance the *Sh* phenotype. Similarly, deficiencies of *DSC1* were without noticeable phenotypic effects: a single dose of *DSC1* was not lethal in a *nap*^{ts} background and did not suppress *Sh*.

These results are in accord with the sequence comparison of *para* and *DSC1*, which indicated that although the two genes are related they are clearly very distinct from one another. The genetic studies support the conclusion that the sodium channels encoded by *DSC1* are functionally distinct from those encoded by *para* and/or that the time and place of expression of the two genes have little, if any, overlap. It will be of interest to elucidate the functional roles of the different types of sodium channels in *Drosophila* and the mechanisms that differentially regulate the genes encoding them.

CONCLUSIONS

Using a combination of genetic, electrophysiological and molecular techniques, we have identified a sodium channel structural gene in *Drosophila*. So far *para* is the only sodium channel gene in any organism that has been mutated in situ. Thus, the *para* locus provides a unique opportunity for detailed genetic and molecular analyses of sodium channel function, regulation and expression in an intact organism.

ACKNOWLEDGEMENTS

Supported by fellowships from the American Heart Association of Wisconsin to K.L., the Damon Runyon Walter Winchell Cancer Research Fund and the National Institutes of Health to M.S. and by grants NS15390 and GM35099 from the National Institutes of Health and a Klingenstein fellowship to B.G. Paper no. 3011 from the Laboratory of Genetics, University of Wisconsin, Madison.

REFERENCES

Baumann A, Krah-Jentgens I, Muller R, Muller-Holtkamp F, Seidel R, Kecskemethy N, Casal J, Ferrus A, Pongs O (1987) Molecular organization of the maternal effect region of the *Shaker* complex of *Drosophila*: characterization of an I_A channel transcript with homology to vertebrate Na^+ channel. EMBO J 6:3419-3429

Barchi RL (1988) Probing the molecular structure of the voltage-dependent sodium channel. Ann Rev Neurosci 11:455-495

Benshalom G, Dagan D (1981) Electrophysiological analysis of the temperature-sensitive paralytic *Drosophila* mutant, $para^{ts}$. J Comp Physiol 144:409-417

Burg MG, Wu CF (1986) Differentiation and central projections of peripheral sensory cells with action-potential block in *Drosophila* mosaics. J Neurosci 6:2968-2976

Catterall WA (1986) Molecular properties of voltage-sensitive sodium channels. Ann Rev Biochem 55:953-985

Engels WR (1983) The P family of transposable elements in *Drosophila*. Ann Rev Genet 17:315-344

Ganetzky B (1984) Genetic studies of membrane excitability in *Drosophila*: Lethal interaction between two temperature-sensitive paralytic mutations. Genetics 108:897-911

Ganetzky B, Wu CF (1982a) *Drosophila* mutants with opposing effects on nerve excitability: Genetic and spatial interactions in repetitive firing. J Neurophysiol 47:501-514

Ganetzky B, Wu CF (1982b) Indirect suppression involving behavioral mutants with altered nerve excitability in *Drosophila melanogaster*. Genetics 100:597-614

Ganetzky B, Loughney K, Wu CF (1986) Analysis of mutations affecting sodium channels in *Drosophila*. In: Kao CY, Levinson SR (eds) Tetrodotoxin, Saxitoxin and the Molecular Biology of the Sodium Channel, New York Academy of Sciences, New York, p325

Ganetzky B, Wu CF (1983) Neurogenetic analysis of potassium currents in *Drosophila*: Synergistic effects on neuromuscular transmission in double mutants. J Neurogenet 1:17-28

Ganetzky B, Wu CF (1986) Neurogenetics of membrane excitability in *Drosophila*. Ann Rev Genet 20:13-44

Jackson FR, Wilson SD, Strichartz GR, Hall LM (1984) Two types of mutants affecting voltage-sensitive sodium channels in *Drosophila melanogaster*. Nature 308:189-191

Jan YN, Jan LY, Dennis MJ (1977) Two mutations of synaptic transmission in *Drosophila*. Proc R Soc Lond B 198:87-108

Kamb A, Iverson LE, Tanouye MA (1987) Molecular characterization of *Shaker*, a *Drosophila* gene that encodes a potassium channel. Cell 50:405-413

Kauvar LM (1982) Reduced [^{3}H]-Tetrodotoxin binding in the *nap*^{ts} paralytic mutant of *Drosophila*. Mol Gen Genet 187:172-173

Kidwell M (1986) P-M mutagenesis. In: Roberts, D. B. (ed) *Drosophila* a practical approach. IRL Press, Oxford Washington DC, p 59

Miller C (1988) Shaker shakes out potassium channels. Trends Neurosci 11:185-186

Noda M, Ikeda T, Kayano T, Suzuki H, Takeshima H, Kurasaki, M, Takahashi H, Numa, S (1986) Existence of distinct sodium channel messenger RNAs in rat brain. Nature 320:188-192

Noda M, Shimizu S, Tanabe T, Takai T, Kayano T, Ikeda T, Takahashi H, Nakayama H, Kanaoka Y, Minamino N, Kangawa K, Matsuo, Raftery MA, Hirose T, Inayama S, Hayashida H, Miyata, T, Numa S (1984) Primary structure of *Electrophorus electricus* sodium channel deduced from cDNA sequence. Nature 312:121-127

O'Dowd DK, Aldrich RW (1986) Characterization of sodium channels in cultured *Drosophila* neurons from wild type and temperature sensitive paralytic mutant embryos. Abstr Soc Neurosci 12:43

Okamoto H, Sakai K, Goto S, Takasu-Ishikawa E, Hotta Y (1987) Isolation of *Drosophila* genomic clones homologous to the eel sodium channel gene. Proc Japan Acad Ser B 63:284-288

Salkoff L, Butler A, Wei A, Scavarda N, Giffen K, Ifune C, Goodman R, Mandel G (1987) Genomic organization and deduced amino acid sequence of a putative sodium channel gene in *Drosophila*. Science 237:744-749

Salkoff L, Wyman R (1981) Genetic modification of potassium channels in *Drosophila Shaker* mutants. Nature 293:228-230

Schmidt JW, Catterall WA (1986) Biosynthesis and processing of the α subunit of the voltage-sensitive sodium channel in rat brain neurons. Cell 46:437-445

Siddiqi O, Benzer S (1976) Neurophysiological defects in temperature-sensitive mutants of *Drosophila melanogaster*. Proc Natl Acad Sci USA 73:3253-3257

Suzuki DT, Grigliatti T, Williamson R (1971) Temperature-sensitive mutants in *Drosophila melanogaster*, VII. A mutation (*para*^{ts}) causing reversible adult paralysis. Proc Natl Acad Sci USA 68:890-893

Suzuki N, Wu CF (1984) Altered sensitivity to sodium channel-specific neurotoxins in cultured neurons from temperature-sensitive paralytic mutants of *Drosophila*. J Neurogenet 1:225-238

Tsubota S, Ashburner M, Schedl P (1985) P-element induced control mutations at the *r* gene of *Drosophila melanogaster*: transposon induced control mutations. Mol Cell Biol 5:2567-2574

Tanouye MA, Ferrus A, Fujita, S (1981) Abnormal action potentials associated with the *Shaker* complex locus of *Drosophila*. Proc Natl Acad Sci USA 78:6548-6552

Tempel BL, Papazian DM, Schwarz TL, Jan YN, Jan LY (1987) Sequence of a probable potassium channel component encoded at *Shaker* locus of *Drosophila*. Science 237:770-775

Wu CF, Ganetzky B (1980) Genetic alteration of nerve membrane excitability in temperature-sensitive paralytic mutants of *Drosophila melanogaster*. Nature 286:814-816

Wu CF, Ganetzky B, Haugland FN, Liu AX (1983) Potassium currents in *Drosophila*: Different components affected by mutations in two genes. Science 220:076-1078

Wu CF, Ganetzky B, Jan YN, Jan LY, Benzer S (1978) A *Drosophila* mutant with a temperature-sensitive block in nerve conduction. Proc Natl Acad Sci USA 75:4047-4051

Wu CF, Haugland FN (1985) Voltage clamp analysis of membrane currents in larval muscle fibers of *Drosophila*: Alteration of potassium currents in *Shaker* mutants. J Neurosci 5:2626-2640

A FAMILY OF POTASSIUM CHANNELS FROM THE SHAKER LOCUS OF DROSOPHILA.

T. L. Schwarz[1], R.Carretto, D. Papazian[2], B. Tempel[3], L. Timpe, Y.-N. Jan, and L.Y. Jan
Howard Hughes Medical Institute and Department of Physiology
University of California, San Francisco
San Francisco, California 94143
USA

1. Present address: Dept. of Molecular and Cellular Physiology, Beckman Center, Stanford University, Stanford CA 94305 USA
2. Present address: Dept. of Physiology, UCLA School of Medicine, Los Angeles CA 90024-1751 USA
3. Present address: Dept. of Pharmacology, University of Washington School of Medicine, Seattle WA 98195 USA

Potassium channels are found in nerve, muscle, epithelial, endocrine, exocrine and immune cells. Despite their importance in the control of excitation and ion transport, biochemical understanding of the channels has grown slowly because there are no high abundance sources of these channels and because high affinity ligands for them have only recently been found (Carbone et al., 1982; Miller et al, 19854; Seagar et al., 1986; Sternsfeld et al., 1987). An alternative route to studying these molecules has been provided by the fruit fly, Drosophila melanogaster. Many neurological mutations of this organism have been found that define loci that may encode important neuronal proteins (Ganetzky and Wu, 1986). One of these, the Shaker locus, appeared likely from electrophysiological and genetic studies (Jan et al., 1977; Salkoff and Wyman, 1983; Timpe and Jan, 1987) to contain

NATO ASI Series, Vol. H 32
Molecular Biology of Neuroreceptors and Ion Channels. Edited by A. Maelicke

the structural gene for a type of potassium channel, called the A channel, that opens in response to depolarization and inactivates thereafter. The gene was mapped to band 16F on the X-chromosome (Tanouye et al., 1981). An entry point to cloning the region was provided by a cDNA, cloned for independent reasons (Wollfner, 1980); the cDNA is not related to Shaker but hybridizes to the same band on the X-chromosome.

The nearby cDNA was used to screen genomic libraries for clones from the 16F region. The ends of these clones were used to rescreen the library for overlapping clones. By repeating this process our lab and others (Papazian et al., 1987; Kamb, et al., 1987; Baumann et al., 1987) "walked" in either direction from the starting point and obtained a set of overlapping genomic clones from the region. By comparing DNA from wildtype and mutant Shaker flies we located the positions of 4 breakpoint mutations and an insertional mutation (Papazian et al., 1987). These mutations were spread over approximately 60kb of the genome and this region was then used to screen cDNA libraries. Several clones were found that represent a complex transcriptional unit with exons from the region in which the mutations were found (Papazian et al., 1987). Since the sequences of numerous clones have been reported by ourselves (Tempel et al., 1987; Schwarz et al., 1988) and others (Kamb et al, 1987; Kamb et al., 1988; Baumann et al, 1987; Pongs et al., 1988), we will not repeat them here but will concentrate instead on two features: describing the homology of the Shaker products to other voltage-gated channels and determining the legitimacy of the variant sequences reported by this lab and others. Thereafter we will summarize recent results using antibodies to the Shaker products and electrophysiological studies on the variants expressed in oocytes.

The kinship of voltage-gated channels.

Upon sequencing the first Shaker clones and comparing them with the Dayhoff protein sequence database (Tempel et al., 1987), a significant similarity of a portion of the Shaker sequence to the voltage-dependant Na^+ channel was noted (Noda et al., 1984). The homology was centered on the region of the Na^+ channel called "S4", and the essential feature of this region was preserved - positively charged residues at every third position with predominantly hydrophobic residues between. Modellers of the sodium channel had already suggested that this region might be involved in the voltage-dependent gating of the channel (Catterall, 1986; Guy and Seetheramulu, 1986) and an S4 region had also been found in a putative voltage-gated calcium channel, the dihydropyridine receptor (Tanabe et al., 1987). For us, its significance was three-fold. First, it indicated that the genetic predictions were correct and these cDNA were very likely to be structural components of a K^+ channel. Second, that this region should be found to be conserved among three different types of voltage-gated channels provided further support for the hypothesis that it served an essential function. Third, it indicated that all of these voltage-gated channels might be derived from a common ancestor that first evolved the S4 structure and acquired the ability to change conformation in response to a depolarization of the membrane (Tempel et al., 1987).

How similar are the voltage-gated channels channels? The largest predicted Shaker product is 656 amino acids, whereas the eel Na^+ channel and the dihydropyridine receptor have 1,820 and 1,873 residues. In addition, the Shaker products contain only a single S4-like region. The Na^+ and putative Ca^{++} channels contain 4 such regions, each within a larger hydrophobic domain. Thus, while the larger channels appear to consist of 4 homologous "pseudo-subunits" linked to one another by cytoplasmic loops (Noda et al., 1984; Tanabe et al., 1987), the Shaker products appear to be individual subunits. Perhaps, by analogy to the other channels, 4 Shaker products come together to make a single channel (Tempel et al., 1987). A multimeric channel would be consistent with earlier genetic evidence that the products of the locus interact with one another (Salkoff and Wyman, 1983; Timpe and Jan, 1987).

How similar are the sequences of the channels? Away from the S4 region, the similarity falls of rapidly. If the S4-like structure is removed from the Shaker sequence, no significant similarities were found by a computer search (Tempel et al., 1987). Although both the Shaker products and the Na+ channel contain many hydrophobic stretches, the actual sequences of these regions are highly divergent. Some of the similarities that have been reported, for example that of the third hydrophobic region of Shaker to the sixth hydrophobic of the Electrophorous Na^+ channel (Baumann et al., 1987) are likely to be fortuitous. Potentially more significant are 7 conserved residues within the hydrophobic stretches which were identified by Tanabe et al. (1987) when they compared the "pseudo-subunits" of the Na^+ channel genes to one another and to those of the dihydropyridine receptor. Five of these residues also occur in analogous positions in the Shaker sequence, while a sixth has undergone a conservative lysine to arginine change (Schwarz et al., 1988). The conservation of these residues suggests that they may play important functional or structural roles. Their appearance in the Shaker products may indicate that the overall design of K^+ channels and Na^+ and Ca^{++} channels are similar despite the great divergence of their sequences. The divergence is not surprising: since protozoans contain distinct voltage-gated Ca^{++} and K^+ channels, these two classes may have been diverging for as long as a billion years (Hille, 1984).

Multiple products arise from alternative splicing.

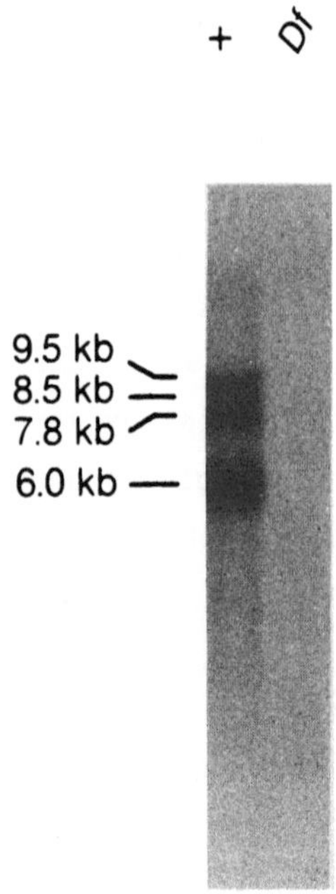

Figure 1. Transcripts from the Shaker locus. Polyadenylated RNA was extracted from the heads of wildtype (OrR) flies (+) and flies with the genotype $T(1;Y)B55^D/T(1;Y)W32^P$ which are deficient for the Shaker region (Df). The blotted RNA was probed with a portion of a ShB cDNA that contains much of the core that is present in all the splicing variants (see Schwarz et al., 1988). The absence of the transcripts in the mutant indicates that these are authentic Shaker transcripts. As a control, the blot was later reprobed to visualize the 7.2 kb transcript of the gene sevenless and was found to contain undegraded large transcripts in equal amounts in both the wildtype and mutant lanes (data not shown).

Northern blots (Schwarz et al., 1988; Fig.1) and analysis of cDNA (Schwarz et al., 1988; Pongs et al., 1988; Kamb et al., 1988) indicate that multiple transcripts arise from the Shaker locus by alternative splicing. Some of these give rise to different predicted proteins while others differ only in their untranslated portions (Schwarz et al., 1988). Because of the large size of the transcripts (6 to 9.5 kb) and the complexity of the transcription unit, it is hard to say just how many different transcripts arise: a Northern blot such as that shown in figure 1 reveals 4 bands, a broad band of 6.0kb and three of 7.8 to 9.5 kb, but the number of transcripts is probably larger, although the individual bands can not be resolved.

By sequencing 7 cDNAs that represent correctly spliced transcripts and contain complete open reading frames, we derived the sequences of 4 predicted proteins that we called ShA, ShB, ShC, and ShD. They share a common core but differ at either the amino or carboxy terminus. They all contain a single S4 region and six hydrophobic regions that we have called H1 - H6. The variant amino terminals seem unrelated to one another, but the variant carboxy ends have large regions of identical amino acid sequences (Schwarz et al., 1988). From our lab and others (Kamb et al., 1987; Baumann et al., 1987; Pongs et al., 1988; Kamb et al., 1988) the sequences of 11 predicted products have now been reported. Each group has adopted its own nomenclature, and sometimes refer to a single clone by several names. To simplify this, Table 1 correlates the different names for predicted products.

Do all of these clones represent real splicing variants that give rise to protein products? We think this is unlikely. Polyadenylated RNA from Drosophila can contain many transcripts that have not been fully processed, i.e. transcripts may contain introns that have not been excised (see the articles by Gundelfinger and Ganetzky in this volume for examples). Naturally, these incompletely processed transcripts are also represented in cDNA libraries. The cDNA that have been reported from the Shaker locus fall into two classes: those that contain the full complement of an S4 region and 6 hydrophobic stretches, and those that do not (see Table 1). The clones that have the full complement of hydrophobics all appear to be correctly spliced - where they differ from one another, they do so by splicing away to distinct exons and the sequences flanking the

Table 1. Nomenclature of Predicted Shaker Proteins.

	JAN LAB (1)	TANOUYE LAB (2)	PONGS LAB (3)
FULL LENGTH CLONES	ShA		Sh-alpha (ShA)
	ShB	H4 (II.iv)	
	ShC		
	ShD		Sh-beta (ShB)
		H37 (III.i)	
ENDS IN INTRON AFTER H1		P1 (F-c) (I.u)	
ENDS IN INTRON AFTER H3			Sh-delta (ShD) (gt141)
		E1 (I.iii)	
		H2 (III.iii)	
			Sh-epsilon (ShE)
ENDS IN INTRON AFTER H5		H29 (I.ii)	

Table 1. For each sequence that has been reported, the name(s) used by the laboratory that reported it are shown. Clones that predict the same protein have been placed on the same line. (Where only 2 or 3 residues differ, the products are treated as the same because the differences are likely to be due to strain variations.) The classification into full length clones and those that terminate in introns is based on genomic sequencing, exon mapping, and a comparison of published cDNA sequences with genomic sequences (see text and Schwarz et al., 1988). The references that employ these terminologies are: (1) Papazian et al., 1987; Tempel et al., 1987; Schwarz et al., 1988; Timpe et al., 1988a and 1988b; Tempel et al., 1988. (2) Kamb et al., 1987; Iverson et al., 1988; Kamb et al., 1988. (3) Baumann et al., 1987; Pongs et al., 1988; Baumann et al., 1988.

splice sites conform to the consensus sequence for eukaryotic splice sites (Schwarz et al., 1988). In contrast, the other clones diverge not by splicing away but by proceeding through a splice site into the adjacent intron. Clone P1 (Kamb et al., 1987) terminates in the intron that follows H1. Clones for Sh-delta and Sh-epsilon (Baumann et al., 1987; Pongs et al., 1988), and for I.iii and III.iii (Kamb et al., 1988) terminate in an intron following H3. The clones for I.ii (Kamb et al., 1988) contain the intron that follows H5. Upon entering the intron, the open reading frame may close after a single codon (e.g. Sh-epsilon) or may continue for approximately 180 bases (I.ii). We have not been able to detect on Western blots (see below) any of the truncated products that these cDNA predict. Furthermore, whereas each of the full-length clones can direct expression of functional channels when expressed in Xenopus oocytes (see below), 2 of the truncated products have been tried but failed to produce channels (Iverson et al., 1988).

Thus it seems likely that only the clones that contain the full complement of an S4 and six hydrophobic regions represent legitimate splicing variants. This reduces the bewildering array of Shaker products to only 5 sequences that have been reported so far: ShA, ShB, ShC, and ShD (Tempel et al., 1987; Schwarz et al., 1988) and a clone called H37 or III.i (Kamb et al., 1988). Together, these clones have 4 different amino terminals and two carboxy terminals. It would not be surprising if additional legitimate products are found. These may be more combinations of the already known sequences or additional variants (the amino terminal of Sh-delta, for example, has not yet been reported in a full-length clone).

Antibodies to predicted proteins.

Because many of the interesting questions now facing us will require protein chemistry to answer, we have raised antibodies to the predicted K^+ channel components. Some of these will recognize all the known variants, others are specific for a single type of amino or carboxy region. They were made to either a fusion protein (F.P.) or to synthetic peptides (p3-S, pN/AB, pC/AC). The location of these antigens within the ShA sequence is outlined in figure 2.

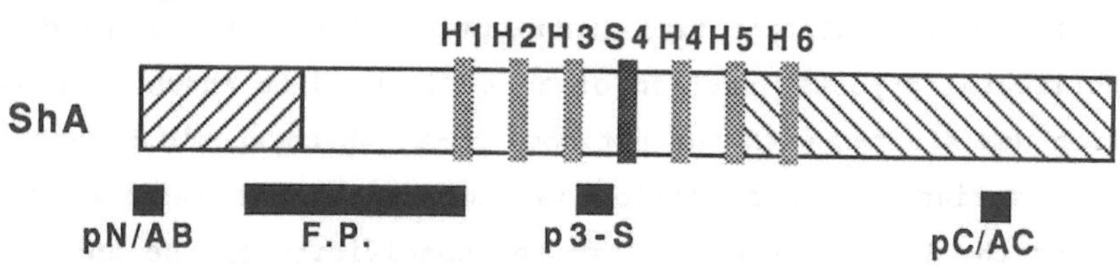

Figure 2. A cartoon of ShA beneath which is marked the positions of the peptides and fusion protein from that have been used as immunogens. The positions of the S4-like region and the six hydrophobic stretches, H1 - H6, are indicated above. The shaded portions are those that are contributed by regions that differ in the splicing variants.

We have data of two sorts: Western blots and immunocytochemistry on fly brains. From Western blots (Schwarz et al., manuscript in preparation) we have been able to determine the sizes of the protein products of the Shaker locus. As described above, this has been important in sorting out the legitimate and artifactual cDNA. We have compared protein from wildtype flies and a mutant that is deficient for most of the Shaker coding region (genotype: $T(1;Y)B55^D$, $T(1;Y)W32^P$) in order to discriminate background staining from the products of the Shaker gene. A set of bands of M_r 65 - 85 kD is observed in wildtype heads that is missing in the control. The full-length cDNAs, i.e. those that include an S4 region and H1-6, are in good agreement with this measurement since they predict proteins of M_r 64 - 74 kD. The bands that indicate a larger mass may be due to glycosylation of the products or to larger splicing variants that have not yet been sequenced. No specific staining was observed lower on the gel that might correspond to the truncated products that some have predicted (Baumann et al., 1987; Pongs et al., 1988; Kamb et al., 1988). We cannot exclude the possibility that these proteins exist at levels that are a few percent of those of the full-length products.

We have undertaken cytochemistry with these antibodies for two reasons. The first is to learn something of the basic biology of the channels whose gene we have cloned - are they evenly distributed in all cells, are they found in axons, in synaptic regions? The second was to clarify the significance of the multiple splicing variants. These variants might represent different subunits of a multimeric channel like the 4 different gene products of the muscle ACh receptor. If so, we would expect all the variants to co-localize. Alternatively, the splicing variants might give rise to different channel subtypes - a hypothesis strongly suggested by the electrophysiological evidence described below. In this case we would expect a differential distribution of the variant products.

A horizontal cryostat section of an adult head is shown in Figure 3. This section was stained with the antibody p3-S, which should detect all of the splicing variants. This section is compared with a section of a fly deficient for the Shaker locus so that the specificity of the antibody for the products of this gene can be seen. The Shaker products are not uniformly distributed in the tissue: while much of the neuropil is stained, there are regions of much greater labelling than others. These include a set of stripes that appear to be a tract connecting the lobula with the lateral protocerebrum and a second set of stripes that represent

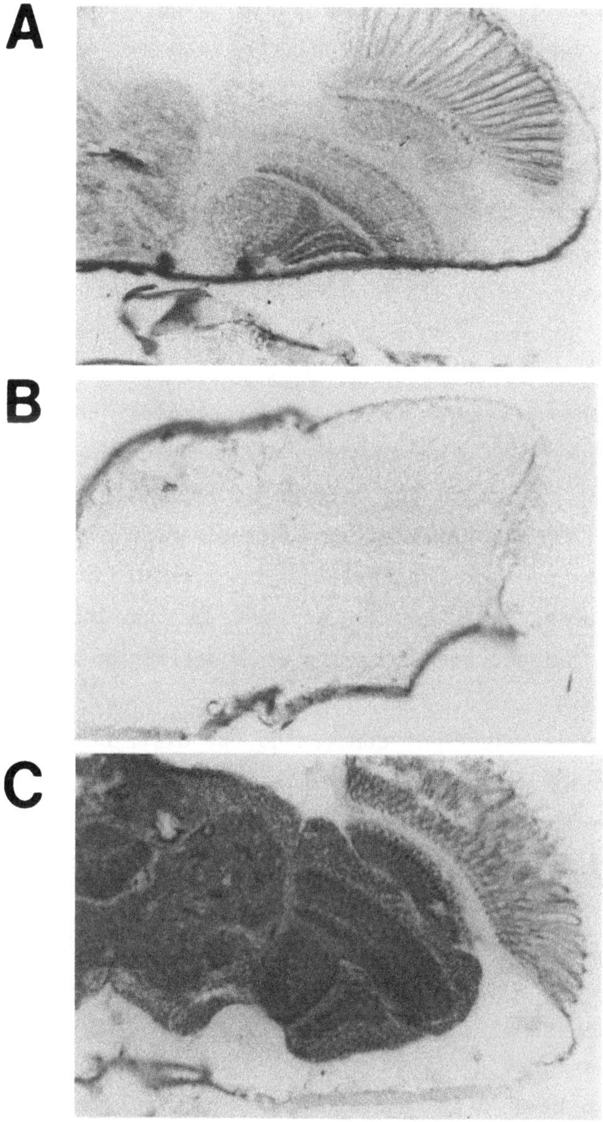

Figure 3. Immunocytochemistry of adult fly heads. A. A horizontal section containing the eye, optic ganglia, and a portion of the protocerebrum of a wild type fly stained with the antiserum p3-S that sees all the known variants. B. A section of a fly deficient for the Shaker locus, at the same plane of section, stained with the same antiserum. C. A similar section of a wild type fly stained with an antibody to horseradish peroxidase to label neuronal membranes (Jan and Jan, 1982).

strata in the neuropil of the lobular plate. That these are not simply the sites with the greatest membrane density can be seen in Figure 3 by comparing Shaker staining with staining by an antibody (anti-HRP) that appears to label all neuronal membranes uniformly (Jan and Jan, 1982). This antibody gives a relatively even stain; there is no sign of the prominent features of the Shaker pattern described above. In the medulla, for example, there is a uniform staining of the neuropil whereas the Shaker antigen is concentrated in the portion of the neuropil that borders on the secondary optic chiasm. From these observations we conclude that the expression and localization of the Shaker products is tightly regulated. Some of the characteristic membrane properties that distinguish the excitability of one cell from another, and perhaps one part of a cell from another part, may be due to the pattern of expression of this class of K^+ channels. Some of the labelled structures, such as the antennal nerve, strongly suggest that the channel may be present in axons, while other labelled structures, such as the stripes in the lobular plate, may be synaptic zones. An accurate subcellular localization must, however, await electron microscopy.

For the second question, concerning the distribution of the different splicing variants, we have obtained a partial answer by comparing the staining of the antibody pN/AB that recognizes one of the variant amino terminal ends (that shared by ShA and ShB) to that obtained with the antibodies that are predicted to see all the known splicing variants. The variant specific antibody labels a subset of the structures stained by the antibodies to the core; the stripe from the lobula to the lateral protocerebrum is labeled, for example, but the conspicuous stripes in the lobullar plate are not (Schwarz et al., manuscript in preparation). Thus there appears to be a differential distribution of the splicing variants. This supports the hypothesis that channel subtypes are produced by the different products. The data do not indicate, however, whether only a single variant is present in a given cell or whether heteromultimeric channels are formed.

Expression of Distinct Functional Channels in Oocytes.

The oocytes of Xenopus laevis can be used as an expression system for testing the functional properties of a cloned channel gene (Methfessel et al., 1986). Initially we expressed 2 of the Shaker cDNA (ShA and ShB) in oocytes and found that each on its own is sufficient to induce the expression of channels in oocytes (Timpe et al., 1988a). These channels are K^+ selective and produce currents that are similar in many ways to those that have been described in fly muscle. The currents produced by ShA and ShB differed from one another, however, in their kinetic properties (Timpe et al., 1988a). Since then we and others have successfully expressed each of the 5 known, full-length products in oocytes and a detailed comparison of their properties is underway (Iverson et al., 1988; Timpe et al., 1988b).

Timpe et al. (1988a) noted that ShA differed from ShB in both the time course of its macroscopic inactivation and, much more dramatically, in its ability to recover from inactivation. These two products differ only in the carboxy terminal portion of the proteins and so the structural features that are responsible for the differences in these properties must reside in that portion of the molecule. As additional variants have been examined, this correlation has been maintained. ShD shares the same carboxy terminal as ShB and, like ShB, it can recover from inactivation within a few hundred milliseconds (Timpe et al, 1988b). In contrast, ShA and ShC (Timpe et al, 1988b), and H37 (Iverson et al., 1988) which share a different carboxy terminal, recover from inactivation very poorly. The current produced by ShC, for example, was only partially recovered even after 5 minutes at -120 mV (Timpe et al., 1988b). Do currents with this unusual property of extremely slow recovery from inactivation exist in vivo or are they an artifact of expression in an abnormal system? None have been described as yet in the nervous system, but a slowly recovering, inactivating K^+ current has been seen in the thrombocyte of the newt (Kawa, 1987). Antibodies that recognize these Shaker proteins may indicate other sites for electrophysiological studies in the fly. Such a channel could produce long lasting changes in the functioning of a neuron; a depolarization sufficient to inactivate the channel would leave the cell in a more excitable for several minutes (Timpe et al., 1988b).

Variations at the amino terminal end also affect the kinetics of the channel (Timpe et al., 1988b). ShB and ShD, which differ only at the amino end, were compared with regard to both their macroscopic time course of inactivation and the time course of recovery from inactivation. Each process could be fitted best as the sum of two exponentials, according to the equations below.

$$1)\ I(t) = A_1\ e^{-t/tau1} + A_2\ e^{-t/tau2}$$

$$2)\ I_p/I_{p-max} = 1 - B_1\ e^{-t/tau1} - B_2\ e^{-t/tau2}$$

In equation 1, which governs the rate of inactivation, A_1 and A_2 are the amplitudes of the two components and tau1 and tau2 are their time constants. In equation 2, which governs the recovery from inactivation, I_p/I_{p-max} represents the fraction of the peak current that is not in the inactivated state at time t after a first pulse, B_1 and B_2 represent the amplitudes of the two components, and tau1 and tau2 are their time constants. Table 2 compares the values for the amplitudes and time constants for each of these processes for each of the currents (Timpe et al., 1988b). From this table it can be seen that both ends of the protein influence the kinetics of inactivation and recovery.

Table 2. Inactivation and Recovery of 4 Variants

		ShA	ShB	ShC	ShD
From measurements of inactivation in currents evoked by steps to 40 mV	tau1 (ms)	11+3	9+1	20+3	11+1
	tau2 (ms)	33+9	181+99	45+10	90+8
	A1/(A1+A2)	0.73+0.15	0.94+0.05	0.65+0.09	0.76+0.12
From measurements of the recovery from inactivation at -80 mV	tau1 (ms)	(after 5 min only 35% recovery)	47	(after 5 min no detectable recovery)	60
	tau2 (ms)		255		500
	B1/(B1+B2)		0.69		0.42

In contrast, the voltage-dependence of activation does not appear to differ among ShA, ShB, and ShD (Timpe et al., 1988a,b). (ShC, because of its prolonged inactivation, could not be studied in detail). Where it could be studied (in ShB and ShD), the voltage- dependence of inactivation also appeared to be the same. Therefore these properties may be governed by the invariant portions of the molecule. Interestingly, homologs of the *Shaker* proteins that have been isolated from mouse (Tempel et al., 1988) and rat (Baumann et al., 1988) brain cDNA libraries are very similar to the fly products in the invariant portions of the molecule, but show little homology at the amino terminal or in the variable portions of the carboxy terminal. The vertebrate homologs have not yet been expressed in oocytes.

Summary

Through a genetic approach, we and others have uncovered a family of potassium channels encoded at the Shaker locus of Drosophila. Although the picture has been clouded by many cDNA that are likely to be incompletely spliced transcripts, it is clear that at least 5 legitimate products arise from this locus by alternative splicing. These are ShA, ShB, ShC, ShD, and H37. They differ at either end of the protein, but share a central core that contains an S4 region and 6 hydrophobic stretches. Additional variants probably remain to be found. From immunocytochemistry we have learned that some of the variant products appear to be differentially distributed; this suggests that the gene gives rise to different channel subtypes in different regions. Electrophysiological studies on each of these 5 variants has shown that a single variant is sufficient to induce expression of K^+ channels in oocytes, that the channels are all of the inactivating type known as A-channels, and that they differ from one another in particular kinetic properties.

References

Baumann A, Krah-Jentgens I, Mueller R, Mueller-Holtkamp F, Seidel R, Kecskemethy N, Casal J, Ferrus A Pongs O (1987) Molecular organization of the maternal effect region of the Shaker complex of Drosophila: characterization of an I_A channel transcript with homology to vertebrate Na^+ channel. EMBO J 6:3419-3429

Baumann A, Grupe A, Ackermann A, Pongs O (1988) Structure of the voltage-dependent potassium channel is highly conserved from Drosophila to vertebrate central nervous systems. EMBO J 7:2457-2464

Carbone E, Wanke E, Prestipino G, Possani L, Maelicke, A (1982) Selective blockage of voltage-dependent K^+ channels by a novel scorpion toxin. Nature 296:90-91

Catterall W (1986) Molecular properties of voltage-sensitive sodium channels. Ann Rev Biochem 55:953-985

Ganetzky B, Wu C-F (1986) Neurogenetics of membrane excitability in Drosophila. Ann Rev Genet 20:13-44

Guy H, Seetharamulu P (1986) Molecular model of the action potential sodium channel. Proc Natl Acad Sci USA 83:508-512

Hille B (1984) Ionic Channels of Excitable Membranes. Sinauer Assoc, Sunderland Massachusettes

Iverson L, Tanouye M, Lester H, Davidson N, Rudy B (1988) Expression of A-type potassium channels from Shaker cDNAs. Proc Nat Acad Sci USA 85:5723-5727

Kamb A, Iverson L, Tanouye M (1987) Molecular characterization of Shaker, a gene that encodes a potassium channel. Cell 50:405-413

Kamb A, Tseng-Crank J, Tanouye M (1988) Multiple products of the Drosophila Shaker gene contribute to potassium channel diversity. Neuron 1:421-430

Kawa K (1987) Transient outward currents and changes of their gating properties after cell activation in thrombocytes of the newt. J Physiol 385:189-205

Jan Y-N, Jan L, Dennis M (1977) Two mutations of synaptic transmission in Drosophila. Proc Roy Soc London 198:87-108

Jan L Jan Y-N (1982) Antibodies to horseradish peroxidase as specific neuronal markers in Drosophila and grasshopper embryos. Proc Natl Acad Sci USA 72:2700-2704

Methfessel C, Witzemann V, Takahashi T, Mishina M, Numa S and Sakmann B (1986) Patch clamp measurements on Xenopus laevis oocytes: currents through endogenous channels and implanted acetylcholine receptor and sodium channels. Pflugers Arch. Ges. Physiol. 407:577-588

Miller C, Moczydlowski E, Latorre R, Phillips M (1985) Charybdotoxin, a protein inhibitor of single Ca^{++}-activated K^+ channels from mammalian skeletal muscle. Nature 313:316-318

Noda M, Shimizu S, Tanabe T, Takai T, Kayano T, Ikeda T, Takahashi H, Nakayama H, Kanaoka Y, Minamino N, Kangawa K, Matsuo H, Raftery M, Hirose T, Inayama S, Hayashida H, Miyata T, Numa S (1984) Primary structure of Electrophorus electricus sodium channel deduced from cDNA sequence. Nature 312:121-127

Papazian D, Schwarz T, Tempel B, Jan Y-N, Jan L (1987) Cloning of genomic and complementary DNA from Shaker, a putative potassium channel gene. Science 237:749-753

Pongs O, Kecskemethy N, Mueller R, Krah-Jentgens I, Baumann A, Kiltz H, Canal I, Llamazares S, Ferrus A (1988) Shaker encodes a family of putative potassium channel proteins in the nervous sytem of Drosophila. EMBO J 7:1087-1096

Salkoff L, Wyman R (1983) Ion currents in Drosophila flight muscles. J Physiol London 337:687-708

Schwarz T, Tempel B, Papazian D, Jan Y-N, Jan L (1988) Multiple potassium channel components are produced by alternative splicing at the Shaker locus in Drosophila. Nature 331:137-142

Seagar M, Labbe-Julie C, Granier C, Goll A, Glossmann H, Van Rietschoten J, Courand F (1986) Molecular structure of rat brain apamine receptor: differential photoaffinity labelling of a putative K^+ channel subunits and target size analysis. Biochem 25:4051-4057

Sternsfeld C, Marsh S, Parcej D, Dolly J, Brown D (1987) Mast cell degranulating peptide and dendrotoxin selectively inhibit a fast-activating potassium current and bind to common neuronal proteins. Neurosci 23: 893-902

Tanabe T, Takeshima H, Mikami A, Flockerzi V, Takahashi H, Kangawa K, Kojima M, Matsuo H, Hirose T, Numa S (1987) Primary structure of the receptor for calcium channel blockers from skeletal muscle. Nature 328:313-318

Tanouye M, Ferrus A, Fujita S (1981) Abnormal action potentials associated with the Shaker locus of Drosophila. Proc Nat Acad Sci (USA) 78:6548-6552

Tempel B, Papazian D, Schwarz T, Jan Y-N, Jan L (1987) Sequence of a probable potassium channel component encoded at the Shaker locus of Drosophila. Science 237:770-775

Tempel B, Jan Y-N, Jan L (1988) Cloning of a probable potassium channel gene from mouse brain. Nature 232:837-839

Timpe L, Jan L (1987) Gene dosage and complementaion analysis of the Shaker locus in Drosophila. J Neurosci 7:1307-1317

Timpe L, Schwarz T, Tempel B, Papazian D, Jan Y-N, Jan L (1988) Expression of functional potassium channels from Shaker cDNA in Xenopus oocytes. Nature 331:143-145

Timpe L, Jan Y-N, Jan L (1988b) Four cDNA clones from the Shaker locus of Drosophila induce kinetically distinct A-type potassium channels in Xenopus oocytes. Neuron 1:659-667

Wolfner M (1980) Ecdysone responsive genes of the salivary gland of Drosophila melanogaster, Ph.D. Thesis. Stanford University, Stanford California

MOLECULAR BASIS OF BIOLOGICAL DIVERSITY AT THE SHAKER LOCUS OF DROSOPHILA

A. Baumann*, I. Krah-Jentgens*, R. Müller*, F. Müller-Holtkamp*, I. Canal+, J. Galceran+, Alberto Ferrús+, and O. Pongs*

*Ruhr-Universität Bochum, Lehrstuhl für Biochemie
4630 Bochum, FRG

+Instituto Cajal, C.S.I.C. Velazques 144, 28006 Madrid, Spain

Potassium channels contribute to the most ubiquitous group of membrane ion channels and seem to be present in almost every eukaryotic cell. Voltage-sensitive potassium channels in excitable membranes are involved (Hille, 1984) in the regulation of action potential duration, in neuron bursting and in cardiac pacemaking. Recent studies indicated that the Shaker gene complex encodes a family of potassium channel proteins in the nervous system of Drosophila (Schwarz et al., 1988, Pongs et al., 1988).

The Shaker (Sh) mutant phenotype captured the attention of physiologists more than ten years ago (2,3) because of its phenotype, the continuous trembling of appendages Although this phenotype has frequently been described in connection with the state of anesthesia, it is now clear that anesthetics do not cause this behavioural abnormality and Sh mutants show their physiological defects either anesthetized

NATO ASI Series, Vol. H 32
Molecular Biology of Neuroreceptors and Ion Channels. Edited by A. Maelicke

or awake (Tanouye a. Ferrus, 1985). The study of the larval neuromuscular junction (Jan a. Jan, 1976) indicated that Sh mutants release an excess of neurotransmitter (Jan et al., 1977). The measurement of action potentials at the adult interneuron cervical giant fiber (Koto et al,., 1981) demonstrated that the membrane repolarization is delayed and that this effect is axon autonomous (Tanouye et al., 1981) (Figure 1). Voltage clamp analysis of adult muscles showed that a voltage dependent potassium current (I_A) is abolished or diminished in some alleles or its closing kinetics is delayed, in another allele (Salkoff a. Wyman, 1981; Wu a. Haugland, 1985; Wu et al., 1983)) respectively.

In addition to all this previous electrophysiological work, the Shaker locus has been the subject of extensive genetic and molecular analysis (Tanouye et al., 1981; Wu et al., 1983; Papazial et al., 1987; Baumann et al., 1987; Schwarz et al., 1988, Pongs et al., 1988; Kamp et al., 1988). A large transcription unit, the largest cloned so far from Drosophila, has been characterized in the central region of the Sh-gene complex. The corresponding Shaker transcripts originate by multiple, differential as well as alternative splicing mechanisms into functionally related proteins with varying 5' and 3'ends. The strucutre of the Sh proteins, derived from the corresponding cDNA sequences, indicate a close homology to other ionic channels such as the Na^+ channel and the dihydropyridine receptor (Noda et al., 1983; Noda et al., 1986; Tanabe et al., 1987). Moreover, the injection of Sh cRNAs into Xenopus oocytes seems to give the final proof that functional potassium channels are encoded in the Sh locus (Timpe et al., 1988; Iverson et al., 1988).

How many products are encdoded in Shaker?

The available molecular data refer to a strech of DNA of 120 kb encompassing three chromosomal breakpoints identified as the X:Y translocations B55 and W32 and the X:3 translocation

Sh^{Lc}, respectively (Tanouye et al., 1981) (Fig. 2). A total of 12 different cDNAs have been isolated and sequenced including their genomic correlates (Papazian et al., 1987; Baumann et al., 1987; Tempel et al., 1987; Schwarz et al., 1988, Pongs et al., 1988; Kamp et al., 1988). The major features emerging from these DNA sequences and the inferred protein structures are: All products analysed so far contain amino acids sequences compatible with proteins which filter ions across the membrane. A common structural property of these ion channel proteins is that they have 6 membrane spanning helices (Fig. 3). In analogy to the Na^+ channel structure of vertebrates we will refer to them as domains S1 to S6 (Noda et al., 1986; Pongs et al., 1988). The structural similarities between the voltage-sensitive sodium, calcium and potassium channels (Tanabe et al., 1987; Pongs et al., 1988) are likely to be the molecular basis of similar electrophysiological behaviour. The primary structures of voltage-dependent sodium and calcium channels show that they contain four internal repeats with similar topology (Tanabe et al., 1987). In contrast Drosophila Shaker proteins are members of a protein family, with variant primary structures, which have only one repeat. It is probable that the potassium channels are not built by one polypeptide, but by homo or heteromultimeric aggregates offering a molecular basis for potassium channel diversity (Schwarz et al., 1988; Timpe et al., 1988; Pongs et al., 1988; Kamp et al., 1988; Iverson et al., 1988).

How is potassium channel diversity regulated?

Alignment of the cDNA sequences with the genomic DNA counterparts has revealed a minimum number of 21 exons distributed over 120 kb of genomic DNA. The cDNAs presently known share some of these exons and differ in others. If we establish the criteria that all Sh proteins will contain, at least, domains S1 to S3, then the available data indicate that a theoretical number of 18 products would be generated

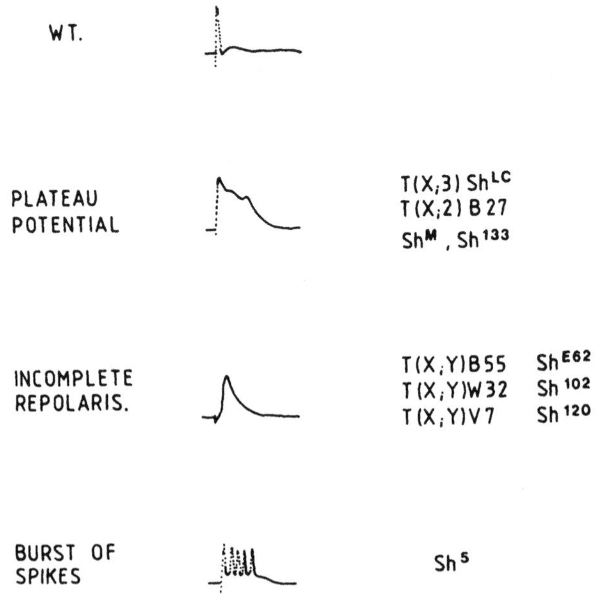

Fig. 1

Action potential phenotypes of the Shaker gene complex. The action potentials have been measured in the cervical giant fiber interneuron from several mutant constitutions (Tanouye <u>et al</u>., 1981, Tanouye a. A. Ferrus, 1985) as indicated.

Fig. 2

Cytogenetic and molecular map of the <u>Shaker</u> gene complex (ShC).
(Top) Schematic drawing of the salivary gland chromosome for this region. The breakpoints of five chromosomal rearrangements used to localize <u>Sh</u> cytogenetically (Tanouye <u>et al</u>. 1981) are indicated by brackets. The centromere is to the right. The genetic organization of the ShC into maternal effect (ME), viable (V) and haplolethal (HL) regions is indicated by open boxes. The location of some translocation breakpoints is indicated by vertical arrows on the physical map. The coordinates on the scale are as defined in Pongs <u>et al</u>. (1988). Calibration is in Kb.
(Bottom) Alignment of exons 1-21 of <u>Sh</u> transcripts to the physical map of the genomic <u>Sh</u> region from which they derive (Pongs <u>et al</u>., 1988). Exons are indicated as small boxes. Open boxes are untranslated regions, bars correspond to derived open reading frames of <u>Sh</u> cDNAs. Exons are numbered from right to left in the direction of transcription. The dot

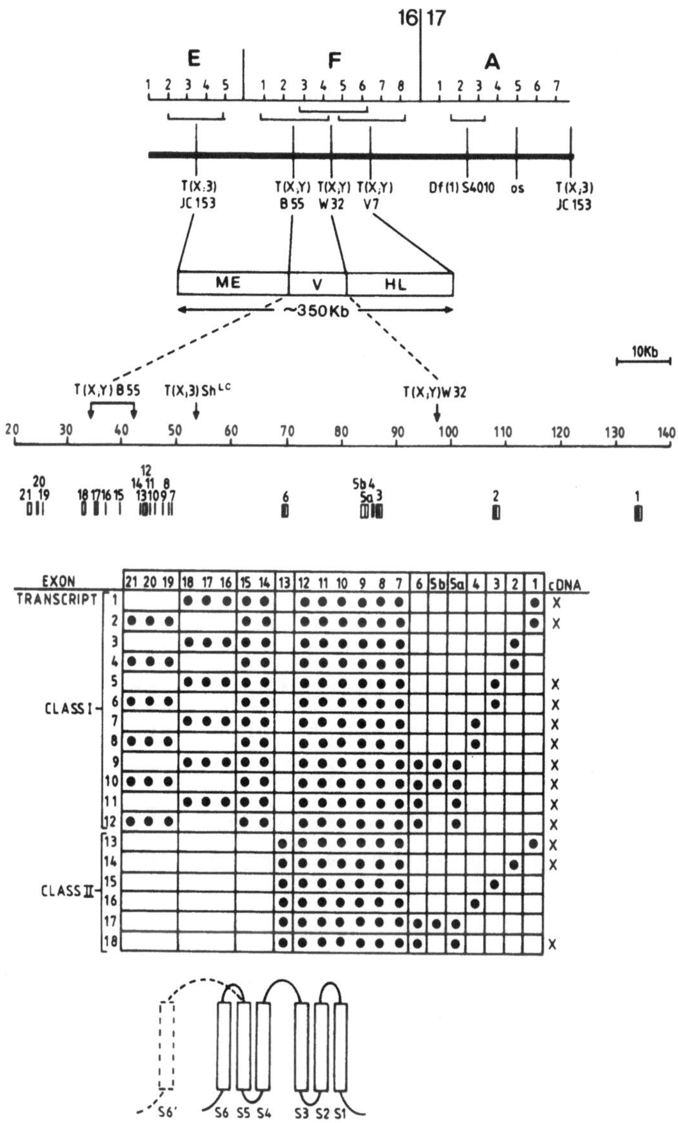

matrix indicates possible combinations of exons 1-21. Exon combinations which have been found in *Sh* cDNAs (Schwarz *et al.*, 1988; Kamb *et al.*, 1988; Pongs *et al.*, 1988) are marked at the right hand side by crosses. *Sh* proteins derived from class I transcripts contain six proposed membrane spanning helices (S1 to S6) while *Sh* proteins derived from class II transcripts contain three proposed transmembrane helices (S1 to S3) (Baumann *et al.*, 1987). As indicated below the dot matrix, exon 10 encodes helix S1, exon 11 helix S2, exon 12 helix S3, exon 14 helix S4 and exon 15 helix S5, respectively. Helix S6 is encoded alternatively in exon 16 (S6) or in exon 19 (S6'). The amino acid sequences of both helices differ only by one conservative exchange of Val into Ala.

from the combinatorial use of the 21 exons by mechanisms of differential or alternative splicing (Fig. 2). Any additional exon to be found in this region would increase the potential number of Sh products. Most exons are spliced out to generate transcripts with alternative exons. Exons 5a/5b and exons 12/13, however, are not separated by intervening sequences. In this case, the differential use of a splice site generates either exon 5a or exon 5a/5b, and exon 12 or exon 12/13, respectively. That is, a fragment of exons 5a/5b and/or 12/13 is spliced out in some cDNAs but not in others. We refer to this modality of RNA processing as differential splicing. The enormous combinatorial potential of the exons in the structure of the Sh transcription unit is unprecedented. Although not all the possible combinations have been injected as cRNAs into Xenopus oocytes for expression studies, the presently available data indicates that the class I transcripts express potassium channel subunits which give rise to A-type channels with diverse inactivation kinetics (Timpe et al., 1988; Iverson et al., 1988). The results of these expression studies thus offer a molecular basis for A-type channel diversity.

It seems that the Sh transcription unit encodes a family of potassium channels whose differences in structure could be adapted to the needs of individual neurons. We suspect that part of the complexity found in this locus is required to have cell specific potassium channels, which contribute to fast membrane repolarization, modulation of neurotransmitter release, neuronal firing in bursts or single spikes, delayed rectification, some forms of electrical coupling, etc. Probably the hardware of different neurons includes a setup of different potassium channels for modulating electrical activities. Apart from possible posttranscriptional regulatory mechanisms, cell specific expression of diverse potassium channels requires cell specific promoters in the Shaker transcription unit. As illustrated in Fig. 2, the mutation T(X;Y)W32 interrupts the Sh transcription unit. It separates exons 1 and 2 from the remaining exons. I_A current is absent in muscle cells of T(X;Y)W32 (Salkoff, 1984) and neuronal action potentials are affected (Tanouye et al., 1981)). In situ hybridizations with exon specific probes in the central nervous system of this mutant as compared to wild type showed that (X;Y)W32 abolishes the synthesis of some but not all Sh RNAs (unpublished observations). This suggests the existence of promoter(s) which are distal as well as proximal to the T(X;Y)W32 mutation. It remains to be shown, however, that these promoters are involved in the regulation of cell-specific expression of Sh RNAs.

Mutant phenotypes in view of ShC diversity

The Sh products known today are expressed in the mature cells of the nervous system and their presence is kept during metamorphosis (Pongs et al., 1988) when most neurons are undergoing developmental changes. These observations are compatible with the relevance of Sh products in the biology of the neural tissue including the subtle morphological changes in arborization found in some Sh mutants (A. Ferrus unpublished results).

Yet, Shaker mutant phenotypes are not easily reconciled with the available molecular and physiological data. The mutation T(X;Y)B55 causes more severe vibration of appendages than T(X;Y)W32 (Tanouye et al., 1981), however it is reported that T(X;Y)B55 has a residual I_A while T(X;Y)W32 is deprived totally of it (Salkoff, 1984). Other Shaker mutations, e.g. T(X;Y)V7 that cause behavioural and physiological Sh phenotypes (Tanouye et al., 1981), reside more than 100 kb away from the region currently analyzed. They might not alter the products presently known. Almost no attention has been paid so far, to the stoichiometry of Sh products in the normal vs the various mutant constructions and this is perhaps an essential feature to understand the biology of

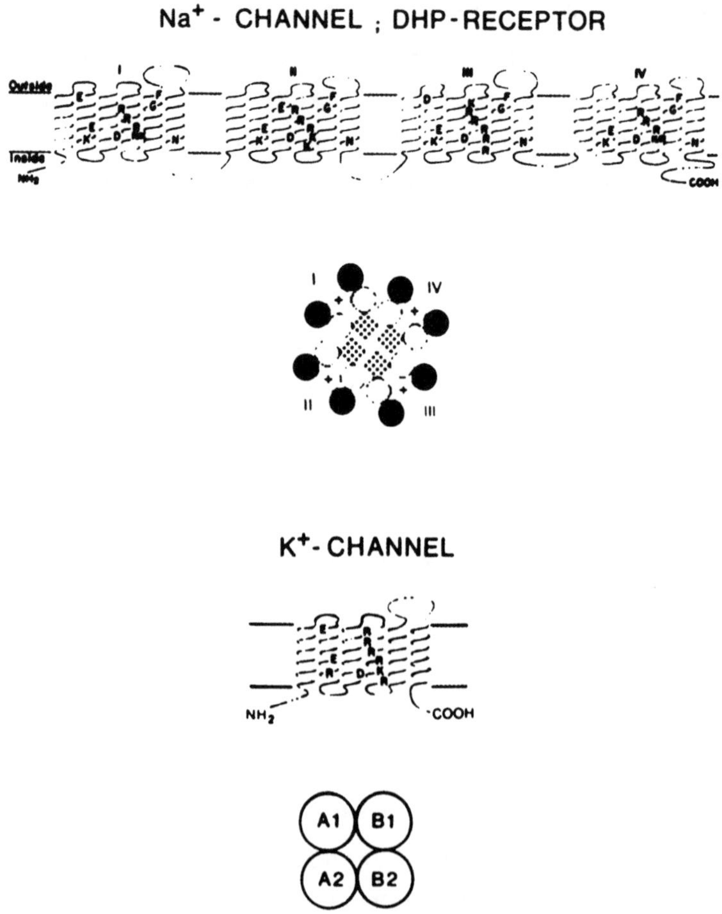

Fig. 3

Schematic representation of structural characteristics common to the dihydropyridine receptor, the voltage gated sodium channel and <u>Sh</u> proteins. Data for vertebrate ion channels are taken from Tanabe <u>et al.</u>, (1987). The proposed transmembrane alfa helical segments S1-S6 (from left to right) are displayed linearly. The dihydropyridine receptor and the sodium channels have four units of homology spanning the membrane, which are assumed to surround the ionic channel. <u>Sh</u> proteins have only one such unit. Amino acids conserved in all segments of the dihydropyridine receptor, the sodium channels and the <u>Sh</u> voltage sensitive potassium channels are shown in one letter code. Roman numbers indicate the four units of homology. Black circles are proposed membrane spanning helices S5 and S6, cross-hatched circles S1, stippled circles S2, hatched circles S3. The proposed S4 is indicated by a plus sign.

these products. The mutations T(X;Y)B55, T(X;Y)W32, Df(1)B55^P/W32^D have very different behavioural and physiological phenotypes (see Fig. 1). Distally to the breakpoint of T(X;Y)B55, relevant functions to the membrane repolarization must be coded, since the mutated element T(X;Y)B55^D causes, by itself, a Sh phenotype. Yet, the more astonishing observation is that Df(1)B55^D/W32^P flies are viable in spite of the fact that no I_A is present and that a multiplicity of products are eliminated. Propably, other regions of the genome, functionally related with the one presently studied, compensate the absence of these Sh products.

Structure of potassium channels

The derived Shaker protein sequences have the properties of integral membrane proteins containing six identical, presumably membrane spanning segments (Papzian et al, 1987; Schwarz et al., 1988; Pongs et al., 1988; Kamp et al., 1988). Their topology as well as amino-acid sequences are similar to that of vertebrate voltage sensitive sodium and calcium channels (Figure 3). In analogy to the model proposed for the voltage-sensitive sodium channel (Noda et al., 1986), the six membrane spanning α-helices of the potassium channel subunits are similarly oriented in a pseudosymmetric fashion across the membrane. The location of charged residues in segments S2, S3 and S4 is remarkably similar to that in the similar segments of the individual repeats of the vertebrate voltage-sensitive sodium channels and of the DHP-receptor (Figure 3). The proposed transmembrane-spanning sequences are encoded in separate exons: segment S1 in exon 10, S2 in exon 11, S3 in exon 12, S4 in exon 14, S6 in exons 16 or 19 (Pongs et al., 1988). Segment S5, interestingly, is encoded in exons 14 and 15. Since the proximal T(X;Y)B55 breakpoint was located between exons 14 and 15, Shaker-ClassI-proteins (Figure 2) cannot contain an intact segment S5 and lack segment S6 altogether.

The alignment of positive charges in S4 is thought to carry the gating current. It has been proposed that segment S4 contribute to the voltage sensor in voltage-sensitive channels. The conservative alignment of charged residues in voltage-sensitive ion channels must fulfil an important role in channel function related to functional properties which all the voltage-sensitive ion channels share. However, the common location of charged residues in the transmembrane helices does probably not contribute to the molecular basis of channel specific ion selectivity.

The amino and carboxyl termini are located in the model of Figure 3 at the cytoplasmic side of the membrane. The respective potassium channel sequences, are not similar to that of the other known voltage sensitive ion channels. Therefore, they might contribute to ion selectivity, pharmacology and inactivation kinetics which are different from those measured for the sodium and calcium channels. Most notably, diversity of _Shaker_ potassium channels is brought about by variations in the amino- and carboxyterminal sequences (Timpe _et al_., 1988; Iverson _et al_., 1988). Indeed, it has been found that injection of variant _Shaker_ cRNAs in to frog oocytes leads to the expression of variant A-type potassium currents which differ in their inactivation kinetics and in their rates of recovery from inactivation (Timpe _et al_., 1988; Iverson _et al_., 1988). But, more extensive studies are needed before a definitive structure-function relationship might become elucidated.

Conservation of general potassium channel structure from Drosophila to rat brain

Voltage-sensitive potassium channels have been identified in coelenterate (Hille, 1984) and in yeast (Saimi _et al_., 1988). Thus, voltage-sensitive potassium channels probably envolved well before the separation of the eukaryotic world into protostomes and deuterostomes. Accordingly, vertebrate voltage-sensitive potassium channels should be homologous to that envolved in the *Shaker* gene complex. Indeed, low stringency hybridization protocols could be utilized to isolate mouse (Tempel *et al.*, 1988) and rat brain cDNAs (Baumann *et al.*, 1988) homologous to *Drosophila Shaker* cDNAs. The derived aminoacids sequences are ~70% identical with the derived *Shaker* potassium channel sequences. Most notably, the isolated potassium channel cDNAs of vertebrate brain encode proteins which also have the potential to span the lipid bilayer six times. The respective transmembrane α-helices are nearly identical in sequence to those of *Shaker* potassium channel. The six transmembrane α-helices have been conserved to different degrees. Sequences of segments S4 and S6 are completely identical between Drosophila and vertebrate brain. The sequence of S5 contains one conservative amino acid sustitution (Val Ile). The sequence of segment S1 contains three conservative and one non-conservative amino acid substitution. The sequences of segments of S2 and S3 contain additional non-conservative amino acid substitutions, including the introduction of one additional negative charge (Glu-272) into segment S3 which is not found in *Shaker* potassium channels at that position. It is presently not known how this affects channel function. However, the availability of rat cDNA clones encoding diverse potassium channels opens an important avenue for the detailed analysis of potassium channels in the vertebrate brain.

ACKNOWLEDGEMENT

The work of our laboratories was supported by the Volkswagenstiftung, the German-Spanish cultural exchange program -acciones integradas-, the Deutsche Forschungsgemeinschaft (O. Pongs), the CAYCIT, and R. Pascual Fund (A. Ferrus).

References

Baumann, A., Grupe, A., Ackermann, A., and Pongs, O. (1988) Structure of the voltage-dependent potassium channel is highly conserved from Drosophila to vertebrate central nervous systems. EMBO J. 7, 2457-2463.

Hille, B. (1984) Ionic channels in excitable membranes, Sinauer, Sunderland MA.

Iverson, L.E., Tanouye, M.A., Lester, H.A., Davidson, N., a. Rudy, B. (1988) A-type potassium channels expressed from Shaker locus cDNA. Proc. Natl. Acad. Sci. USA 85, 5723-5727.

Jan, L.Y., a. Jan, Y.N. (1976) Properties of the larval neuromuscular junction in Drosophila melanogaster. J. Physiol. (London) 262, 215-236.

Jan, Y.N., Jan, L.Y., a. Dennis, M.J. (1977) Two mutations of synaptic transmission in Drosophila. Proc. R. Soc. London (Biol.) 198, 87-108.

Kamb, A., Tseng,-Crank, J. a. Tanouye, M.A. (1988) Multiple Products of the Drosophila Shaker gene may contribute to potassium channel diversity. Neuron 1, 421-430.

Koto, M., Tanouye, M.A., Ferrus, A., Thomas, J.B., a. Wyman, R. (1981) The morphology of the cervical giant fiber neuron of Drosophila. Brain Res. 221, 213-217.

Noda, M., Shimizu, S., Tanabe, T., Takai, T., Kayanao, T., Ikeda, T., Tkahashi, H., Nakayama, H., Kanoka, Y., Miniamino, N., Kangawa, K., Natsuo, H. Raftery, M.A., Hirose T., Inayama S., Hayashida, H., Miyata, T. and Numa, S. (1984) Primary structure of Electrophorus electricus sodium channel deduced from cDNA sequence. Nature 312, 121-127.

Noda, M. Ikeda, T., Kayano, T., Suzuki, H., Takeshima, H., Kurasaki, M., Takasashi, H. and Numa, S. (1986) Existence of distinct sodium channel messenger RNAs in rat brain. Nature 320, 188-192.

Papazian, D.M., Schwarz, R.L., Tempel, B.L., Jan. Y.N., a. Jan. L.Y. (1987) Cloning of genomic and complementary DNA from Shaker., a putative potassium channel gene from Drosophila., Science 237, 749-753.

Pongs, O., Kecskemethy, N., Müller, R., Krah-Jentgens, I., Baumann, A., Kiltz, H.H., Canal, I., Llamazares, S., and A. Ferrus (1988) Shaker encodes a family of putative potassium channel proteins in the nervous system of Drosophila. EMBO J., 7, 1087-1096.

Saimi, Y., Martinac, B., Gustin, M.C., Cubertson, M.R., Adler, J. a. Kung, C. (1988) Ion channels in Paramecium, yeast and Escherichia coli. Trends Biochem. 13, 304-309.

Salkoff, L. (1983) Genetic and voltage clamp analysis of a Drosophila potassium channel. Cold Spring Harbor Symp. Quant. Biol. 48, 221- 231.

Salkoff, R. a. Wyman, R. (1981) Genetic modification of potassium channels in Drosophila Shaker mutants. Nature 293, 228-230.

Schwarz, T.L., Tempel, B.L., Papazian, D.M. Jan, Y.N., a. Jan, L.Y. (1988) Multiple potassium-channel components are produced by alternative splicing at the Shaker locus of Drosophila. Nature 331, 137-142.

Tanabe, T., Takeshima, H., Mikami, A., Flockerzi, V., Takahashi, H., Kangawa, K., Kojima, M., Matsuo, H., Hirose, T. and Numa, S. (1987) Primary structure of the receptor for calcium channel blockers from skeletal muscle. Nature 328, 313-318.

Tanouye, M., Ferrus, A., a. Fujita, S.C. (1981) Abnormal action potentials associated with Shaker locus of Drosophila. Proc. Natl. Acad. Sci. USA 78, 6548-6552.

Tanouye, M. and Ferrus, A. (1985) Action potentials in normal and Shaker mutant Drosophila. J. Neurogenetics 2, 253-271.

Tempel, B.L., Papazian., D.M., Schwarz, T.L., Jan, Y.N., a. Jan. L.Y. (1987) Sequence of a probable potassium channel component encoded at Shaker locus of Drosophila. Science 237, 770-775.

Timpe, LC., Schwarz, T.L. Tempel, B.L., Papazian, D.M., Jan, Y.N., a. Jan, L.Y. (1988) Expression of functional potassium channels from Shaker cDNA in Xenopus oocytes. Nature 331, 143-145.

Wu, C.-F., Ganetzky, B., Haugland, F., a. Liu, A.-X. (1983) Potassium currents in Drosophila: Different components affected by mutations of two genes. Science 220., 1076-1078.

Wu, C.-F. a. Haugland, F.N. (1985) Voltage clamp analysis of membrane currents in larval muscle fibers of Drosophila: Alteration of potassium current in Shaker mutants. J. Neurosci. 5, 2626-2640.

COMMON PROPERTIES OF RECEPTORS COUPLED TO GTP BINDING REGULATO PROTEINS

L. Emorine, S. Marullo, M.-M. Sutren, C. Delavier, Y. Eshdat, G. Raposo, A.D. Strosberg

CNRS, University Paris VII and Pasteur Institute
Laboratory of Molecular Biology of Receptors
25, rue du Docteur Roux
75724 Paris Cedex 15 - France

Introduction

A number of membrane receptors coupled to GTP binding regulatory proteins have recently been shown to share structural propertie at the level both of the genes and of the corresponding proteins (Fig. 1 and 2). By comparing results obtained for α_2-, β_1- and β_2-adrenergic catecholamine receptors, M1 to M5 muscarinic acetylcholine receptors as well as receptors for the neuropeptides substance K and angiotensin, and for dopamine and serotonin (see references in legend of figure 1), a number of common characteristics have been identified.

The genes

The genes coding for these receptors display sizeable homology and are approximatively of similar length with most proteins constituted of 410 to 500 amino-acid residues, although some may be as short as 325 residues, such as the MAS oncogen which was recently identified as an Angiotensin II receptor (Jackson et al. 1988) and others

NATO ASI Series, Vol. H 32
Molecular Biology of Neuroreceptors and
Ion Channels. Edited by A. Maelicke

may be as long as 590 residues, such as the M4 and M5 muscarinic receptors (Peralta et al 1987, Bonner et al 1988). Glycosylation brings the molecular weight from 47 kDa to up to 80 kDa.

Despite their similarity, the chromosomal location of the genes vary extensively : the human β2 -adrenergic receptor gene is on chromosome 5, the α2A adrenergic receptor gene on chromosome 10, the α2B on chromosome 4; the genes coding for the five different subtypes of muscarinic receptors are located on four different chromosomes.

Notwithstanding their evolutionary relationship with the rhodopsin gene which has four introns (see figure 2), most of the coding regions of the genes for the adrenergic and muscarinic receptors are intronless, irrespective of the animal species in which they were studied.This absence of introns was, until very recently, considered as a characteristic of this multigene family but several reports now suggest that introns are present in the genes coding for the HT_{1C} (Lübbert et al.1988) and HT_2 (Pritchett and Seeburg 1988) subtypes of serotonin receptors, for the D2 dopamine receptor (Bunzow et al 1988) and for the neuropeptide substance K receptor (Nakanishi 1988).

The domains in the receptors

The receptor proteins are constituted of single polypeptide chains composed of alternative hydrophilic and hydrophobic stretches. The abundance of hydrophobic regions explains the strict requirement for detergents to solubilize the receptors.

By analogy to bacterio-rhodopsin and rhodopsin (which is also coupled to a GTP binding protein, "transducin"), and from *in situ* proteolytic studies, it

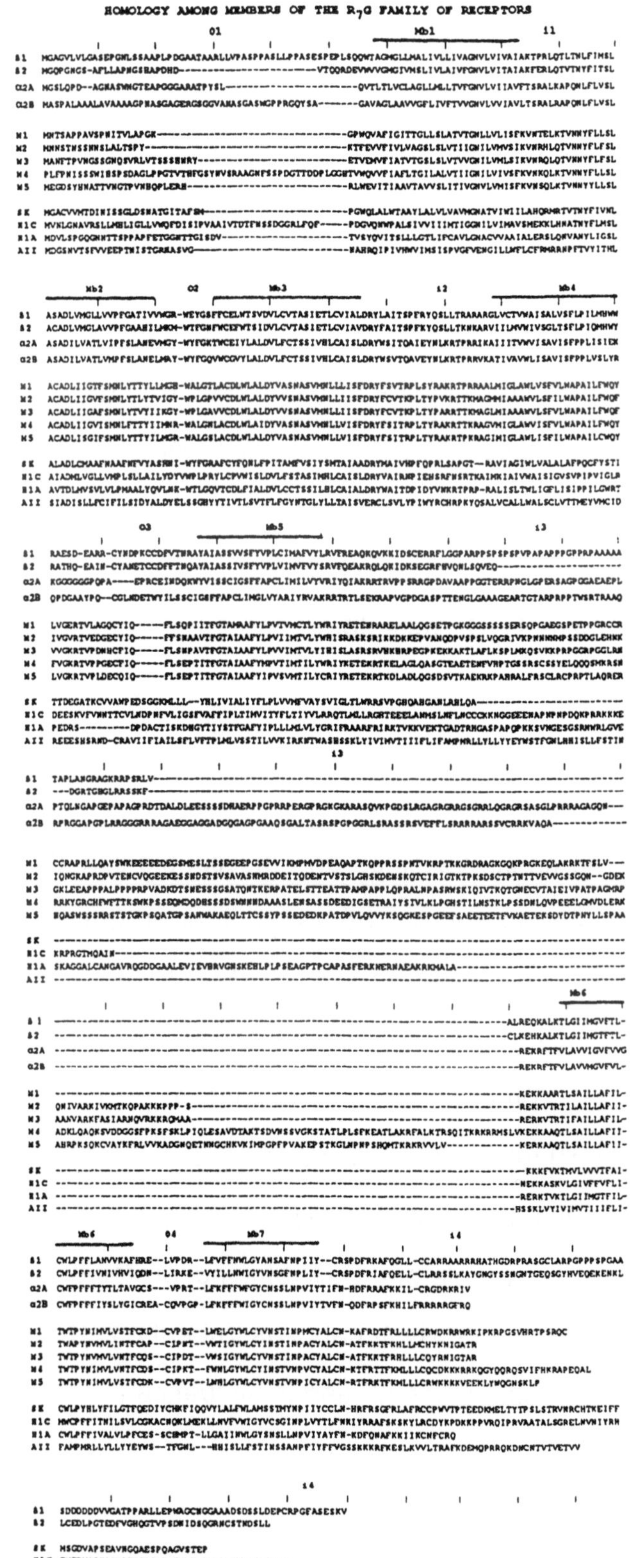

Figure 1:Sequence homology of human receptors coupled to GTP-binding proteins

We have lined up here all the available sequences of human membrane receptors with seven hydrophobic segments coupled to GTP binding regulatory proteins (the "R_7G" proteins).

References are as follows: β1: Frielle et al 1987, β2:Emorine et al 1987, α **2A** : Kobilka et al 1987,α**2B** : Regan et al 1988, **M1** to **M4** Peralta et al 1987, **M5** :Bonner et al 1988 . We have included the bovine receptor for the neuropeptide substance K (**SK**) (Masu et al 1987), the sequence of the G21 gene (Fargin et al 1988) which was recently shown to correspond to the $5HT_{1A}$receptor , hence the symbol **H1A**, and the sequence of the human oncogene "mas" (Young et al. 1986) which was recently identified as an angiotensin II receptor (Jackson et al 1988), hence the symbol **AII**. The serotonin receptor $5HT_{1C}$ has been presented under the symbol **H1C** (Julius et al 1988). Deletions have been inserted to maximize homology. Proteins have been grouped in families which display high internal homology: two such readily apparent groups are the adrenergic and muscarinic receptors.

appears that the N-terminal **O1** region which contains several potential N-glycosylation sites is extracellular whereas the seven **Mb-1** to **Mb-7** hydrophobic stretches are transmembranar. The existence of the seven transmembrane domains has been recognized as a hallmark of this new family of membrane receptors for which we propose the collective name of "$\mathbf{R_7G}$" proteins.

The C-terminal region **i4,** which in most receptors contains serine and threonine-rich regions , is probably intracellular and might be the target of protein kinases involved in receptor-desensitization. Several receptors also possess such phosphorylation consensus sites in the **i3** intracellular loop. We present in Figure 2 a schematic drawing which shows the alternated intra-and extracellular loops for these $\mathbf{R_7G}$ receptors.

The binding sites

The binding sites for the catecholamines and muscarinic ligands are composed of residues belonging to at least four different transmembranar domains **Mb-2, Mb-3**, **Mb-6** and **Mb-7** (Fig. 1). Among these residues are, in the β2 receptors, aspartic acids 79 and 113, asparagine 318 and tyrosine 330. Two half cystines located at positions 106 and 184 (Fig. 1 and 2) constitute a disulfide bond essential for the binding activity (Vauquelin et al 1980, Dixon et al 1987) .

- the site of interaction with the regulatory GTP-binding proteins is probably located in intracellular domain **i3** which happens to be the part of the $\mathbf{R_7G}$ proteins which varies most both in length and in sequence .

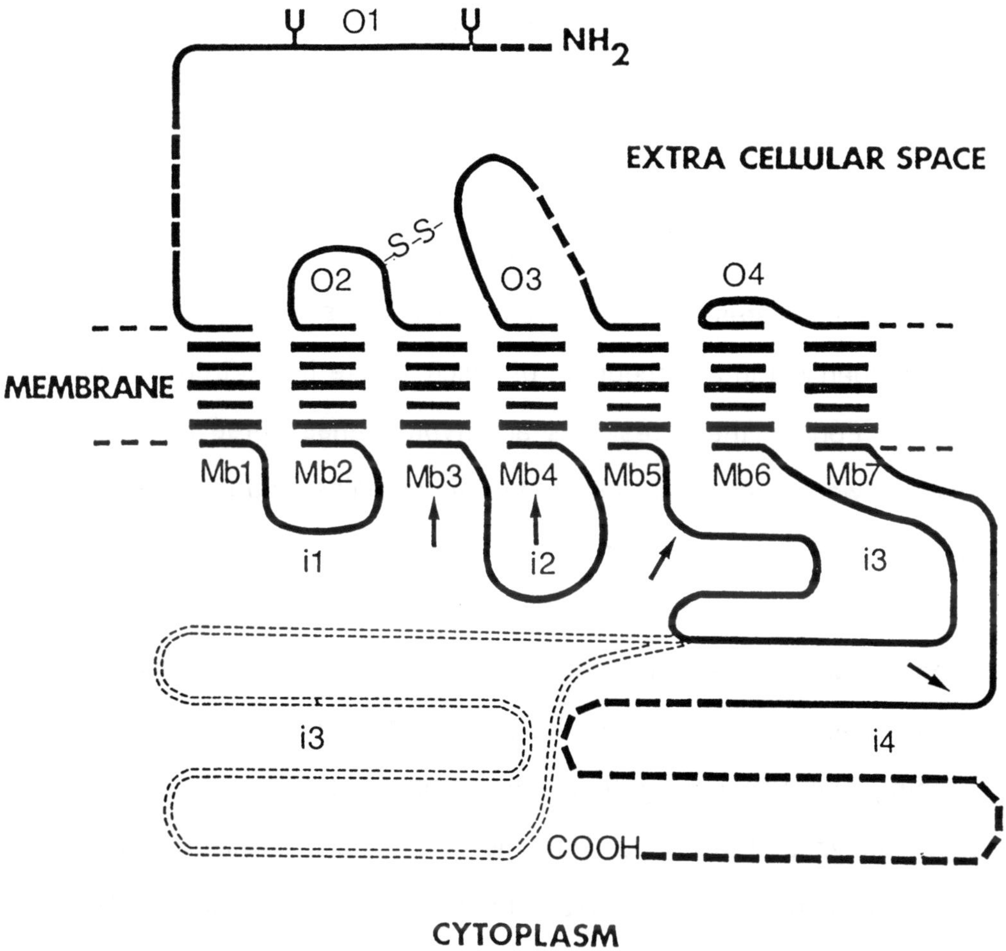

Figure 2 : Possible model for insertion of the R_7G receptors in the plasma membrane.
The seven membrane spanning segments correspond to the hydrophobic sequences. These transmembrane domains are connected through extra and intracellular loops, one of which (the 3rd intracellular) is very variable in length possibly to insure the possibility for the sixth and seventh domain to physically interact with the second and third domain as inferred by site- directed mutagenesis (Dixon et al 1987, Fraser et al 1988), affinity labeling (Wong and al 1988) and chimeric receptor studies (Kobilka et al 1988, Marullo et al 1988). We have indicated the disulfide bond (-S-S) which in β-adrenergic receptors is essential for ligand binding (Vauquelin et al, 1980, Dixon 1987).and the sites of insertions of introns (arrows) in the rhodopsin gene.

Sequence homologies and pharmacologic subtypes.

Although the homologies among the **R7G** proteins are striking, the number of actual identities is highest, as expected, among receptors that bind the same ligands with similar pharmacologic profiles. These groups of proteins thus constitute subfamilies (see figure 1).

When one compares the first twelve proteins in this figure one find only 21 residues identical in all receptors, of which 15 are in the transmembrane domains. Four of these residues are tryptophanes, three are prolines, including two in presumed transmembrane domains, and two are aspartic acids, also in transmembrane domains. The Asp-Arg-Tyr tripeptide is conserved in all these proteins. If we include in the comparison the Mas oncogene, which is an Angiotensin II receptor, only seven residues are conserved and four of these are located in **Mb-6** and **Mb-7.**

Cloning and sequencing has considerably increased the number of known subtypes of receptors. Thus the two muscarinic receptors identified by their high (M1) and low (M2) affinity for the antagonist pirenzepine became five by the isolation of three additional homologous genes. Expression of each of these revealed selective binding and effector activation properties (Peralta et al 1988, Bonner et al 1988). The α2 adrenergic receptors were also subdivided in α2A and α2B after several genes were identified by homology cloning (Kobilka et al 1987, Regan et al 1988).

Nevertheless, pharmacological subtypes are not necessarily always very homologous: the HT_{1C} subtype (Julius et al 1988) for instance is much more similar to the HT_2 receptor than to the HT_{IA} subtype (Pritchett and Seeburg 1988).

Expression of mammalian receptors in *E. coli*

In view of the high degree of conservation in the **R_7G** receptor genes, we set out to develop a common expression system that would be simple and straightforward and most importantly, applicable to all members of the family.

Based on the structural homology with bacterio-rhodopsin, we reasoned that similar mammalian proteins could also be embedded in the bacterial membrane. The coding region of the gene for the human β2-adrenergic receptor was fused to the gene coding for β-galactosidase in the λgt11 expression vector. Escherichia coli bacteria were lysogenized with this modified vector and transcription of the fused genes was induced . Production of functional receptors was shown by detection, on intact bacteria, of binding sites for catecholamine agonists and antagonists possessing a typical β2-adrenergic pharmacological profile (Marullo et al 1988). Binding and photoaffinity labeling studies performed on intact E.coli and membrane fractions revealed that these binding sites were located in the inner membrane of the bacteria.

Functional expression in bacteria of other β–adrenergic receptors such as the human β1 and the turkey erythrocyte receptors as well as of a muscarinic M2 receptor (Koman et al in preparation) confirmed that glycosylation is not essential for binding nor for insertion in membranes and supports the hypothesis that all these receptors with a common distribution of hydrophobic and hydrophilic domains share a well conserved topology.

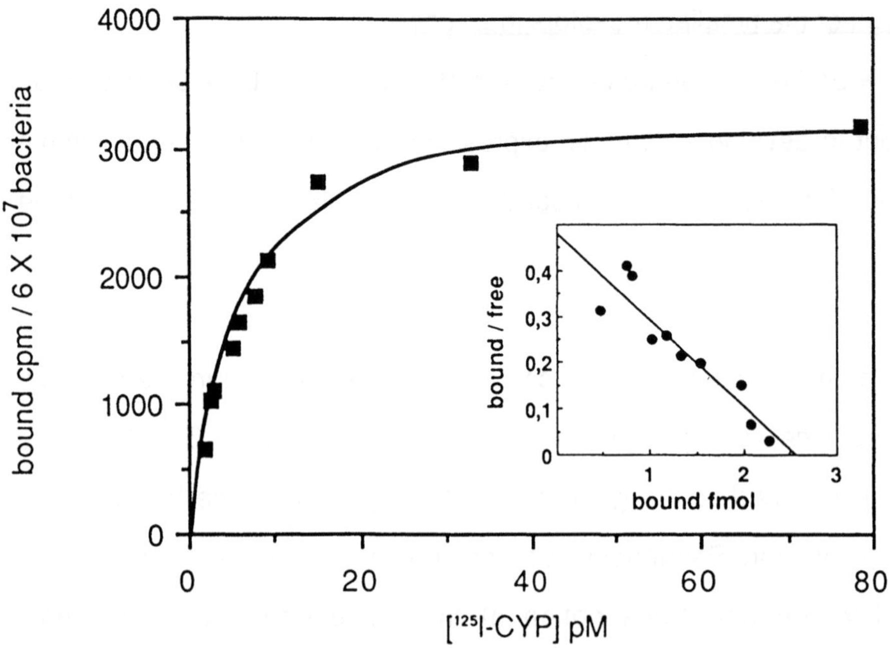

Compound	pharmacological property	K_i (nM)
(-) norepinephrine	agonist	9100 ± 1100
(-) epinephrine	agonist	4700 ± 1400
(+) isoproterenol	agonist	1400 ± 100
(-) isoproterenol	agonist	110 ± 10
procaterol	agonist ß2	40 ± 8
CGP 20712-A	antagonist ß1	2500 ± 60
practolol	antagonist ß1	790 ± 220
CGP 12.177	antagonist ß1	4.8 ± 1.3

Figure 3: Functional expression of human β-adrenergic receptor in E.coli

Binding of β-adrenergic ligands to intact E.coli bacteria containing the (β-galactosidase β-adrenergic receptor) fusion gene (Marullo et al 1988). Figure 3A : Saturation binding curve of Iodo-cyano-pindolol with the Scatchard analysis of the binding isotherm in inset. Figure 3B: Inhibition constants (K_i) of various ligands. Controls include bacteria infected with wild-type λgt-11 or phages containing the receptor gene in the opposite orientation .

Regulation of the R-G family members: redistribution and internalization

ß-adrenergic and muscarinic receptor-responsive systems become desensitized upon exposure to excess agonists. This diminution of pharmacologic responsiveness has been linked to a decrease of the number of ligand binding sites at the cell surface as well as an uncoupling from the GTP regulatory proteins. Although it has been suggested previously that receptors actually become inserted in "light vesicles",neither the precise nature nor the intracellular location of these structures has been further discussed.

By using monoclonal anti-receptor antibodies raised against affinity-purified muscarinic (Andre et al 1985) and ligand -complexed $ß_2$-adrenergic receptors (Kaveri et al 1987), we have shown by immunofluorescence and electron microscopic studies that in excess agonist treated-cells, receptors become incorporated in smooth pits which later coalesce into smooth vesicles at which time they leave the surface of the cell to ultimately form endosomes (Raposo et al 1987, 1988).

This common pathway of smooth pits and vesicles for muscarinic and ß-adrenergic receptors is in stark contrast with the clathrin-coated pits and vesicles identified for receptors with tyrosine kinase activity such as the EGF and Insulin receptors.

A "smooth" pathway has also been proposed for the receptors for the yeast α and β mating factors which also contain seven transmembrane domains (Payne et al. 1988). It is likely that other members of the $\mathbf{R_7G}$ family which display seven transmembrane domains will also be internalized via a smooth pathway.

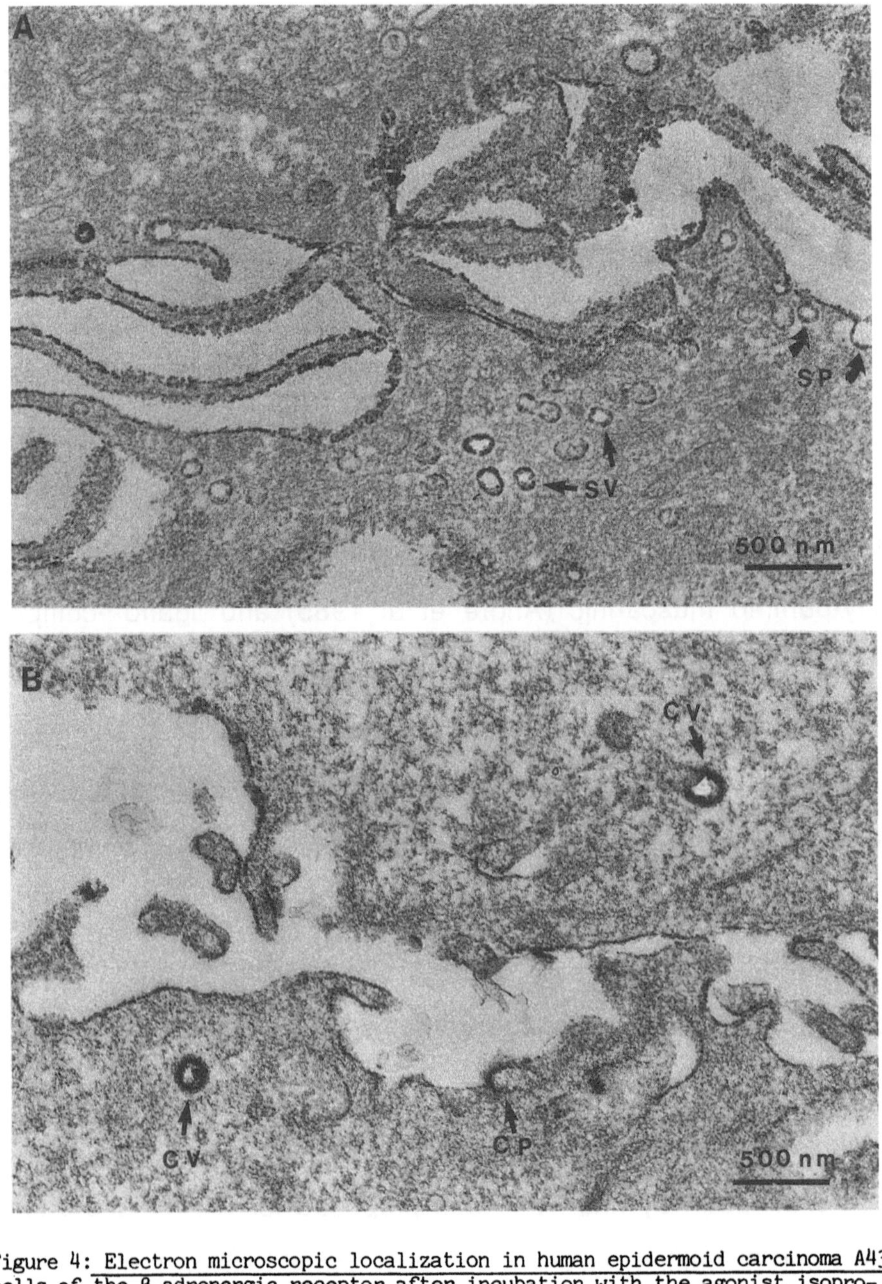

Figure 4: Electron microscopic localization in human epidermoid carcinoma A431 cells of the ß-adrenergic receptor after incubation with the agonist isoproterenol (A) and of the EGF receptor after incubation with EGF (B).
A431 cells were treated with the agonists before fixation in paraformaldehyde glutaraldehyde. The receptors were localized using specific anti-receptor antibodies and the indirect immunoperoxidase labeling method. The ß-adrenergic receptor is internalized by smooth pits (SP) and smooth vesicles (SV). The EGF receptor is internalized by coated vesicles (CV). (CP) points to coated pit formations (Raposo et al. 1987, 1988).

Conclusion :

We have summarized here a few properties of the membrane receptors which belong to a new family which we propose to designate under the term **R_7G** to emphasize that they display seven hydrophobic domains and are coupled to GTP-binding regulatory "G" proteins. In addition to the structural similarities of the genes and corresponding proteins, we have reported here that several of these receptors may be expressed in E.coli bacteria with full retention of their pharmacological binding properties despite the absence of glycosylation. We have shown furthermore that both the muscarinic M2 and the adrenergic β2 receptors after prolonged exposure to their respective antagonist become internalized via smooth, non-clathrin coated pits and vesicles rather than through the clathrin-coated pits and vesicles utilized for instance by receptors for growth factors. This new mode of internalization for membrane receptors is likely to be shared by other members of the **R_7G** family.

Acknowledgements: We thank Dr.J-G.Guillet and E.Dassa for useful discussions .This work was supported by grants from the Centre National de la Recherche Scientifique, the Institut National de la Santé et de la Recherche Médicale, the Ministère de la Recherche et de l'Enseignement Supérieur, the Tobacco Research Council, the Association pour la Recherche sur le Cancer, la Ligue contre le Cancer, and the Fondation pour la Recherche Médicale Française.

References:

André, C., Guillet, J.G., De Backer, J.P.,Vanderheyden, P., Hoebeke, J. and Strosberg, A.D. "Monoclonal antibodies against muscarinic acetylcholine receptors recognize active and denatured forms". EMBO J. 3, 17-21 (1984)

Bonner, T.I., Young, A.C., Brann, M.R., and Buckley, N.J. (1988) "Cloning and Expression of the Human and Rat m5 Muscarinic Acetylcholine Receptor Genes".Neuron 1 403-410.

Bunzow, J., Albert, P., Christies, M.,Van Tol, H., Salon, J., Machida, C., Grandy, D.E., and Civelli, O."Cloning and Expression of a New G-Protein Coupled Receptor". These Proceedings.

Dixon, R.A.F., Sigal, I.S., Candelore, M.R., Register, R.B., Scattergood, W., Rands, E. and Strader, C. (1987) : "Structural features required for ligand binding to the ß-adrenergic".EMBO 6 3269-3275.

Emorine, L.J.,Marullo, S.,Delavier-Klutchko, C., Kaveri, S.V.,Durieu-Trautmann, O. and A.D. Strosberg (1987) "Structure of the gene for the human $ß_2$-adrenergic receptor: expression and promoter characterization". Proc. Natl. Acad.Sci USA 84 6995-6999.

Fargin, A., Raymond, J.R., Lohse, M.J., Kobilka, B.K., Caron, M.G., and Lefkowitz; R.J. (1988) "The genomic clone G-21 which resembles a ß-adrenergic receptor sequence encodes the $5\text{-}HT_{1A}$ receptor".Nature 335, 358-360.

Fraser, C.M., Fu-Zon Chung, Cheng-Dian Wang, and Venter, J.C. (1988) "Site-directed mutagenesis of human ß-adrenergic receptors: Substitution of aspartic acid-130 by asparagine produces a receptor with high-affinity agonist binding that is uncoupled from adenylate cyclase" Proc.Natl.Acad.Sci.USA 85 5478-5482.

Frielle, T., Collins, S., Daniel, K.W., Caron, M.G., Lefkowitz, R.J. and Kobilka, B.K.(1987) "Cloning of the cDNA for the human ß1-adrenergic receptor" Proc.Natl.Acad.USA .84 7920-7924.

Jackson, T.R., Blair, L.A.C., Marshall, J., Goedert, M., and Hanley, M.R. (1988) " The mas oncogene encodes an angiotensin receptor". Nature 335, 437-440.

Julius, D., MacDermott, A.B., Axel, R., Jessel, T.M. (1988) "Molecular characterization of a functional cDNA encoding the serotonin 1c receptor". Science 241, 558-564.

Kaveri, S.V., Cervantes-Olivier, P., Delavier-Klutchko, C. and Strosberg, A.D. (1987) "Monoclonal antibodies directed againt the human A431 $ß_2$-adrenergic receptor recognize two major polypeptide chains".Eur.J. Biochem. 167, 449-445.

Kobilka, B.K., Matsui, H., Kobilka, T.S., Yang-Feng, T.L., Francke, U., Caron, M.G., Lefkowitz, and R.J.and Regan, J.W. (1987) " Cloning, Sequencing, and Expression of Gene Coding for the Human Platelet α_2-Adrenergic Receptor".Science 238 650-656

Kobilka, B.K., Kobilka, T.S. , Daniel, K., Regan, J.W., Caron, M.G., Lefkowitz, R.J. (1988) "Chimeric α_2-$ß_2$-adrenergic receptors : Delineation of domains involved in effector coupling and ligand binding specificity" (1988) Science 240, 1310-1316.

Lübbert, H., Hoffman, B.J.,Nguyen, M.H.T. and Le, IM.H.T. (1988) :"Isolation by functional expression and analysis of serotonin 5-HT1C receptor cDNA and genomic clones".Proc.Eur.Neurosci. Assoc. Zürich 1988, Abstr. 58.1, p.225.

Marullo,S., Delavier-Klutchko,C., Eshdat, Y., Strosberg, A-D,. and Emorine, L. (1988):"Human β_2-adrenergic receptors expressed in Escherichia coli membranes retain their pharmacological properties" Proc.Natl.Acad.Sci.USA 85, in press

Masu, Y., Nakayama, K., Tamaki, H., Harada, Y., Kuno, M., and Nakanishi, S. (1987) :"cDNA cloning of bovine substance-K receptor through oocyte expression system" Nature 329 836-838.

Payne, G.S., Baker, D., Van Tuinen, E., Schekmann, R. (1988) "Protein transport to the vacuole and receptor-mediated endocytosis by clathrin heavy chain-deficient yeast" J.Cell Biol. 106 1453-1461.

Peralta, E.G., Ashkenazi, A., Winslow, J.W., Smith, D.H., Ramachandran, J., and Capon, D.J. (1987) "Distinct primary structures, ligand-binding properties and tissue-specific expression of four human muscarinic acetylcholine receptors" EMBO J. 6 3923-3929.

Raposo, G. Dunia I., Marullo S., Guillet J.G., André C., Strosberg A.D.., Benedetti E.L. and Hoebeke J. (1987) "Redistribution of muscarinic acetylcholine receptors on CCL137 fibroblasts induced by regulatory ligands" Biol.Cell 60,117-124.

Raposo, G., Dunia, I., Benedetti, E.L. and Strosberg, A.D. (1988) : Agonist-induced internalization of β–adrenergic receptors and EGF-accompanied endocytosis of EGF-receptor occur through different and independent mechanisms. Eur.J.Cell.Biol. Submitted for publication.

Regan, J.W., Kobilka, T.S., Yang-Feng, T.L., Caron, M.G., Lefkowitz, R.J., and Kobilka, B.K. (1988) :" Cloning and expression of a human kidney cDNA for an α_2-adrenergic receptor subtype" Proc.Natl.Acad.Sci.USA 85 6301-6305

Vauquelin, G., Bottari, S., Kanarek, L., and Strosberg, A.D. (1979) "Evidence for essential disulfide bonds in β_1-adrenergic receptors of turkey erythrocyte membranes"J.of Biol.Chem.254 4462-4469.

Wong, S.K.F., Slaughter, C., Ruoho, A.E. and Ross, E.M.(1988) "The catecholamine binding site of the β-adrenergic receptor is formed by juxtaposed membrane-spanning domains" J.of Biol.Chem.263 7925-7928

Young, D., Waitches, G., Birchmeier, C., Fasano, O., and Wigler, M. (1986) : "Isolation an characterization of a new cellular oncogene encoding a protein with multiple potential transmembrane domains" Cell 45 711-719.

CLONING OF A RAT D_2-DOPAMINE RECEPTOR cDNA

Olivier Civelli, James Bunzow, Hubert van Tol, David Grandy, Paul Albert, John Salon, Curtis Machida and Kim Neve
Vollum Institute for Advanced Biomedical Research
Oregon Health Sciences University
3181 SW Sam Jackson Park Road
Portland, OR 97201

Recent structural analyses have revealed that neurotransmitter receptors can be divided into two families: the G protein-coupled receptors and the ligand-gated ion channels (Stevens, 1987; Dohlman, 1987; Hall, 1987).

The ligand-gated ion channels are made of several subunits each spanning the membrane four or five times. The subunits aggregate to form a channel which can be opened or closed upon interaction of specific extracellular ligands (Stevens, 1987). Each subunit might have its corresponding ligand. Each subunit of one receptor complex has a high degree of sequence similarity with the other subunits from the same complex. Moreover, it has been shown that subunits from different receptor complexes do also share a significant degree of sequence similarity, suggesting that these receptors are members of a large gene family. The prototypical receptor of this gene family is the nicotinic acetylcholine receptor (Noda et al., 1982).

The G protein-coupled receptors are made of one single subunit which is expected to span the membrane seven times. These receptors share also a significant degree of sequence similarity, suggesting again that they are part of a large gene family. The first cloned receptor of this type was the rhodopsin receptor, since then the adrenergic receptors have been extensively studied (Dohlman et al., 1987). We are interested in the G protein-coupled receptors.

THE G PROTEIN-COUPLED RECEPTORS

The G protein-coupled receptors owe their name to their mode of action: they activate second messenger systems by first interacting with one of the

NATO ASI Series, Vol. H 32
Molecular Biology of Neuroreceptors and Ion Channels. Edited by A. Maelicke

transducing proteins, the G proteins (Berridge, 1984). This family of receptors comprise some important neurotransmitters and most of the neuropeptide receptors. Several members of this family have been cloned and sequenced. These experiments have showed three structural characteristics which are found in all the thus-far cloned G protein-coupled receptors (Dohlman et al., 1987; Masu et al., 1987; Julius et al., 1988). First, seven hydrophobic stretches of about 20 amino acid residues are present in these receptors, indicating that these receptors might span seven times the membrane. Second, several amino acid residues are positioned at conserved sites on these molecules, such as those for potential glycosylation in the N-terminal part of the molecule, the potential phosphorylation sites in the cytoplasmic loops and about twenty more conserved residues. Third, these receptors share a significant level of sequence similarity with the other members of this gene family. It should be stressed that all the G protein-coupled receptors which will be discovered in the future might not fulfill these criteria. We only want to argue that a protein which fulfills these criteria can be accepted as part of the G protein-coupled receptor family.

These common features are of importance because they open the possibility of isolating new G protein-coupled receptors on the basis of their similarities to the other receptors part of the same gene family.

CLONING OF THE D_2-DOPAMINE RECEPTOR

We have developed a strategy based on nucleotide sequence homologies to isolate genes coding for new G protein-coupled receptors. We show that, by applying this strategy, we have isolated the cDNA encoding the rat D_2-dopamine receptor. This receptor is characterized on the basis of three criteria (1) its sequence which has all the characteristics expected from a G protein-coupled receptor, (2) its tissue distribution which overlaps with its of the D_2-dopamine receptor and (3) its pharmacological profile as measured on the membranes of Ltk^- cells transformed with this cDNA.

Cloning Strategy

Our cloning strategy uses the DNA coding one of a G protein-coupled receptor, the β_2-adrenergic receptor (β_2AR), as an hybridization probe in screening a rat genomic library under low stringency conditions. These conditions have been shown to detect a large array of DNA fragments in analyses of rat genomic DNA.

A partial hamster genomic library (EcoRI digested DNA size 4 -7 kb) was constructed in the phage λgt10. The hamster β_2-adrenergic receptor gene was isolated using specific oligonucleotides designed according to the published sequence (Dixon et al., 1986). A 1.3 kb DNA fragment, known to contain most of the receptor coding sequence, was used as hybridization probe to screen a rat genomic library under low stringency hybridization. Several clones were found to hybridize to the hamster probe under these conditions, in particular, one positive clone, RGB-2, was found to contain an hybridizable 0.8 kb EcoRI-PstI fragment. This fragment was sequenced and shown to contain a stretch of nucleotides with a high degree of identity (31 out of 40) with the nucleotide sequence of transmembrane VII of the hamster β2AR. In addition, the peptide sequence corresponding to this DNA fragment contained a significant degree of sequence similarity with transmembrane domains VI and VII of the β2AR. Further analysis of this fragment revealed, however, that it contains a 3′ intron splice site and approximately 500 bp of putative intronic sequences. The 0.8 kb EcoRI-PstI fragment was used to probe a rat brain cDNA library. Under high stringency conditions, two positive clones were identified. Sequence and restriction analysis suggest that the two clones were replicas of the same sequence and contain the sequence of the genomic fragment used as probe.

The RGB-2 Receptor Clone

The RGB-2 clone was totally sequenced. It contains 2455 bases. This cDNA can code for a 415 amino acid protein (Mr=47064). Determination of the start of initiation was defined by the presence of a Kozak consensus sequence and because we have shown that our cDNA has approximately the same size as the mRNA from which it is derived. Worth mentioning, the 5′ untranslated part of the mRNA contains an in-frame dipeptide 42 bases upstream of the initiation start. The 3′ untranslated region of the mRNA contains one poly A addition signal.

The 415 amino acid long polypeptide was analyzed for the criteria known to belong to G protein-coupled receptors. By carrying on an hydrophobicity plot analysis, we found out that this protein contains seven hydrophobic stretches. The first five are located in the same position on the molecule as are the first five hydrophobic stretches found in the β_2-adrenergic receptor. In contrast, the two last are separated from the others by a long peptide sequence, indicating that this receptor would have a large third cytoplasmic loop.

It was also found that the RGB-2 protein contains three potential glycosylation sites in its N-terminus and one cAMP-dependent protein kinase phosphorylation site in its presumed third cytoplasmic loop. In addition, all of the 21 amino acid residues expected to be conserved among the G protein-coupled receptors have been found in this protein, in particular, the well conserved LA.ADL sequence in the second transmembrane domain and the aspartic residue known to be important in cationic amine binding.

Finally, the RGB-2 protein was shown to share sequence similarities with G protein-coupled receptors. The regions of greatest amino acid similarity are clustered in the putative transmembrane domains. Within these domains the RGB-2 sequence is 49%, 45%, 41%, 39%, 37%, 30% and 24% respectively with the human α_2-adrenergic-, the human G21-, the hamster β_2-adrenergic-, the human β_1-adrenergic-, the porcine m2 muscarinic-, the porcine ml muscarinic- and the bovine substance K receptors (Dohlman et al., 1987).

Interestingly, the RGB-2 protein as the α_2-adrenergic receptor contains a large cytoplasmic loop between transmembrane domains V and VI and a short C-terminus. This structural organization appears to be characteristic of receptors which interact with a Gi protein complex. All together these data indicate that the RGB-2 gene can code for a G protein-coupled receptor.

LIGAND SPECIFICITY OF THE RGB-2 RECEPTOR

The next question was which receptor is the RGB-2 receptor. It is clear that the approach we took can result in the isolation of clones which might code for known or thus far undescribed neurotransmitters or neuropeptides. We knew from our sequence comparisons that RGB-2 was apparently not part of one of the G protein-coupled receptor subfamilies, such as the adrenergic or the muscarinic families, since the level of identity between RGB-2 and these cloned receptors was too low.

Tissues Distribution of the RGB-2 mRNA

As the first approach in determining the ligand specificity of the RGB-2 receptor, we undertook a study of its site of synthesis. The RGB-2 cDNA served as hybridization probe in Northern blot analysis of mRNA isolated from selected rat tissues. The RGB-2 mRNA is about 2500 bases long and is differentially expressed in a different part of the brain, the highest concentration being in the basal ganglia (data not shown). Very interestingly, RGB-2 mRNA sequences are also present in high level in the neurointermediate lobe of the pituitary and in lower concentration in the anterior lobe. This particular tissue distribution of the RGB-2 mRNA points to the localization of the D_2-dopamine receptor as determined by receptor autoradiography (Boyson et al., 1986). In addition, the RGB-2 mRNA was also found to be present or absent in peripheral tissues and cell lines where the dopamine receptor can or cannot be detected. Since no other receptors were known to display this particular tissue pattern, we set up to demonstrate if RGB-2 was or was not a dopamine receptor. We, therefore, needed to determine the ligand specificity of the RGB-2 protein.

Pharmacological Profile of the RGB-2 Receptor

To determine the ligand specificity of RGB-2, the cDNA was cloned into the eukaryotic expression vector pZem3 which contains the mouse metallothionein promoter. This plasmid was cotransfected into mouse fibroblast Ltk$^-$ cells with the selective neomycin transferase gene (pRSVneo). Stable transformants were isolated and analyzed for their content in RGB-2 mRNA. The parent cells did not express this mRNA.

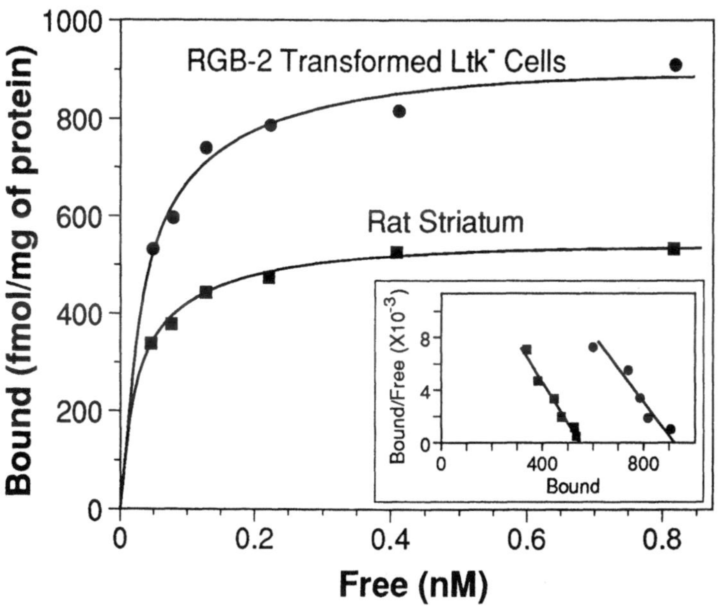

Fig 1. Saturation isotherms of the specific binding of spiperone to membranes prepared from L-RGB2Zem-1 cells and rat striatum. (+)-butaclamol was used to define non-specific binding. Insert: Scatchard transformation of the data.

Membranes were prepared from RGB-2 transformed and non-transformed Ltk$^-$ cells and were tested for their ability to bind specifically the D_2-dopamine ligand, ^{3}H-spiperone by displacement against (+)-butaclamol or sulpiride. Control Ltk$^-$ cell membranes were unable to bind spiperone. Binding of spiperone to membranes prepared from a clone of transfected cells, L-RGB2Zem-1, was saturable with a Kd value of 48 pM (Figure 1). The density of binding sites as determined in four experiments was 945 fmol/mg of protein in L-RGB2Zem-1 membranes and 454 fmol/mg of protein in rat striatal membranes. The binding of spiperone to membranes of L-RGB2Zem-1 cells was inhibited by a number of drugs and the resulting K_i values (Cheng and Prusoff, 1973) closely matched those obtained using striatal membranes (Figure 2 and Table 1). The D_2 antagonists (+)-butaclamol and haloperidol were the most potent inhibitors followed by

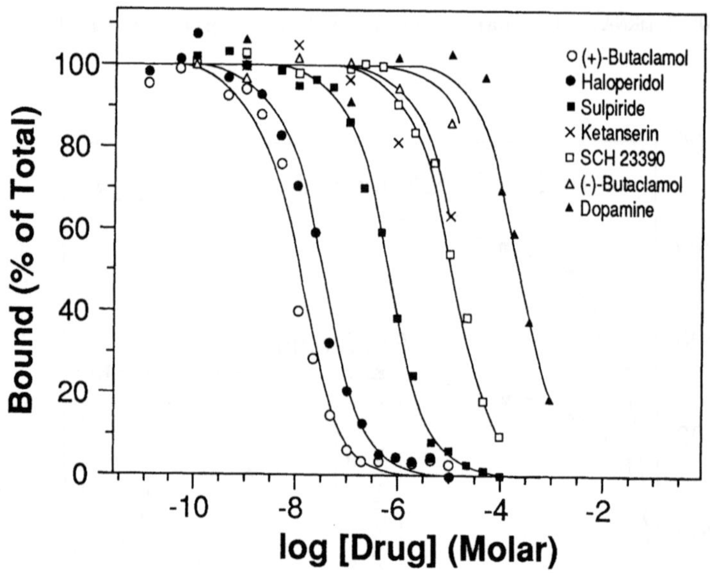

Figure 2. Competition curves using L-RGB2Zem-1 membranes. Representative curves are shown for inhibition of specifically bound spiperone by drugs in membranes from L-RGB2Zem-1 cells. Each drug was tested three times.

sulpiride. The D_1 antagonist SCH 23390 and the serotonin 5HT2 antagonist ketanserin were much less potent at blocking spiperone binding. The binding appeared to be stereoselective as (+)-butaclamol was much more potent than (-)-butaclamol at inhibiting binding. In these experiments the absolute affinities of dopaminergic antagonists and the rank order of potency of the drugs (spiperone > (+)-butaclamol > haloperidol > sulpiride >> (-)-butaclamol) agree closely with the known values for the D_2-dopamine receptor.

Drug	Ki (nM)		
	RGB-2 Transformed Ltk⁻ Cells	Rat Striatum	
(+)-Butaclamol	0.83	1.0	
(-)-Butaclamol	>1,000	>1,000	
Haloperidol	3.0	5.3	
Dopamine + GTP	17,000	6,300	
Sulpiride			
high affinity	80	67	(87%)
low affinity	——	>10,000	(13%)
SCH 23390			
high affinity	——	35	(16%)
low affinity	1,000	780	(84%)
Ketanserin			
high affinity	——	27	(25%)
low affinity	>1,000	>1,000	(75%)

Table 1. K_i values for L-RGB2Zem-1 and rat striatum. The IC 50 values calculated in Figure 2 were converted to K_i values (Cheng and Prusoff, 1973). Results are geometric means of 3 experiments in which spiperone binding was inhibited by various concentrations of unlabelled drug. The presence of two binding sites in rat striatal membranes is consistent with the low affinity binding of spiperone to D_1 dopamine receptor sites and to serotonin 5HT2 receptor sites.

All binding data for L-RGB2Zem-1 were fit best by assuming the presence of only one class of binding sites. On the other hand, inhibition by several drugs of spiperone binding to rat striatal membranes was fir best by assuming the presence of two classes of binding sites. Thus, SCH 23390 and ketanserin inhibited 10-20% of spiperone binding to rat striatal membranes with high affinity. It seems likely that this binding, which is not displaced by sulpiride, represent binding of spiperone to 5HT2 serotonin receptors. The pharmacological data presented here proves that the binding profile of the D_2-dopamine receptor is found in Ltk^- after transformation with the RGB-2 cDNA.

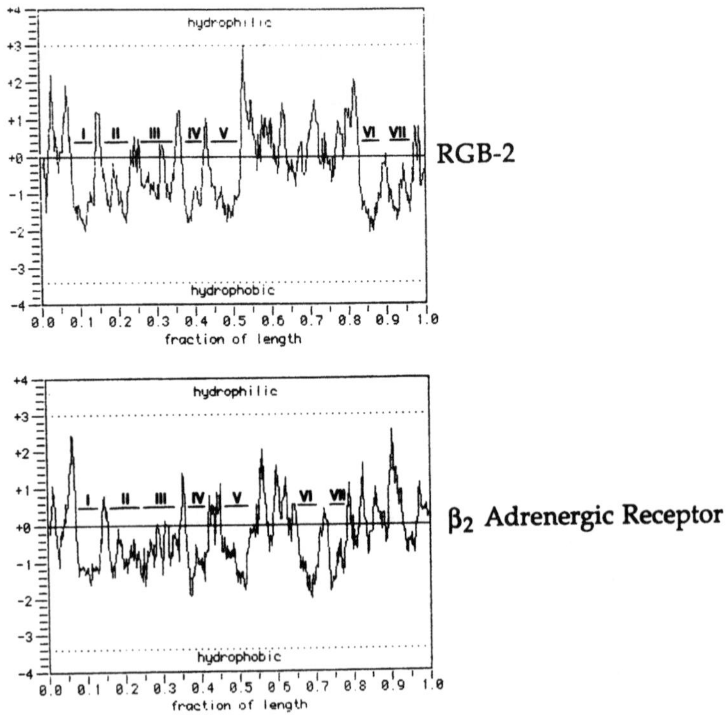

Fig 3. Hydrophobicity profiles of D_2 dopamine and the β_2-adrenergic receptors. The Hopp-Woods analysis program was employed.

CONCLUSION: RGB-2 IS THE D_2-DOPAMINE RECEPTOR

In conclusion, we have cloned a rat cDNA which, when transfected into eukaryotic cells induces the expression of a D_2-dopamine binding protein. In addition we have shown that its corresponding mRNA is localized in the tissues where the D_2-dopamine receptor is known to be present. Since this mRNA codes for a protein which has all the expected characteristics of a G protein-coupled receptor, in which the D_2-dopamine receptor is thought to be part of, we conclude that we have cloned the rat D_2-dopamine receptor. Of particular interest is the presence in the gene sequence of at least one intron upstream to the sixth transmembrane domain. This finding is important for our understanding of the evolution of the G protein-coupled receptors since most of them do not have intron within their coding sequence. Finally, it is worth mentioning that the cloning strategy we have followed can lead to the isolation of numerous receptor cDNAs and that the successful cloning of the D_2-dopamine receptor will open new research directions in the studies of this important receptor.

ACKNOWLEDGEMENTS

This work is dedicated to the memory of Ed Herbert and is supported by a NIH grant and a grant from Cambridge NeuroScience Research, Inc. to OC.

REFERENCES

Berridge MJ (1984) The molecular basis of communication within the cell. Sci Am 66:142-152

Boyson SJ, McGonigle P, Molinoff PB (1986) Quantitative autoradiographic localization of the D_1 and D_2 subtypes of dopamine receptors in rat brain. J of Neurosci 6:3177-3188

Cheng YC, Prusoff WH (1973) Relationship between the inhibition constant (K_i) and the concentration of inhibitor which causes 50 percent inhibition (I50) of an enzymatic reaction. Biochem Pharmacol 22:3099-3108

Dixon RAF, Koblika BK, Strader DJ, Benovic JL, Dolhman HG, Frielle T, Bolanowski M, Bennett C, Rands E, Diehl R, Mumford R, Slater E, Sigal I, Caron M, Lefkowitz R, Strader C (1986) Cloning of the gene and cDNA for mammalian β-adrenergic receptor and homology with rhodopsin. Nature 321:75-79

Dohlman HG, Caron MG, Lefkowitz RJ (1987) A family of receptors coupled to guanine nucleotide regulatory proteins. Biochemistry 26:2657-2664

Hall ZA (1987) Three of a kind: the β-adrenergic receptor, the muscarinic acetylcholine receptor, and rhodopsin. Trends in Neuro Sci 10:99-100

Julius D, MacDermott AB, Axel R, Jessell TM (1988) Molecular characterization of a functional cDNA encoding the serotonin 1c receptor. 241:558-564

Masu Y, Nakayama K, Tamaki H, Harada Y, Kuno M, Nakanishi S (1987) cDNA cloning of bovine substance-K receptor through oocyte expression system. Nature 329:836-838

Noda M, Takahashi H, Tanabe T, Toyosato M, Furutani Y, Hirose T, Asai M, Inayama S, Miyata T, Numa S (1982) Primary structure of α subunit precursor of Torpedo californica acetylcholine receptor deduced from cDNA sequence. Nature 299:793-797

Stevens CF (1987) Channel families in the brain. Nature 328:198-199

Molecular characterization of neuropeptide receptors and of a novel membrane protein involved in K^+ permeation.

Shigetada Nakanishi, Hiroaki Ohkubo, Yasuo Masu, Toru Takumi, Hiromi Tamaki and Yoshifumi Yokota
Institute for Immunology, Kyoto University Faculty of Medicine, Yoshida, Kyoto 606, Japan

Peptide receptors and ion channels generally represent minor cellular components and are tightly embedded in the plasma membrane. Therefore, the conventional biochemical approach is not readily available for the characterization of these proteins. For the past few years, we have been working on the mammalian neuropeptide tachykinin system to understand the regulatory mechanisms underlying interneuronal communications. These investigations have indicated that the neuropeptide system is diversified not only at the level of peptide production but also at the level of receptor interactions (1, 2). However, the nature of the tachykinin receptors remained to be clarified. We therefore developed a new strategy to characterize the receptors at the molecular level by combining the molecular cloning method with electrophysiological techniques. Because this strategy is also applicable to molecular cloning of ion channels and their

NATO ASI Series, Vol. H 32
Molecular Biology of Neuroreceptors and Ion Channels. Edited by A. Maelicke

associated modulatory proteins, we extended it to investigate the molecular nature of these proteins. This article describes our studies of the tachykinin receptors and of a novel membrane protein that is involved in inducing selective K^+ permeation by membrane polarization.

TACHYKININ RECEPTORS

The mammalian tachykinin system consists of three distinct peptides: substance P, substance K, and neuromedin K (1, 2). Substance P is one of the best characterized neuropeptides and probably acts as a neuromediator involved in the transmission of pain impluses by primary sensory neurons. Substance K/ neuromedin K, though possessing biological activities characteristic of the tachykinin peptides, differs in its biological potencies from those of substance P (2, 3). These pharmacological studies as well as ligand-binding studies suggested that there are at least three distinct receptors, each specific for the three mammalian tachykinins (3, 4). However, their molecular nature remained to be clarified. The brain and stomach mRNAs induce functional substance P and substance K receptors, respectively, when injected into Xenopus oocytes, and these expressions can be assayed electrophysiologically by application of the respective receptor ligand (5). We extended this observation and developed a new cloning strategy to isolate a functional cDNA clone for bovine substance K receptor (6). We constructed a cDNA library from a size-fractionated stomach

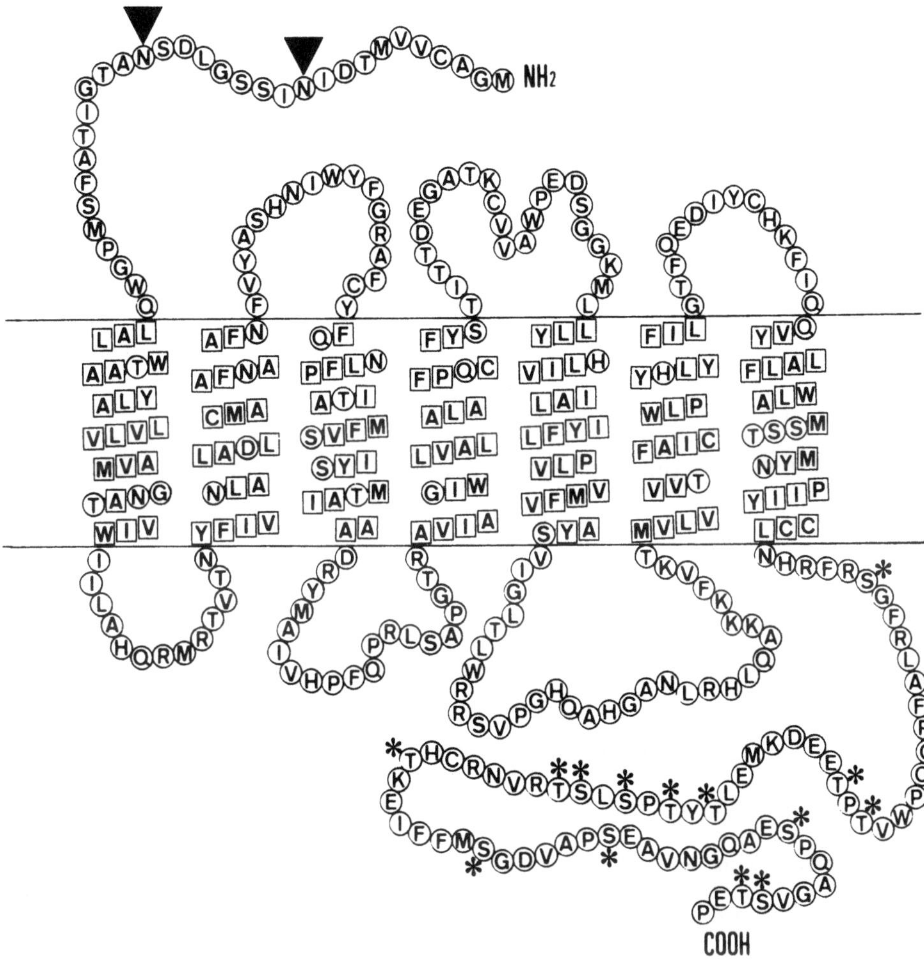

Fig. 1. A transmembrane model of substance K receptor. Triangles, potential glycosylation sites; asterisks, possible phosphorylation sites; squared amino acids, hydrophobic amino acids in the putative transmembrane domains. Data taken from ref. 6,

mRNA. The vector DNA used for this cloning contained an RNA polymerase promoter next to the insertion site of the cDNA, thus allowing the *in vitro* mRNA synthesis from the cloned cDNA by a specific RNA polymerase. We electrophysiologically tested for the receptor expression following injection of the *in vitro* synthesized mRNA into the oocyte system and identified a cloned cDNA mixture containing a receptor cDNA clone. A response-evoking cloned cDNA mixture was then repeatedly fractionated and we finally obtained a single cDNA clone that encodes the substance K receptor (6).

The bovine substance K receptor consists of 384 amino acid residues possessing seven hydrophobic segments and shares significant sequence similarity with rhodopsin-type receptors (7). Figure 1 shows a transmembrane model for the substance K receptor, based on the structural similarity to rhodopsin-type receptors. The seven putative transmembrane domains consist of 20-24 uncharged amino acid residues. The transmembrane segments II, V and VI each contains a charged amino acid. The Asp in segment II is conserved for all rhodopsin-type receptors, whereas the His residues in segments V and VI are characteristic of the substance K receptor. The amino-terminal region with no signal sequence contains two potential N-glycosylation sites, while the carboxyl-terminal region has many serine and threonine residues as possible phosphorylation sites. Thus, these terminal regions of the substance K receptor also show a pattern similar to those of the rhodopsin-type receptors. The results obtained thus

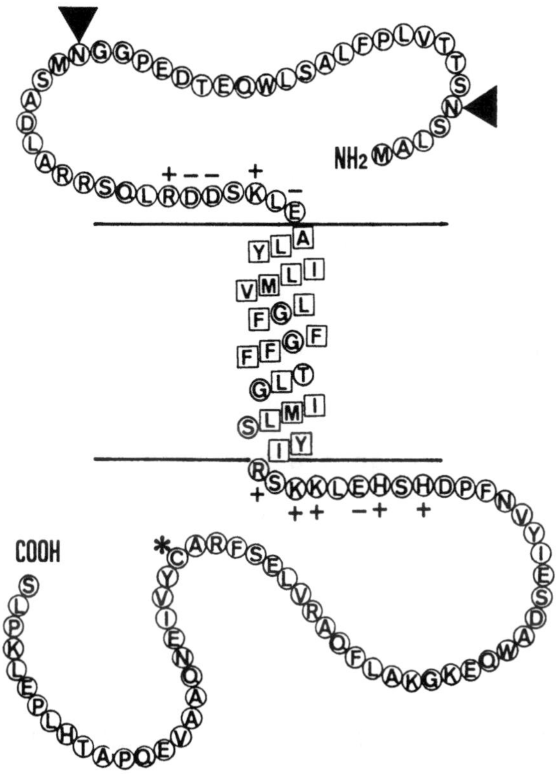

Fig. 2. A transmembrane model of a protein involved in voltage-dependent, selective permeation of potassium ions. Triangles, potential N-glycosylation sites; + and -, positively and negatively charged residues at the membrane-flanking regions; asterisk, a cysteine residue; squared amino acids, hydrophobic amino acids in the putative transmembrane domain. Data taken from ref. 10.

demonstrate that the structure and perhaps the function of the neuropeptide receptor are fundamentally similar to those of the receptors for classical small molecule transmitters such as catecholamine and acetylcholine. Furthermore, these structural characteristics suggest that the substance K receptor is associated with a G protein to effect the second messenger system involved in the substance K function. In fact, an electrophysiological examination indicated that Ca^{++}-dependent activation of a chloride ion channel serves as an ionic mechanism for the substance K receptor reaction in the oocyte system (5). Thus, the substance K receptor expressed in the oocyte acts through a G protein to cause production of IP_3, which in turn elevates cytoplasmic Ca^{++}. This Ca^{++} then activates a Ca^{++}-dependent chloride channel in oocyte plasma membrane.

A NOVEL MEMBRANE PROTEIN INVOLVED IN SELECTIVE K^+ PERMEATION

Ion channels exhibit a high degree of diversity, varying in their electrophysiological and pharmacological properties (8). However, the molecular nature of the ion channels with distinct properties and their associated modulatory proteins remained to be determined. The kidney epithelial cells are known to control various types of ion permeations to maintain the electrolyte homeostasis in the body (9). Because rat kidney mRNAs were found to induce a very slowly activating outward current by membrane depolarization after being expressed in *Xenopus* oocytes, we isolated a functional cDNA clone for the mRNA responsible for this induction by using the cloning method described above (10).

Electrophysiological characterizations revealed that the voltage-dependent outward current is carried selectively by potassium ions. The sequence determination showed that the cloned cDNA encodes a polypeptide consisting of 130 amino acids. The hydropathicity analysis indicated a clear hydrophobic segment consisting of 23 continuous uncharged amino acids in the middle portion of the protein. Consistent with this finding, the translation product derived from the mRNA was located in the membrane fraction of oocytes. This protein, however, did not show any sequence similarities to known protein sequences including the ion channel proteins. Thus, despite the fact that the protein identified is capable of inducing selective permeation of K^+ ions by membrane polarization, it represents a novel membrane protein that differs from the conventional ion channel proteins, as schematically illustrated in Figure 2.

The novel membrane protein differs from the known ion channel proteins not only in its structural characteristics but also in its electrophysiological properties. The protein is considerably small and contains only a single putative transmembrane domain. The K^+ current induced by this protein is unusually slow in both activation and deactivation after membrane polarization. These characteristics resemble those of channel-forming peptide ionophores (11). Gramicidin and alamethicin are examples of the channel-forming peptide ionophores. Although these peptides are small and are made up solely or almost solely of uncharged amino acids, they form a pore that allows selectively monovalent cations to flow down along their electrochemical gradient (11). Thus, the simplest and most likely explanation for the function of the novel membrane protein is that it functions as a discrete K^+-conducting ion channel. Alternatively, it remains possible that the protein may subserve as a modulatory protein that activates endogenous potassium channels.

The mRNA for this protein is distributed in tissues comprising glandular epithelial cells. Various types of epithelial cells are able to permeate potassium ions into the interstitial space as well as into the lumen, when intracellular K^+ concentrations are increased by the sodium-potassium ATPase pump (9). These K^+ permeations are thought to be mediated through K^+ channels. Thus, an intriguing hypothesis for the role of this protein is its involvement in the permeation of potassium ions in epithelial cells to control the K^+ homeostasis. In any event, K^+ channels are known to exhibit a high degree of diversity, and the protein reported here will provide a new insight into the molecular nature of the K^+ channels and their modulatory proteins.

ACKNOWLEDGMENTS

Our studies were partly supported by research grants from the Ministry of Education, Science and Culture of Japan, the Institute of Physical and Chemical Research, and the Science and Technology Agengy of Japan.

REFERENCES

1. Nakanishi, s. (1986) Trends Neurosci., 9, 41-44.
2. Nakanishi, S. (1987) Physiol. Rev., 67, 1117-1142.
3. Nawa, H., Doteuchi, M., Igano, K., Inouye, K. and Nakanishi, S. (1984) Life Sci., 34, 1153-1160.
4. Quirion, R. (1985) Trends Neurosci., 8, 183-185.
5. Harada, Y., Takahashi, T., Kuno M., Nakayama, K., Masu, Y. and Nakanishi, S. (1987) J. Neurosci., 7, 3265-3273.
6. Masu, Y., Nakayama, K., Tamaki, H., Harada, Y., Kuno, M. and Nakanishi, S. (1987) Nature, 329, 836-838.
7. Dohlman, H.G., Caron, M.G. and Lefkowitz, R.J. (1987) Biochemistry, 26, 2657-2664.
8. Kaczmarek, L.K. and Levitan, I.B. (1987) Neuromodulation: the Biochemical Control of Neuronal Excitability. Oxford University Press, New York, Oxford.
9. Schultz, S.G. (1986) In: Andreoli, T.E., Hoffman, J.F., Fanestil, D.D. and Schultz, S.G. (eds) Membrane Transport Processes in Organized System. Plenum, New York, London, pp. 135-150.
10. Takumi, T., Ohkubo, H. and Nakanishi, S. (1988) Science, in press.
11. Eisenman, G. and Dani, J.A. (1987) Annu. Rev. Biophys. Biophys. Chem., 16, 205-226.

A 64 kDa protein is a candidate for a TRH receptor in prolactin-producing rat pituitary tumour cells.

Marianne Wright, Jan O.Gordeladze, Anders Høgset,
Peter Alestrøm and K.M.Gautvik
Institute of Medical Biochemistry
University of Oslo, Oslo, Norway

Introduction

Binding of the hypothalamic tripeptide TRH to specific membrane receptors stimulates Prl secretion and synthesis. The action of TRH (Fig.1) has been extensively studied in clonal strains of pituitary tumour cells (Table 1) which have retained their ability to secrete hormone and are sufficiently differentiated to generate and express hormone receptor signal systems analogous to normal anterior pituitar cells (for review see Tashjian and Hoyt 1972, Gautvik et al.1984). TRH induces the activation of adenylate cyclase (AC) and phospholipase C (PLC) through G_S- and G_{plc}-proteins, respectively (Gordeladze et al.1987). $[Ca^{2+}]_i$ peaks rapidly within seconds due to the enhanced production of inositolpolyphosphates while cAMP elevation occurs within minutes (Fig.2, K.M.Gautvik et al.1988). The TRH stimulated adenylate cyclase is rapidly and totally desensitized by TRH, while phospholipase C is only partially down regulated (Table 2). The two signal pathways are crosslinked by the inactivation of G_i by protein kinase C, thereby enhancing cAMP synthesis further (Gordeladze et al.1988).

The binding characteristics of the TRH receptor has been described in detail (Hinkle and Tashjian 1973, Gourdji 1980, Gautvik et al.1983) and can be summarized as follows:
I. $[^3H]$-TRH is stereospecific and the binding affinities of TRH structure analogs parallel their biological potencies; II. The binding of $[^3H]$-TRH shows the existence of two independent binding sites with K_D s of about 1 nM and 25 nM, respectively, 1/3 of the calculated 10^5 receptors per cell represent the higher affinity binding site; III. The binding

NATO ASI Series, Vol. H 32
Molecular Biology of Neuroreceptors and
Ion Channels. Edited by A. Maelicke

is reversible within 24 hours and IV. The binding shows complete homologous desensitization.

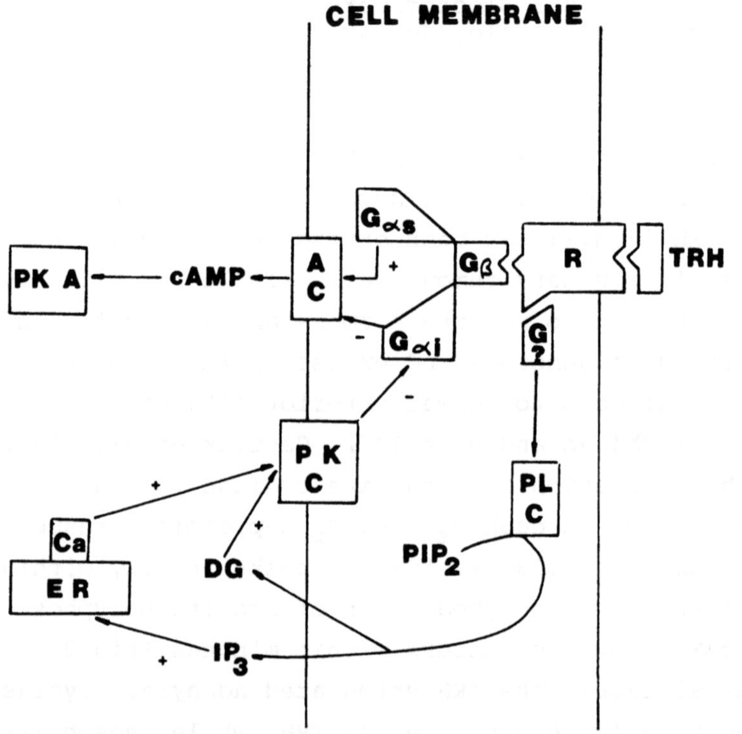

Figure 1: Schematic illustration of the TRH induced second messenger system

	produce Prl	GH	TRH-R	G-proteins G_s	$G_{i/k}$
GH_12C_1	–	++	–		
GH_3	+	+	$5 \cdot 10^4$	+	++
GH_4C_1	++	–	$10 \cdot 10^4$	+++	+

Table 1: Characteristics of GH - cell clones

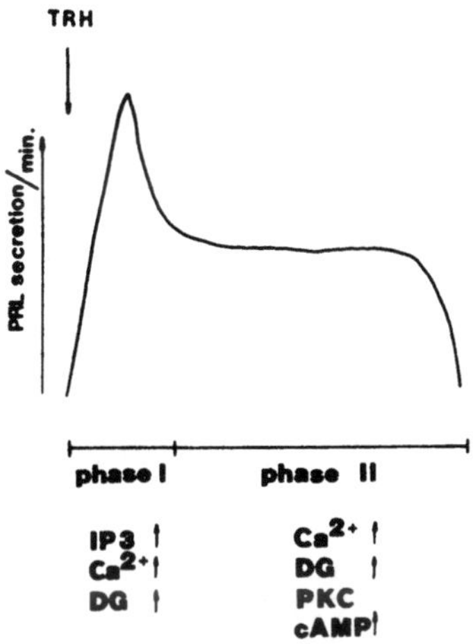

Figure 2: Time course of Prl secretion after TRH stimulation

Preincubation (30 min)	AC (20 min)	IP_3-accumulation (5 sec)
Control	2.0	12.4
TRH (10^{-6}M)	1.0	6.3
VIP (10^{-6}M)	1.9	13.9
TPA (10^{-6}M)	1.9	10.1
Ionomycin (20μM)	2.0	5.4

Table 2: Desensitation of TRH(10^{-6}M)-stimulated AC and PL-C activities (relative to basal) in intact GH_4C_1 cells.

Identification of a TRH receptor protein

[^{3}H]-TRH has been crosslinked to rat pituitary tumour cells by UV- illumination. The total cell protein was subsequently fractionated by SDS polyacrylamide gel electrophoresis and blotted over to nitrocellulose filter. Exposure of the scintillator-impregnated filters to x-ray film revealed a 64 kDa [^{3}H]-labelled protein. The incorporation of [^{3}H]-TRH to this protein is UV-dose dependent and is inhibited by an excess of cold TRH . In addition, this 64 kDa protein can be visualized by antiserum raised against surface antigens of the receptor-positive cell strain (GH_4C_1). In a related cell line (GH_12C_1) which does not bind [^{3}H]-TRH, the 64 kDa protein cannot be demonstrated by UV - crosslinking or on immunoblots using the same antiserum (Wright et al.).

References

Gautvik, K.M., Gordeladze J.O., Jahnsen,T., Haug E., Hansson, V. and Lystad, E.(1983) J.Biol. Chem. 258,10304-10311

Gautvik, K.M., Haug, E., Sand, O., Iversen, J.-G., Johansen, P.W., Sletholt, K. and Gordeladze, J.O. (1984) Regulation of Prolactin secretion by thyrolibrin (TRH) in cultured rat pituitary tumor cells. In: Regulation of Target Cells Responsiveness, eds. McKerns, K.W., Aakvaag, A. and Hansson, V., Vol.2, Plenum Publishing Corporation, New York

Gautvik, K.M., Bjøro, T., Sletholt, K., Østberg, B.Chr., Sand, O., Torjesen, P. Gordeladze, J.O.,Iversen, J.-G. and Haug, E.(1988) Molecular Mechanism in Secretion, Alfred Benzon Symposium 25, 211-227; eds. Thorn, N.A., Treiman, M., Petersen, O.H., Munksgaard, Copenhagen

Gordeladze, J.O., Bjøro, T., Østberg, B.Chr., Sand, O., Torjesen, P., Haug, E., and Gautvik, K. M.(1987) Bio-chem.Pharmacol.37,(16),3133-3138

Gordeladze J.O. et al.(1988) Eur.J.of Biochem.(in press)

Gourdji, D.,(1980) In: Synthesis and release of adenophysop-hyseal hormones, eds. Utisz,M.J and Kerns, K.W.,pp.463-491, Plenum Press, New York

Hinkle, P.M. and Tashjian, A.H.Jr.(1973) J.Biol.Chem.248, 6180-6186

Tashjian, A.H.Jr. and Hoyt, R.F.Jr. (1972) In: Molecular and General Developmental Biology, ed.Sussman, M., pp. 353-387. Prentiss-Hall, Eaglewood Cliffs, New Jersey, USA

Wright, M., Høgset, A., Aleström, P. and Gautvik K.M. (1988) Biochem.Biophys.Res.Comm.(in press)

COMPARISON OF THE PRESYNAPTIC VESICLE COMPONENT SYNAPTOPHYSIN AND GAP JUNCTION PROTEINS: A CLUE FOR NEUROTRANSMITTER RELEASE?

Leo Thomas, Petra Knaus and Heinrich Betz
Zentrum für Molekulare Biologie, Universität Heidelberg, Im Neuenheimer Feld 282, D-6900 Heidelberg, Federal Republic of Germany

INTRODUCTION

Synaptophysin is a major integral membrane protein of presynaptic vesicles (Wiedenmann and Franke, 1985; Jahn et al., 1985). Elucidation of its primary structure by different groups in combination with biochemical work on its structural organization revealed a transmembrane topology and quarternary structure similar to that of gap junction proteins. This led to speculations that synaptophysin might form a channel which spans the synaptic vesicle membrane, a hypothesis which we have now confirmed by reconstitution experiments. The identification of a gap junction protein-like channel in presynaptic vesicles may prove crucial for our understanding of the molecular mechanisms of neurotransmitter release (Reichardt and Kelly, 1983).

SYNAPTOPHYSIN, A STRUCTURAL ANALOGUE OF GAP JUNCTION CHANNEL PROTEINS

Synaptophysin is a homo-oligomer of a glycosylated 38 kD subunit (Rehm et al., 1986) whose primary structure has recently been determined from cloned cDNA (Südhof et al., 1987; Leube et al., 1987; Buckley et al., 1987). A model of the transmembrane organization deduced from hydropathy plots proposes four transmembrane domains for synaptophysin (Fig. 1A) and locates both its carboxy and amino termini on the cytoplasmic face of the vesicular membrane. This topology is reminiscent of that of liver gap junction channel proteins (Paul,

NATO ASI Series, Vol. H 32
Molecular Biology of Neuroreceptors and Ion Channels. Edited by A. Maelicke

1986; Kumar and Gilula, 1986) or "connexins" (Beyer et al., 1987), which form a hexameric transmembrane structure, the connexon or gap junction "hemi-channel" (Unwin and Zampighi, 1980). Although synaptophysin shares no overall sequence homology to any other known protein, there is a common motif of hydrophobic amino acids (Ala/Val - Ile/Val - Phe - Ala/X - Phe - Leu/X) in all predicted transmembrane regions of synaptophysin that reappears in one of the hydrophobic segments of identified and putative connexins. This may relate to common functional features of these membrane proteins.

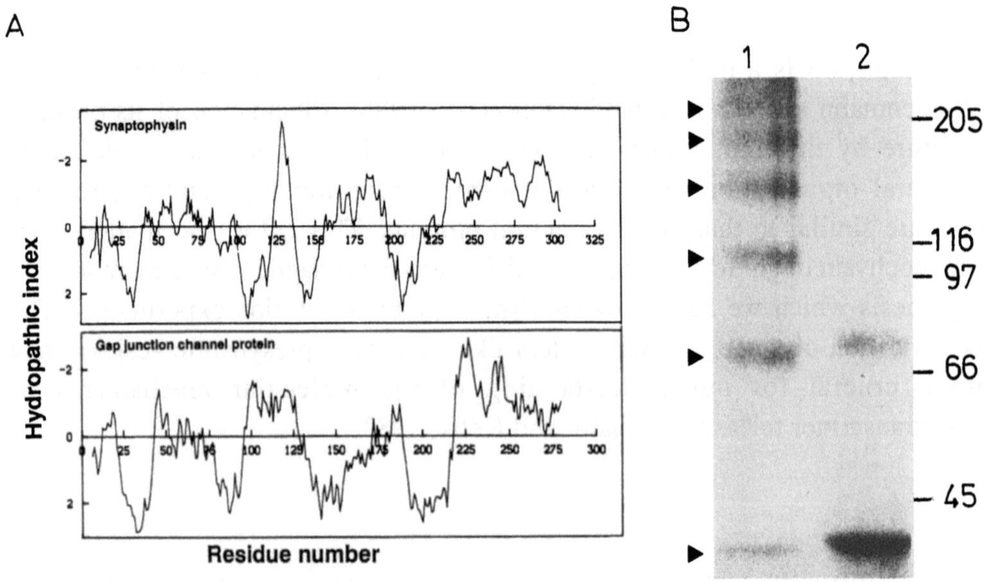

Fig. 1: (A) Hydropathy profiles of synaptophysin (top) and the liver gap junction channel protein (bottom) according to Kyte and Doolittle. Note the striking similarity concerning the four hydrophobic domains as well as an extended hydrophilic carboxy terminal region. (B) SDS-PAGE under non-reducing conditions of purified synaptophysin before (lane 2) and after (lane 1) crosslinking with 3,3'-dithiobis(sulfosuccinimidylpropionate). The gel was silver stained. Molecular masses of marker proteins are given on the right. Synaptophysin monomers and oligomers are indicated by triangles. Dimer formation due to intramolecular disulfide bonding is visible in lane 2.

The quarternary structure of synaptophysin was examined by hydrodynamic measurements, chemical crosslinking and ultrastructural analysis (Thomas et al., 1988). Upon gel exclusion chromatography, we found a Stokes radius of 6.5 nm; sedimentation on H_2O and D_2O sucrose density gradients yielded a sedimentation coefficient $S_{20,w}$ of 8.9 S. A molecular mass of 230 kD was calculated from these values for native synaptophysin, pointing to a hexameric assembly of the 38 kD monomer. Consistent with this interpretation , treatment of synaptic vesicles and purified synaptophysin preparations with different crosslinking reagents (N-sulfosuccinimidyl(4-azidophenyl)-1,3'-dithiopropionate, dimethylsuberimidate, dimethyl-3,3'-dithiobispropionimidate and 3,3'-dithiobis(sulfosuccinimidyl-propionate)) produced up to five homo-oligomeric adducts of 72 kD, 110 kD, 145 kD, 180 kD, and 220 kD (Fig. 1B). Also, electron microscopy of negatively stained synaptophysin preparations showed particles of mean diameter of 7.8 nm, consistent with a total molecular mass of about 200 to 250 kD.

Based on these data, synaptophysin is concluded to be a homo-oligomeric hexamer, as are gap junction connexons. Thus, although these membrane proteins are barely related in primary structure, both share a similar deduced transmembrane topology of their respective subunits as well as a common quarternary structure.

SYNAPTOPHYSIN, A FUNCTIONAL ANALOGUE OF GAP JUNCTION CHANNEL PROTEINS?

In electron micrographs, the 7.8 nm synaptophysin particles display a rosette-like appearance, showing a central 1 to 2 nm cavity (Thomas et al., 1988). These features are reminiscent of the negative staining pictures of other transmembrane channels, including the nicotinic acetylcholine receptor, the voltage-gated sodium channel, the so-called "mediatophore" of the presynaptic plasma membrane of Torpedo and in particular hepatic gap junction proteins. The structural analogy and similar ultrastructural appearance suggested a possible function of synaptophysin, i. e. providing a channel across the synaptic vesicle membrane.

To detect pore-forming properties of synaptophysin, the purified protein was reconstituted into liposomes and fused to planar lipid bilayers (Thomas et al., 1988). Thereafter, channel activity could indeed be observed which exhibited an

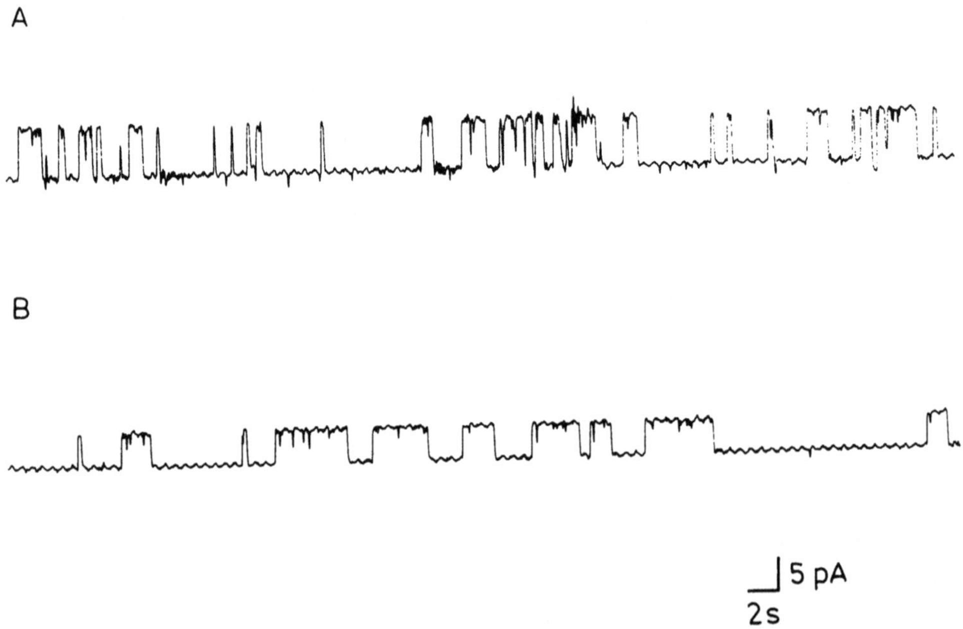

Fig. 2: (A) Channel activity of reconstituted synaptophysin at +50 mV under conditions of physiological ionic strength. Channel openings are depicted as upward deflections. (B) Alteration of channel properties after addition of monoclonal antibody SY38. Note decrease of elementary conductance and frequency of channel events and increase in the mean channel opening time.

average conductance of about 150 pS (Fig. 2A). This value is similar to that found for reconstituted gap junction channels, indicating that synaptophysin can form transmembrane pores of similar size. Since a monoclonal antibody (SY38) against the carboxy terminal region of synaptophysin specifically altered the conductance properties of synaptophysin-containing vesicles, channel formation by contaminating proteins can be excluded (Fig. 2B). Interestingly, positive voltages were required for synaptophysin channel opening. Assuming a right-side-out orientation of synaptophysin in our reconstituted vesicles (which usually holds true for the nicotinic acetylcholine receptor, e.g. McCrea et al., 1987), this would correspond to a depolarization event in vivo. A large conductance channel, among others, has very recently been disclosed in whole synaptic vesicles from Torpedo electric organ by patch clamp analysis (Rahamimoff et al., 1988).

Thus, synaptophysin, as the connexons, is capable of forming transmembrane channels. However, there are considerable differences between the two channel proteins. In particular, gap junction channels exhibit a different voltage dependence of opening. In any case, our data clearly show that the similar transmembrane topology of synaptophysin and gap junction channel monomers reflects not only a common quarternary structure, but also similar functional properties. This analogy extends the notion that channels of similar function share a common protein architectural design (Unwin, 1986). In other words, voltage-gated ion channels are "tetrameric", ligand-gated receptors with intrinsic ion channels pentameric, and large conductance channels of the connexon type hexameric, membrane proteins.

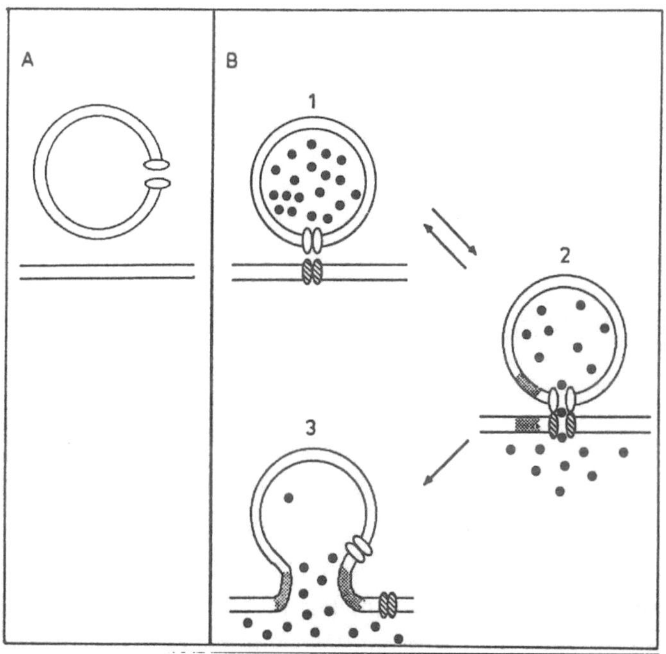

Fig. 3: Possible functions of synaptophysin. (A) Synaptophysin may serve as a transmembrane channel connecting cytoplasm and vesicle interior. (B) Synaptophysin participates in fusion pore formation during transmitter release. Docking of synaptophysin (○) to a channel protein (⊘) in the presynaptic plasma membrane (1) causes formation of a gap junction-like pore structure (2) followed by fusion of vesicle and plasma membranes (3) Membrane areas fusing incidentally are dotted. Note that membrane fusion is not required for release.

What may be the physiological role of synaptophysin in the synaptic vesicle membrane? One possibility is that synaptophysin serves as a transmembrane channel which allows exchange of low molecular mass components between cytoplasm and vesicle interior (Fig. 3A). Alternatively, the analogy between synaptophysin and the connexons may extend much further. Formation of gap junction channels is known to involve coupling of two connexons located in different cell membranes, thus creating a pore through two apposed lipid bilayers (Unwin and Zampighi, 1980). Could such a two-membrane coupling process also be involved in the in vivo function of synaptophysin? Some light is shed on this question by the recent identification of defined channels and fusion pores during the initial stage of exocytosis (Breckenridge and Almers, 1987). Fusion pores are thought to consist of proteinaceous structures of diameter and conductance similar to that of gap junction channels. Correspondingly, synaptophysin might "dock" to a related channel protein in the presynaptic membrane. Concurrent formation of a pore traversing both synaptic vesicle and synaptic plasma membranes then may constitute the initial step in the release of small neurotransmitter molecules (Fig. 3B). As synaptophysin has been shown to have a cytoplasmic Ca^{2+}-binding site (Rehm et al., 1986), docking might be triggered by depolarization-induced Ca^{2+} influx known to initiate quantal release.

ACKNOWLEDGEMENTS

We thank Drs. E. Bamberg, W. W. Franke and K. Hartung for experimental help at various stages of this work, Dr. B. Wiedenmann for continous collaboration and supply of the SY38 antibody, D. Langosch and Dr. H. Rehm for helpful discussions and C. Grund and E. Harbich for expert technical assistance. This work was supported by the Bundesministerium für Forschung und Technologie (BCT 381-5) and, in part, the Deutsche Forschungsgemeinschaft (SFB 317 and 169).

REFERENCES

Beyer EC, Paul DL, Goodenough DA (1987) Connexin43: a protein from rat heart homologous to a gap junction protein from liver. J. Cell Biol. 105: 2621-2629.

Breckenridge LJ, Almers W. (1987) Currents through the fusion pore that forms during exocytosis of a secretory vesicle. Nature 328: 814-817.

Buckley KM, Floor E, Kelly RB (1987) Cloning and sequence analysis of cDNA encoding p38, a major synaptic vesicle protein. J. Cell Biol. 105: 2447-2456.

Jahn R, Schiebler W, Ouimet C, Greengard P (1985) A 38,000-dalton membrane protein (p38) present in synaptic vesicles. Proc. Natl. Acad. Sci. USA 82: 4137-4141.

Kumar NM, Gilula NB (1986) Cloning and characterization of human and rat liver cDNAs coding for a gap junction protein. J. Cell Biol. 103: 767-776.

Leube RE, Kaiser P, Seiter A, Zimbelmann R, Franke WW, Rehm H, Knaus P, Prior P, Betz H, Reinke H, Beyreuther K, Wiedenmann B (1987) Synaptophysin: molecular organization and mRNA expression as determined from cloned cDNA. EMBO J. 6: 3261-3268.

McCrea PD, Popot JL, Engelman DM (1987) Transmembrane topography of the nicotinic acetylcholine receptor delta subunit. EMBO J. 6: 3619-3626.

Paul DL (1986) Molecular cloning of cDNA for rat liver gap junction protein. J. Cell Biol. 103: 123-134.

Rahamimoff R, DeRiemer SA, Sakmann B, Stadler H, Yakir N (1988) Ion channels in synaptic vesicles from *Torpedo* electric organ. Proc. Natl. Acad. Sci. USA 85: 5310-5314.

Rehm H, Wiedenmann B, Betz H. (1986) Molecular characterization of synaptophysin, a major calcium-binding protein of the synaptic vesicle membrane. EMBO J. 5: 535-541.

Reichardt LF, Kelly RB (1983) A molecular description of nerve terminal function. Ann. Rev. Biochem. 52: 871-926.

Südhof TC, Lottspeich F, Greengard P, Mehl E, Jahn R (1987) A synaptic vesicle protein with a novel cytoplasmic domain and four transmembrane regions. Science 238: 1142-1144.

Thomas L, Hartung K, Langosch D, Rehm H, Bamberg E, Franke WW, Betz H (1988) Identification of synaptophysin as a hexameric channel protein of the synaptic vesicle membrane. Science, in press.

Unwin N (1986) Is there a common design for cell membrane channels? Nature 323: 12-13.

Unwin PNT, Zampighi G (1980) Structure of the junction between communicating cells. Nature 283: 545-549.

Wiedenmann B, Franke WW (1985) Identification and localization of synaptophysin, an integral membrane glycoprotein of M_r 38,000 chracteristic of presynaptic vesicles. Cell 41: 1017-1028.

USE OF SYNTHETIC PEPTIDES AND HIGH AFFINITY PROTEIN LIGANDS FOR STRUCTURAL STUDIES OF CENTRAL AND PERIPHERAL NICOTINIC RECEPTORS.

Bianca M. Conti-Tronconi[1], Gregor Fels[2], Katya McLane[1], Fen Tang[1], Matteo Bellone[1], Anna Kokla[3], Socrates Tzartos[3], Robert Milius[4] and Alfred Maelicke[2].
[1]Department of Biochemistry, CBS, University of Minnesota, St. Paul, MN 55108, USA

INTRODUCTION

Neurotransmitter receptors are key components of neuronal function. A major breakthrough in their study has been provided by cloning and sequencing of their genes and deduction of the amino acid sequence of their precursors.

By this approach the amino acid sequence of the precursors of several receptors which function through coupling to different second messenger systems (ion channels, adenylate cyclase, etc.) have been cloned and sequenced. Different GABA receptor subunits (Schofield et al., 1987, Levitan et al., 1988), the glycine receptor (Grenningloh et al., 1987), β adrenergic receptors (Kubo et al., 1986a, Kubo et al., 1986b, Bonner, et al., 1987) and muscarinic receptors (Yarden et al., 1986, Dixon et al., 1986, Kobilka et al., 1987, Fukuda et al., 1988) and different nicotinic receptors (McCarthy et al., 1986, Boulter et al., 1986, Hermans-Borgmeyer et al., 1986, Goldman et al., 1987, Baldwin et al., 1988, Deneris et al., 1988) have been sequenced.

Receptors activated by different ligands, and even functioning through different second messengers, can have similar structure. The amino acid sequence of receptors which function via an ion gating unit can be very similar irrespective of their ion specificity. The sequences of constituent subunits of the GABA and the glycine receptors, which both contain an anion (Cl^-) gating unit are very similar to the subunit sequences of the nicotinic receptors (AcChR), which contain a cation (Na^+) channel (Schofield et al., 1987, Grenningloh et al., 1987, Levitan et al., 1988). Elucidation of the molecular structure of these receptors is therefore necessary to understand the structural basis for their function and for the ion discrimination operated by these very similar molecules. Regrettably, direct studies of receptor structure by X-ray

[2]Max-Planck-Institut fur Ernahrungphysiologie, Rheinlanddamm 201, D-4600 Dortmund 1, West Germany.
[3]Hellenic Pasteur Institute, 127, Vas. Sofias Av., Athens 115 21, Greece.
[4]Department of Biochemistry and Molecular Biology, Mayo Clinic and School of Medicine, Rochester, MN 55904, USA.

NATO ASI Series, Vol. H 32
Molecular Biology of Neuroreceptors and Ion Channels. Edited by A. Maelicke

crystallographic approaches are unsuitable for neurotransmitter receptors, with the possible exception of the nicotinic receptor from Torpedo electric organ, because of their scarcity. New experimental strategies need to be devised to obtain structural information on these molecules, and the available sequence information could be the key towards this endeavour.

The nicotinic receptors are an excellent system to develop new approaches interfacing molecular biology and protein chemistry, because considerable information regarding their primary structure, function and pathology is available, at least in peripheral tissues like muscle and fish electroplax. Sequences of different constituent subunits of neuronal AcChRs are also available, and it is reasonable to expect that the approaches developed for peripheral AcChRs will be easily adapted and applied to neuronal AcChRs.

Large numbers of peptides can be synthesized in parallel, in amounts large enough for their full characterization and use in controlled binding studies (Houghton, 1985). It is therefore possible to synthesize peptide panels screening the complete sequence of large proteins and identify constituent peptide loops of functionally important surface domains, by the binding of protein probes like toxins and antibodies. These bind with high affinity to a relatively large areas of the protein surface and therefore they can be expected to recognize even portions of their binding sites, when they are contained within the sequence of a synthetic peptide.

Snake neurotoxins, like α-bungarotoxin (α–BTX) and α-naja toxin (α-NTX) for peripheral AcChRs and k-bungarotoxin (k-BGT) for neuronal AcChRs are good probes for this purpose, because they bind with very high affinity and by multipoint attachment, involving different segments of the toxin sequence (Martin et al., 1983; Chibber et al., 1983). Antibodies directed against functionally important or unique parts of the AcChR molecule can also be used for this approach. Antibodies which compete with antagonist and/or agonists binding can be used to localize their binding site(s); antibodies which inhibit ion flux without interfering with agonist/antagonist binding can be used to identify sequence segments involved in forming ion gating structures. Antibodies known to bind to the surface of the AcChR can identify sequence loops and individual residues which are exposed on the surface of the molecule. Finally, antibodies can be made against synthetic peptides corresponding to unique segments of the amino acid sequence of a receptor. Such antibodies can be used to probe the presence of a particular subunit or of a particular receptor by immunoblotting or immunohistochemistry.

The strategies outlined above are suitable for both peripheral and neuronal AcChRs as well as for any receptors for which the primary sequence is known and for which high affinity ligands are available. In this regard, monoclonal antibodies (mAbs) can be regarded as custom-made high affinity protein ligands and construction of a mAb library is generally feasible even with the scarce neuronal receptors.

Because neurotransmitter receptors are the functional interface between the extracellular environment and the cells interior, the surface of these molecules which is the target for extracellular messengers and intracellular modulating systems is of particular interest. In the present paper we summarize the results of studies aimed at defining the structure of surface domains on AcChRs from different tissues. In particular, we identified sequence segments forming binding sites for cholinergic ligands in both Torpedo and neuronal AcChRs. We also localized the Main Immunogenic Region (MIR), which is the immunodominant epitope in the human disease Myasthenia Gravis (MG). The structure of these surface domains has been identified by the ability of toxins and monoclonal antibodies (mAbs) to bind to synthetic peptides corresponding to different parts of the AcChR sequence. We also summarize the results of a study aimed at directly characterizing the structural properties of a chick optic lobe AcChR (Conti-Tronconi et al., 1985), which is representative of the several attempts made at directly studying the structural features of different neuronal AcChRs (Schmidt-Nielsen et al., 1977, Norman et al., 1972, Betz et al., 1982, Schneider et al., 1982, Breer et al., 1985, Whiting and Lindstrom, 1986), some of which are reported in other chapters of this book, and of a study (Conti-Tronconi and Raftery, 1986) which demonstrated that multiple binding sites are present on Torpedo AcChR. These results are used to interpret the existing conflicting data on the binding of nicotine, α-BTX and other cholinergic ligands in different brain areas.

THE BRAIN α–BGT BINDING COMPONENTS ARE MEMBERS OF THE NICOTINIC RECEPTOR FAMILY.

Some regions of the vertebrate brain contain nicotinic AcChR(s) with pharmacological characteristics similar but not identical to muscle AcChR (reviewed in Curtis and Crawford, 1969, Morley et al., 1981, Lindstrom et al., 1987). In addition, some brain areas and peripheral ganglia contain high-affinity binding sites for α-BTX and similar snake toxins, which are known to bind to peripheral AcChR. However, the identity of the neuronal αBTX-binding component with a nicotinic AcChR has been disputed (for reviews, see Schmidt et al., 1980, Oswald and Freeman, 1981, Lindstrom et al., 1987) because i) α-BTX seldom blocks cholinergic function in neuronal systems, ii) there is a differential distribution in rat and mouse brain of high affinity binding sites for ^{3}H-nicotine/^{3}H-acetylcholine, and high affinity α–BTX sites (Marks and Collins, 1982, Clarke et al., 1985, Marks et al., 1986), and iii) neuronal protein(s) which bind α-BTX can be physically separated from the component(s) able to bind antibodies which recognize neuronal AcChR (Patrick and Stallcup, 1977; Whiting et al., 1987), or ^{3}H-acetylcholine (Schneider et al., 1985).

The α-BTX binding protein has been purified from avian optic lobe (Norman et al., 1982). This protein displays the ligand-binding characteristics of a nicotinic receptor and it can be covalently labeled by the affinity reagent bromoacetylcholine in a manner similar to muscle and electric organ AcChR. It also shows distinct immunological crossreactivity with an antiserum to mammalian muscle AcChR and with some mAbs raised against chicken muscle AcChR (Norman et al., 1982, Meheraban et al., 1984). In spite of these indications, identification of this and other neuronal α-BTX binding proteins as true AcChRs is under debate. To resolve this issue and to determine if and how neuronal α-BTX binding components relate to the gene family formed by the peripheral AcChRs and by neuronal AcChRs not recognized by α-BTX requires information on the amino acid sequence of their subunits.

The avian central nervous system contains α-BTX binding in the optic lobe, where α-BTX blocks the cholinergic responses, and in other brain areas where on the contrary α-BTX, although it binds in a competitive fashion with cholinergic ligands, does not seem to interfere with any cholinergic function. We isolated the α-BTX binding component(s) from both the optic lobe and the rest of the chick brain using α-BTX affinity chromatography (Conti-Tronconi et al., 1985). The purified α-BTX-binding component from optic lobe showed high-affinity binding of ^{125}I-α-BTX with specific activities of 4000-6000 nmol/g of protein. The component purified from the rest of the brain had similar α-BTX-binding characteristics and specific activity. Upon SDS gel electrophoresis the α-BTX-binding proteins from either chick optic lobe or the rest of the brain had very similar peptide compositions. Four or five major components were consistently present, of apparent molecular weights between 48,000 and 72,000. All the peptides contained in purified optic lobe α-BTX-binding protein preparations were isolated and submitted to amino-terminal amino acid sequencing. The lowest molecular weight component gave a readily identifiable single sequence. Comparison of this sequence with the known sequences of AcChR subunits and with the sequence of chick muscle AcChR subunits (Conti-Tronconi, Dunn, Barnard and Raftery, unpublished) revealed that the optic lobe component was highly homologous to the subunits of the other known sequence of AcChRs. The highest degree of homology was with the α subunits, and particularly with the α subunit of Torpedo. The other subunits isolated from optic lobe α-BTX-binding protein had blocked amino termini, which prevented determination of their aminoterminal sequence.

Because of the very low AcChR concentration in the other parts of chick brain, isolation of individual subunits was not feasible. Intact α-BTX binding protein preparations from the rest of the brain were therefore submitted to simultaneous amino-terminal analysis (Conti-Tronconi et al., 1985). Only one signal was present above a high background, which indicated the existence of other proteins with blocked amino termini. The sequence obtained was identical to that of the lowest molecular weight component of optic lobe AcChR. These data, the similar gel patterns and the indistinguishable reactivities to polyclonal and monoclonal antibodies against muscle AcChR (Norman et al., 1982, Meheraban et al., 1984) suggest that the AcChRs from the optic lobe and from other regions of the chick brain are very similar or even identical complex proteins, in spite of their somewhat different pharmacological properties. The different properties of these two receptors could arise from differentiation post-translational modifications of their constituent subunits. These results support the identification of the α-BTX-binding proteins from avian optic lobe and brain as true AcChRs, structurally homologous to the peripheral AcChRs, and they indicate that the divergence of peripheral and central nicotinic receptors happened very early during vertebrate evolution.

Further evidence that other polypeptides are components of this AcChR was obtained by virtue of their immunoreactivity (Conti-Tronconi et al., 1985). A monoclonal antibody raised against chick muscle AcChR crossreacted with the α-BTX binding protein purified from chick optic lobe (Meheraban et al., 1984) and precipitated the polypeptides of M_r 48,000, 56,000, and 69,000. Note that these peptides include the subunit for which the sequence data reveal a similarity with the α-subunit of peripheral AcChRs, and the subunit of M_r 56,000 which can be labeled by bromoacetylcholine (Norman et al., 1982).

In the absence of amino termini available for sequencing it was not possible to determine the stoichiometry of the receptor subunits, or to ascertain whether the other two polypeptides present in $NaDodSO_4$ gels of purified brain receptors (M_r 62,000 and 72,000) are contaminants or AcChR subunits or different proteolytic products or post-translational modifications of one subunit. We can exclude that the M_r 48,000 subunit is a breakdown product of any of the others, since the amino terminus is different from theirs (not blocked) and is in sequence homologous to the true amino terminus of peripheral AcChR α subunits. Actin, which is found complexed to calf muscle AcChR (Gotti et al., 1982) and which migrates in the M_r 45,000-48,000 region and has a blocked amino terminus was absent, as shown by comparison with a chicken gizzard actin standard and by the lack of change of the pattern upon treatment of the AcChR with anti-actin antibody (Conti-Tronconi et al., 1985).

The common ancestry of peripheral and central AcChRs well explains the pharmacological characteristics of these receptors, which only partially overlap. Similarly, the partial structural identity explains why polyclonal and monoclonal antibodies raised against peripheral AcChRs may fail to recognize central acetylcholine receptors (Morley et al., 1983, Wonnacot et al., 1982) and conversely antibodies against chick optic lobe receptor may not bind to peripheral AcChR from chick or from Torpedo californica (Betz and Pfeiffer, 1984).

Similar results have been obtained in studies on rat brain α-BTX binding protein(s). Although these studies on the structural properties of brain AcChRs and AcChR-like proteins (Schmidt-Nielsen et al., 1977, Norman et al., 1982, Betz et al., 1982, Schneider et al., 1982, Breer et al., 1985, Whiting and Lindstrom, 1986) were careful and labor-intensive, they gave rather non-detailed insights into the molecular structure of these neuronal AcChRs.

In the case of mammalian neuronal systems, a further complication arises from the possibility that α–BTX recognizes different AcChRs, and perhaps blocks only some of them. This is suggested by the finding that in mammalian brain two populations of α-BTX binding sites exist, characterized by different binding affinity (Kd of 5 and 400 nM respectively) (Lukas 1984a,b). In addition, α-BTX and k-bungarotoxin (k-BTX), a neurotoxin from Bungarus multicinctus venom which specifically blocks the function of several neuronal AcChRs, partially compete with each other. This may indicate that α-BTX can recognize cholinergic proteins to which k-BTX cannot bind, and others which are also recognized by k-BTX.

In support of this possibility, we have isolated the α–BTX binding component purified from different murine neuronal cell lines. The purified α–BTX binding protein(s) were submitted to SDS-gel electrophoresis and probed with ^{125}I-α-BTX, ^{125}I-k-BTX and antisera specific for unique sequence segments of murine muscle α subunit (α_1) or different neuronal α subunits (α_3 and α_4) (McLane, Kemp and Conti-Tronconi, unpublished). Consistently, both α–BTX and k-BTX recognized peptide components, of approximate molecular weight 55,000 and 45,000 respectively. In the immunoblots, in addition to the expected recognition of a peptide component by the anti-α_1 antibodies, binding of anti-α_4 antibodies was also present. These results are well explained if α–BTX recognizes two different cholinergic proteins, containing two different α subunits, and one of them can be recognized also by the k-BTX.

TORPEDO AcChR CONTAINS MULTIPLE LIGAND BINDING SITES.

Identification of the ligand binding events leading to activation and inactivation of the ion channel contained in the AcChR molecule and definition of the number, location and function of the binding sites for cholinergic ligands is necessary to elucidate the mechanism of AcChR function. Electrophysiological (Dionne et al., 1978, Dreyer et al., 1978) and stopped-flow (Raftery et al., 1983) studies indicated that channel opening results from acetylcholine binding to two low-affinity sites, which operate in a cooperative fashion. Ligand binding also causes the AcChR to slowly become desensitized to a state characterized by inability of the channel to open and by high affinity for ligands (Katz and Thesleff, 1957; Rang and Ritter, 1970). Because each of the two α-subunits has a high-affinity binding site for cholinergic ligands, several models of the linear type have been proposed in which both activation and desensitization of the receptor are controlled by ligand binding to these two sites, which triggers physiological effects by sequential mechanisms involving multiple conformational changes with concomitant alterations in ligand affinity (reviewed in Conti-Tronconi and Raftery, 1982). However, the relatively early divergence of the genes encoding the AcChR subunits, as well as the extremely high degree of homology between corresponding subunits from distant animal species (reviewed in McCarthy et al., 1986) suggests that the AcChR subunits might have evolved separately to perform different functions and that each subunit could in principle have an independent ligand binding site.

The long α-neurotoxins from snake venoms are a group of polypeptide cholinergic antagonists useful for study of AcChR binding sites, since they bind in a specific and slowly reversible fashion (Lee, 1972, Banks et al., 1974). α-Neurotoxins that have been most extensively used are α-NTX from the venom of Naja naja species and α-BTX from Bungarus multicinctus. α-Dendrotoxin (α-DTX), a polypeptide neurotoxin from the venom of Dendroaspis viridis, is a cholinergic antagonist closely related to α-NTX and α-BTX, to which it is highly homologous (Banks et al., 1974). These toxins are powerful blockers of muscle and electric organ AcChR (Lee, 1972, Banks et al., 1974) and they bind in a competitive fashion with acetylcholine, carbamoylcholine, curare and other classic cholinergic ligands (Neubig and Cohen, 1979, Sine and Taylor, 1980). They are good ligands to study the binding site stoichiometry of AcChRs from peripheral tissues, because the complexes with the AcChR dissociate very slowly and, due to their hydrophilic nature, the unbound toxins can be efficiently washed away, resulting in minimal nonspecific binding.

We have studied the stoichiometry of the binding of α-DTX and α-NTX to membrane-bound AcChR from Torpedo californica electric organ (Conti-Tronconi and Raftery, 1986) by simultaneous-quantitative gas-phase microsequencing of all the amino acid sequences present in AcChR/α-neurotoxin complexes. This method permits the use of homogeneous (nonradiolabeled) preparations of native toxins to obtain molar ratios of neurotoxin-receptor complexes. The stoichiometry obtained for α-NTX was 2.1 ± 0.2 ($n = 4$), in agreement with the accepted view that α-NTX, like α-BTX, binds to the two α subunits of the AcChR molecule. α-DTX bound with a stoichiometry of 4.1 ± 0.5 ($n = 12$). The AcChR molecule therefore contains four binding sites for this α-neurotoxin, two of which are recognized by α-NTX and α-BTX. In support of this contention we found that when the AcChR is saturated with α-BTX, addition of α-DTX markedly accelerates the dissociation of the bound α-BTX (Conti-Tronconi and Raftery, 1986). This demonstrates that occupancy of the additional two sites by α-DTX influences and decreases the affinity of α-BTX for its two binding sites. Because the AcChR subunits are highly homologous, as many as five binding sites for cholinergic ligands could be present, one on each subunit.

α-DTX blocks AcChR function in neuronal systems that are not sensitive to α-BTX blockade, such as the frog spinal cord (Miledi and Szczepaniak, 1975) and the snail neurons (Szczepaniak, 1974). α-DTX binds to twice as many sites as α-BTX in the sympathetic cell line PC12 (Patrick et al., 1980). These findings support the notion of a broader binding specificity of α-DTX, which might bind to sites recognized by α-NTX and α-BTX with much lower affinity or not recognized at all. It is therefore possible that neuronal AcChR could also carry multiple, homologous binding sites for cholinergic ligands, possibly one on each subunit. Each such site could recognize a cholinergic ligand with different affinity, and each receptor would therefore be characterized by the spectrum of affinities for the different cholinergic agonists and antagonists, so that the binding of the same ligand could have different functional consequences in different AcChRs. At low-ligand concentration only high-affinity sites are revealed, both in terms of their number and of their effect(s). The use of ligands with broad specificity, such as α-DTX, or suitably high concentrations of the other ligands, can reveal the low-affinity sites which, at least in the peripheral AcChRs, are associated with channel activation.

Mapping of the binding of different cholinergic agonists and antagonists in the brain has been the subject of careful and detailed studies, with conflicting results. The most obvious discrepancy is the differential distribution in mouse and rat brain of high affinity sites for nicotine/acetylcholine and α-BTX (Marks and Collins, 1982, Marks et al., 1986, Clarke et al., 1985, Goldman et al., 1987), in spite of the fact that high concentrations of nicotine and other cholinergic ligands can inhibit α-BTX binding in the brain (Marks and Collins, 1982, Clarke et al., 1985). This dilemma would be reconciled if each of the structurally related AcChRs have multiple binding sites for ligands, one on each of their homologous subunits.

STRUCTURE OF CHOLINERGIC BINDING SITE(S) ON THE α-SUBUNIT OF PERIPHERAL AcChRs.

α-BTX binds to the α subunits (Haggerty and Froener, 1981, Gershoni et al., 1983, Mishina et al., 1985, Pedersen et al., 1986), and because α-BTX competitively inhibits the binding of carbamoylcholine, curare and other classic cholinergic ligands (Neubig and Cohen, 1979, Sine and Taylor, 1980) as well as affinity labeling of the α-subunit (Damle and Karlin, 1978, Moore and Raftery, 1979, Wolosin et al., 1980), it is believed that this toxin binds to cholinergic high affinity sites on the α subunits. The binding site for α-BTX seems to contain one or both of the cysteinyl residues at position 192 and 193 on the α-subunit (Wilson et al., 1984, Neumann et al., 1986), which are labelled by cholinergic affinity labels (Kao and Karlin, 1986). To identify further structural features of the binding site for cholinergic ligands on the α-subunit, we studied the binding of ^{125}I-labelled α-BTX and of three different monoclonal antibodies (mAbs), raised against native Torpedo AcChR and able to competitively inhibit α-NTX and α-BGT binding (Watters and Maelicke, 1983, Fels et al., 1986, Conti-Tronconi, Tang and Maelicke, unpublished), to a panel of synthetic peptides corresponding to the complete Torpedo α-subunit (Conti-Tronconi, Fels, Spencer, Tang and Maelicke, unpublished). The peptides were 18-20 residues long and overlapped each other by 4-6 residues. The mAbs used in this study are specific for the α subunit, can completely inhibit α-BTX and α-NTX binding but do not completely compete with each other for binding to native AcChR (Watters and Maelicke, 1983, Fels et al., 1986). The binding of all these mAbs is full inhibited by α-NTX and by the cholinergic agonist carbamoylcholine. The binding of one of these mAb (mAb #6) is not inhibited by cholinergic antagonists (Fels et al., 1986), which on the other hand efficiently inhibit binding of the other two mAbs (#2 and #5) (Watters & Maelicke, 1983). The different patterns of binding inhibition by cholinergic ligands and competition between mAbs indicate that these mAbs bind to different, yet overlapping epitopes which together comprise a region of the AcChR surface containing the α-neurotoxin and cholinergic ligand binding sites. Within this area, two subsets of binding sites may exist. Only agonists bind at one site while the second may also be recognized by antagonists. While mAb #6 would primarily recognize the former, mAbs #2 and #5 may recognize primarily the latter. They are therefore ideal tools to complement α-BTX in the search of peptide loops forming the cholinergic binding site on the α subunit. When tested in dot-blot assays, at concentrations below 4 μM, α-BTX recognized two peptides. One of them corresponded to the sequence segment flanking the cysteinyl residues 192 and 193, which general consensus indicates as very close to the cholinergic ligand binding site (Wilson et al., 1984, Neumann et al., 1986, Kao and Karlin, 1986). α-BTX also recognized, although to a lesser extent, a second peptide corresponding to the sequence α55-74. At higher (>5 μM) concentration, α-BTX bound to a third peptide, which contained the cysteinyl residue at position 128 and the flanking sequence segments. The mAbs were tested against the peptides by dot-blot assays and by an enzyme-linked immunoassay (ELISA). The results of the binding of the three mAbs to the α subunit peptides, can be summarized as follows. mAbs #6 recognized primarily peptide α181-200, which contains the cysteinyl residues 192 and 193 and to a minor extent peptide α43-60. The other two mAbs recognized primarily either both the overlapping peptides α43-60 and α55-74, or peptide α43-60 alone. They also recognized, to a much minor extent, peptide α181-200. We suggest therefore that the binding site for α-BTX includes three peptide loops. One is contained within the sequence α181-200, and may include one or both the cysteinyl residues 192 and 193. The second sequence segment is within the sequence α43-74. Because the Main Immunogenic Region is formed by a 10-residue loop which is partially included in the carboxyterminal site of peptide α55-74 (see below and Tzartos et al., 1988), and anti-MIR antibodies do not interfere with α-BTX binding (Tzartos and Lindstrom, 1980), we propose that the second constituent peptide loop forming the binding site for α-BTX and cholinergic ligands is within the sequence α51-66, i.e. on the aminoterminal side of peptide α55-74, which largely overlaps the sequence of peptide α43-60. The third loop should contain the cysteinyl residue 128 and some of the flanking residues. In support of this possibility, site-specific mutation experiments have demonstrated that when either Cys α128 or Cys α142, which are believed to form together a disulfide bridge, were changed to a different residue the ability to bind α-BTX was completely lost. On the other hand, mutations of either Cys 192 or Cys 193 reduced but did not eliminate α-BTX binding (Mishina et al., 1985).

The preferential recognition of the sequence segment α43-66 by mAbs #2 and #5, whose binding to native AcChR can be inhibited by curare, supports the possibility that residues contained within this segment contribute to form the site preferentially recognized by antagonists.

LOCALIZATION OF CHOLINERGIC BINDING SITES ON NEURONAL AcChRs.

We are using a strategy similar to the one described above to achieve structural definition of cholinergic binding sites on neuronal AcChRs (McLane, Tang and Conti-Tronconi, unpublished results). Peptides 18-20 residues long, overlapping each other by five residues, corresponding to the known sequences of the neuronal subunits α_3 (Boulter et al., 1986) were synthesized, and the binding of radiolabelled k-BTX (Ravdin and Berg, 1979, Quik and Lamarca, 1982, Loring et al., 1984, 1986, Grant and Chiappinelli, 1985) to these peptides was studied. ^{125}I-k-BTX consistently recognized peptides corresponding to the sequence segment $\alpha_3$51-70 and the binding was efficiently inhibited by unlabelled k-BTX. This sequence segment is homologous to the segment α51-66, which, as it was discussed above, should be part of the binding site for cholinergic ligands and α-BTX on Torpedo AcChR.

LOCALIZATION OF THE MAIN IMMUNONOGENIC REGION TO RESIDUES α67-76 OF HUMAN MUSCLE α-SUBUNIT

Myasthenia Gravis is a human paralysis due to an autoimmune response against the AcChR at the neuromuscular junction (Engels, 1979, Lindstrom, 1985). Anti-AcChR antibodies cause accelerated destruction and functional impairment of the AcChR and failure of neuromuscular transmission (reviewed in Lindstrom, 1985). Myasthenic symptoms can be induced in a variety of animals by immunization with purified AcChR (Patrick and Lindstrom, 1973) or by injection of anti-AcChR antibodies (Richman et al., 1980, Lennon and Lambert, 1980).

Most (>60%) anti-AcChR antibodies both in myasthenic patients and in rats immunized with AcChR are directed against an extracellular area of the α subunit called the Main Immunogenic Region (MIR) (Tzartos and Lindstrom, 1980, Tzartos et al., 1981). Because of the important role of the MIR in the pathogenisis of Myasthenia Gravis, its localization on the muscle α subunit sequence has been the subject of an intense investigation (Ratnam et al., 1986, Barkas et al., 1987, 1988, Tzartos et al., 1988).

We have studied the binding of a panel of anti-MIR monoclonal antibodies to 26 synthetic peptides corresponding to the hydrophilic parts of the human AcChR α subunit (Tzartos et al. 1988). Eight such mAbs bound to one peptide corresponding to residues 63-80 of the human α subunit. A second panel of peptides corresponding to the various parts of the α–subunit segment

63-80 was synthesized. Anti-MIR mAbs bound to a peptide that contained the α–subunit sequence 67-76. A main constituent loop of the MIR on the human muscle AcChR is therefore localized between residues 67 and 76 of the α subunit. Because anti-MIR antibodies bind with higher affinity to native AcChR than to denatured α subunit (Tzartos and Lindstrom, 1980, Tzartos et al., 1981), other residues could contribute to form the native MIR.

The binding of sera from myasthenic patients to synthetic peptides corresponding to the complete human α-subunit gave ambiguous results (Bellone & Conti-Tronconi, unpublished). Peptides containing the sequence segment α 63-80 seemed to be recognized more frequently than the other segments of the α subunit sequence. However when peptide binding is studied using total polyclonal myasthenic sera the background noise is high, and peptides are recognized at a low level because most antibodies present in these sera should be primarily directed against the native form of the AcChR. These problems make it very difficult to be confident about the specificity of the binding.

The segments of α subunit corresponding to residues 63-80 is highly conserved among species. Peptide α 63-80 is particularly conserved in its middle segment, which corresponds to peptide α 67-76. This explains the species crossreactivity of anti-MIR antibodies (Tzartos and Lindstrom, 1980, Tzartos et al., 1981).

STRUCTURAL PREDICTIONS OF THE MAIN IMMUNOGENIC REGION

The sequence segments α 63-80 and 67-76 from the human α subunit were analyzed for potential to form secondary structure with the Sequence Analysis Software Package of the Genetics Computer Group at the University of Wisconsin (UWGCC) (Devereux et al. 1984). PEPPLOT (Gribskov et al., 1986) was used to obtain Chou-Fasman (CF) and Garnier-Osgorthurpe-Robson (GOR) secondary structure profiles (Chou & Fasman, 1978, Garnier et al., 1978), Eisenberg's hydrophobic moment (Eisenberg et al., 1984), hydropathy analysis of Kyte & Doolittle (Kyte & Doolittle, 1982), and transbilayer helix prediction of Goldman, Engelman, and Steitz (GES)(Engelman et al., 1986). PEPTIDESTRUCTURE (Wolf et al, 1987) provided CF and GOR secondary structure profiles, hydropathic profiles of Hopp-Woods (Hopp & Woods, 1981) and Kyte-Doolittle (Kyte & Doolittle, 1982), surface probability according to Emini (Emini et al., 1985), Darplus-Schultz chain flexibility analysis (Karplus & Schultz, 1985), and antigenicity index according to Jameson-Wolf (Jameson & Wolf, 1988). Molecular modelling was carried out using two programs from Polygen Corp. (Waltham, MA). Initial model building and all energy evaluations and minimizations were carried out using CHARMM (Brooks et al, 1983) version 20.3 implemented on a DEC VAX 8600. Graphic display and further manipulations of the modes were carried out using HYDRA with an Evans & Sutherland PS300 and DEC MicroVAX II as the host system.

These structural predictions indicated a high segmental mobility for the sequence Asp-Asp-Tyr-Gly-Gly-Val-Lys-Lys and a very strong turn potential centered around residues 68-71 (Asn-Pro-Asp-Asp). The segment 67-76 should form a region that is flexible, hydrophilic, negatively charged, and possibly of convex shape. All these structural properties have been proposed as important features of antibody-binding epitopes (Westhof et al., 1984, Tainer et al., 1984, Novotny et al., 1986, Geysen et al., 1987) and they support the identification of this segment as part of a strongly immunogenic region. The minimized structure of the segment α63-80 is shown in Fig. 1, and has a calculated potential energy of -594.76 kcal/mole. This sequence can form a β strand, followed by a Type I β turn centered around the residues described above, and β strand again. Its potential for tertiary structure is to form a hairpin loop, whose two arms are β strands. The

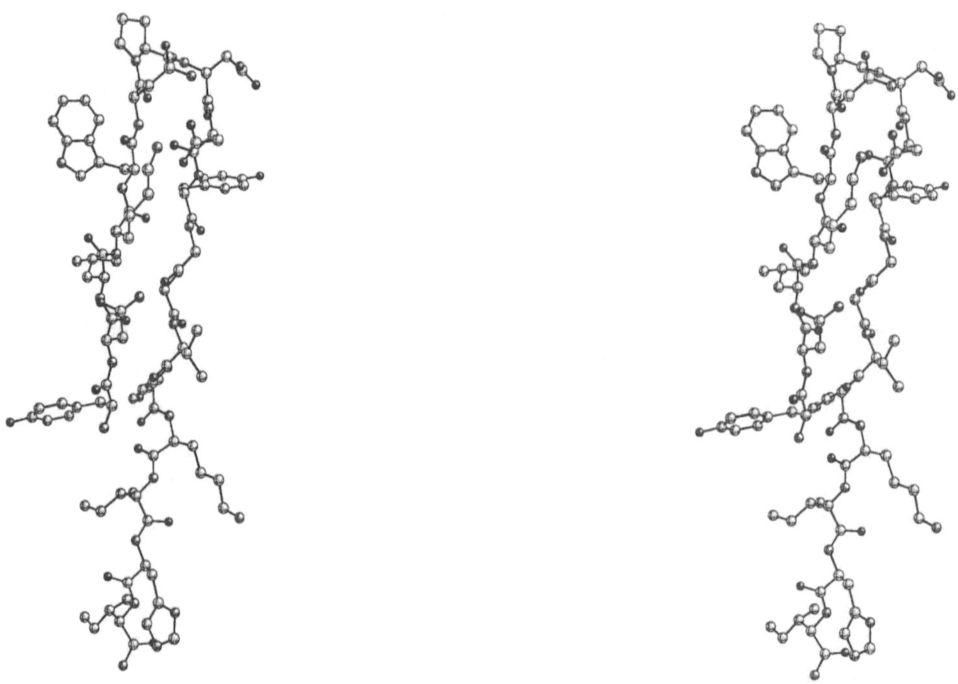

Figure 1. Stereo drawings of the minimized structure of the sequence segment 63-80 of human AcChR α-subunit. The hydrogen atoms are not shown in order to simplify the drawings. The amino terminus is located on the left side of each drawing. The chain extends up, forms a Type I β-turn, and then extends downwards, towards the carboxyl-terminus. The side chains of the amino acid residues are predominantly perpendicular to the plane of the paper. The loop formed by the two sheets is stabilized by the interaction between the side chains of Lys 66 and Asp 71.

interaction between the two sheets is stabilized by the interaction between the side chains of Lys 66 and Asp 71. Peptides corresponding to the sequence segment α67-76 can form a very similar structure (Fig 2). The aminoterminal of Trp 67 instead of the ε aminogroup of Lys 66 stabilizes the loop, by interacting with the side chain of Asp 71. The apex of this loop is charged and hydrophilic, because it contains residues Asn 68, Asp 70 , Asp 71 and Tyr 72. These residues should therefore play an important role in anti-MIR antibody binding.

This possibility is supported by the finding that in the Xenopus AcChR, which is the only peripheral AcChR not recognized by anti-MIR antibodies, positions 68, 70 and 71 on the α subunit are substituted by non-conservative residues (Baldwin et al., 1988). In particular, at position 68 a negatively charged Asp residue substitutes the neutral Asn residue present in both Torpedo and human AcChR, at position 71 the positively charged Lys substitutes the negative charge of the invariant Asp residue, and at position 70 Xenopus has the hydrophobic Ala residue, instead of Asp in human. In addition, at position α61 Xenopus has the hydrophylic, positively charged Arg residue instead of an invariant hydrophobic residue (Val or Ile) present in all the other known α-subunits. The residues 65 (Leu), 73 (Gly), 74 (Gly) and 75 (Ile) can form a hydrophobic surface which could make contact with the hydrophobic core of the molecule.

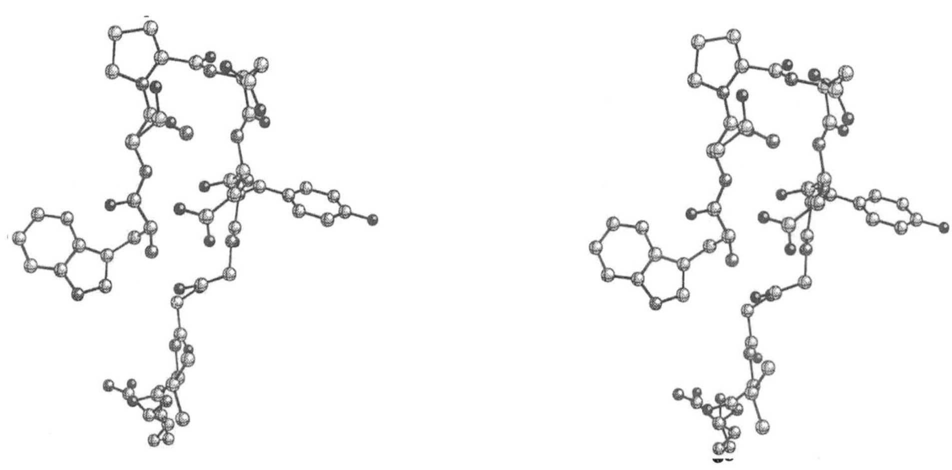

Figure 2. Stereo drawings of the predicted minimized structure of the sequence segment 67-76 of human muscle AcChR α-subunit. The hydrogen atoms are not shown. The peptide forms a hairpin loop, whose two arms are β sheets, joined by a Type I β turn. The interaction between the two sheets is established by the interaction between the aminoterminal of Try 67 and the side chain of Asp 71. The side chains of the aminoacid residues extend predominantly either towards or away from the viewer.

ACKNOWLEDGMENTS. Supported in part by the U.S. NINCDS grant NS 29919 (B.M.C.-T.), a grant from the Muscular Dystrophy Association of America (B.M.C.-T.), the U.S. NSF grant BNS-8607 289 (B.M.C.-T.), grants SFB 168-B2 and Ma 599/10-1 from the Deutsche Forshungsgemeinschaft (A.M.) and a grant from the Fonds der Chemischen Industrie (A.M.).

REFERENCES

Amit, A.G., Mariuzza, R.A., Phillips, S.E.V. and Poljak, R.J. Three-dimensional structure of an antigen-antibody complex at 2.8 A resolution. (1986) Science 233, 747-753.

Baldwin, T.J., Yoshihara, C.H., Blacmer, K., Kintner, L.R. and Burden, S.J. Regulation of acetylcholine receptor transcript expression during development in xenopus laevis. (1988) J. Cell Biol. 106, 469-478.

Banks, B.E.C., Miledi, R. and Shipolini, R.A. The primary sequences and neuromuscular effects of three neurotoxic polypeptides from the venom of Dendroaspis viridis. (1974) Eur. J. Biochem. 45, 457-468.

Barkas, T., Gabriel, J-M., Mauron, A., Hughes, G.J., Roth, B., Alliod, C., Tzartos, S.J. and Ballivet, M. Monoclonal antibodies to the main immunogenic region of the nicotinic acetylcholine receptor bind to residues 61-76 of the α-subunit. (1988) J. Biol. Chem. 263, 5916-5920.

Barkas, T., Mauron, A., Roth, B., Alliod, C., Tzartos, S.J. and Ballivet, M. Mapping the main immunogenic region and toxin-binding site of the nicotinic acetylcholine receptor. (1987) Science 235, 77-80.

Betz, H., Graham, D. and Rehm, H. Identifcation of polypeptides associated with a putative neuronal nicotinic acetylcholine receptor. (1982) J. Biol. Chem. 257, 11390-11394.

Betz, H. and Pfeiffer, F. Monoclonal antibodies against the α-bungarotoxin-binding protein of chick optic lobe. (1984) J. Neurosci. 4, 2095-2105.

Bonner, T.I., Buckley, N.J., Young, A.C. and Brann, M.R. Identification of a family of muscarinic acetylcholine receptor genes. (1987) Science 237, 527-531.

Boulter, J., Evans, K., Goldman, D., Martin, G., Treco, D., Heinemann, S. and Patrick, J. Isolation of a cDNA clone coding for a possible neural nicotinic acetylcholine receptor α-subunit. (1986) Nature 319, 368-375.

Breer, H., Kleene, R. and Hinz, G. Molecular forms and subunit structure of the acetylcholine receptor in the central nervous system of insects. (1985) J. Neuroscience 5, 3386-3392.

Brooks, B.R., Bruccoleri, R.E., Olafson, BD., States, D.J., Swaminathan, S. and Karplus, M. CHARMM: A program for macromolecular energy, minimization, and dynamics calculations. (1983) J. Comp. Chem. 4, 187-217.

Chibber, B.A., Martin, B.M., Walkinshaw, M.D., Saenger, W. and Maelicke, A. The sites of neurotoxicity in alpha-cobratoxin. (1983) In: Toxins as Tools in Neurochem., pp. 141-150.

Chou, P.Y. and Fasman, G.D. Prediction of the secondary structure of proteins from their amino acid sequence. (1978) Adv. Enzymol. 47, 45-147.

Clarke, P.B.S., Schwartz, R.D., Paul, S.M., Pert, C.B. and Pert, A. Nicotinic binding in rat brain: autoradiographic comparison of [^{3}H] acetylcholine, [^{3}H]nicotine and [^{125}I]-α-bungarotoxin. (1985) J. Neurosci. 5, 1307-1315.

Colman, P.M., Laver, W.G., Varghese, J.N., Baker, A.T., Tulloch, P.A., Air, G.M. and Webster, R.G. Three-dimensional structure of a complex of antibody with influenza virus neuraminidase. (1987) Nature 326, 338-363.

Conti-Tronconi, B.M., Dunn, S.M.J., Barnard, E.A., Dolly, J.O., Lai, F.A., Ray, N. and Raftery, M.A. Brain and muscle nicotinic acetylcholine receptors are different but homologous proteins. (1985) Proc. Natl. Acad. Sci. USA 82, 5208-5212.

Conti-Tronconi, B.M., Gotti, C., Paggi, P. and Rossi, A. Acetylcholine receptors in the ciliary ganglion and in the iris muscle of the chick: specific binding and effect on the synaptic transmission of the neurotoxin from Naja naja siamensis. (1979) Br. J. Pharmac. 66, 33-38.
Conti-Tronconi, B.M. and Raftery, M.A. The nicotinic cholinergic receptor: correlation of molecular structure with functional properties. (1982) Annu. Rev. Biochem. 51, 491-530.
Conti-Tronconi, B.M. and Raftery, M.A. Nicotinic acetylcholine receptor contains multiple binding sites: evidence from binding of α-dendrotoxin. (1986) Proc. Natl. Acad. Sci. USA 83, 6646-6650.
Curtis, D.R. and Crawford, J.M. Central synaptic transmission - microelectrophoretic studies. (1969) Annu. Rev. Pharmacol. 9, 209-240.
Damle, V. and Karlin, A. Affinity labeling of one of two α-neurotoxin binding sites in acetylcholine receptor from Torpedo californica. (1978) Biochemistry 17, 2039-2045.
Deneris, E.S., Connolly, J., Boulter, J., Wada, E., Wada, K., Swanson, L.W., Patrick, J. and Heinemann, S. Primary structure and expression of β2: A novel subunit of neuronal nicotinic acetylcholine receptors. (1988) Neuron 1, 45-54.
Devereux, J., Haeberli, P. and Smithies, O. A comprehensive set of sequence analysis analysis programs for the VAX. (1984) Nucl. Acids Res. 12, 387-395.
Dionne, V.E., Steinbach, J.M. and Stevens, C.F. An analysis of the dose-response relationship at voltage-clamped frog neuromuscular junctions. (1978) J. Physiol. 281, 421-444.
Dixon, R.A.F., Kobilka, B.K., Strader, D.J., Benovic, J.L., Dohlman, H.G., Frielle, T., Bolanowski, M.A., Bennett, C.D., Rands, E., Diehl, R.E., Mumford, R.A., Slater, E.E., Sigal, I.S., Caron, M.G., Lefkowitz, R.J. and Strader, C.D. Cloning of the gene an cDNA for mammalian beta-adrenergic receptor and homology with rhodopsin. (1986) Nature 321, 75-79.
Dreyer, F., Peper, R. and Sterz, R. Determination of dose-response curves by quantitative ionophoresis at the frog neuromuscular junction. (1978) J. Physiol. 281, 395-419.
Eisenberg, D., Sweet, R.M. and Tersilliger, T.C. The hydrophobic moment detects periodicity in protein hydrophobicity. (1984) Proc. Natl. Acad. Sci. USA 81, 140-144.
Emini, E.A., Hughes, J.V., Perlow, D.S. and Boger, J. Induction of hepatitis A virus neutralizing antibody by a virus-specific synthetic peptide. (1985) J. Virol. 55, 836-839.
Engel, A. Myasthenia Gravis. In Handbook of Clinical Neurology, ed. P. Vinken, G. Bruyn, 41(II), 95-145. New York:North Holland (1979).
Engelman, D.M., Steitz, T.A. and Goldman, A. Identifying nonpolar transbilayer helices in amino acid sequences of membrane proteins. (1986) Ann. Rev. Biophys., Biophys. Chem. 15, 321-353.
Fels, G., Plumer-Wilk, R., Schreiber, H. and Maelicke, A. A monoclonal antibody interfering with binding and response of the acetylcholine receptor. (1986) J. Biol. Chem. 261, 15746-15754.
Fukuda, K., Higashida, H., Kubo, T., Maeda, A., Akiba, I., Bujo, H., Mishina, M. and Numa, S. Selective coupling with K^+ currents of muscarinic acetylcholine receptor subtypes in NG108-15 cells. (1988) Science 335, 355-358.
Garnier, J., Osguthorpe, D.J. and Robson, B. Analysis of the accuracy and implications of simple methods for predicting the secondary structure of globular proteins. (1978) J. Mol. Biol. 120, 97-120.
Gershoni, J.M., Hawrot, E. and Lentz, T.L. Binding of α-bungarotoxin to isolated α-subunit of the acetylcholine receptor of Torpedo californica: Quantitative analysis with protein blots. (1983) Proc. Natl. Acad. Sci. USA 80, 4973-4977.
Geysen, H.M., Meloen, R.H. and Barteling, S.J. Use of peptide synthesis to probe viral antigens for epitopes to a resolution of a single amino acid. (1984) Proc. Natl. Acad. Sci. USA 81, 3998-4002.
Geysen, H.M., Tainer, J.A., Rodda, S.J., Mason, T.J., Alexander, H., Getzoff, E.D. and Lerner, R.A. Chemistry of antibody binding to a protein. (1987) Science 235, 1184-1190.
Gotti, C., Conti-Tronconi, B.M. and Raftery, M.A. Mammalian muscle acetylcholine receptor purification and characterization. (1982) Biochemistry 21, 3148.

Grant, G.A. and Chiapinelli, V.A. κ-bungarotoxin: complete amino acid sequence of a neuronal nicotinic receptor probe. (1985) Biochemistry 24, 1532-1537.
Grenningloh, G., Riemitz, A., Schmitt, B., Methfessel, C., Zensen, M., Beyreuther, K., Gundelfinger, E.D. and Betz, H. The strychnine binding subunit of the glycine receptor shows homology with nicotinic acetylcholine receptors. (1987) Nature 328, 216-219.
Gribskov, M., Burgess, R.R. and Devereux, J. PEPPLOT, a protein secondary structure analysis program for the UWGCC sequence analysis software package. (1986) Nucl. Acids Res. 14, 327-334.
Haggerty, J.G. and Froehner, S.C. Restoration of ^{125}I-α-bungarotoxin binding activity to the α-subunit of Torpedo acetylcholine receptor isolated by gel electrophoresis in sodium dodecyl sulfate. (1981) J. Biol. Chem. 256, 8294-8297.
Hermans-Borgmeyer, I., Zopf, D., Ryseck, R-P., Hovemann, B., Betz, H. and Gundelfinger, E.D. Primary structure of a developmentally regulated nicotinic acetylcholine receptor protein from Drosophila. (1986) EMBO J. 5, 1503-1508.
Hopp, T.P. and Woods, K.R. Prediction of protein antigenic determinants from amino acid sequences. (1981) Proc. Natl. Acad. Sci. USA 78, 3824-3828.
Houghten, R.A. General method for the rapid solid-phase synthesis of large numbers of peptides: specificity of antigen-antibody interaction at the level of individual amino acids. (1985) Proc. Natl. Acad. Sci. USA 82, 5131-5135.
Jacob, M.H. and Berg, D.K. The ultrastructural location of α-bungarotoxin binding sites in relation to synapses on chick cilary ganglion neurons. (1983) J. Neurosci. 3, 26-271.
Jameson, B.A. and Wolf, H. The antigenic index: a novel algorithm for antigenic determinants. (1988) CABIOS 4, 181-186.
Jemmerson, R., Morrow, P.R., Klinman, N.R. and Paterson, Y. Analysis of an evolutionarily conserved antigenic site on mammalian cytochrome c using synthetic peptides. (1985) Proc. Natl. Acad. Sci. USA 82, 1508-1512.
Kao, P.N. and Karlin, A. Acetylcholine receptor binding site contains a disulfide crosslink between adjacent half-cystinyl residues. (1986) J. Biol. Chem. 261, 8085-8088.
Karplus, P.A. and Schulz, G.E. Prediction of chain flexibility in proteins. (1985) Naturwissenschaften 72, 212-213.
Katz, B. and Thesleff, S. A study of the "desensitization" produced by acetylcholine at the motor end-plate. (1957) J. Physiol. 138, 63-80.
Kobilka, B.K., Dixon, R.A.F., Frielle, T., Dohlman, H.G., Bolanowski, M.A., Sigal, I.S., Yang-Feng, T.L., Francke, U., Caron, M.G. and Lefkowitz, R.J. cDNA for the human beta-2 adrenergic receptor: a protein with multiple membrane-spanning domains and a chromosomal location shared with the PDGF receptor gene. (1987) Proc. Natl. Acad. Sci. USA 84, 46-50.
Kubo, T., Fukuda, K., Mikami, A., Maeda, A., Takahashi, H., Mishina, M., Haga, T., Haga, K., Ichiyama, A., Kangawa, K., Kojima, M., Matsuo, H., Hirose, T. and Numa, S. Cloning, sequencing and expression of complementary DNA encoding the muscarinic acetylcholine receptor. (1986) Nature 323, 411-416.
Kubo, T., Maeda, A., Sugimoto, K., Akiba, I., Mikami, A., Takahashi, H., Haga, T., Haga, K., Ichiyama, A., Kangawa, K., Matsuo, H., Hirose, T. and Numa, S. Primary structure of porcine cardiac muscarinic acetylcholine receptor deduced from the cDNA sequence. (1986b) Fed. Eur. Biochem. Soc. Lett. 209, 36-372.
Kyte, J. and Doolittle, R.F. A simple method for displaying the hydropathic character of a protein. (1982) J. Mol. Biol. 157, 105-132.
Lee, C.Y. Chemistry and pharmacology of polypeptide toxins in snake venoms. (1972) Annu. Rev. Pharmacol. 12, 265-298.
Lennon, V.A. and Lambert, E.H. Myasthenia Gravis induced by monoclonal antibodies to acetylcholine receptors. (1980) Nature 285, 233-236.
Levitan, E.S., Schofield, P.R., Burt, D.R., Rhee, L.M., Wisden, W., Kohler, M., Fujita, N., Rodriguez, H.F., Stephenson, A., Darlison, M.G., Barnard, E.A. and Seeburg, P.H. Structural and functional basis for $GABA_A$ receptor homogeneity. Nature 335, 76-79.

Lindstrom, J. Immunobiology of Myasthenia Gravis, experimental autoimmune Myasthenia Gravis, and Lambert-Eaton syndrome. (1985) Ann. Rev. Immunol. 3, 109-131.
Lindstrom, J., Schoepfer, R. and Whiting, P. Molecular studies of the neuronal nicotinic acetylcholine receptor family. (1978) Molecular Neurobiology 1, 281-337.
Loring, R.H., Andrews, D., Lane, W., Zigmond, R.E. Amino acid sequence of toxin F, a snake venom toxin that blocks neuronal nicotinic receptors. (1986) Brain Research 385, 30-37.
Loring, R.H., Chiapinelli, V.A., Zigmond, R.E. and Cohen, J.B. Characterization of a snake venom neurotoxin which blocks nicotinic transmission in the avian ciliary ganglion. (1984) J. Neurosci. 11, 989-999.
Lukas, R.J. Properties of Curaremimetic Neurotoxin Binding Sites in the Rat Central Nervous System. (1984) Biochemistry 23, 1152-1160.
Lukas, R.J. Detection of low-affinity α-bungarotoxin binding sites in rat central nervous system. (1984) Biochemistry 23, 1160-1164.
Lukas, R.J. Immunochemical and pharmacological distinctions between curaremimetic neurotoxin binding sites of central, autonomic, and peripheral origin. (1986) Proc. Natl. Acad. Sci. USA 83, 5741-5745.
Luzzatto, A.C., Conti-Tronconi, B.M., Paggi, P. and Rossi, A. Binding of Naja naja siamensis α-toxin to the chick ciliary ganglion: A light-microscopy autoradiographic study. (1980) 5, 313-318.
Marks, M.J. and Collins, A.C. Characterization of nicotine binding in mouse brain and comparison with the binding of α-bungarotoxin and quinuclidinyl benzilate. (1982) Molecular Pharmacology 22, 554-564.
Marks, M.J., Stitzel, J.A., Romm, E., Wehner, J.M. and Collins, A.C. Nicotinic binding sites in rat and mouse brain: Comparison of acetylcholine, nicotine, and α-bungarotoxin. (1986) Molecular Pharmacology 30, 427-436.
Martin, B.M., Chibber, B.A. and Maelicke, A. The sites of neurotoxicity in alpha-cobratoxin. J. Biol. Chem. 258, 8714-8722.
McCarthy, M.P., Earnest, J.P., Young, E.F., Choe, S. and Stroud, R.M. The molecular neurobiology of the acetylcholine receptor. (1986) Ann. Rev. Neurosci. 9, 383-413.
Mehraban, F., Kemshead, J.T. and Dolly, J.O. Properties of monoclonal antibodies to nicotinic acetylcholine receptor from chick muscle. (1984) Eur. J. Biochem. 138, 53-61.
Miledi, R., Szczepaniak, F.R.S. and Szczepaniak, A.C. Effect of Dendroaspis neurotoxins on synaptic transmission in the spinal cord of the frog. (1975) Proc. R. Soc. Lond. B. 190, 267-274.
Mishina, M., Tobimatsu, T., Imoto, K., Tanaka, K-i., Fujita, Y., Fukuda, K., Kurasaki, M., Takahashi, H., Morimoto, Y., Hirose, T., Inayama, S., Takahashi, T., Kuno, M. and Numa, S. Location of functional regions of acetylcholine receptor α-subunit by site-directed mutagenesis. (1985) Nature 313, 364-369.
Moore, H.P. and Raftery, M.A. Studies of reversible and irreversible interactions of an alkylating agonist with Torpedo californica acetylcholine receptor in membrane-bound and purified states. (1979) Biochemistry 18, 1862-1867.
Morley, B.J., Dwyer, D.S., Strang-Brown, P.F., Bradley, R.J. and Kemp, G.E. Evidence that certain peripheral anti-acetylcholine receptor antibodies do not interact with brain BTX binding sites. (1983) Brain Res. 262, 109-116.
Neubig, R.R. and Cohen, J.B. Equilibrium binding of [^{3}H] tubocurarine and [^{3}H] acetylcholine by torpedo postsynaptic membranes: stoichiometry and ligand interactions. (1979) Biochemistry 18, 5464-5475.
Neumann, D., Barchan, D., Safran, A., Gershoni, J.M. and Fuchs, S. Mapping of the α-bungarotoxin binding site within the α-subunit of the acetylcholine receptor. (1986) Proc. Natl. Acad. Sci. USA 83, 3008-3011.
Norman, R.I., Mehraban, F., Barnard, E.A. and Dolly, J.O. Nicotinic acetylcholine receptor from chick optic lobe. (1982) Proc. Natl. Acad. Sci. USA 79, 1321-1325.
Novotny, J., Handschumacher, M., Haber, E., Bruccoleri, R.E., Carlson, W.B., Fanning, D.W., Smith, J.A., Rose, G.D. Antigenic determinants in proteins coincide with surface

regions accessible to large probes (antibody domains). (1986) Proc. Natl. Acad. Sci. USA 83, 226-230.

Oswald, R.E. and Freeman, J.A. Alpha-bungarotoxin binding and central nervous system nicotinic acetylcholine receptors. (1981) Neuroscience 6, 1-14.

Patrick., J. and Lindstrom, J. Autoimmune response to acetylcholine receptors. (1973) Science 180, 871-873.

Patrick, J. and Stallcup, W.B. Immunological distinction between acetylcholine receptor and the α-bungarotoxin-binding component on sympathetic neurons. (1977) Proc. Natl. Acad. Sci. USA 74, 4689-4692.

Patrick, J., Stallcup, W.B., Zavanelli, M. and Ravdin, P. Binding properties of a neurotoxin from the venom of the green mamba, Dendroaspis viridis. (1980) J. Biol. Chem. 255, 526-533.

Pedersen, S.E., Dreyer, E.B. and Cohen, J.B. Location of ligand-binding sites on the nicotinic acetylcholine receptor α-subunit. (1986) J. Biol. Chem. 261, 13735-13743.

Quik, M. and Lamarca, M.V. Blockade of transmission in rat sympathetic ganglia by a toxin which co-purifies with α-bungarotoxin. (1982) Brain Res. 238, 385-399.

Raftery, M.A., Dunn, S.M.J., Conti-Tronconi, B.M., Middlemas, D.S. and Crawford, R.D. The n-acetylcholine receptor: subunit structure, functional binding sites and ion transport properties. (1983) Cold Spring Harbor Symp. Quant. Biol. 48, 21-33.

Rang, M.P. and Ritter, J.M. On the mechanism of desensitization at cholinergic receptors. (1970) Mol. Pharmacol. 6, 357-383.

Ratnam, M., Sargent, P.B., Sarin, V., Fox, J.L., Nguyen, D.L., Rivier, J., Criado, M. and Lindstrom, J. Location of antigenic determinants on primary sequences of subunits of nicotinic acetylcholine receptor by peptide mapping. (1986) Biochemistry 25, 2621-2632.

Ravdin, P.M. and Berg, D.W. Inhibition of neuronal acetylcholine sensitivity by α-toxins from bungarus multicinctus venom. (1979) Proc. Natl. Acad. Sci. USA 76, 2072-2076.

Richman, D.P., Gomez, C.M., Berman, P.W., Burres, S.A., Fitch, F.W. and Arnason, B.G.W. Monoclonal anti-acetylcholine receptor antibodies can cause experimental myasthenia. (1980) Nature 286, 738-739.

Rodda, S.J., Geysen, H.M., Mason, T.J. and Schoofs, P.G. The antibody response to myoglobin. I. Systematic synthesis of myoglobin peptides reveals location and substructure of species-dependent continuous antigenic determinants. (1986) Molec. Immunol. 23, 603-610.

Schmidt-Nielsen, B.K., Gepner, J.I., Teng, N.N.H. and Hall, L.M. Characterization of an α-Bungarotoxin binding component from Drosophila melanogaster. (1977) J. Neurochem. 29, 1013-1029.

Schmidt, T., Hunt, S. and Polz-Tejera, G. Nicotinic receptors of the central and autonomic nervous system (1980) In: Essman, W.B. (ed). Neurotransmitters, Receptors, and Drug Action. Spectrum Publications, Inc. New York, 1-45.

Schneider, M., Adee, C., Betz, H. and Schmidt, J. Biochemical characterization of two nicotinic receptors from the optic lobe of the chick. (1985) J. Biol. Chem. 260, 1505-14512.

Schofield, P.R., Darlison, M.G., Fujita, N., Burt, D.R., Stephenson, F.A., Rodriguez, H., Rhee, L.M., Ramachandran, J., Reale, V., Gelcorse, T., Seeburg, P.H. and Barnard, E.A. Sequence and functional expression of the $GABA_A$ receptor shows a ligand-gated receptor super-family. (1987) Nature 328, 221-224.

Sheriff, S., Silverton, E.W., Padlan, E.A., Cohen, G.H., Smith-Gill, S.J., Finzel, B.C. and Davies, D.R. Three-dimensional structure of an antibody-antigen complex. (1987) Proc. Natl. Acad. Sci. USA 84, 8075-8079.

Sine, S.M. and Taylor, P. The relationship between agonist occupation and the permeability response of the cholinergic receptor revealed by bound cobra α-toxin. (1980) J. Biol. Chem. 255, 10144-10156.

Szczepaniak, A.C. Effect of α-bungarotoxin and dendroaspis neurotoxin on acetylcholine response of snail neurons. (1974) J. Physiol., Lond. 241, 55-56.

Tainer, J.A., Getzoff, E.D., Paterson, Y., Olson, A.J., Lerner, R.A. The atomic mobility component of protein antigenicity. (1985) Ann. Rev. Immunol. 3:501-35.

Toldi, J., Joo, F., Adam, G., Feher, O. and Wolff, J.R. Inhibition of synaptic transmission in the rat superior cervical ganglion by intracarotid infusion of bungarotoxin. (1983) Brain Res. 262, 323-327.

Tzartos, S.J., Kokla, A., Walgrave, S.L. and Conti-Tronconi, B.M. Localization of the main immunogenic region of human muscle acetylcholine receptor to residues 67-76 of the α subunit. (1988) Proc. Natl. Acad. Sci. USA 85, 2899-2903.

Tzartos, S.J. and Lindstrom, J.M. Monoclonal antibodies used to probe acetylcholine receptor structure: localization of the main immunogenic region and detection of similiarities between subunits. (1980) Proc. Natl. Acad. Sci. USA 77, 755-759.

Tzartos, S.J., Rand, D.E., Eirnason, B.L. and Lindstrom, J.M. Mapping of surface structures of electrophorus acetylcholine receptor using monoclonal antibodies. (1981) J. Biol. Chem. 256, 8635-8645.

Watters, D. and Maelicke, A. Organization of ligand binding sites at the acetylcholine receptor: a study with monoclonal antibodies. (1983) Biochemistry 22, 1811-1819.

Westhof, E., Altschuh, D., Moras, D., Bloomer, A.C., Mondragon, A., Klug, A., Van Regenmortel, M.H.V. Correlation between segmental mobility and the location of antigenic determinants in proteins. (1984) Nature 311, 123-126.

Whiting, P.J. and Lindstrom, J.M. Purification and characterization of a nicotinic acetylcholine receptor from chick brain. (1986) Biochemistry 25, 2082-2093.

Whiting, P.J., Schoepfer, R., Swanson, L.W., Simmons, D.M. and Lindstrom, J.M. Functional acetylcholine receptor in PC12 cells reacts with a monoclonal antibody to brain nicotinic receptors. (1987) Nature 327, 515-518.

Wilson, P.T., Gershoni, J.M., Hawrot, E. and Lentz, T.L. Binding of α-bungarotoxin to proteolytic fragments of the α-subunit of Torpedo acetylcholine receptor analyzed by protein transfer on positively charged membrane filters. (1984) Proc. Natl. Acad. Sci. USA 81, 2553-2557.

Wolf, J., Modrow, S., Motz, M., Jameson, B., Hermann, G. and Fortsch, B. An integrated family of amino acid sequence analysis programs. (1987) CABIOS

Wolosin, J.M., Lyddiatt, A., Dolly, J.O. and Barnard, E.A. Stoichiometry of the ligand-binding sites in the acetylcholine-receptor oligomer from muscle and from electric organ. (1980) Eur. J. Biochem. 109, 495-505.

Wonnacott, S., Harrison, R. and Lunt, G. Immunological cross-reactivity between the α-bungarotoxin-binding component from rat brain and nicotinic acetylcholine receptor. (1982) J. Neuroimmunol. 3, 1-13.

Yarden, Y., Rodriguez, H., Wong, S.K.F., Brandt, D.R., May, D.C., Burnier, J., Harkins, R.N., Chen, E.Y., Ramachandran, J., Ullrich, A. and Ross, E.M. The avian beta-adrenergic receptor: primary structure and membrane topology. (1986) Proc. Natl. Acad. Sci. USA 83, 6795-6799.

FUNCTIONAL DOMAINS OF THE NICOTINIC ACETYLCHOLINE RECEPTOR

F. Hucho, R. Hilgenfeld§, G. Bandini, S. Jaweed

Institute of Biochemistry
Free University of Berlin
Thielallee 63
1000 Berlin 33
W. Germany

Introduction

Type I receptors contain two main functional domains as integral domains of its protein structure: the binding site (s) for the signal molecules and the ion channel. Ion channels are transmembrane structures whose purpose it is to lower the energy barrier for charged particles diffusing through the lipid bilayer of biological membranes. Made from proteins, they have to be hydrophobic enough to form a stable membrane complex and, simultaneously, they have to be hydrophilic to interact with polar permeant molecules. Furthermore, they have to be regulated to permit chemical or electrical signals to trigger or terminate the transmembrane flow. The hypothesis generally accepted is that this dual property - being hydrophobic and hydrophilic at the same time - is brought about by polypeptide chains having hydrophobic sequences (about 24 amino acids in length) which traverse the lipid bilayers as α-helices. At least one of these helices possibly has clusters of polar amino acids which in the context of a quaternary structure contribute to the wall of a polar pore. Individual charged amino acid side chains may represent the gate of such a pore.

---

§ - Hoechst AG, Frankfurt am Main

NATO ASI Series, Vol. H 32
Molecular Biology of Neuroreceptors and Ion Channels. Edited by A. Maelicke

In the absence of any tertiary structure of a macromolecular ion channel, solved, for example, by X-ray analysis of a crystalline protein, this hypothesis is based on the analogy to antibiotic channels and to X-ray structures of crystalline water-soluble proteins, which are similar to the very few membrane proteins crystallized so far. It is in fact based on pure plausibility, i.e., on pure speculation. The nicotinic acetylcholine receptor (AChR) may become the channel protein enabling us to separate fact from fiction. It is intensively investigated (1, 2) as a model for the family of chemically-gated ion channels which includes, among others, the glycine- (3) and $GABA_A$-channels (4).

The Helix-M2 model

In 1986, at this symposium's predecessor, we proposed the Helix-M2 model of the ion channel of the AChR (5). We had observed that UV irradiation makes the channel-blocking cation triphenylmethylphosphonium ($TPMP^+$) react with homologous amino acid sequences of several of the receptor subunits. These sequences represented the predicted transmembrane helices M2 of all the otherwise disagreeing hypothetical folding patterns. The reaction site of $TPMP^+$ in one of the subunits (δ) was identical to the reaction site of the allosteric, voltage-independent channel blocker chlorpromazine (6). We postulated that the helices M2 contributed by possibly all five receptor subunits of its α2β δ quaternary structure form the channel wall. Their properties seem to favor this hypothesis (fig. 1).

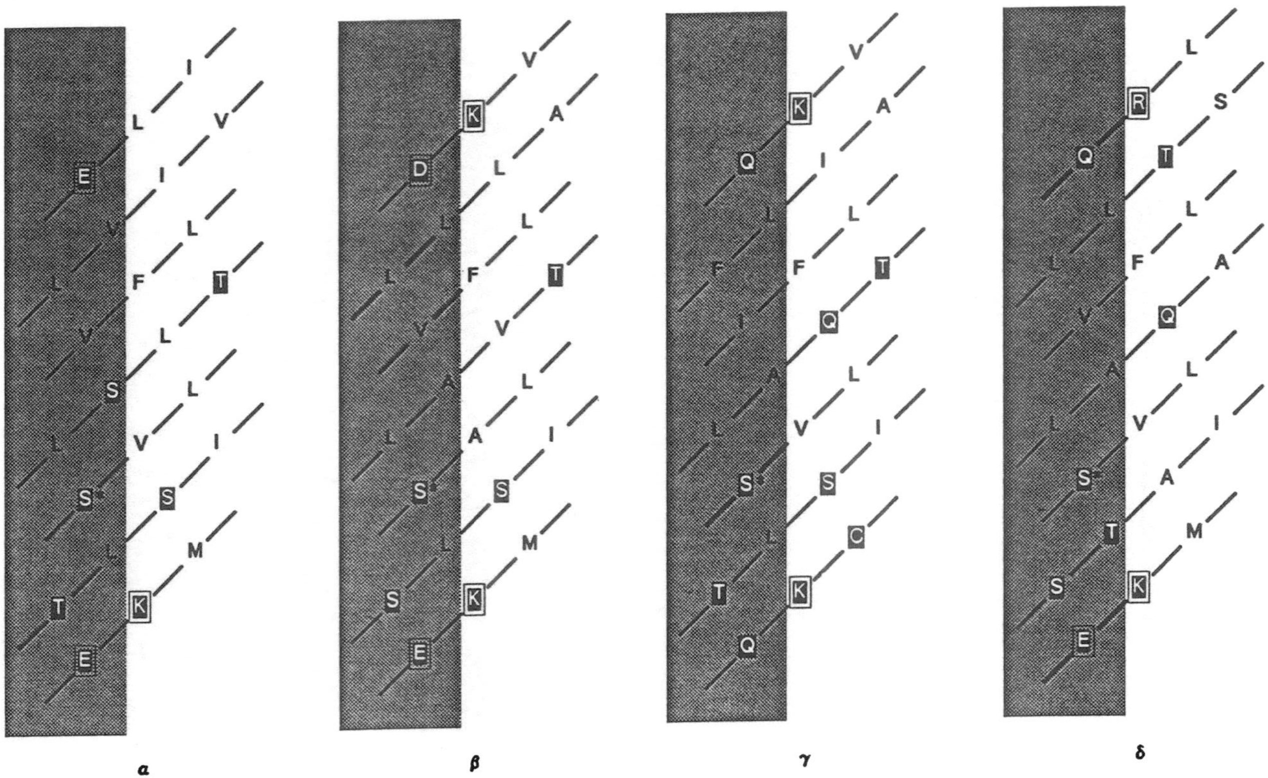

Fig. 1: The Helix-M2 sequences of the receptor subunits. Shaded areas: amino acid residues lining the channel. Boxed letters: polar amino acids. Boxed and framed letters: charged amino acids (one letter code). The asterix marks the serin residues identified in α, β, and δ as the sites labeled by TPMP$^+$.

Although hydrophobic in the two-thirds of their sequences facing the extracellular side, they contain several polar side chains in the sequences from the reaction site of TPMP to the cytoplasmic end of the helices. The known dimensions of the channel entrance (30 A, as deduced from electron microscopy (7)) of the diameter of the channel blocker $TPMP^+$ (11.5 A, X-ray crystallography (8)) and of the selectivity filter (6.4 A, evaluated by electrophysiologists (9)) made a funnel-shaped model plausible (fig. 2), with the narrowest part of the channel at or near its cytoplasmic exit.

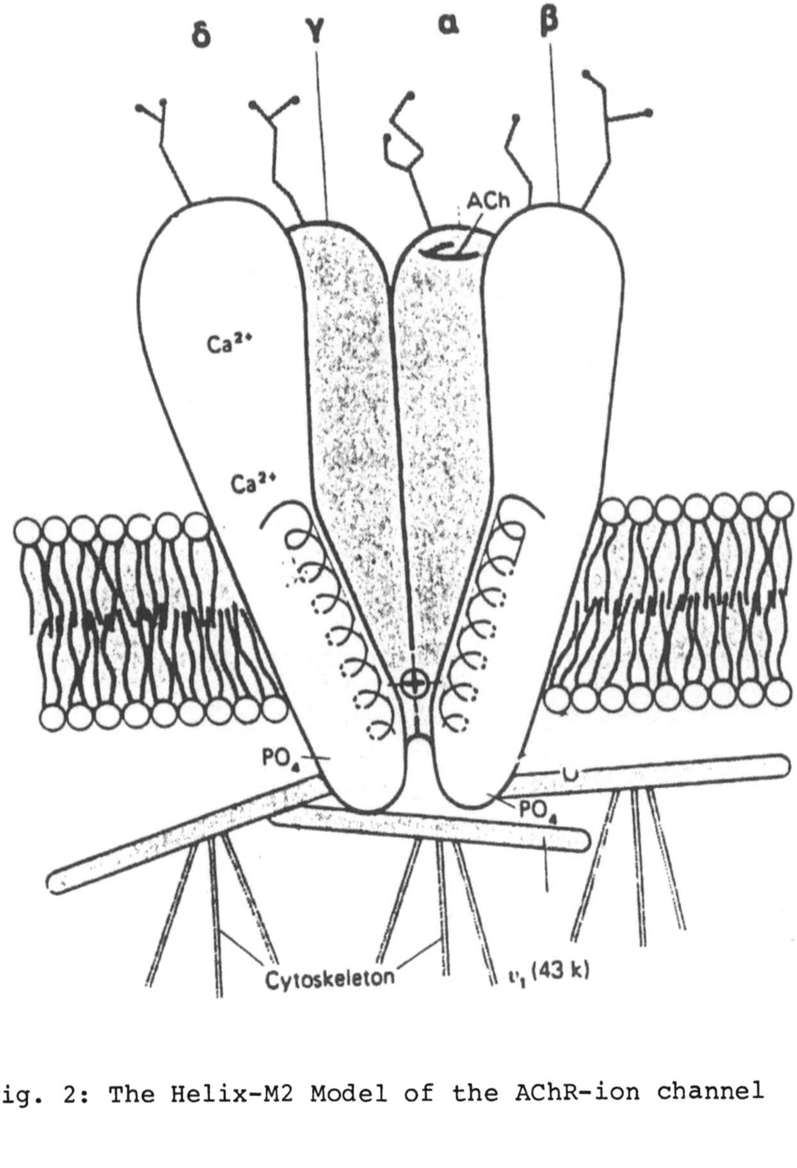

Fig. 2: The Helix-M2 Model of the AChR-ion channel

The Helix M2 model is especially compatible with investigations applying electrophysiology (patch clamp) with recombinant DNA techniques (10). By constructing chimeric receptors from calf and Torpedo proteins and expressing them in Xenopus oocytes it was clearly shown that the channel conductivity is determined by the helix M2 segment (and the neighboring sequence connecting the M2 and M3 helices). More recently it has been shown by site-directed mutagenesis experiments (11) that the charged amino acid side chains at both ends of the channel-lining helices M2 of the five receptor subunits are of special significance. They appear to form three negatively-charged rings. The middle ring seems to be especially narrow, and is therefore a good candidate for the selectivity filter.

Computer modeling

Computer modeling (fig. 3) of the channel with five helices M2 makes possible some refinement of this model. First, it shows that gaps arise between the helices at the upper (extracellular) end of the channel. They are a consequence of the tilt of the helices forming the funnel with the three diameters mentioned (30 A, 11.5 A, and 6.4 A). One has to postulate that parts of other transmembrane helices fill these gaps, although there is no evidence from labeling studies for other sequences contributing to the channel wall. Second, computer modeling reveals special structural requirements at the narrow cytoplasmic end of the channel. The charged glutamate side chains (glu α_{241}, glu β_{247}, glu δ_{255}) would jam the channel exit if they were in a trans/trans conformation. This problem may be overcome by chosing appropriate side chain torsion angles. We propose that the χ_1 and χ_2 torsion angles of the glutamate or glutamine side chains at the cytoplasmic ends of the M2 helices have to be adjusted to about +60° and +180°, respectively. The resulting gauche (+)/trans conformation is the most prominent found for these side chains in X-ray structures of proteins (12). In this conformation the glu residues of the α, β, and δ chains might favorably interact via one or more H_2O molecules with the positively charged lys side chains in positions α_{242}, β_{248}, and

a)

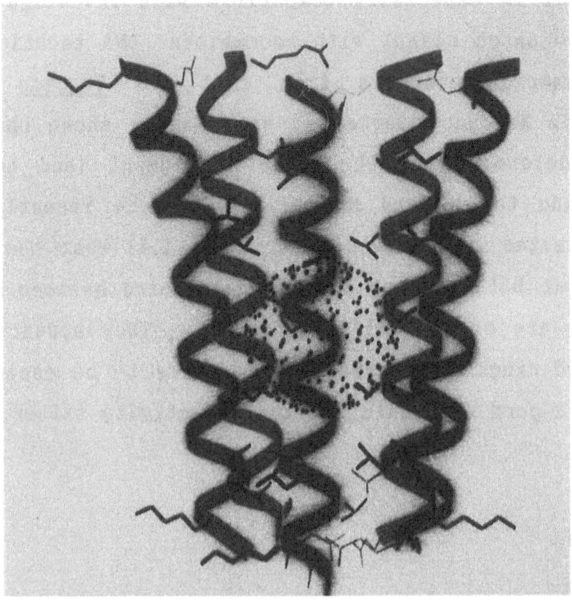

b)

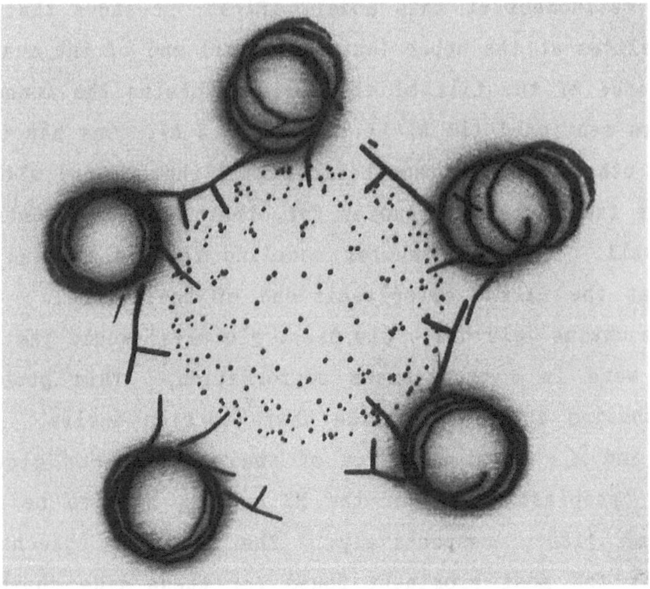

Fig. 3: Computer model of the ion channel

a) Side view with the channel blocker $TPMP^+$ represented as a sphere

b) View from the extracellular side. The sphere represents $TPMP^+$ which is in touch with the serine-OH groups identified as the site of photoaffinity labeling.

δ_{256}. It has recently been shown that such i/i+1 pairs of glu/lys occur more often in α-helices than expected (13).

The diameter of the ring formed by the carboxylate (α, β, δ) and carboxylamide groups is approximately 10 A, which is wider than the 6.5 A suggested by ion permeability experiments (9). However upon ion transfer through the channel the glx side chains might undergo conformational changes from gauche(+)/trans to gauche(-)/transfer its χ_1 and χ_2 torsion angles. This conformation is usually much less populated but equally permitted (14). The diameter of the channel opening would then reduce to approximately 7 A, a value not too far from that observed in permeability experiments with organic cations. Of course, such a relatively close approach of the glu side chains would only be possible if their unfavorable charge interaction would be neutralized by binding of the cation being transported. It is also conceivable that these negatively charged side chains form one of the many Ca^{2+} binding sites of the AChR. The permeant sodium or potassium ions would then have to remove the cation bound at this site.

In conclusion, our model is compatible with the biochemical and electrophysiological data available. It is especially compatible with the notion that the selectivity filter of the AChR ion channel is formed by a ring of glu and gln residues which are located very close to the cytoplasmic end of the channel.

Of course models obtained by computer modelling are highly speculative too, especially in absence of sufficiently exact data. Such data can be obtained by X-ray crystallography. We therefore set out to crystallize the nicotinic actylcholine receptor from Torpedo electric tissue. We developed a rapid and efficient method for preparing two-dimensional crystals of the membrane-bound receptor (15). But the structural resolution obtainable with these is limited. Three dimensional crystals would be more helpful. For this purpose we developed a large-scale purification procedure enabling us to obtain 5 - 10 mg of pure AChR within two days. It starts out with receptor-rich membranes from which

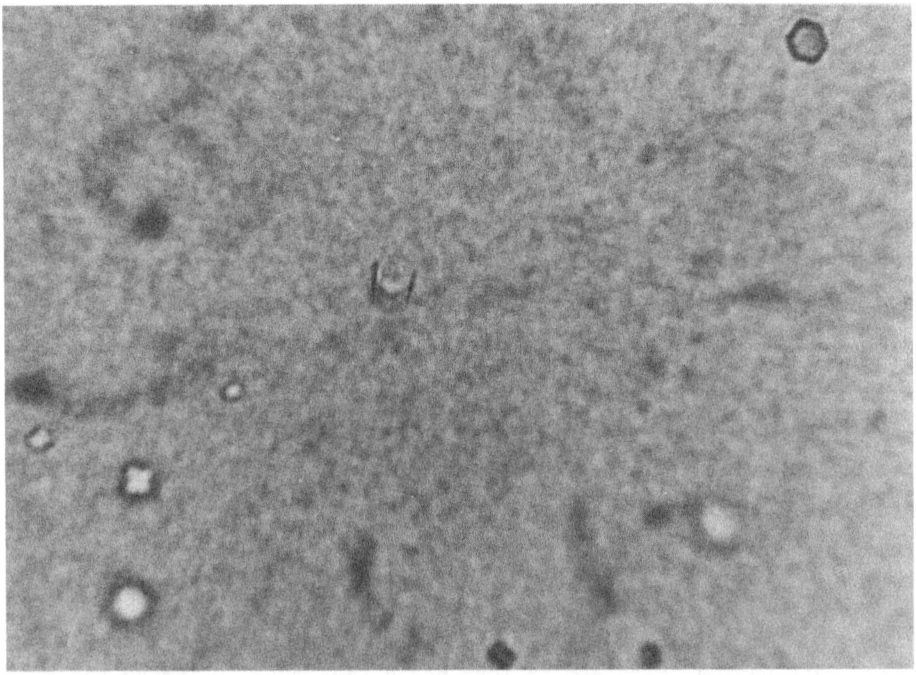

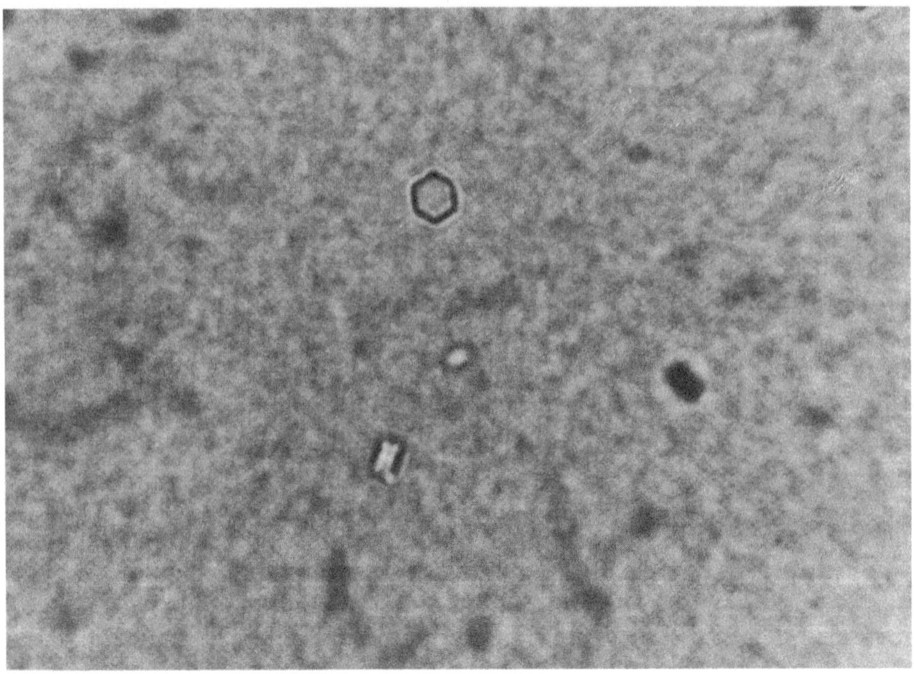

Fig. 4: Crystals obtained with an AChR preparation at 5 mg/ml

peripheral proteins have been extracted by treatment with lithium iodosalicylate. After dissolving these membranes in β-octylglycoside, further purification is performed by chromatography with a Blue Column, which retains most of the contaminating proteins. The AChR elutes from this column in an active, stabile state, together with its membrane lipids. Avoiding delipidation seems to be important in obtaining a native receptor preparation. Extensive crystallization attempts are under way. Fig. 4 shows some crystals obtained by the "hanging drop method" with a protein solution of 5 mg/ml in 1% β-octylglycoside.

Acknowledgements: Financial support from the Deutsche Forschungsgemeinschaft (DFG, SFB 312) and from the Fonds der Chemischen Industrie is gratefully acknowledged.

References

1) Popot, J.-L., & Changeux, J.-P., Physiol. Rev. 64, 1164 - 1239 (1984).

2. Hucho, F., Eur. J. Biochem. 158, 211-226 (1986).

3. Grenningloh, G., Rienitz, A., Schmitt, B., Methfessel, Ch., Zensen, M., Beyreuther, K., Gundelfinger, E., & Betz, H. Nature 328, 215 - 220 (1987).

4. Barnard, E., Darlison, M.G., & Seeburg, P., TINS 10, 502 - 509 (1987).

5. Hucho, F., Oberthür, W., Lottspeich, F., & Wittmann-Liebold, B., NATO ASI Series, Vol. H3, 115 - 127 (1986).

6. Giraudat, J., Dennis, M., Heidmann, T., Chang, J.-Y., & Changeux, J.-P., Proc. Natl. Acad. Sci. USA 83, 2719 - 2723 (1986).

7. Brisson, A., & Unwin, P.N.T., Nature 215, 474 - 477 (1985).

8. McPhail, A.T., Semenink, G.M., & Chesnut, D.B., J. Chem. Soc. (A) 2174 - 2180 (1971).

9. Huang, L.-Y., Catterall, W.A., & Ehrenstein, G., J. Gen. Physiol. 71, 394 - 410 (1978).

10. Imoto, K., Methfessel, Ch., Sakmann, B., Mishina, M., Mori, Y., Kormo, T., Fukuda, K., Kurasaki, M., Bujo, H., Fujita, Y., & Numa, S., Nature 324, 670 - 674 (1986).

11. Imoto, K., Busch, Ch., Sakmann, B., Mishina, M., Kormo, T., Nakai, J., Bujo, H., Mori, Y., Fukuda, K., & Numa, S. Nature, in press.

12. Janin, J., Wodak, S., Levitt, M. & Maigret, B., J. Mol. Biol. 125, 357 - 386 (1978).

13. Sundaralingam, M., Sakharudu, Y.C., Yatkindra, N., Ravichandran, V., Proteins Vol. 2, 64 - 71 (1987).

14. Pormuswamy, P.K., Sasisekkaran, V., Int. J. Prot. Pept. Res. Vol. 3, 9 - 18 (1971).

15. Giersig, M., Kunath, W., Pribilla, I., Bandini, G. & Hucho, F., submitted for publication.

THE LIMITED SEQUENCE SPECIFICITY OF ANTI-PEPTIDE ANTIBODIES MAY INTRODUCE AMBIGUITY IN TOPOLOGICAL STUDIES.

Alfred Maelicke*, Rita Plümer-Wilk*, Gregor Fels*, Sandra R. Spencer&, Martin Engelhard*, Detlef Veltel# and Bianca Conti-Tronconi&

*) Max-Planck-Institut für Ernährungsphysiologie, Rheinlanddamm 201, D-4600 Dortmund 1, FRG
#) Abteilung Zellbiologie, Universität Münster, 3 Domagkstrasse, D-4400 Münster, FRG
&) Department of Biochemistry, University of Minnesota, 1479 Gortner Avenue, St. Paul, Minnesota 55108, USA

Our present view of the membrane topology of nicotinic acetylcholine receptors (nAChR) heavily depends on hydropathy plots (Kyte & Doolittle, 1982; Noda et al., 1982; Claudio et al., 1983) and epitope mapping (Lindstrom et al., 1984; Maelicke et al., 1984; Neumann et al., 1984; Ratnam & Lindstrom, 1984; Plümer et al., 1984; Young et al., 1985; Fuchs & Safran, 1986; Ratnam et al., 1986a,b). The latter is usually performed with antibodies raised against rather long synthetic peptides. We have criticized this approach (Maelicke et al., 1984; Plümer et al., 1984) as it may lead to ambiguous results: Recent structural data have established that the antigen recognition site of antibodies is quite large and that antibody-antigen interaction probably involves multipoint attachment (Amit et al., 1986). Thus, only under limiting conditions will the charge density pattern defining an antibody binding site be directly correlated to the primary structure of this particular region of the antigen. The following results of a topological study of the nAChR from *Torpedo* electric organ (Maelicke et al., 1989) may exemplify this problem.

The synthetic peptides P3 (FDQQNC-Y) and P4 (FDWQNC-Y) matching the *Torpedo* nAChR sequence regions α137-142 and βγδ"137-142" differ in only one position. We have raised rabbit antisera against conjugates of each of these peptides with a carrier

NATO ASI Series, Vol. H 32
Molecular Biology of Neuroreceptors and Ion Channels. Edited by A. Maelicke

protein and have purified the peptide-specific antibodies. Their binding properties were then tested in solid phase assays with the free peptides. We found that anti-P3 serum also bound to peptide P4, and anti-P4-serum also bound to peptide P3. In both cases, the serum titre against the immunizing peptide was 50-100 fold higher than against the analogous peptide. Thus, either there existed in the two sera antibodies which do not use the exchanged sequence position as epitope determinant, or antigen-antibody interaction is by multipoint attachment with a missing attachment point reducing affinity but not completely eliminating binding. In view of the available structural data on antigen-antibody complexes (Amit et al., 1986), the latter explanation appears more likely. It agrees with the notion that antibodies in general recognize three-dimensional rather than linear epitopes.

As it refers to highly conserved sequence regions within the same molecule, the above example may be considered an unlikely case for most epitope mapping studies. This reservation does not apply to the following data. The synthetic peptide P7 (KSDEES-Y) matches the sequence region α387-392 of *Torpedo* nAChR; there is no other match with more than two aminoacids within this subunit. However, when ELISAs were performed with purified rabbit anti-P7 antiserum and a library of synthetic peptides covering overlappingly the whole primary structure of nAChR α-subunit from *Torpedo californica*, the serum recognized an additional peptide (Fig. 1). As shown in Fig. 2, peptide E58 (α184-203) does not display any sequence homology with the one (E67, α374-394) containing the sequence of P7. Notwithstanding this fact, our results suggest that parts of peptide E58 have a three-dimensional folding and charge pattern closely similar to that of peptide P7 and the related part of peptide E67.

In addition to peptide E58 of nAChR α-subunit, anti-P7 serum recognized also a sequence region of another subunit of *Torpedo* receptor. This was shown by immunoblotting (the α and β subunits but not the γ and δ subunits were labelled) and by

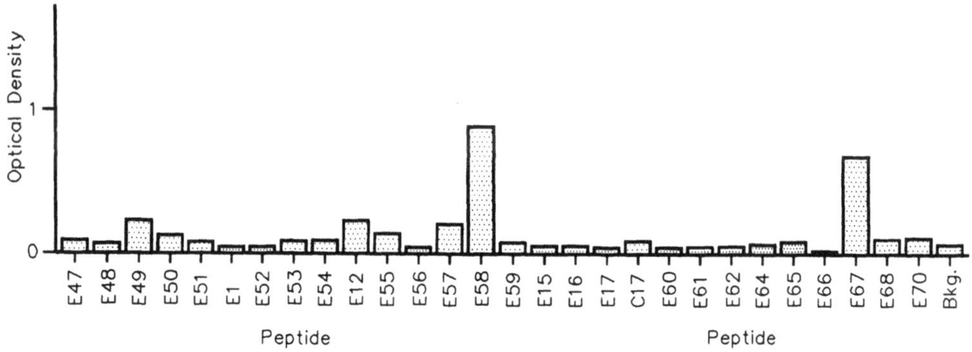

FIGURE 1: ELISA of anti-P7 antiserum, tested using synthetic peptides corresponding to the complete α-subunit sequence. After 15 minute incubation, peptide E67, which contains the immunizing sequence segment KSDEES, and peptide E58, which corresponds to the apparently unrelated sequence α181-200 were recognized to a similar extent. Upon more prolonged incubations (up to 60 minutes) also the recognition of peptide E57, which correspond to the sequence α165-184, became evident (not reported in this figure).

ELISAs employing appropriate libraries of synthetic peptides. Anti-P7 serum recognized peptides E50 (which is a shorter analog of E58) and E52 (the related sequence of *Torpedo* β-subunit) but not E54 and E56 (the related sequences of γ and δ subunit) (Fig. 2). These results may be considered as internal controls as E52 contains a sequence portion (..RSDDPSY..) quite homologous to that of the immunizing peptide while E54 and E56 display little or no sequence homology, respectively.

Our results show that the widely accepted assumption of sequence specificity of anti-peptide antibodies is not warranted. Further evidence against this assumption is provided by the demonstration of "discontinuous" epitopes (Amit et al., 1986) and the general observation that antibody affinity often increases with increasing size of the immunizing peptide. Another limitation of antibodies raised against long peptides comes from studies with anti-nAChR monoclonal antibodies (Fels

E67 (α): S A I E G V K Y I A E H M *K S D E E S* S N

E58 (α): Y R G W K H W V Y Y T C C P D T P Y L D

E50 (α): W K H W V Y Y T C C P D T P Y L D I T Y

E52 (β): H K P S R K N W - - - - - *R S D D P S Y* E D V T F

E54 (γ): K K N Y N W Q L T *K - D D* T D F Q E I I F

E56 (δ): K K N I - Y P D K F P N G T N Y Q D V T F

Fig. 2: Sequences of synthetic peptides studied with rabbit anti-P7 immuneserum.
Sequence positions in italics refer to the sequence of the immunizing peptide and homologous regions.

et al., 1986), some of which have been shown to being capable of allosteric action, i.e. they induce conformational changes of their antigens. Antibodies raised against peptides may therefore induce or favor antigen conformations that do not significantly contribute to the equilibrium state of conformations of the antigen under study [for further examples, see Getzoff et al., 1987)]. Taken together, these are serious reservations against all epitope mapping data (and predictions of membrane topology based on them) obtained with antibodies suggested, but not proven by appropriate methods, to have specificity for matching sequences (Young et al., 1985; Ratnam et al., 1986a,b; Lindstrom, 1986).

As a consequence of our studies, we suggest careful analysis of the epitope specificity of anti-peptide antibodies prior to their application in topological studies. One way of establishing the epitope (sequence) specificity of antibodies is toa test them against a library of peptides as performed in the present study.

A.M. acknowledges grants SFB 168-B2 and Ma 599/10-1 from the Deutsche Forschungsgemeinschaft and a grant from the Fonds der Chemischen Industrie. B.M.C.-T. acknowledges grants NS 28919 (NINCDS) and BNS 8607289 (NSF).

References:

Amit AG, Mariuzza RA, Phillips SEV, Poljak RJ (1986) Three-dimensional structure of an antigen-antibody complex at 2.8 Å resolution. Science (Washington, D.C.) 232:747-753

Claudio T, Ballivet M, Patrick J, Heinemann S (1983) Nucleotide and deduced amino acid sequences of *Torpedo californica* acetylcholine receptor gamma subunit. Proc Natl Acad Sci USA 80:1111-1115

Fels G, Plümer-Wilk R, Schreiber M, Maelicke A (1986) A monoclonal antibody interfering with binding and response of the acetylcholine receptor. J Biol Chem 261:15746-15754

Fuchs S, Safran A (1986) What can we learn about the acetylcholine receptor from synthetic peptides? In: Maelicke A (ed) Cell Biology. Springer Berlin Heidelberg, pp 49-59

Getzoff ED, Geysen HM, Rodda SJ, Alexander H, Tainer JA, Lerner RA (1987) Mechanisms of antibody binding to a protein. Science (Washington, D.C.) 235:1191-1196

Kyte J, Doolittle RJ (1982) A simple method for displaying the hydropathic character of a protein. J Mol Biol 157:105-132

Lindstrom J, Criado M, Hochschwender S, Fox JL, Sarin V (1984) Immunochemical tests of acetylcholine receptor subunit models. Nature 311:573-575

Lindstrom J (1986) Probing nicotinic acetylcholine receptors with monoclonal antibodies. Trends Neurosci 9:401-407

Maelicke A, Watters D, Fels G, Plümer R (1984) Antibodies Directed against functional sites on the acetylcholine receptor from Torpedo marmorata. Journ of Receptor Research 4:671-679

Maelicke A, Plümer-Wilk R, Fels G, Spencer SR, Engelhard M, Veltel D, Conti-Tronconi B.M. (1989) Epitope mapping employing antibodies raised against short synthetic peptides: a study of the nicotinic acetylcholine receptor. Biochemistry 28:1396-1405

Neumann D, Fridkin M, Fuchs S (1984) Anti-acetylcholine receptor response achieved by immunization with a synthetic peptide from the receptor sequence. Biochem Biophys Res Commun 121:673-679

Noda M, Takahashi H, Tanabe T, Toyosato M, Furutani Y, Hirose T, Asai M, Inayama S, Miyata T, Numa S (1982) Primary structure of alpha-subunit precursor of *Torpedo californica* acetylcholine receptor deduced from cDNA sequence. Nature 299:793-797

Plümer R, Fels G, Maelicke A (1984) Antibodies against preselected peptides to map functional sites on the acetylcholine receptor. FEBS Letters 178:204-208

Ratnam M, Lindstrom J (1984) Structural features of the nicotinic acetylcholine receptor revealed by antibodies to synthetic peptides. Biochem Biophys Res Commun 122:1225-1233

Ratnam M, Sargent PB, Sarin V, Fox JL, Le Nguyen DL, Rivier J, Criado M, Lindstrom J (1986a) Location of antigenic determinants on primary sequences of subunits of nicotinic acetylcholine receptor by peptide mapping. Biochemistry 25:2621-2632

Ratnam M, Le Nguyen DL, Sargent PB, Lindstrom J (1986b) Transmembrane topology of nicotinic acetylcholine receptor: Immunochemical tests contradict theoretical predictions based on hydrophobicity profiles. Biochemistry 25: 2633-2643

Young EF, Ralston, E, Blake J, Ramachandran J, Hall ZW, Stroud RM (1985) Topological mapping of acetylcholine receptor: evidence for a model with five transmembrane segments and a cytoplasmic COOH-terminal peptide. Proc Natl Acad Sci USA 82:626-630

STRUCTURE, CONFORMATION AND IMMUNOLOGICAL ANALYSIS OF PEPTIDE 188-201, THE PUTATIVE BINDING SITE FOR α-BUNGAROTOXIN IN THE α SUBUNIT OF ACETYLCHOLINE RECEPTOR.

C. Gotti°, R. Longhi#, F. Frigerio*, M. Bolognesi*, D. Fornasari°, G. Racchetti+, G. Mazzola# and F. Clementi°

°CNR Center of Cytopharmacology, Department of Medical Pharmacology, University of Milano, via Vanvitelli 32, 20129 Milano; #CNR Institute of Hormone Chemistry, via M. Bianco, 20133, Milano; *Department of Genetics, University of Pavia; +Section of Biochemistry and Immunology, Recordati Spa, Milano.

* * * * *

The nicotinic acetylcholine receptor (AChR) is a transmembrane protein made up of five subunits (2α, β, γ, δ). The α subunits bind cholinergic agonists and antagonists and α-Bungarotoxin (αBgtx) (1). The primary structure of the receptor subunits has been deduced from the cDNA sequences (2) and from partial aminoacid sequences of the protein (3). On the basis of sequence and monoclonal antibody binding studies, several models have been proposed for describing the probable orientation of the subunits in the membrane and for localizing the most relevant functional sites of these proteins (4).

Identification of the acetylcholine (ACh) binding sites on the receptor α subunits is a key point for understanding the molecular mechanism underlying receptor activation.

Chemical modification and affinity labeling studies of both native and

NATO ASI Series, Vol. H 32
Molecular Biology of Neuroreceptors and Ion Channels. Edited by A. Maelicke

detergent-solubilized AChR indicate that a disulphide bridge is present on the α subunit chains within 10 A from the acetylcholine binding site (5), and it has been identified between residues 192 and 193 (6). Peptide mapping studies of the α subunit have indicated that αBgtx binding sites is located in regions between residues 153-241 (7), 152-313 (8), 179-207 (9), 152-180 (10). Furthermore it has been shown that synthetic peptides corresponding to the sequences 185-196 (8), 173-204 (11), 184-200 (12), 188-201 (13) and 185-199 (14) bind αBgtx. Sequences comparison of the AChR subunits has shown that these subunits are highly homologous except in those regions containing the sequence 192-193, putative toxin binding regions. These findings suggest that this sequence is the site which is mainly responsible for the binding of αBgtx and cholinergic agents in the α subunit.

However, other reports claim that synthetic or proteolytic peptides of the α subunits, that do not contain the sequence flanking residues 192-193, bind αBgtx and that this binding is prevented by specific cholinergic agents. The suggested alternative sites for αBgtx binding are the regions between residues 23-49, 100-115 and 125-148 (15).

1 - Site of αBgtx binding

1.1 Synthetic peptides

We addressed the problem of finding the αBgtx binding site by using synthetic peptides. We synthetized three peptides of the extracellular region of the Torpedo α-subunit corresponding to the sequences 125-143, 158-167, 188-201 and one corresponding to sequence 188-201 of human AChR (see Table 1) (16). Peptides were synthetized manually or by using a Beckman system 990 synthesizer in the stepwise solid-phase method proposed by Merrifield. Final deprotection and cleavage of the peptides from the resin was performed by the low-high HF treatment proposed by Tam and Merrifield (16).

TABLE 1 - Sequences of the synthetic peptides of the α-subunit of the Torpedo californica (T) and human (H) acetylcholine receptor

peptide αT 125-143:	KSYCEIIVTHFDQNCT
peptide αT 158-167:	ISPESDRPDL
peptide αT 188-201:	KVYYTCCPDTPYLDI
peptide αT 188-194:	GVYYTCCP
peptide αT 191-198:	GTCCPDTPY
peptide αT 195-201:	GDTPYLDI
peptide αH 188-201:	KVTYSCCPDTPYLDI

Peptide purification was achieved by gel filtration, ion exchange and semipreparative reverse-phase high performance liquid liquid chromatography (RP-HPLC); purity was ascertained by RP-HPLC in different eluting systems and by aminoacid analysis. Both human and Torpedo 188-201 sequences contained an additional Lys, and the peptides overlapping the Torpedo 188-201 sequence contained an additional Gly at the NH_2 terminus as a spacer and in order to improve the yield of immobilization. Peptides were immobilized on CNBr-activated Sepharose 4B (Pharmacia) and the amount of peptides coupled was determined by aminoacid analysis of the HCl hydrolisate. Toxin bindings were performed on peptides bound to Sepharose with αBgtx purified from the venom of Bungarus multicinctus according to Gotti et al. (18), and labeled with ^{125}I to a specific activity of 30 μCi/nmol by the chloramine T method (19).

We found that the greatest αBgtx binding was with the peptide corresponding to Torpedo sequence 188-201, while peptides 158-167 and 125-143 did not bind ^{125}I-αBgtx (Table 2). Peptide corresponding to human 188-201 significantly bound ^{125}I-αBgtx but this binding was only 10% of the binding to the corresponding peptide of Torpedo AChR (16).

TABLE 2 - ^{125}I-αBgtx Binding to 2 mmol different peptides bound to Sepharose

peptide 8-22 calcitonin	190± 28	191± 47
peptide 12-26 endorphin	196± 61	142± 33
peptide 125-143 AChR	330± 83	342± 93
peptide 158-167 AChR	214± 47	192± 34
peptide αT 188-201 AChR	43234± 670	3047± 446
peptide αH 188-201 I	4260± 292	654± 116

1.2 Characteristics of αBgtx binding to peptide 188-201

The specific binding of αBgtx to peptide 188-201 was linear over a wide range of concentrations of peptide. The toxin binding was fast (70% of binding was reached after only 2 min and maximum effect was seen after 30 min). The kinetics of αBgtx binding was temperature independent. Scatchard analysis of binding data indicated that all the sites belong to a single class, characterized by low affinity for the ligand (K_d=1.03±0.1 μM) and B_{max} was 16.5 pmol/nmol of peptide (Fig.1).

The characteristics of αBgtx binding to the peptide are different from those of the binding to the native Torpedo receptor. The rate of association is faster and the affinity is 3 orders of magnitude lower. This last finding is consistent with the concentration of antagonists necessary for the inhibition of αBgtx binding to the peptide which is also 3 orders of magnitude greater than that normally used in competition experiments with native AChR. Probably for the same reason, we were not able to show any specific or consistent binding of $[^3H]$ACh to the peptide (data not shown). The binding data of αBgtx to the peptide are not very different from those found studying αBgtx binding to isolated and denatured Torpedo subunits.

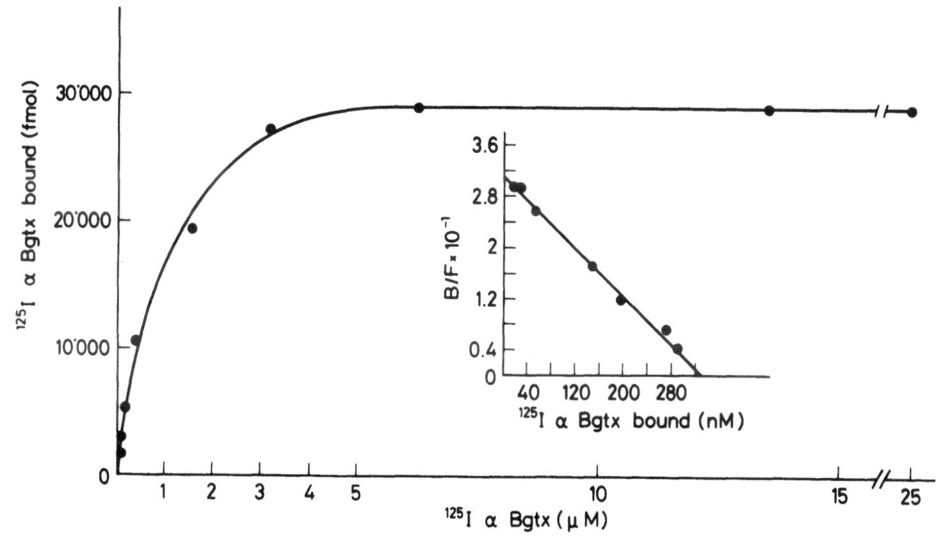

Fig.1 - Specific binding of ^{125}I-αBgtx to peptide 188-201. The unspecific binding was measured in the presence of excess of unlabeled αBgtx and did not exceed 20% of the total. The inset shows a Scatchard analysis of the binding data which are in agreement with only one site of binding for αBgtx.

The specific binding of ^{125}I-αBgtx to peptide 188-201 was not inhibited at high concentrations (80μM final concentration) by the typical basic proteins cytochrome c, soybean inhibitor or chick lysozyme.

1.3 Pharmacology of αBgtx binding

αBgtx binding was inhibited by cholinergic drugs: d-tubocurarine was the most potent, followed by decamethonium and hexamethonium. No inhibition by carbamylcholine could be detected, even at 300 mM concentration. The profile of inhibition with these ligands is quite similar to that obtained with intact AChR although affinities are several orders of magnitude lower (13).

However, we did not observe any inhibition of αBgtx binding with Naja Naja siamensis (NNS) nor with P15 toxin (13) used up to 80 μM. These results were unexpected because these toxins bind to muscle and Torpedo AChR, and NNS shares with αBgtx a very high degree of aminoacid homology and a number of pharmacological and electrophysiological properties (18). Similar results have been obtained by Atassi et al. (15), who found that cobratoxin binds to peptide 182-198 with as low affinity as to other parts of the α subunit sequences.

Our findings support the hypothesis that sequence 188-201 comprises a site relevant for the binding of cholinergic agents. Particularly similar to our results are those of Neumann et al., who found that toxin binds to peptide 185-196 (21). Comparing the sequence of this peptide with ours we can see that the difference between them lies in the N-terminus and C-terminus portion. The overlapping sequence 188-196 is the following: Val-Tyr-Tyr-Thr-Cys-Cys-Pro-Asp-Thr. It is thus possible that these 9AA are required for αBgtx binding.

However, the low affinity binding of αBgtx peptide 188-201 in comparison with the native receptor, together with the fact that Naja Naja toxin binds to native receptors as efficiently as αBgtx but not to peptide 188-201, and the low binding of αBgtx to human 188-201 in comparison with native receptors, suggests that sequence 188-201 is a privileged site for determining toxin binding, but that it is not the only one. The other sites proposed by Noda et al. (2), and Atassi et al (15), and by Boulter et al. (22) could play a role in determining the affinity and kinetics of binding to this peculiar and specific region of the sequence.

2 - Requirements of peptide 188-201 for toxin binding

2.1 S-S bridge between cysteines 192-193.

Karlin et al. (5,6,23) have shown that a disulphide bond was within 10 A from the cholinergic binding site and that, in native receptors cysteines 192-193 are disulphide cross bridged. We tested this hypothesis using two approaches. First, we synthetized a short peptide 195-201 (Table 1) which did not contain the two cysteines and we found that it does not bind ^{125}I-αBgtx (Fig.2d), confirming that the two cysteines are relevant for toxin binding. Then, in order to evaluate the relevance for toxin binding of the integrity of the S-S bridge between the two cysteines we modified it by reduction and by reduction plus alkylation (16). We found that after both reduction alone and reduction plus alkylation the peptide had partially lost its toxin binding property (Fig.2 e,f). Restoring the disulphide bridge by mild oxydation completely restored toxin binding capacity to peptide (Fig.2).

Our findings strongly indicate that the free sulphydryl group is relevant for the binding of αBgtx. We do not know if the S-S bridge is important because it is a site for toxin binding or because it imposes a physical constraint on the peptide. We favour the first hypothesis since the integrity of the S-S bridge has been found relevant also in native AChR. In fact, Karlin found that a mild reduction with DTT causes an inhibition in the response of electroplax to acetylcholine and to cholinergic analogs. This activity was fully restored by subsequent oxidation (23). It was further observed that alkylation by N-ethylmaleimide or other S-blockers irreversibly inhibits the response of electroplax to acetylcholine (5).

2.2 Peptide length

We investigated what could be the minimum length of the sequence flanking residues 192-193 required for toxin binding. From the data present in the literature one can argue that nine aminoacids are relevant for toxin binding in this region. We synthesized two short peptides, both containing the disulphide bond, 188-194 and 191-198 (Table 1). We found that both of them was able to bind ^{125}I-αBgtx but the binding capacity was very low (Fig.2, b,c). This indicate that a certain length of the chain is necessary on both sides of the residue 192-193 for the proper structure of the binding site.

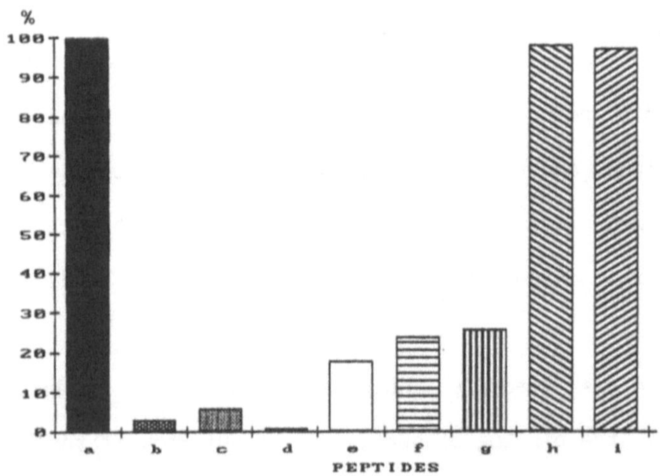

Fig.2 - Binding of ^{125}I-αBgtx to peptides corresponding to the aminoacid sequences 188-201 of Torpedo AChR and to chemically modified and shortened peptides. (a) Peptide 188-201 of Torpedo AChR in the oxidized form (expressed as 100%); (b) peptide 188-194; (c) peptide 191-198; (d) peptide 195-201; (e) peptides 188-201 reduced with DTT; (f) peptide 188-201 reduced and alkylated with N-ethylmaleimide; (g) peptide 188-201 with O-acylated Tyr; (h) peptide 188-201 with deacylated O-acetyl Tyr; (i) peptide 188-201 with β-carboxyl of aspartic acids and COOH terminal amidated.

2.3 Chemical modification of side chains

In order to verify if other aminoacids or side chain residues could be relevant for toxin binding to peptide 188-201, peptide 188-201 conjugate to sepharose was chemically modified in order to obtain o-acyl Tyr peptide, and δ-carboxylamide peptide (16). A decrease in the binding properties of the peptide was observed when the Tyr were O-acylated, but β-carboxyl amidated peptide maintained the binding activity. Restoring the phenolic hydroxyl of Tyr by hydroxylamine completely restored binding capacity (Fig.2 g,h,i). The relevance of Tyr for αBgtx binding is further underlined by the data found with the sequence 188-201 of the α-subunit of human AChR. This sequence differs from that of Torpedo in that Tyr 189 is substituted with a Thr and Thr 192 is substituted with a Ser. This peptide still specifically binds αBgtx but this binding is only 10% of the binding of αBgtx to the corresponding peptide of the Torpedo sequence (Table 2). These findings indicate that Tyr in position 189 is really relevant for αBgtx binding.

In conclusion, the data we report indicate that the requirements of peptide 188-201 for toxin binding are the disulphide bridge between 192-193, a chain of appropriate length on both sides of the S-S bridge, and a Tyr in position 189.

3 - Molecular model of peptide 188-201

We approached this part of our work by using three methods: circular dichroism (CD), computer modeling and immunological studies.

3.1 Circular dichroism.

CD of peptide 188-201 at neutral pH showed negative bands centered at 205 and 198 nm in the far UV region. Treatment with organic solvent, non-ionic and ionic surfactants, natural phospholipids and different pH's left the peptide 188-201 profile unchanged. No differences were observed in the spectra of reduced and oxidized 188-201 peptides. Using the Chou and Fasman's model (24), the analysis of the region around residues 192-193 predicts a tendency for a β-turn at residues 193-196 and β-sheet conformation in both side chains. This suggested that this sequence may be stabilized in the protein by means of an antiparallel β-sheet conformation and our CD data appear to support this idea (see ref.13 for more details).

3.2 Computer modeling

In order to confirm the conformation postulated by CD and in order to have a more precise analysis of the peptide structure we performed a computer modeling of the peptide 188-201. For the details used for establishing the secondary structure and for molecular modeling see ref.16. Atomic coordinates for the C-α positions of αBgtx were those available from the Brookhaven Protein Data Bank file.

Consideration of the secondary structure potentials for tetrapeptides in the 188-201 segment suggests a significant propensity for β structure (24). On the other hand the presence of an accessible disulphide bridge linking two adjacent Cys, followed by a Pro, imposes structural constraints on the conformation attainable by the whole polypeptide. In particular, in order to attain proper S-S covalent bond formation, the peptide bond between adjacent Cys 192 and Cys 193 must adopt the unusual cis instead of trans conformation. Isomerization of the peptide bond may also occur at Pro residues, such as Pro 194. In this case, because of the five membered side chain ring constraint, the trans Pro structure is favoured by only 2 kcal/mole compared with the cis isomer. Under particular circumstances, such as the occurrence of a -Pro-Pro- segment within longer polypeptides or proteins, one of the two residues has a cis peptide bond.

The 188-201 polypeptide model was built according to these criteria as an extended structure, with chain reversal at Cys 192-Cys 193, and taking into consideration the two alternative structures for Pro 194 peptide bond. Independent refinement of the two models indicated that lower overall energy can be attained by the structure displayed in Fig.3, in which both 192-193 and 193-194 are <u>cis</u> peptide bonds. This structure could be refined to an overall energy of -27 Kcal/mole, while isomerization of Pro 194 (and subsequent rearrangement of the following residues) yielded a polypeptide conformation which could be refined to -14 Kcal/mole. As can be seen from Fig.3 the conformation proposed for the polypeptide is extended but not regular, especially at the turn where considerable constraints are present. Nevertheless the phi-psi backbone conformational parameters fall well within the accepted limits of the Ramachandran plot. Three hydrogen bonds connect the two antiparallel strands, but only Tyr 189 O-Thr 196 N shows adequate donor-acceptor orientation.

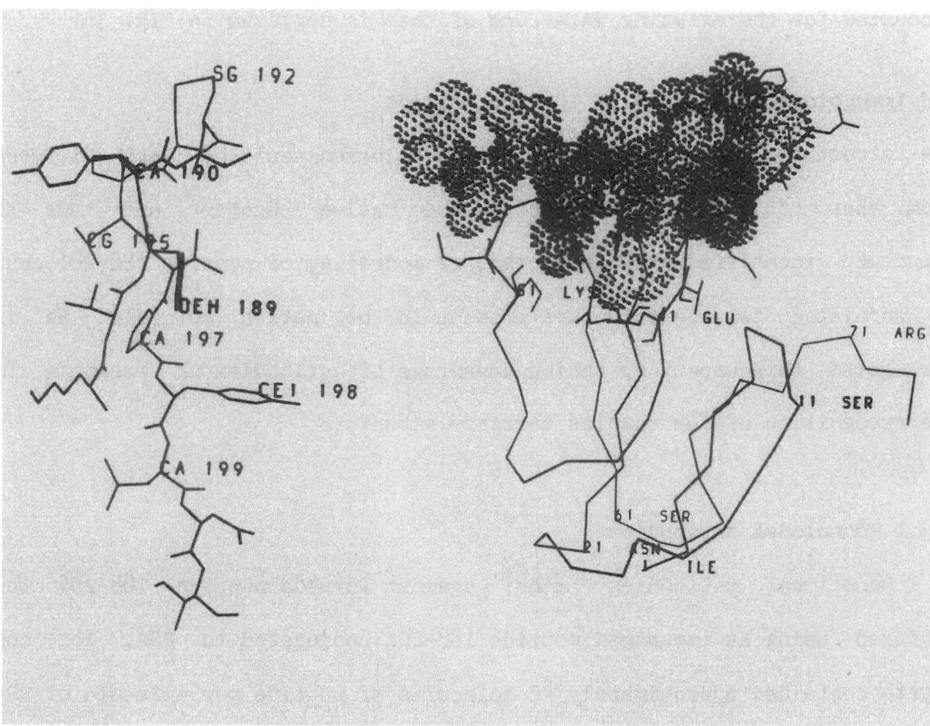

Fig.3 - (A) Computer generated picture of the model 188-201 peptide. The disulphide bridge between cysteines 192-193 is in the upper part of the picture. Aminoacids are labeled according to their sequence position. (B) A computer generated picture of the interaction of peptide 188-201 (Hatched-spheres) with the concave face of the αBgtx (lower part).

The conformation proposed is consistent with a protein loop pointing to the receptor surface at the site of chain reversal. This is in agreement with solution studies showing the ready accessibility of Cys 192-Cys 193 disulphide to reducing reagents (6). The structure here proposed for the 188-201 peptide of the α-subunit of nicotinic acetylcholine receptor however does not allow the extension of the prediction of its binding mode to the surfaces of snake neurotoxins, for which detailed molecular models are available. Despite convincing evidences that the receptor peptide binds to the concave face of α-neurotoxins (25), and that specific toxin residues have been identified as functionally important for the binding (26), we could model at least two peptide-neurotoxin complexes which equally well accounted for the existing data. One of them is depicted in Fig. 3b.

3.3 Immunological studies of 188-201 peptide

The structure of synthetic peptides in solution could be very different from that of the same sequence in the native receptor and thus the structure identified by CD and computer modelling of peptide 188-201 could be unrelated to the structure present in the native molecule. We investigated this aspect by taking advantage of antibodies as reagents for the recognition of the complex tertiary structure.

3.3.1 Monoclonal antibodies

Monoclonal antibodies (mAbs) against Torpedo peptide 188-201 were produced using as immunogen peptide 188-201 conjugated to BSA. The conjugate contained approximately 40 molecules of peptide per molecule of BSA. We found that a mAb (Mcl T1) raised against Torpedo AChR 188-201 sequence recognized, in an ELISA assay previously decribed (27), peptide 188-201 (data not shown). Furthermore, three mAbs (Mc 5,7,8) against peptide 188-201 are able to recognize Torpedo AChR (Fig.4). However, these mAbs were unable to precipitate Torpedo-AChR which had been preincubated with α-Bgtx indicating that the antigenic site recognized by mAbs was hidden by αBgtx (data not shown).

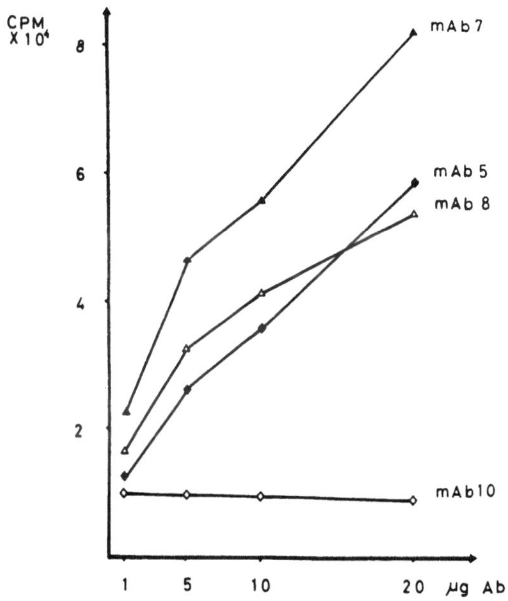

Fig. 4 - Immunoprecipitation of ^{125}I-Torpedo AChR by monoclonal antibodies raised in mouse against peptide 188-201 bound to KLH. ^{125}I-Torpedo AChR was incubated, overnight at 4°C with the indicated concentrations of purified mAbs against 188-201 (mT7, mT5, mT8) or control mAb (mT10). The complexes were then precipitated by addition of 50 μl of goat and anti-mouse IgG. Precipitates were washed twice and counted.

3.3.2 Antibodies of myasthenic animals and patients

We were also interested in the immunological study of the peptide 188-201 because it is already known that sera from animals with Experimental Myasthenia Gravis and sera from patients with Myasthenia Gravis are both able to inhibit the binding of αBgtx to AChR both in solution and membrane bound. This αBgtx inhibition could be due either to a direct competition between the antibodies and the αBgtx for the cholinergic site, i.e. the epitope of the antibodies overlaps the αBgtx binding site; or to an allosteric modification of the αBgtx binding site, as a result of the binding of the antibodies to a different site.

Given that peptide 188-201 has been demonstrate to be a privileged site for the αBgtx binding, it seemed reasonable to expect that polyclonal antibodies obtained from animals immunized with AChR or sera from myasthenic patients would bind to peptide 188-201. In order to demonstrate the latter point, we analyzed 3 sera obtained from rabbits immunized with purified Torpedo AChR, 3 sera obtained from rabbits immunized with veal AChR, 100 sera obtained from myasthenic patients and 20 sera obtained from normal subjects. The direct binding of antibodies to peptide 188-201 was measured by an Elisa assay (27).

We found that 2 sera of rabbits immunized with Torpedo AChR and 2 sera of rabbits immunized with veal AChR were able to specifically recognize the peptide 188-201 (Fig.5). The amount of anti-peptide antibodies varied in the tested sera and, as expected, was higher for sera obtained from animals immunized against Torpedo AChR. No relationship was found between the amount of anti-peptides antibodies and the anti-AChR antibody titer.

In the case of myasthenic patients we found that only 5 over 100 sera and none of the control sera contained anti-peptide antibodies. Also in this case no correlation was found between the amount of anti-peptide antibodies and the anti-AChR titer. There was no correlation between the clinical severity of the desease and the presence of anti-peptide antibodies. The reason why we found a small number of sera of myasthenic patients which recognize peptide 188-201 can be that this epitope in humans has a low antigenicity and/or that the different sequences of Torpedo and human 188-201 peptides (Table 1) make the tertiary structure of the peptides considerably different.

All these immunological studies suggest that the synthetic peptide 188-201 in solution can have a conformation similar to that shown in native AChR since it is recognized by pAbs against AChR. In addition, mAbs against the peptide and sera from myasthenic patients and from myasthenic animals also recognize AChR. Thus the structure that we have suggested for peptide 188-

201 may be considered valid also for the toxin binding region of native AChr. This peptide is a relevant part of the binding site of αBgtx in the native α subunit since mAbs against the synthetic peptide do not recognize AChR when it has bound αBgtx.

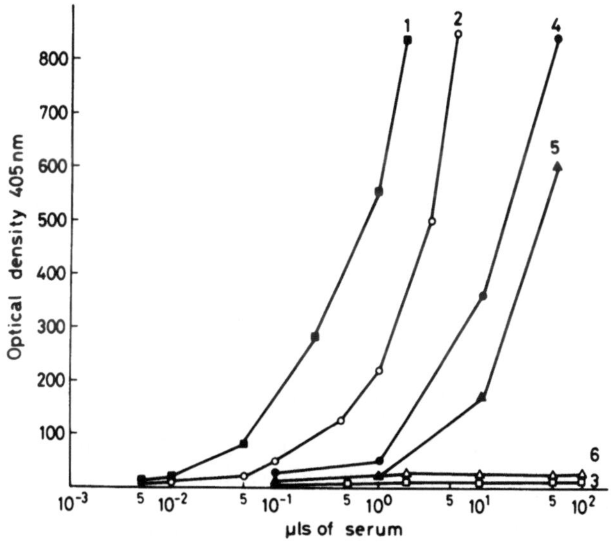

Fig.5 - Absorbance profile of sera of rabbits immunized with purified Torpedo AChR (no.1,2,3) and sera of rabbits immunized with veal AChR (no. 4,5,6) as a function of µls of serum tested.
Wells of microtiter plates were coated with 100 µl of 1 µg/ml of Torpedo peptide 188-201 conjugated to BSA and incubated overnight at 4°C.
Plates were then washed and rabbit sera at the indicated concentration were added and left for 2 h at 37°C. The Elisa assay was then performed as previously descirbed (27).

4 - CONCLUSIONS

Our studies of the αBgtx binding site in the α-subunit of AChR by means of synthetic peptides have shown that peptide 188-201 is the most important peptide for the in binding of αBgtx. The requirements for toxin binding are: the integrity of the disulphide bridge between cysteines 192-193, a minimum length of the flanking cysteines chains, and a Tyr in position 189.

The peptide has a well defined structure that fits with the presumed binding site present in αBgtx. The immunological experiments confirm that the synthetic peptide has epitopes in common with the native α-subunit.

These findings strongly support the hypothesis that sequence 188-201 is determinant for toxin binding also in native receptor. However, our data and other data present in the literature suggest that further binding sites are important for the perfect fitting of αBgtx to native receptor.

REFERENCES

1. Changeux JP, Devillers-Thiery A, Chemanilli P (1984) Acetylcholine receptor: an allosteric protein. Science 225:1335-1345.
2. Noda M, Takahashi H, Tanabe T, Toyosato M, Furutani Y, Hirose T, Asai M, Nayama S, Miyata T, Numa S (1982) Primary structure of α-subunit precursor of Torpedo californica acetylcholine receptor deduced from cDNA sequence. Nature 299: 793-797
3. Conti-Tronconi BM, Raftery MA (1982) The nicotinic cholinergic receptor: correlation of molecular structure with functional properties. Ann Rev Biochem 51: 491-530.
4. McCarthy MP, Earnest JP, Young EF, Choe S, Stroud M (1986) The molecular neurobiology of the acetylcholine receptor. Ann Rev Neurosci 9:383-413.
5. Karlin A (1969) Chemical modification of the active site of the acetylcholine receptor. J Gen Physiol 54:245-264.
6. Kao PN, Karlin A (1986) Acetylcholine receptor binding site contains a disulphide cross-link between adjacent half cystinil residues. J Biol Chem 261:8085-8088.
7. Wilson PT, Lentz TL, Hawrot E (1985) Determination of the primary aminoacid sequence specifying the α-Bungarotoxin binding site on the subunit of the acetylcholine receptor from Torpedo californica. Proc. Natl. Acad. Sci. USA 82:8790-8794
8. Neumann D, Barchan D, Safran A, Gershoni JM, Fuchs S (1986) Mapping of the α-Bungarotoxin binding site within the α-subunit of the acetylcholine receptor. Proc Natl Acad Sci USA 83: 3008-3011.
9. Dennis M, Giraudat J, Kotzyba-Hibert F, Goeldner M, Hirth C, Chang JY, Changeux JP (1986) A photoaffinity ligand of acetylcholine binding site predominantly labels the region 179-207 of the α-subunit on native acetylcholine receptor from Torpedo marmorata. FEBS Lett 207: 243-249
10. Oblas B, Singer RH, Boyd ND (1986) Localization of a polypeptide sequence within the α-subunit of the acetylcholine receptor containing the cholinergic binding site. Mol Pharmacol 29:649-656.
11. Wilson PT, Leutz T (1988) Binding of α-Bungarotoxin to synthetic peptides corresponding to residues 173-204 of the α-subunit of Torpedo, calf and human acetylcholine receptor and restoration of high affinity binding by sodium dodecyl sulphate. Biochemistry 27:6667-6674.

12. Aronheim A, Eshel Y, Moskovitz R, Gershoni JM (1988) Characterization of the binding of α-Bungarotoxin to bacterially expressed cholinergic binding sites. J Biol Chem 263:9933-9937.
13. Gotti C Mazzola G, Longhi R, Fornasari D, Clementi F (1987) The binding site for α-Bungarotoxin resides in the sequence 188-201 of the α-subunit of acetylcholine receptor: structure, conformation and binding characteristics of peptide [Lys] 188-201. Neurosci Lett 82:113-119
14. Ralston S, Sarin V, Lam Thamh H, Rivier J, Fox L, Lindstrom J (1987) Synthetic peptides used to locate the α-Bungarotoxin binding site and immunogenic regions on α-subunit of the nicotinic acetylcholine receptor. Biochemistry 26: 3261-3266.
15. Atassi M, Mulac-Jericevic B, Yokoi T, Manshouri T (1987) Localization of the functional sites on the α chain of acetylcholine receptor. Fed Proc 46:2538-2547.
16. Gotti C, Frigerio F, Bolognesi M, Longhi R, Racchetti G, Clementi F, (1988) Nicotinic acetylcholine receptor: a structural model for α-subunit peptide 188-201, the putative binding site for cholinergic agents. FEBS Lett 228:118-122.
17. Tam JP, Merrifield RB (1985) Solid phase synthesis of gastrin. Int. J. Peptide Protein Res 26:262-273.
18. Gotti C, Omini C, Berti F, Clementi F (1985) Isolation of a polypeptide from the venom of Bungarus multicinctus that binds and blocks ganglionic transmission in mammals. Neuroscience 15: 563-575
19. Greenwood FC, Hunter WM, Glover JC (1963) The preparation of ^{131}I-labeled human growth hormone of high specific radioactivity. Biochem J 89:114-117
20. Karlson E (1979) Chemistry of protein toxins in snake venoms, in: Handbook of experimental pharmacology, Lee Ch Y (ed), vol.52:159-212, Springer Verlag Berlin.
21. Neumann D, Barchan D, Fridkin M, Fuchs S (1986) Analysis of ligand binding to the synthetic dodecapeptide 185-196 of the acetylcholine receptor α-subunit. Proc Natl Acad Sci USA 83:9250-9253.
22. Boulter J, Luyten W, Evans K, Mason B, Ballivet M, Goldman D, Stengelin S, Martin G, Heinemann S, Patrick J (1985) Isolation of a clone coding for the α-subunit of a mouse acetylcholine receptor. J Neurosci 5:2545-2552
23. Karlin A and Bartels E (1966) Effect of blocking sulphidryl groups and of reducing disulphide bands on the acetylcholine activated permeability system of the electroplax. Biochim Biophys Acta 126:525-535
24. Chou PY, Fasman GD (1978) Empirical predictions of protein conformation. Annn Rev Biochem 47:252-276
25. Love RA and Stroud RM (1986) Prot. Engineering 1,37-46.
26. Endo T, Nakanishi M, Furukawa S, Joubert F, Tomiya N, Hoyoshi K (1986) Role of C terminal tail of long neurotoxins from snake venoms in molecular conformation. Biochemistry 25:4592-4598
27. Comitti R, Racchetti G, Gnocchi P, Morandi E and Galante Y (1987) A monoclonal-based, two site enzyme immunoassay of human insulin. J Imm Meth 99:25-37.

TOWARDS AN UNDERSTANDING AND USE OF THE CHOLINERGIC BINDING SITE

J. M. Gershoni, A. Aronheim,
R. Mosckovitz, B. Ohana, I. Rusu
Department of Biophysics
The Weizmann Institute of Science
Rehovot
Israel

INTRODUCTION

Systematic analysis of the α-subunit of the nicotinic acetylcholine receptor (AChR) has led to the identification of the cholinergic binding site. By probing protein blots of proteolysed α-subunits with: α-bungarotoxin (BTX), lectins, sequence-specific antibodies and biotinylated SH reagents, the disulfide arrangement of this subunit has been determined (Mosckovitz and Gershoni 1988) and the α-neurotoxin binding domain has been mapped to the region of α180-200 (Criado et al. 1986, Neumann et al. 1985; 1986, Pedersen et al. 1986, Wilson et al. 1984; 1985). Indeed, synthetic polypeptides corresponding to this area specifically and directly bind BTX (Gotti et al. 1987; 1988, Mulac-Jericevic and Atassi 1986, Neumann et al. 1986, Ralston et al. 1987, Wilson et al. 1985) as do similar recombinant fusion proteins (Aronheim et al. 1988, Barkas et al. 1987, Gershoni 1987). Moreover, competition assays reveal that these sites bind d-tubocurarine as well. Recently, selective T1 NMR relaxation measurements have shown that the sequence α184-200 can bind both nicotine and acetylcholine specifically, proving that this region is an essential element of the cholinergic binding site (Fraenkel et al. 1988).

However, can we now honestly say that we have a better comprehension of why or how this site actually works? Surprisingly, with all the data that has accumulated we are still only able to speculate possible mechanisms for the recognition process responsible for ligand binding. The major significance of these elaborate studies has therefore, been to teach us how to pursue and more critically define the next experiments that should be performed.

NATO ASI Series, Vol. H 32
Molecular Biology of Neuroreceptors and
Ion Channels. Edited by A. Maelicke

This report describes some of the work we have been doing during the last year directed towards better understanding and possible use of the cholinergic binding site.

QUESTIONS ON HAND

The 17 amino acid sequence: WKHWVYYTCCPDTPYLD (α184-200, *Torpedo californica*, Noda et al. 1982), binds BTX with a K_D=3x10^{-7}M (Aronheim et al. 1988). Which of these residues actually form points of contact with the toxin? How are they oriented in space? What dictates and maintains their required three dimensional conformation? These are primarily questions of structure and function.

A second type of question however, can be formulated and tested. Independent of the degree of our understanding of the binding process, a pragmatic question arises: could α184-200 provide *in vivo* protection against α-neurotoxins? Can a toxin/binding-site complex be formed *in vivo*? If so, "mimic binding sites" could serve as molecular decoys, suppressing the effect of the toxin before it reaches its target*.

STRATEGIES

In order to address the question of the identification of relevant residues, an extremely flexible experimental system has been devised.

The system consists of constructing oligonucleotides which correspond to a variety of selected amino acid sequences. These sequences are then expressed as fusion proteins in bacteria and finally, biochemically tested for toxin binding capacity. Two series of binding sites have been prepared and analysed.

THE MINIMAL BINDING SITE

The 17 amino acid sequence, α184-200, has been systematically chopped down, one residue at a time, from both the amino and the carboxy termini (Fig. 1).

----

* The term "decoy" was originally selected from the Herbrew: דכוי (pronounced dee-cuy) meaning "suppression", and thus has dual meaning.

	184			187													200
1737	.			W	V	Y	Y	T	C	C	P	D	T	P	Y	L	D
2137	.		H	W	V	Y	Y	T	C	C	P	D	T	P	Y	L	D
2337	.	K	H	W	V	Y	Y	T	C	C	P	D	T	P	Y	L	D
4137	W	K	H	W	V	Y	Y	T	C	C	P	D	T	P	Y	L	D
4134	W	K	H	W	V	Y	Y	T	C	C	P	D	T	P	Y	L	.
4131	W	K	H	W	V	Y	Y	T	C	C	P	D	T	P	Y		.
4128	W	K	H	W	V	Y	Y	T	C	C	P	D	T	P			.
4125	W	K	H	W	V	Y	Y	T	C	C	P	D	T				.
4122	W	K	H	W	V	Y	Y	T	C	C	P	D					.
	184											195					200

Fig. 1: A series of binding sites was prepared by systematically removing N-terminal and C-terminal residues from the sequence α184-200 (4137). The peptides were produced as bacterial fusion proteins and were analysed as described in Aronheim et al 1988.

In this manner the most proximal and critical residues can be identified. Thus for example, residues W-184 or D-200 are not essential for efficient binding. Once the binding sites are fully analysed the minimal binding sequence which continues to bind BTX with reasonable affinity will be discovered.

AN EVOLUTIONARY SURVEY

Over the past seven years the AChR's of a variety of organisms, representing different taxonomic groups, have been cloned and sequenced. Thus the sequence of α184-204 of Torpedo (Noda et al. 1982), xenopus (Baldwin et al. 1988), chick (Ballivet et al. 1983), mouse (Boulter et al. 1985), calf and human (Noda et al. 1983) as well as the invertabrate; Drosophila (Bossy et al. 1988) have been compared (Fig. 2).

	184																				204
Drosophila	V	R	N	E	K	F		S			-	E	E						V	F	N
Torpedo	W	K	H	W	V	Y	Y	T	C	C	P	D	T	P	Y	L	D	I	T	Y	H
Xenopus				.		.		.				.	K								
Chick				.		.		A				.									
Mouse				.		F		S				T									
Calf				.		F		A				S									
Human				S		T		S													

Fig. 2: A series of binding sites corresponding to the sequence α184-204 derived from different animals is given. Only the variant residues are shown.

Each of these peptides has been prepared as a fusion protein and they are currently being biochemically analysed. Interestingly, one cannot make the simple conclusion that maximal homology with Torpedo, promises preferred BTX binding. The Drosophila site, for example, binds BTX markedly better than that of human and a distinct difference exists between chick binding as compared to that for xenopus, or even mouse as compared to calf. By systematically comparing the sites and producing novel combinations with defined structures it is hoped that the qualitative contribution of individual residues will be learned.

MOLECULAR DECOYS

Mixtures of α-cobratoxin (CTX) with either a toxin binding fusion protein (R4137, Aronheim et al. 1988), a nontoxin binding bacterial protein (truncated *trpE* derived from pATH transformed E. coli HB101, Gershoni 1987) or affinity purified AChR (Froehner and Rafto 1979) were injected into mice. Then the mice were monitored and the number of deaths per time was recorded as described in figure 3.

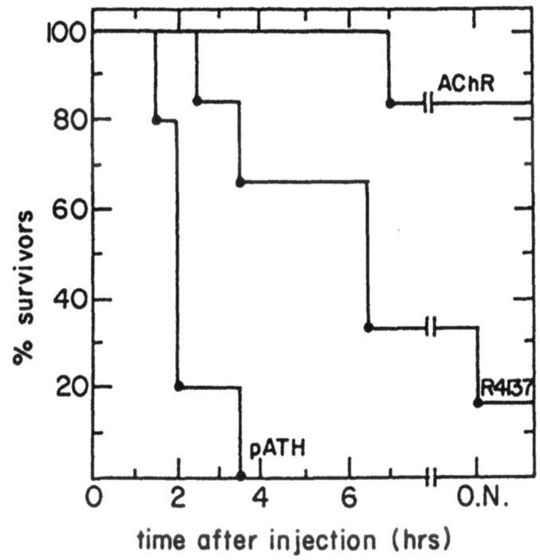

Fig. 3: CTX preincubation experiment, see text.

AChR is extremely efficient in reducing the toxicity of CTX while the pATH containing mixture is still very toxic. As can be seen, the R4137 actually acts as a genuine AChR mimic. This simple experiment demonstrates that complexed CTX is dramatically suppressed for toxicity.

Clearly, however, the question arises could such a complex be formed *in vivo.* Therefore, mice were injected with CTX or d-tubocurarine subcutaneously in the neck and with R4137 or with pATH (serving as a placebo) intraperitoneally. R4137 was found to render protection against these toxins in males and females of inbred (Balb/C) or outbred (CD1) mice (Gershoni and Aronheim 1988).

These experiments provide the first *in vivo* validation of the concept of "decoyance". However, in order to generate efficient decoys, considerable work must still be done. Questions pertaining to the pharmacokinetics of such a system or the potential induction of autoimmunity must be answered, and the most effective delivery schemes must be devised. The cholinergic decoy/mouse animal system will allow the solution of these problems.

In the meantime, the concept of decoyance has taken on a new perspective. Soluble CD4 (the T4 lymphocyte antigen which serves as the receptor for the human immuno-defficienty virus (HIV) (Dalgleish et al. 1984, Klatzmann et al. 1984) has been found to suppress HIV infectivity *in vitro* (Deen et al. 1988, Fisher et al. 1988, Hussey et al. 1988, Smith et al. 1987, Traunecker et al. 1988). This discovery implies that CD4-based decoys could be constructed, and employed in the treatment of Acquired Immuno Deficiency Syndrome (AIDS).

In conclusion, once the cholinergic binding site had been discovered, we have continued our effort to learn the function/structure relationships which govern its capacity to bind α-neurotoxins. This knowledge will undoubtedly provide the basis for the design of more effective decoys. It is believed that through the cholinergic decoy/mouse animal model, better insight to more efficient therapeutic regimens may be realized.

REFERENCES

Aronheim A, Eshel Y, Mosckovitz R, Gershoni JM (1988) Characterization of the binding of α-bungarotoxin to bacterially expressed cholinergic binding sites. J. Biol. Chem. 263:9933-9937.

Baldwin TJ, Yoshihara CM, Blackmer K, Kintner CR, Burden SJ (1988) Regulation of acetylcholine receptor transcript expresion during development in *Xenopus laevis* J. Cell Biol. 106:469-478.

Ballivet M, Nef P, Standler R, Fulpius B (1983) Genomic sequences encoding for the α-subunit of acetylcholine receptor are conserved in evolution. Cold Spring Harbor Symp. Quant. Biol. 48:83-87.

Barkas T, Mauron A, Roth B, Alliod C, Tzartos SJ, Ballivet M (1987) Science 235:77-80.

Boulter J, Luyten W, Evans K, Mason P, Ballivet M, Goldman D, Stengelin S, Martin G, Heinemann S, Patrick J (1985) Isolation of a clone coding for the α-subunit of a mouse acetylcholine receptor. J. Neurosci. 5:2545-2552.

Bossy B, Ballivet M, Spierer P (1988) Conservation of neural nicotinic acetylcholine receptors from *Drosophila* to vertebrate central nervous systems. EMBO J. 7:611-618.

Criado M, Virender S, Lawrence Fox J, Lindstrom J (1986) Evidence that the acetylcholine binding site is not formed by the sequence α127-143 of the acetylcholine receptor. Biochemistry 25:2839-2846.

Dalgleish AG, Beverley PCL, Clapham PR, Crawford DH, Greaves MF, Weiss RA (1984) The CD4 (T4) antigen is an essential component of the receptor for the AIDS retrovirus. Nature 312:763-767.

Deen KC, McDougal JS, Inacker R, Folena-Wasserman G, Arthos J, Rosenberg J, Maddon PJ, Axel R, Sweet RW (1988) A soluble form of CD4 (T4) protein inhibits AIDS virus infection. Nature 331:82-84.

Fisher RA, Bertonis JM, Meier W, Johnson VA, Costopoulos DS, Liu T, Tizard R, Walker BD, Hirsch MS, Schooly RT, Flavell RA (1988) HIV infection is blocked *in vitro* by recombinant soluble CD4. Nature 331:76-86.

Fraenkel Y, Navon G, Aronheim A, Gershoni JM (1988) Selective and non-selective T1 measurements of binding constants of acetylcholine and its antagonists to synthetic and genetically engineered peptides of the acetylcholine receptor. 13th Int. Conf. Magnetic Resonance in Biological Systems. 14-19 August, Madison, Wisconsin

Froehner SC, Rafto S (1979) Comparison of the subunits of *Torpedo californica* acetylcholine receptor by peptide mapping. Biochemistry 18:301-307.

Gershoni JM (1987) Expression of the α-bungarotoxin binding site of the nicotinic acetylcholine receptor by *Escherichia coli* transformants. Proc. Natl. Acad. Sci. USA 84:4318-4321

Gershoni JM, Aronheim A (1988) Molecular decoys: ligand-binding recombinant proteins protect mice from curarimimetic neurotoxins. Proc. Natl. Acad. Sci. USA 85:4087-4089.

Gotti C, Mazzola G, Longhi R, Fornasari D, Clementi F (1987) The binding site for α-bungarotoxin residesin the sequence 188- 201 of the α-subunit of acetylcholine receptor:structure, conformation and binding characteristics of peptide [Lys] 188-201. Neurosci. Letters 82:113-119.

Gotti C, Frigerio F, Bolognesi M, Longhi R, Racchetti G, Clementi F (1988) Nicotinic acetylcholine receptor: a structural model for α- subunit peptide 188-201, the putative binding site for cholinergic agents. FEBS Letters 228:118-122.

Hussey RE, Richardson NE, Kowalski M, Brown NR, Chang H-C, Siliciano RF, Dorfman T, Walker B, Sodroski J, Reinherz EL (1988) A soluble CD4 protein selectively inhibits HIV replication and syncytium formation. Nature 331:78-81.

Klatzmann D, Champagne E, Chamaret S, Gruest J, Guetard D, Hercend T, Gluckman J-C, Montagnier L (1984) T-lymphocyte T4 molecule behaves as the receptor for human retrovirus LAV. Nature 312:767-768.

Mosckovitz R, Gershoni JM (1988) Three possible disulfides in the acetylcholine receptor α-subunit. J. Biol. Chem. 263:1017-1022.

Mulac-Jericevic B, Zouhair Atassi M (1986) Segment α182-198 of *Torpedo californica* acetylcholine receptor contains a second toxin-binding region and binds anti-receptor antibodies. FEBS Letters 199:68-74.

Neumann D, Gershoni JM, Fridkin M, Fuchs S (1985) Antibodies to synthetic peptides as probes for the binding site on the α subunit of the acetylcholine receptor. Proc. Natl. Acad. Sci. USA 82:3490-3493

Neumann D, Barchan D, Safran A, Gershoni JM, Fuchs S (1986) Mapping of the α-bungarotoxin binding site within the α subunit of the acetylcholine receptor. Proc. Natl. Acad. Sci. USA 83:3008-3011

Noda M, Takahashi H, Tanabe T, Toyosato M, Furutani Y, Hirose T, Asai M, Inayama S, Miyata T, Numa S (1982) Primary structure of α-subunit precursor of *Torpedo californica* acetylcholine receptor deduced from cDNA sequence. Nature 299:793-797.

Noda M, Furutani Y, Takahashi H, Toyosato M, Tanabe T, Shimizu S, Kikyotani S, Kayano T, Hirose T, Inayama S, Numa S (1983) Cloning and sequence analysis of calf cDNA and human genomic DNA encoding α-subunit precursor of muscle acetylcholine receptor. Nature 305:818-823.

Pedersen SE, Dreyer EB, Cohen JB (1986) Location of ligand-binding sites on the nicotinic acetylcholine receptor α-subunit. J. Biol. Chem. 261:13735-13743.

Ralston S, Sarin V, Thanh HL, Rivier J, Fox L, Lindstrom J (1987) Synthetic peptides used to locate the α-bungarotoxin binding site and immunogenic regions on α subunits of the nicotinic acetylcholine receptor. Biochemistry 26:3261-3266.

Smith DH, Byrn RA, Marsters SA, Gregory T, Groopman JE, Capon DJ (1987) Blocking of HIV-1 infectivity by a soluble, secreted form of the CD4 antigen. Science 238:1704-1707.

Traunecker A, Luke W, Karjalainen K (1988) Soluble CD4 molecules neutralize human immunodeficiency virus type I. Nature 331:84-86.

Wilson PT, Gershoni JM, Hawrot E, Lentz TL (1984) Binding of α-bungarotoxin to proteolytic fragments of the α subunit of *Torpedo* acetylcholine receptor analyzed by protein transfer on positively charged membrane filters. Proc. Natl. Acad. Sci. USA 81:2553-2557

Wilson PT, Lentz TL, Hawrot E (1985) Determination of the primary amino acid sequence specifying the α-bungarotoxin binding site on the α subunit of the acetylcholine receptor from *Torpedo californica*. Proc. Natl. Acad. Sci. USA 82:8790-8794

EVIDENCE FOR MULTIPOINT ATTACHMENT OF LIGANDS TO THE NICOTINIC ACETYLCHOLINE RECEPTOR.

Alfred Maelicke, Gregor Fels and Bianca M. Conti-Tronconi[&]

Max-Planck-Institut für Ernährungsphysiologie,
Rheinlanddamm 201, D-4600 Dortmund 1, FRG
&) Department of Biochemistry, University of Minnesota,
1479 Gortner Avenue, St. Paul, Minnesota 55108, USA

Binding of ^{125}I-labelled α-bungarotoxin (αBTX) and of several ^{3}H-labelled αBTX-competitive anti-nAChR antibodies suggest that the cholinergic binding site(s) are "discontinuous", i.e. they are formed by several sequence segments which are well separated along the primary structure of the nAChR α-subunit (Conti-Tronconi et al., 1989). Presumably, these non-continuous segments are brought together, by the tertiary folding of the α-subunit, to form the area of the nAChR surface recognized by the neurotoxin and the antibodies. Our data agree with previous reports (Neumann et al., 1986; Gotti et al., 1987; Ralston et al., 1987; Gershoni 1987; Gotti et al., 1989; Gershoni et al., 1989) in that the region around cysteines 192 and 193 forms a major component of the toxin binding site. They result in a different model of the cholinergic binding site, however, in that they suggest a rather intricately structured binding region, with particular sensitivity to conformational changes and the capability of selective binding of classes of ligands (Conti-Tronconi et al., unpublished).

In agreement with our present findings, previous studies have shown that the binding of cholinergic ligands, including α-neurotoxins and antibodies, to the nAChR is rather sensitive to conformational changes (Conti-Tronconi et al., 1982;

NATO ASI Series, Vol. H 32
Molecular Biology of Neuroreceptors and Ion Channels. Edited by A. Maelicke

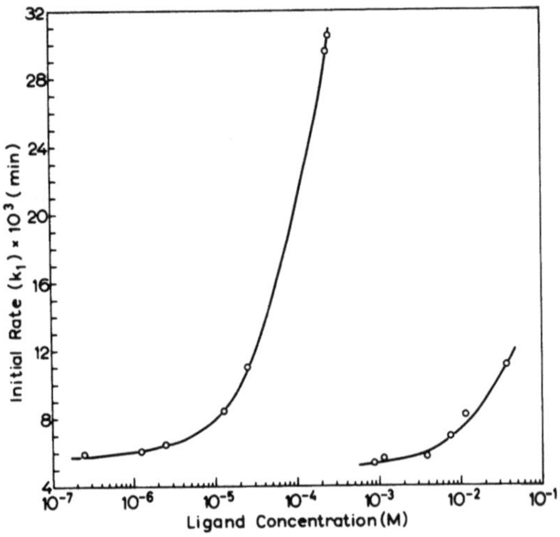

Initial Rate (k1) × 10^3 (min)
32
28
24
20
16
12
8
4
10^-7
10^-6
10^-5
10^-4
10^-3
10^-2
10^-1
Ligand Concentration (M)

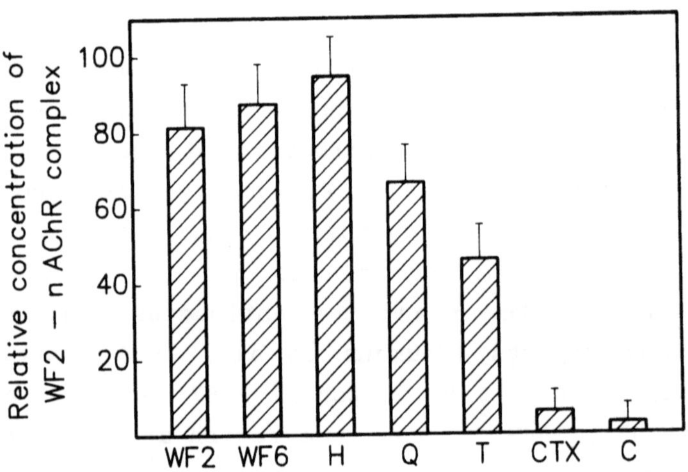

Relative concentration of WF2 – n AChR complex
100
80
60
40
20
WF2
WF6
H
Q
T
CTX
C

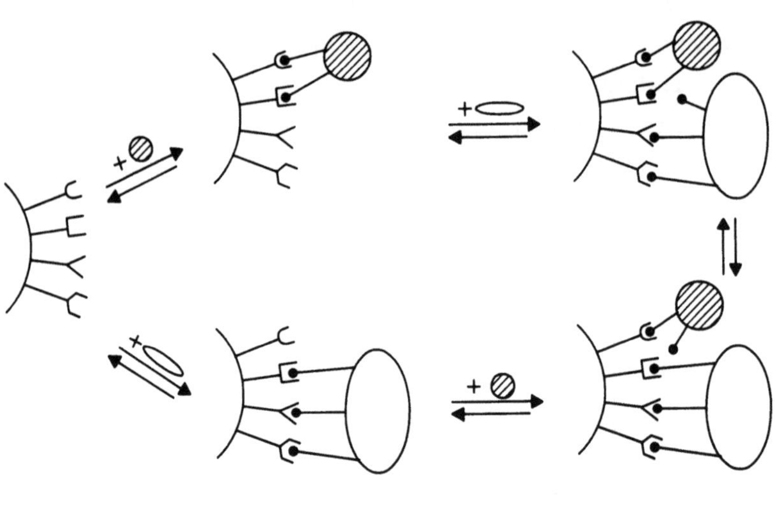

+
+
+
+

Figure 1:

Accelerated dissociation of (a) αCTX-nAChR complexes and of (b) antibody WF2-nAChR complexes in the presence of high concentrations of competing ligand.

(a) Initial rate constants of dissociation of αCTX-nAChR complexes (concentration of toxin and receptor in the reaction mixture were 4.04 nM and 4.88 nM, respectively) in the presence of various concentrations of dimethyl-D-tubocurarine and carbamoylcholine, respectively. Concentrations for half-maximal rate constant were 1.1×10^{-4}M and 5×10^{-2}M, respectively. Original data see (Kang & Maelicke, 1980).

(b) Relative concentration of remaining WF2-nAChR complex after incubation with various ligands. Receptor-rich membrane fragments (2 nM in toxin binding sites) and ^{3}H-labelled WF2 (2 nM) were incubated overnight. The reaction mixture was subdivided into aliquots and the following ligands were added, respectively (final concentration in brackets): buffer; WF2 (500 nM); WF6 (500 nM); Hexamethonium (1 mM); Quinacrine (1 mM); dimethyl-D-tubocurarine (1 mM); αCTX (10μM); and carbamoylcholine (1 mM). After 1 h of incubation, solute and particulate phase were separated by means of an airfuge (Fels et al., 1982), and the amount of unbound radioactivity was determined by radioactive counting of aliquots of the supernatant mixed with a scintillation cocktail. The data show accelerated dissociation of antibody-nAChR complexes in the presence of αCTX or carbamoylcholine, and to a lesser extent in the presence of tubocurarine or Quinacrine.

(c) A plausible model for accelerated dissociation of ligand-nAChR complexes in the presence of competing ligand. Because the two ligands share overlapping attachment points, there may exist mutually exclusive binding in the low concentration range, and low affinity non-competitive binding (ternary complexes) in the high concentration range. The latter would result in an increase in the dissociation rate of secondary complexes in the presence of high concentrations of competing ligand.

Maelicke 1988) and represents an entropy-driven process (Maelicke et al., 1977). Multipoint attachment of cholinergic ligands to the receptor is suggested by the phenomenon of accelerated dissociation of toxin-receptor complexes (Maelicke et al., 1977; Kang & Maelicke, 1980) and of antibody-receptor complexes (Fels, G. and Maelicke, A., unpublished) in the presence of high concentrations of competing ligand (Fig. 1).

TABLE I

Affinities of Binding of α-Cobratoxin and its Derivatives to the Acetylcholine Receptor from *Torpedo marmorata*.

Toxin Derivative	K_I[M]	Rel. Affinity [%]
native α-toxin	2×10^{-10}	100
hexa-N,N-dimethyl	8×10^{-10}	25
hexa-acetyl	4×10^{-7}	0.05
methylcarbamylated	1×10^{-8}	5
exhaustively methylated	1×10^{-7}	0.1
cyclohexanedione treated	2×10^{-8}	1
tetranitromethane treated	5×10^{-10}	40
di-S-pyridylethylated	1×10^{-10}	200
di-S-carboxyamidomethylated	2×10^{-11}	1000
di-S-AEDANS	3×10^{-9}	7
fully reduced, carboxyamido-methylated	3×10^{-6}	0.02

The binding affinities of the derivatives were determined by competition binding employing [^{3}H]monopyridoxamine phosphate labelled α-cobratoxin (Maelicke et al., 1977) and purified receptor (Rüchel et al., 1981). The experiments were performed at room temperature with constant incubation times of 300 min.

Further evidence for the existence of several points of interaction between the nAChR and α-neurotoxins has been provided by binding studies of chemically modified α-cobratoxin (αCTX) to the receptor (Martin et al., 1983; Chibber et al., 1983). These studies (table I) showed that none of the positively charged residues nor tyrosine nor the integrity of the disulfide bridges of αCTX is absolutely essential for receptor binding (and neurotoxicity). As the chemical modification of groups of aminoacid residues of the toxin induced only graded changes in binding affinity to the nAChR (Table

I), several locally separate structural elements of the toxin appear to be involved in binding. These are the long loop containing the extra disulfide bridge, parts of the aminoterminal region and the lower portion of the intermediate loop (Chibber et al., 1983). It is suggestive to assume for neurotoxin-nAChR complexes a similar structural arrangement as was recently found for antibody-antigen complexes (Amit et al., 1986). Amit et al. conclusively showed the interaction of several discontinuous sequence portions of the antibody with similarly discontinuous sequence portions of the antigen.

In conclusion, the results of our binding site(s) mapping studies (Conti-Tronconi et al., 1989) are supported by a large body of additional evidence. Furthermore, the involvement of several structural elements in ligand binding could explain the pharmacological and biochemical specificity of ligand recognition by the nAChR (Conti-Tronconi & Raftery, 1982; Maelicke, 1984; Maelicke, 1988).

Acknowledgement:

A.M. acknowledges grants SFB 168-B2 and Ma 599/10-1 from the Deutsche Forschungsgemeinschaft and a grant from the Fonds der Chemischen Industrie. B.M.C.-T. acknowledges grants NS 28919 (NINCDS) and BNS 8607289 (NSF).

References

Amit AG, Mariuzza RA, Phillips SEV, Poljak RJ (1986) Three-dimensional structure of an antigen-antibody complex at 2.8 Å resolution. Science (Washington DC) 232:747-753

Chibber BA, Martin BM, Walkinshaw MD, Saenger W, Maelicke A (1983) The sites of neurotoxicity in α-cobratoxin. In: Hucho F, Ovchinnikov YA (eds) Toxins as tools in neurochemistry, Walter de Gruyter & Co., Berlin - New York, pp 141-150

Conti-Tronconi BM, Fels G, McLande K, Tang F, Bellone M, Kokla A, Tzartos S, Milius R, Maelicke A (1989) Use of synthetic peptides and high affinity protein ligands for structural studies of central and peripheral nicotinic receptors. This volume, pp 291-309

Conti-Tronconi, B.M. & Raftery, M.A. (1982) The nicotinic cholinergic receptor: correlation of molecular structure with functional properties. Annu. Rev. Biochem. 51, 491-530

Fels G, Wolff EK, Maelicke A (1982) Equilibrium binding of acetylcholine to the membrane-bound acetylcholine receptor. Eur J Biochem 127:31-38

Gershoni JM (1987) Expression of the α-bungarotoxin binding site of the nicotinic acetylcholine receptor by *Escherichia coli* transformants. Proc Natl Acad Sci USA 84:4318-4321

Gershoni JM, Aronheim A, Mosckovitz R, Ohana B, Rusu I (1989) Towards an understanding and use of the cholinergic binding site. This volume, pp. 345-352

Gotti C, Longhi R, Frigerio F, Bolognesi M, Fornasari D, Racchetti G, Mazzola G, Clementi F (1989) Structure, conformation and immunological analysis of peptide 188-201, the putative binding site for α-bungarotoxin in the α subunit of acetylcholine receptor. This volume, pp. 327-343

Gotti C, Mazzola G, Longhi R, Fornasari D, Clementi F (1987) The binding site for α-bungarotoxin resides in the sequence 188-201 of the α-subunit of acetylcholine receptor: structure, conformation and binding characteristics of peptide (Lys)188-201. Neurosci Lett 82:113-119

Kang S, Maelicke A (1980) Fluorescein isothiocyanate-labeled α-cobratoxin: biochemical charakterization and interaction with acetylcholine receptor from Electrophorus electricus. J Biol Chem 255:7326-7332

Maelicke A (1984) Biochemical aspects of cholinergic excitation. Angew Chem Int Ed 23:195-221

Maelicke A (1988) Structure and function of the nicotinic acetylcholine receptor. In: Born GVR, Farah A, Herken H, Welch AD (eds) Handbook of experimental pharmacology. Volume: "The Cholinergic Synapse" Whittaker VP (ed) Vol 86, Springer-Verlag Berlin-Heidelberg-New York-Tokyo, pp 267-313

Maelicke A, Fulpius BW, Klett RP, Reich E (1977) Acetylcholine receptor: responses to drug binding. J Biol Chem 252:4811-4830

Martin BM, Chibber BA, Maelicke A (1983) The Sites of Neurotoxicity in α-Cobratoxin. J Biol Chem 258:8714-8722

Neumann D, Barchan D, Safran A, Gershoni JM, Fuchs S (1986) Mapping of the alpha-bungarotoxin binding site within the alpha-subunit of the acetylcholine receptor. Proc Natl Acad Sci USA 83:3008-3011

Ralston S, Sarin V, Thanh HL, Rivier J, Fox JL, Lindstrom J (1987) Synthetic peptides used to locate the α-bungarotoxin binding site and immunogenic regions on α subunits of the nicotinic acetylcholine receptor. Biochemistry 26:3261-3266

Rüchel R, Watters D, Maelicke A (1981) Molecular forms and hydrodynamic properties of acetylcholine receptor from electric tissue. Eur J Biochem 119:215-223

FINE STRUCTURAL CHARACTERIZATION OF THE MAIN IMMUNOGENIC REGION OF THE NICOTINIC ACETYLCHOLINE RECEPTOR

S. Tzartos*, I. Papadouli*, S. Potamianos*, I. Hadjidakis+, H. Bairaktari+, V. Tsikaris+, C. Sakarellos+, M.T. Cung++ and M. Marraud++

*Hellenic Pasteur Institute, 127, Vas. Sofias Ave., Athens 11521, Greece; +Dept. of Chemistry, University of Ioannina, Box 1186, 45110 Ioannina, Greece; and ++Laboratoire de Chimie-Physique Macromoleculaire, CNRS-UA-494, ENSIC-INPL, BP 451, 54001 Nancy Cedex, France.

INTRODUCTION

The nicotinic acetylcholine receptor (AChR) of fish electric organs and mammalian muscle is a complex transmembrane glycoprotein formed by five homologous subunits in a stoichiometry $\alpha_2\beta\gamma\delta$, of mol. weight 290.000. The two α-subunits carry the two ACh binding sites. The complete amino acid sequence of Torpedo electric organ and mammalian muscle AChR subunits is known (reviewed in Numa et al., 1983; Changeux & Revan, 1987).

In the neuromuscular disease myasthenia gravis (MG), AChR is the target of an autoimmune response. Anti-AChR antibodies cause loss and/or blockage of the function of AChR molecules resulting in impaired neuromuscular transmission. Experimental MG can be induced in animals by injecting AChR, anti-AChR antibodies or sera from MG patients (reviewed in Drachman, 1987; Lindstrom et al., 1988).

By the use of monoclonal antibodies (mAbs) against the AChR as probes, it was found that about two/thirds of the anti-AChR antibodies, from human MG patients or from rats immunized with intact AChR, are directed against an extracellular area of the

NATO ASI Series, Vol. H 32
Molecular Biology of Neuroreceptors and Ion Channels. Edited by A. Maelicke

α-subunit named the main immunogenic region (MIR) (Tzartos & Lindstrom, 1980; Tzartos et al., 1981, 1982, 1983). The MIR is different from the ACh binding site (Tzartos and Lindstrom, 1980; Kubalek et al. 1987) and it is not involved in the control of the ion channel (Wan and Lindstrom, 1981). The MIR has been detected on electric organ and muscle AChRs from almost all studied species (reviewed in Tzartos, 1988).

The pathogenic role of the MIR and of the anti-MIR antibodies has been extensively studied. Anti-MIR mAbs cause AChR loss when added to muscle cell cultures and also cause experimental MG when injected into rats (Tzartos & Lindstrom, 1980; Tzartos et al., 1985; 1987). The anti-MIR antibody fraction from human myasthenic sera is primarily responsible for the ability of these sera to cause AChR loss in cell cultures (Tzartos et al., 1985). Therefore, the MIR plays a critical role in the pathogenesis of MG, and its exact localization is of great importance for understanding the disease.

Localization of the MIR has been recently achieved in considerable detail. In a series of studies, in which AChR fragments (proteolytic peptides, fusion proteins or synthetic peptides) and anti-AChR mAbs were used, the MIR was gradually localized between residues 1-150 (Barkas et al., 1986), 46-127 (Ratnam et al., 1986), 37-85 (Barkas et al. 1987), 61-76 (Barkas et al., 1988), and 67-76 (Tzartos et al., 1988a,b) α-subunit.

We have undertaken to investigate the role of each residue in the segment α67-76 with respect to its three-dimensional structure and the antibody-antigen interactions. Comparison of these two effects is expected, in addition to providing a thorough understanding of the MIR, to suggest ways of producing modified synthetic peptides with a conformation approaching that of the intact MIR, such that the anti-MIR antibodies can bind with high affinity.

Here we present preliminary results on the use of synthetic peptides corresponding to the α67-76 of _Torpedo_ electric organ and human muscle AChR, together with nine peptide analogs, each having one amino acid of the _Torpedo_ decapeptide replaced by Alanine, and two intermediate analogs between _Torpedo_ and human peptide. Studying the binding of these peptides to several mAbs it was found that several residues within α67-76 are indispensable for antibody binding whereas others are not. Some peptide analogs enhanced mAb binding, which suggests that construction of a very antigenic MIR is feasible.

TECHNIQUES

A. Peptide synthesis and characterization.

Peptide synthesis was carried out by the stepwise solid phase method introduced by Merrifield (1965). Deprotection and cleavage of the peptides from the resin were performed by HF treatment.

B. mAbs and mapping techniques.

Production of the mAbs used has been earlier described (Tzartos et al., 1981; 1983). The binding of anti-MIR mAbs to the peptides was tested by a solid phase radioimmunoassay. Peptides were attached to 96 well plates using two different techniques. In the first, the wells were coated with poly-DL-Ala-poly-Lys, and the test peptide was bound covalently to it via glutaraldehyde (Tzartos et al., 1988a). In the second, the peptides were adsorbed directly to the plate. The following steps were common in both techniques: addition of the test mAb, then addition of the anti-antibody and finally labelling of the complex with ^{125}I-protein A (Papadouli et al., in preparation)

RESULTS AND DISCUSSION

Peptide characterization

Synthetic peptides were purified by chromatographic methods and their sequence was confirmed by NMR spectroscopy. Fig. 1 shows the Torpedo and human α67-76 peptides together with nine Torpedo analogs (A^1-A^3, and A^5-A^{10}) and two intermediates between Torpedo and human peptides (-A^4, V^9). Analog -A4 is a nonapeptide missing residue No 4 which contains the only non-conservative substitution between the two species in the α67-76 segment (Ala in Torpedo versus Asp in human AChR) .

Using one and two dimensional NMR spectroscopy techniques the conformation of these peptides in DMSO was elucidated (Cung et al. 1988, 1989). Strong and multiple short and long-range NOEs in the *Torpedo* peptide, and especially a strong long-range NOE between the two Asn^{68}-CαH and Gly^{73}-CαH protons argue in favor of a rigid folded conformation. This is stabilized by three interactions involving the D^{71}, G^{74} and K^{76} amide protons (Cung et al. 1989). It was further shown that the conformation of the decapeptide is significantly affected by any Ala-substitution except of that on residue 75 (analogue A^{9}) (Cung et al. 1988).

Anti-MIR mAb binding to Torpedo and human α67-76 peptide analogs.

Figure 2 shows the binding pattern of six anti-MIR mAbs to the *Torpedo* peptide analogs. Analogs A^{1} and A^{6} (as well as the human peptide HUM in Fig. 3) exhibited very different binding behaviour under the two techniques used. These peptides were

Residues (α67 — α76)	Peptide name
WNPADYGGIK	Torp
ANPADYGGIK	A^{1}
WAPADYGGIK	A^{2}
WNAADYGGIK	A^{3}
WNPAAYGGIK	A^{5}
WNPADAGGIK	A^{6}
WNPADYAGIK	A^{7}
WNPADYGAIK	A^{8}
WNPADYGGAK	A^{9}
WNPADYGGIA	A^{10}
WNPDDYGGVK	Hum
WNP-DYGGIK	$-A^{4}$
WNPADYGGVK	V^{9}

Fig. 1. Amino acid sequence of *Torpedo* and human AChR peptides α67-76 and 11 peptide analogs differing by one amino acid from the original sequences.

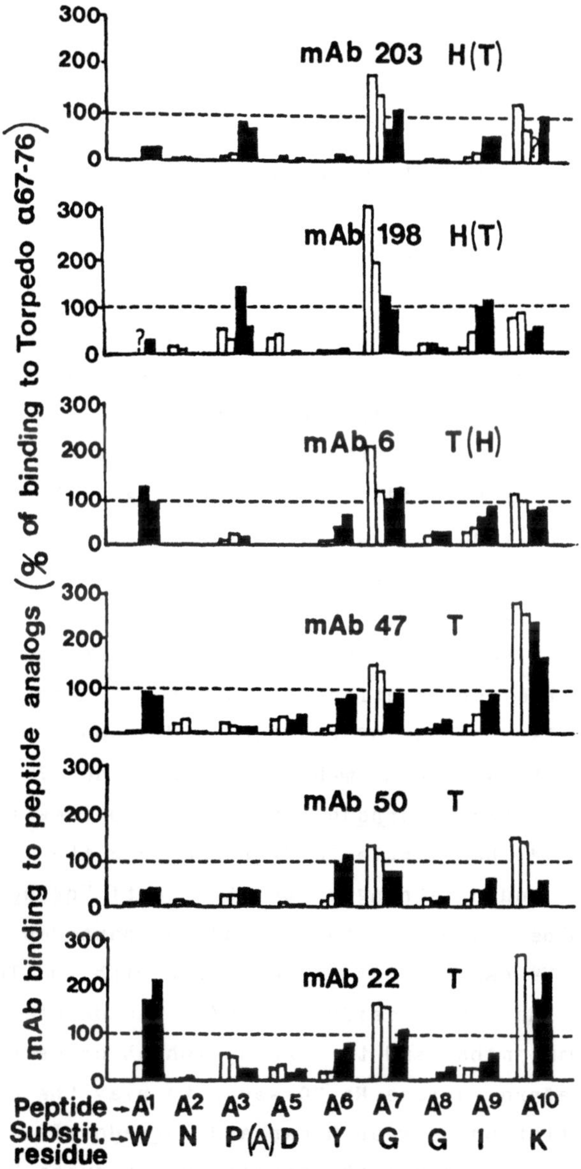

Fig. 2. Binding pattern of anti-MIR mAbs to the *Torpedo* α67-76 peptide analogs. Bars represent the ratios of mAb binding to the peptide analog versus binding to the *Torpedo* peptide. Empty bars: peptides were bound to the plates via poly-DL-Ala-poly-Lys and glutaraldehyde. Black bars: peptides were bound directly to the plate. For each condition two mAb concentrations were used, represented by the two attached bars of the same color. T,H : *Torpedo* or human AChR used as immunogen for the mAb. (T), (H): the mAb cross-reacts with *Torpedo* or human AChR respectively.

almost inactive under the poly-DL-Ala-poly-Lys coating technique, contrary to their behaviour in the direct binding technique. For all the other peptides the two techniques produced usually comparable results.

The epitopes for all mAbs shown in Fig. 2 required residues N^{68}, D^{71} and G^{74}, while P^{69} is required for the anti-<u>Torpedo</u> mAbs only (especially for mAbs 47 and 6), and Y^{72} is part of the epitopes for the anti-human mAbs only (No 198 and 203). Substitution of I^{75} by Ala, only weakly affects the mAb binding for most of the mAbs tested. These results show that the MIR consists of several overlapping epitopes.

Interestingly, substitutions of G^{73} and K^{76} by Ala, resulted in enhancement of binding of some mAbs (Fig. 2). This observation was further analysed by using varying dilutions of peptides and mAbs. It was found that the enhancement of mAb binding was generally stronger when decreasing the mAb and/or peptide concentration (Papadouli et al., in preparation). These results suggest that the construction of a synthetic peptide with binding capacity similar to that of the intact MIR could be feasible.

Binding of anti-MIR mAbs to human α67-76 and to the two intermediates between <u>Torpedo</u> and human peptides is shown in Fig. 3. Peptide V^{9} had little difference from the <u>Torpedo</u> peptide. Actually it bound stronger to the anti-<u>Torpedo</u> than to the anti-human mAbs. 2-D NMR studies of the peptides in DMSO showed that peptide V^{9} has similar conformation with the <u>Torpedo</u> peptide. The nonapeptide $-A^{4}$ (residue A^{70}/D^{70} was deleted) bound only to the anti-human mAbs and to mAb 6 (which crossreacts with human AChR). Therefore residue No 70 seems to play the critical role in the differentiation between human and <u>Torpedo</u> MIR.

Table 1 summarizes the results of Figures 2 and 3 in two groups of mAbs (anti-<u>Torpedo</u> and anti-human) showing the average binding of each peptide to all anti-<u>Torpedo</u> or anti-human mAbs under conditions of direct peptide attachment to the plate. In most cases, only small variations in binding were observed (small SD) within each of the two groups of mAbs. On the contrary, binding between the two groups was very different for several peptide analogs. Therefore the anti-<u>Torpedo</u> and anti-human MIR mAbs constitute two distinct groups of specificities.

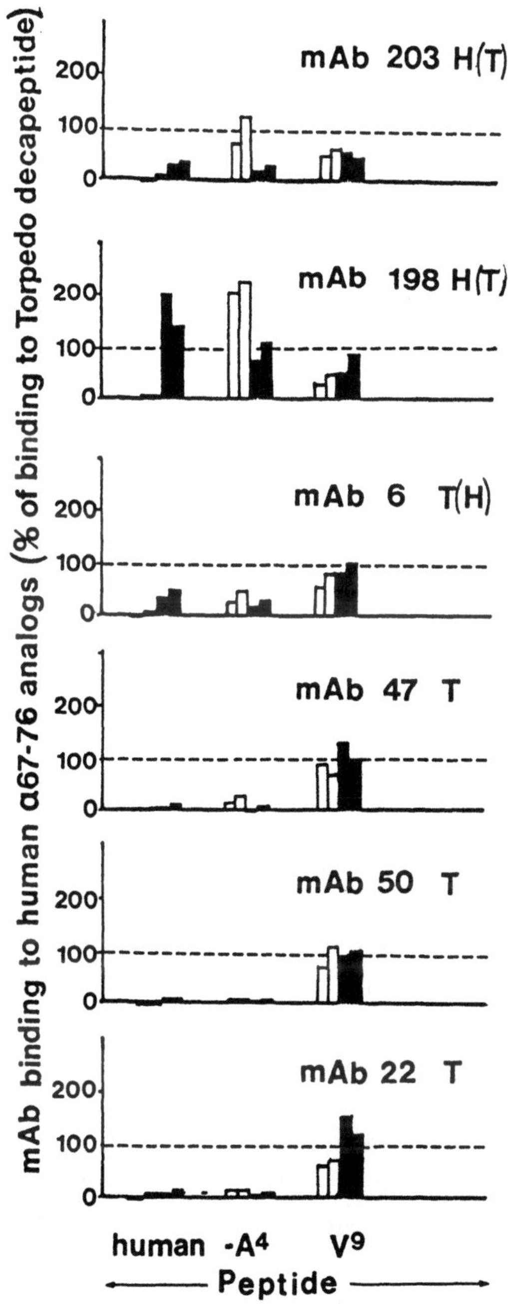

Fig. 3. Binding of anti-MIR mAbs to human AChR peptide α67-76 and to two intermediates between *Torpedo* and human peptides. Symbols as in Fig. 2.

Table 1

Overall contribution of each residue in the Torpedo and human MIR

Peptide analog	Substituted original residue	Ratio of mAb binding to peptide analogs versus binding to Torpedo α67-76 (±SD among mAbs)	
		Anti-Torpedo mAbs	anti-human mAbs
A^1	W^{67}	101 ± 64	29 ± 3
A^2	N^{68}	1 ± 0	0 ± 0
A^3	P^{69}	18 ± 8	89 ± 9
A^5	D^{71}	14 ± 12	3 ± 1
A^6	Y^{72}	71 ± 17	8 ± 2
A^7	G^{73}	85 ± 12	108 ± 3
A^8	G^{74}	23 ± 4	9 ± 3
A^9	I^{75}	60 ± 13	80 ± 20
A^{10}	K^{76}	152 ± 84	130 ± 72
Hum	A^{70}, I^{75}	13 ± 14	98 ± 72
$-A^4$	A^{70} (deleted)	8 ± 9	91 ± 3
V^9	I^{75}	111 ± 20	57 ± 9

CONCLUSIONS

Based on the binding pattern of six anti-MIR mAbs on Ala-analogs of the segment α67-76 of the AChR, the following tentative conclusions can de drawn:

1. Residues N^{68}, D^{71} and G^{74} are indispensable for the epitopes of all tested anti-MIR mAbs. Epitopes for anti-Torpedo mAbs require also residue P^{69}, while those for anti-human mAbs require also residues W^{67} and Y^{72}. Most of these antigenicaly important residues have been found by NMR studies to affect the overall conformation of the peptide.
2. The difference in the binding capacity to the anti-MIR mAbs tested, between the Torpedo and the human decapeptide seems to be determined by residue-70, while residue-75 does not seem to contribute significantly.
3. Substitutions of G^{73} or K^{76} by Ala, resulted in enhancement of

the binding of some mAbs. This suggests that the construction of a peptide with binding capacity similar to that of the intact MIR is likely. Such a product would be invaluable in myasthenia gravis studies.

Acknowledgements: This work was supported by grants from the Greek General Secretariat of Research and Technology, the Muscular Dystrophy Association of America, and the "Stimulation" program of EEC (ST2J-0184).

REFERENCES

Barkas, T., Gabriel, J. M., Juillerat, M., Kokla, A., and Tzartos, S. J. (1986). Localization of the main immunogenic region of the nicotinic Acetylcholine receptor. FEBS Lett. 196, 237-241.

Barkas, T., Mauron, A., Roth, B., Alliod, C., Tzartos, S. J., and Ballivet, M. (1987). Mapping the main immunogenic region and toxin binding site of the nicotinic acetylcholine receptor. Science. 235, 77-80.

Barkas, T., Gabriel, J-M., Mauron, A., Hughes, G.J., Roth, B.,Alliod, C., Tzartos, S.J., Ballivet, M, (1988) Fine localisation of the main immunogenic region of the nicotinic AChR to residues 61-76 of the a subunit. J. Biol. Chem., 263, 5916-5920.

Changeux, J-P., and Revah, F. (1987). The Acetylcholine receptor molecule: allosteric sites and the ion channel. Trends Neurosci. 10, 245-250.

Cung, M.T., Marraud, M., Hadjidakis, I., Bairaktari, H., Tsikaris, V., Sakarellos, C., Papadouli, I., Potamianos, S, & Tzartos, S. (1988). NMR conformational studies of guest-host peptide analogs of the Torpedo acetylcholine receptor α67-76 main immunogenic region. Peptides 1988 (E. Bayer and G. Jung, eds) Walter de Gruyter, Berlin, in press.

Cung, M.T., Marraud, M., Hadjidakis,I., Bairaktari, H., Sakarellos, C., Kokla, A., & Tzartos, S. (1989). 2D-^{1}H NMR study of a synthetic peptide containing the main immunogenic region of the Torpedo acetylcholine receptor. Biopolymers, in press.

Drachman, D. (ed.) (1987). Myasthenia gravis, Biology and treatment. Ann. N.Y. Acad. Sci. Vol. 505.

Kubalek, E., Ralston, S., Lindstrom, J., and Unwin, N. (1987). Location of subunits within the Acetylcholine receptor by electron image analysis of tubular crystals from Torpedo marmorata. J. Cell Biol. 105, 9-18.

Lindstrom, J., Shelton, D. and Fujii, Y. (1988) Myasthenia gravis Adv. Immunol. 42: 233-284.

Merrifield, R.B. (1963) Solid-phase peptide synthesis. I. The synthesis of a tetrapeptide. J. Amer. Chem. Soc. 85, 21-49.

Numa, S., et al (1983). Molecular structure of the Acetylcholine receptor. Cold Spring Harb. Symp. Quant. Biol. 48, 57-69.

Ratnam, M., Sargent, P., Sarin, V., Fox, J. L., Le Nguyen, D., Rivier, J., Criado, M., and Lindstrom, J. (1986). Location of antigenic determinants on primary sequences of the subunits of the nicotinic acetylcholine receptor by peptide mapping. Biochemistry 25, 2621-2632.

Tzartos,J. (1988) Myasthenia gravis studied by monoclonal antibodies to the acetylcholine receptor. In Vivo, 2, 105-110.

Tzartos, S.J. & Lindstrom, J.L. (1980) Monoclonal antibodies to probe acetylcholine receptor structure: Localization of the main immunogenic region and detection of similarities between subunits. Proc. Natl. Acad. Sci. USA 77, 755-759.

Tzartos, S.J., Rand, D.E., Einarson, B.E. & Lindstrom, J.M. (1981) Mapping of surface structures of Electrophorus acetylcholine receptor using monoclonal antibodies,. J. Biol. Chem. 256, 8635-8645.

Tzartos, S.J., Seybold, M. & Lindstrom, J. (1982) Specificities of antibodies to acetylcholine receptors in sera from myasthenia gravis patients measured by monoclonal antibodies. Proc. Natl. Acad. Sci. USA 79, 188-192.

Tzartos, S., Langeberg, L., Hochschwender, S. & Lindstrom, J. (1983) Demonstration of a main immunogenic region on acetylcholine receptors from human muscle using monoclonal antibodies to human receptor,. FEBS Lett. 158, 116-118.

Tzartos, S.J., Sophianos, D. & Efthimiadis, A. (1985) Role of the main immunogenic region of acetylcholine receptor in myasthenia gravis. An Fab mAb protects against antigenic modulation by human sera. J. Immunol. 134, 2343-2349.

Tzartos, S.J., Hochschwender, S., Vasquez, P. & Lindstrom, J. (1987) Passive transfer of experimental autoimmune myasthenia gravis by monoclonal antibodies to the main immunogenic region of the acetylcholine receptor. J. Neuroimmunol. 15, 185-194.

Tzartos, S.J., Kokla, A., Walgrave, S. & Conti-Tronconi, B. (1988a) Localization of the main immunogenic region of human muscle acetylcholine receptor to residues 67-76 of the α-subunit. Proc. Natl. Acad. Sci. USA. 85, 2899-2903.

Tzartos S.J., Loutrari, H.V., Tang, F., Kokla, A., Walgrave S.L., Milius, R. and Conti-Tronconi, B.M. (1988b) The main immunogenic region of Torpedo electroplax and human muscle acetylcholine receptor. Localization and micro-heterogeneity reveiled by the use of synthetic peptides. Submitted.

Wan, K., and Lindstrom, J. (1985). Effects of mAbs on the function of purified acetylcholine receptor from Torpedo Californica reconstituted into liposomes. Biochemistry 24, 1212-1221.

PHOSPHORYLATION OF THE NICOTINIC ACETYLCHOLINE RECEPTOR AND LOCALIZATION OF ITS PHOSPHORYLATION SITES

Anat Safran, Carlo Provenzano, Ronit Sagi-Eisenberg and Sara Fuchs

Department of Chemical Immunology
The Weizmann Institute of Science
Rehovot 76100, Israel.

The nicotinic acetylcholine receptor (AChR) mediates synaptic transmission at the post-synaptic membrane of nicotinic cholinergic synapses. This receptor is a prototype of the chemically-gated ion channel receptors which contain the ligand binding site as well as an intrinsic ion channel responsible for signal transduction (Hucho, 1986; Changeux et al., 1987). Phosphorylation of AChR has been implicated in playing a role in the regulation of its ion channeling. Cyclic AMP-elevating agents which activate the cAMP-dependent protein kinase (PKA), as well as phorbol esters, which activate the Ca^{2+}/phospholipid-dependent protein kinase (PKC), were shown to enhance receptor desensitization (Middleton et al., 1986; Albuquerque et al., 1986; Eusebi et al., 1985). It was postulated that direct phosphorylation of the AChR by these two kinases may result in the observed physiological response. Indeed, phosphorylation of the receptor was demonstrated in intact mammalian muscle cells and this phosphorylation was stimulated by cAMP and Ca^{2+} (Miles et al., 1987; Smith et al., 1987).

The nicotinic AChR from Torpedo electric organ has also been shown to serve as a substrate for phosphorylation by endogenous protein kinases present in Torpedo post-synaptic membranes (Gordon et al., 1977; Teichberg et al., 1977). One of these endogenous kinases was identified as PKA (Huganir and Greengard, 1983). In studying the effect of phosphorylation on receptor function, the rate of desensitization of the purified Torpedo receptor that had been phosphorylated in vitro by PKA and reconstituted into liposomes, was enhanced when

NATO ASI Series, Vol. H 32
Molecular Biology of Neuroreceptors and
Ion Channels. Edited by A. Maelicke

compared with that of the reconstituted non-phosphorylated receptor (Huganir et al., 1986). This result indicates that PKA-mediated phosphorylation of the Torpedo AChR is associated with enhanced rate of desensitization as has been shown for the mammalian receptor.

Phosphorylation of membrane-bound AChR by PKA and PKC

We have employed Torpedo post-synaptic membranes in order to investigate the ability of PKA and PKC to directly phosphorylate the Torpedo receptor within its native environment. Incubation of Torpedo post-synaptic membranes with {γ-^{32}P}-ATP and $MgCl_2$, resulted in the phosphorylation of various proteins which did not correspond to the most abundant proteins present in these membranes. A marked increase in the phosphorylation of some of these proteins was observed when 1 μM cAMP was added to the reaction mixture. This stimulation of phosphorylation by cAMP indicates the presence of an endogenous cAMP dependent protein kinase (PKA) in these membranes. To test whether any of these phosphoproteins are polypeptides related to the acetylcholine receptor, phosphorylated membranes were solubilized and subjected to immunoprecipitation by anti-receptor antibodies. Only a 60 and 65 kDa phosphoproteins which represent the γ and δ subunits of the receptor, were specifically immunoprecipitated. The two subunits were phosphorylated to a comparable extent and the phosphorylation of both was markedly stimulated by cAMP. These results demonstrate that the γ and δ subunits of the nicotinic acetylcholine receptor, in its membrane bound form, undergoes phosphorylation by an endogenous cAMP dependent protein kinase, in agreement with previously reported data (Huganir and Greengard, 1983).

We have then analyzed the phosphorylation of Torpedo post-synaptic membrane proteins by PKC. We could not demonstrate phosphorylation of AChR by an endogenous PKC. This was not surprising as PKC is a soluble protein which binds to the membranes in a Ca^{2+}-dependent manner (Nishizuka, 1986). Considering the fact that our procedure for preparing the post-synaptic membranes involves the use of a high concentration of EGTA, in order to avoid Ca^{2+}-dependent proteolysis of AChR, the membranes are likely to be depleted of, or at best poor in endogenous PKC. However, addition of exogenous PKC resulted in a marked stimulation of the phosphorylation of various membrane proteins. This phosphorylation was dependent on the presence of Ca^{2+} and was sensitive to heat inactivation of the enzyme as well as to its inhibition by polymyxin B. It, therefore, appears that although our membrane preparation does not contain endogenous PKC, post-synaptic membrane proteins do serve as substrates for phosphorylation by this kinase.

Membranes phosphorylated by PKC were solubilized and immunoprecipitated by anti-receptor antibodies. Again, only the 60 and 65 kDa phosphoproteins, representing the γ and δ subunits of AChR were specifically immunoprecipitated. However, while PKA phosphorylated these two subunits to a similar extent, PKC phosphorylated predominantly the δ subunit and the phosphorylation of the γ subunit was considerably lower.

Phosphorylation of affinity purified AChR by PKA and PKC.

The different phosphorylation of the γ subunit by PKA and PKC could result from an intrinsic property such as conformation of the receptor molecule or amino acid sequence of the phosphorylation site. Alternatively, it could result from environmental factors like the activation of Ca^{2+}-dependent phosphatases that specifically dephosphorylate the γ-subunit. In order to distinguish between these possibilities, the phosphorylation characteristics of the soluble, affinity purified receptor, by the two kinases were determined (Souroujon et al., 1986; Fuchs et al., 1986; Safran et al., 1986;1987). A similar pattern of phosphorylation was observed also in the solubilized AChR. Thus, in the affinity purified AChR, both the γ and δ subunits were phosphorylated to a similar extent by PKA while PKC phosphorylated predominantly the δ subunit, and the phosphorylation of the γ subunit was lower. Under these conditions, some phosphorylation of the α subunit by PKC was also detected (Safran et al., 1987). The rate of phosphorylation of the δ and γ subunits by PKA was similar and the maximal stoichiometry was 0.8 and 0.6 moles of phosphate per mole of the δ and γ subunits respectively. On the other hand, the rate of phosphorylation by PKC was considerably different. Phosphorylation of the δ subunit was detectable already following 1 min of incubation, while that of the γ subunit was detected only after longer incubation periods. The maximal stoichiometry of phosphorylation of the δ subunit was 0.6 moles of phosphate per mole of subunit, while that of the γ subunit was less than 0.1. These results clearly suggest that the differences in the phosphorylation of the γ and δ subunits by PKA and PKC result from intrinsic properties of the receptor itself.

Phosphorylation sites for PKA and PKC in the δ subunit

As the δ subunit is phosphorylated by both PKA and PKC to a high stoichiometry, we were interested to localize the respective phosphorylation sites within this subunit. For that purpose, three synthetic peptides were synthesized corresponding to residues 354-367, 364-374, and 373-387 of the δ subunit of Torpedo AChR (Safran et al., 1986). These peptides contains putative phosphorylation sites within the large cytoplasmic domain of this subunit (Huganir et al., 1984).

At the first attempt to localize both PKA and PKC phosphorylation sites within the δ subunit, the ability of the three peptides to be phosphorylated in vitro by both kinases was analyzed. Out of these three peptides, only peptide 354-367 was phosphorylated by both PKA and PKC (Safran et al., 1987).

PKA phosphorylation was inhibited by the specific inhibitor for PKA (PKI), while PKC phosphorylation was Ca^{2+}/PS dependent and was insensitive to the specific inhibitor of PKA (PKI). This peptide contains the sequence Arg-Arg-Ser-Ser-Ser which fulfils the substrate requirement of PKA (Arg-Arg-X-Ser-Y; Kemp et al., 1977), and out of the three adjacent serines (positions 360, 361 and 362 in the δ subunit), serine 361 is the one that should be phosphorylated, as was indeed demonstrated recently (Yee and Huganir, 1987). It still remains to be established which of the three serine residues in this peptide is phosphorylated by PKC.

In further attempts to localize PKA and PKC phosphorylation sites within the δ subunit we have analyzed the phospho-fragments profile obtained by proteolytic digestion of the δ subunit which has been isolated following phosphorylation by either PKA or PKC. Proteolytic digestion of the δ subunit phosphorylated by either kinase revealed identical phosphopeptide profile with a major phospho-fragment of 15 kDa. Furthermore, immunoblotting experiments with antibodies to each of the three synthetic peptides revealed that sequence 354-367 is included within this 15 kDa fragment, whereas the sequence 373-387 is not (Safran et al., 1987). This result further corroborates the finding that both PKA and PKC phosphorylation sites reside within residues 354-367 of the receptor δ subunit. Thus PKA and PKC may phosphorylate adjacent, or identical, serine residues within the δ subunit, and phosphorylation by these kinases may affect biological functioning in an additive, synergistic, or antagonistic manner. This opens the possibility for diverse effects achieved through AChR phosphorylation.

Phosphorylation sites within the γ subunit

The γ subunit of Torpedo AChR contains the sequence Arg-Arg-Arg-Ser-Ser at position 350-354. This sequence is similar, yet not identical to the sequence

Arg-Arg-Ser-Ser-Ser within the δ subunit (residues 358-362), which includes the phosphorylation site for PKA and PKC. The sequence Arg-Arg-Arg-Ser-Ser which is also a consensus sequence for PKA, has been demonstrated to be the site of phosphorylation by PKA within the γ subunit (Yee and Huganir, 1987). As the γ subunit is phosphorylated by both PKA and PKC, we were interested to find out whether, in analogy to the δ subunit, both kinases phosphorylate the same site and whether differences in the amino acid sequence between the δ and γ subunits may account for the observed differences in their phosphorylation by the two kinases. For that purpose we have synthesized a 14-amino acid peptide, (Lys-Pro-Gln-Pro-Arg-Arg-Arg-Ser-Ser-Phe-Gly-Ile-Met-Ile) corresponding to residues 346-359 of the γ subunit. This peptide contains the identified PKA phosphorylation site and is homologous to peptide 354-367 of the δ subunit (Asp-Leu-Lys-Leu-Arg-Arg-Ser-Ser-Ser-Val-Gly-Tyr-Ser-Ile) that has been shown to undergo phosphorylation by both kinases (Safran et al., 1987). Indeed, like peptide 354-367 from the δ subunit, peptide 346-359 of the γ subunit also served as a substrate for phosphorylation by both PKA and PKC. In contrast, several other serine-containing peptides (e.g. peptide 373-387 from the δ subunit which includes the sequence Lys-Ser-Arg-Ser), failed to undergo phosphorylation by any of the two kinases. It should, however, be noted that while PKA phosphorylated both the γ and δ-derived peptides to a similar extent, a difference was observed in the ability of PKC to phosphorylate the two. In both time and dose response analyses, the phosphorylation of the γ peptide was lower when compared with that of the δ peptide. Still, this difference in peptide phosphorylation seems too small to account for the difference in the phosphorylation of the γ subunit within the intact soluble as well as the membrane-bound receptor. Thus, an additional factor, such as the accessibility of this subunit to the two kinases might be involved.

The sequence requirements of PKC within its protein substrates is not completely resolved. Several groups have reported that PKA and PKC phosphorylated a common site which includes the consensus recognition site for PKA (Arg-Arg-X-Ser-Y; Sibley et al., 1988; Kishimoto et al., 1985; Campbell et al., 1986). Our results indicating that sequence 354-367 of the δ subunit (Safran et al., 1987) as well as sequence 346-359 of the γ subunit of AChR serve as substrates for both kinases, provides two more examples to this rule.

In summary, our studies on AChR phosphorylation suggest that observed changes in AChR desensitization rate in intact cells, following treatment with agents which activate various kinases, may result from direct phosphorylation of the receptor by PKA and/or PKC. Our studies also suggest that under conditions which result in activation of PKA, both the γ and δ subunits are phosphorylated to a similar extent. On the other hand, during activation of PKC, these two subunits are phosphorylated to different degrees. This diversity may provide an alternative mode of regulation of receptor function by phosphorylation.

Acknowledgements

This research was supported by grants from the Muscular Dystrophy Association of America, the Los Angeles Chapter of the Myasthenia Gravis Foundation and the United States-Israel Binational Science Foundation (BSF) to S.F., and from the Israel Cancer Research Fund and The Leukemia Research Foundation to R.S.-E. R.S.-E is an incumbent of the Charles H. Revson Career Development Chair. C.P. is a recipient of a long-term EMBO Fellowship.

REFERENCES

Albuquerque EX, Deshpande SS, Aracava Y, Alkondon M, Daly JW (1986) FEBS Letter 199:113-120.

Campbell DG, Hardie DG, Vulliet PR (1986) J Biol Chem 261:10489-10492.

Changeux JP, Giraudat J, Dennis M (1987) Trends in Pharm Sci 8:459-465.

Eusebi F, Molinaro M, Zani BM (1985) J Cell Biol 100:1339-1342.

Fuchs S, Neumann D, Safran A (1986) In: Nato ASI Series, Vol H3 Nicotinic Acetylcholine Receptor, A. Maelicke (ed) Springer-Verlag, Berlin, Heidelberg, pp 49-59.

Gordon AS, Davis CG, Milfay D, Diamond I (1977) Nature 267:539-540.

Hucho F (1986) Eur J Biochem 158:211-226.

Huganir RL, Delcour AH, Greengard P, Hess GP (1986) Nature 321:774-776.

Huganir RL, Greengard P (1983) Proc. Natl. Acad. Sci. USA 80:1130-1134.

Huganir RL, Miles K, Greengard P (1984) Proc. Natl. Acad. Sci. USA 81:6968-6972.

Kemp BE, Graves, DJ, Benjamini E, Krebs EG (1977) J Biol Chem 252:4888-4894.

Kishimoto A, Nishiyama K, Nakanishi H, Uratsuji Y, Nomura H, Takeyama Y, Nishizuka Y (1985) J Biol Chem 260:12492-12499.

Middleton P, Jaramillo F, Schuetze SM (1986) Proc. Natl. Acad. Sci. USA 83:4967-4971.

Miles K, Anthony DT, Rubin LL, Greengard P, Huganir, RL (1987) Proc. Natl. Acad. Sci. USA 84:6591-6595.

Nishizuka, Y (1986) Science 233:305-312.

Safran A, Neumann D, Fuchs S (1986) EMBO J 5:3175-3178.

Safran A, Sagi-Eisenberg R, Neumann D, Fuchs S (1987) J Biol Chem 262:10506-10510.

Sibley DR, Benovic JL, Caron MG, Lefkowitz RJ (1988) Endocrine Rev 9:38-56.

Smith MM, Merlie JP, Lawrence JC (1987) Proc. Natl. Acad. Sci. USA 84:6601-6605.

Souroujon, MC, Neumann D, Pizzighella S, Fridkin M, Fuchs S (1986) EMBO J 5:543-546.

Teichberg VI, Sobel A, Changeux JP (1977) Nature 267:540-542.

Yee GH, Huganir RL (1987) J Biol Chem 262:16748-16753.

A NEW BROMOBIMANE FLUORESCENT LABEL FOR ANION EXCHANGE PROTEINS

N.S. Kosower*, E.M. Kosower**, A.E. Radkowsky** & J. Zipser*
*Department of Human Genetics,
Sackler School of Medicine and
**School of Chemistry,
Tel Aviv University,
Ramat Aviv, Tel Aviv 69978, Israel

ABSTRACT

A new, negatively charged fluorescent labeling agent, sulfobenzoyloxybromobimane (SBBr), selectively labels the membrane anion exchange protein (Band 3) in intact erythrocytes under physiological conditions. The fluorescent product is stable and easily seen after electrophoretic separation of proteins from solubilized membranes (whole membranes or membranes stripped of peripheral proteins). Quantitation is carried out with solubilized membranes or by densitometry of the gel. Chloride ions diminish the rate of labeling. Diisothiocyanatostilbenedisulfonate (DIDS) inhibits labeling by SBBr. The major part of the label remains in the 60kDa fragment after chymotrypsin treatment of SBBr-labeled cells. SBBr-binding site(s) may be different from those attacked by DIDS. SBBr is useful for studying band 3 and related proteins, including Cl^- channels.

INTRODUCTION

Bromobimanes have been used for the fluorescent labeling of thiols, both in small and large molecules, in a variety of biological systems (Kosower et al, 1979; Kosower & Kosower, 1987; Fahey & Newton, 1987). Monobromobimane (mBBr) and bisbromobimane (bBBr) are uncharged molecules which easily penetrate intact, live cells and which label membrane and intracellular thiols under physiological conditions. The positively charged quaternary bimane salt, qBBr, does not penetrate living cells. Bimane-labeled materials are stable to air, to light, to chemical and biochemical procedures, and are resistant to irradiation. The fluorescent moiety is small (two pentacyclic

NATO ASI Series, Vol. H 32
Molecular Biology of Neuroreceptors and
Ion Channels. Edited by A. Maelicke

rings) and would perturb only minimally macromolecular conformation.

We describe here a new, negatively charged bromobimane, sulfobenzoyl-oxy-bromobimane (SBBr), which selectively labels the anion exchange protein, Band 3, in the intact erythrocyte.

MATERIALS AND METHODS

Sulfobenzoyloxy-bromobimane was synthesized by the reaction of the K^+ salt of 4-sulfobenzoic acid with bis-bromobimane in CH_3CN in the presence of a crown ether. The somewhat complex isolation and purification procedure which yields either the K^+ or Li^+ salts will be described in detail elsewhere. (Kosower et al, in preparation).

A 50mM solution of labeling agent (solvents: water for the SBBr-lithium salt and 50% acetonitrile-water for the potassium salt) was prepared just before use and kept in the dark.

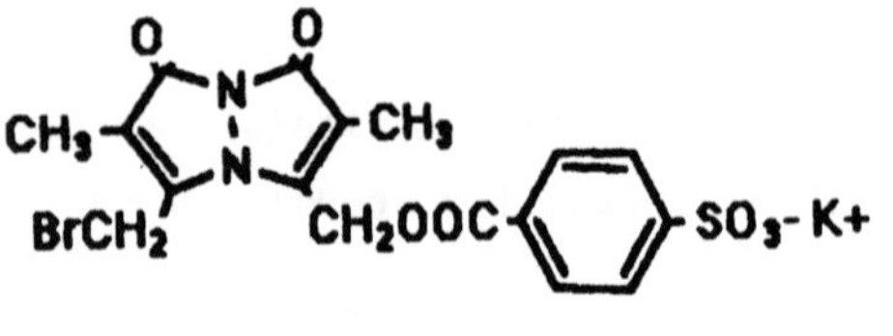

SBBr

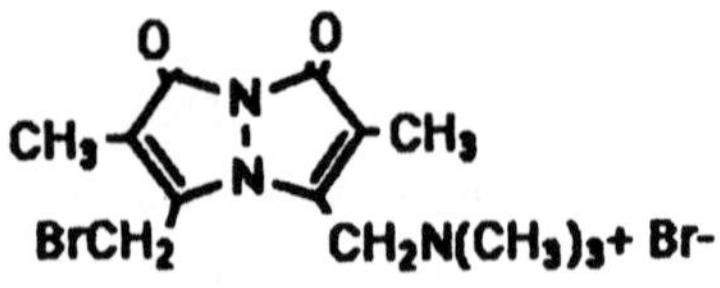

qBBr

Blood samples were obtained from healthy donors and prevented from coagulation by the addition of heparin. Blood was centrifuged, buffy coat removed and the erythrocytes washed and resuspended in isotonic buffer, pH 7.4. The buffers used were a) 8.4mM Na_2HPO_4-1.6mM NaH_2PO_4 (10mM sodium phosphate)-135mM NaCl (PO_4/NaCl buffer) b) 84mM Na_2HPO_4-16mM NaH_2PO_4 (PO_4 buffer) c) 10mM sodium phosphate-280mM sucrose (PO_4/sucrose buffer). In some experiments, Na^+ in the buffer was replaced by either K^+ or Li^+.

Erythrocyte suspensions (8%) are mixed with SBBr solution (final concentration, 0.2-2.0mM SBBr) and incubated at 37^oC for 10-120 min. Reaction vessels containing bromobimanes are covered to avoid photolysis of the reagent. The reaction products are stable to light. Following incubation, cell suspensions are washed free of excess reagent.

In some experiments, aliquots of erythrocyte suspensions were first treated with 50μM of diisothiocyanatostilbenedisulfonate (DIDS), washed, then reacted with SBBr. In other experiments, cell suspensions were allowed to react with 5mM N-ethylmaleimide (NEM) for 60 min prior to SBBr labeling. In additional experiments, SBBr-labeled cells were treated with chymotrypsin (100μg chymotrypsin/ml of 25% cell suspension) for 30min, then washed twice with buffer containing 0.5% albumin and 2mM phenylmethylsulfonyl fluoride (PMSF) and once with pure buffer.

Erythrocyte ghosts were prepared and solubilized by established methods (Steck & Kant, 1974). Spectrofluorimetric measurments (excitation, 380nm, emission, 480nm) were made with solubilized membranes (Kosower et al, 1979). Solubilized membranes were analysed by sodium dodecyl sulfate-polyacrylamide gel electrophoresis (SDS-PAGE)(Fairbanks et al,1971). Fixed gels were viewed and photographed on a transilluminating box with a long wavelength uv source (Kosower & Kosower, 1987). Gels were then stained with Coomassie blue and rephotographed. Densitometric scans were made using the negatives of the gel photographs.

Peripheral proteins were extracted with ice-cold 0.1N NaOH (10 vol/vol ghost suspension), waiting a few seconds, then centrifuging the samples (Steck et al, 1976). The membrane pellets, stripped of peripheral proteins, were solubilised as described above for whole membranes.

RESULTS AND DISCUSSION

After reaction of the intact cells with SBBr, membranes were separated and analysed. Band 3 was selectively labeled in intact cells (Fig.1). The restricted labeling pattern seen in the gels was very different from the pattern obtained after mBBr-labeling, in which many membrane proteins were labeled (Fig.1). SBBr did not penetrate the intact cell, as was evident from the lack of labeling of intracellular glutathione. In erythrocyte ghosts,

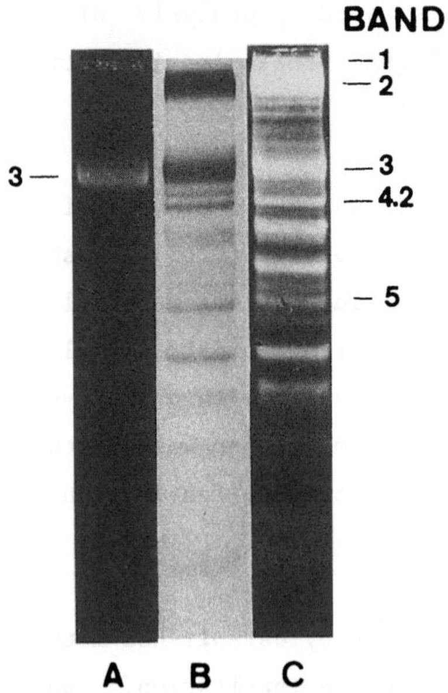

Fig.1. SDS-polyacrylamide gel electrophoresis patterns of erythrocyte membrane proteins.
A, fluorescence (SBBr)
B, dye complexes (Coomassie blue)
C, fluorescence (mBBr)

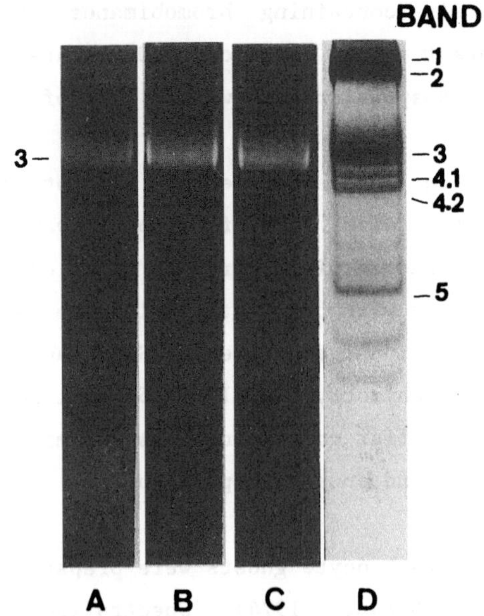

Fig.2A. SDS-polyacrylamide gel electrophoresis patterns of erythrocyte membrane proteins (SBBr)
A, fluorescence (PO_4/NaCl)
B, fluorescence (PO_4)
C, fluorescence (PO_4/sucrose)
D, dye complexes (PO_4)

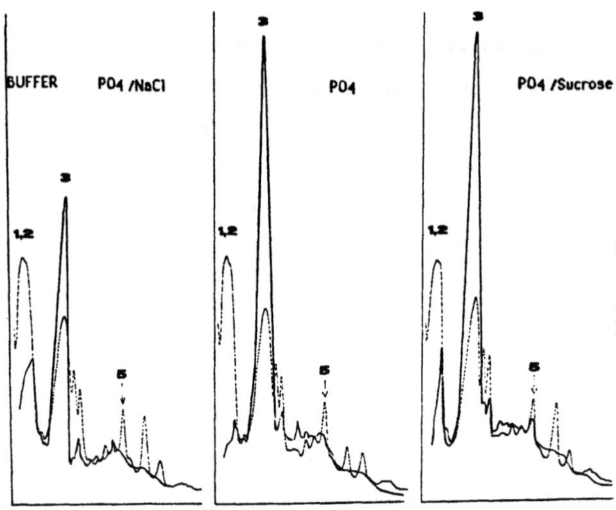

Fig.2B. Densitometric scans of gel patterns detected by either fluorescence or Coomassie blue absorption. (see caption, Fig.2A)
Fluorescence, ———
Dye absorption, ------------

SBBr labeled many proteins, the gel showing a pattern of fluorescent bands similar to that of labeling by mBBr (not shown).

Erythrocytes labeled with SBBr in high Cl^- buffer (PO_4/NaCl) show decreased labeling of Band 3 compared to those labeled in buffers lacking Cl^- (PO_4 or PO_4/sucrose) as is clear in the electrophoretic gel fluorescence patterns (Fig.2A) . Densitometry showed that the Band 3 labeling in high Cl^- buffer was about half that obtained for labeling in buffers lacking chloride in our experiments. (Fig.2B). Replacement of Na^+ in the buffer by Li^+ or K^+ did not affect the labeling.

Pretreatment of erythrocytes with DIDS markedly inhibited SBBr-labeling of Band 3 (Fig.3). The fluorescence of solubilized membranes in such cases was about 30% of controls. Some inhibition of SBBr labeling was observed even when DIDS was added to the cells after 30 min of incubation with SBBr, indicating that the rate of reaction of DIDS may be appreciably greater then that of SBBr. DIDS does not penetrate the intact erythrocyte, but binds to Band 3 and inhibits anion transport (Passow, 1986). High Cl^- concentration in the medium decreases DIDS binding to erythrocytes and other cells (Jessen et al, 1986; Passow, 1986). The effect of DIDS on SBBr labeling suggests that SBBr reacts with site(s) of importance to the functioning of Band 3.

Chymotrypsin treatment of intact erythrocytes leads to the degradation of Band 3 to two major fragments of molecular weights of 60kDa and 35kDa (Passow, 1986). The stained 60kDa fragment can be easily seen on PAGE gels, whereas the 35kDa fragment is scarcely detected for unknown reasons. (Jennings and Passow, 1979; Steck et al, 1976). Treatment of SBBr-labeled cells with chymotrypsin yielded a protein pattern similar to that of unlabeled cells. The 60kDa fragment was found to be labeled by SBBr and stainable by Coomassie blue, with little or no labeling or staining observed in the region of 35kDa proteins(Fig.4A). A major part of the label was found in the 60kDa fragment, based on densitometric comparisons of Band 3 from labeled cells and Band 3 from chymotrypsin-treated labeled cells (Fig.4B). In addition, fluorescence intensities of labeled, solubilised membranes showed that 75-100% of the label was retained in the membranes from chymotrypsin-treated cells. Although we do not have a direct measurement on the extent of SBBr-labeling of the 35kDa fragment, the lack of fluorescence on gels and the

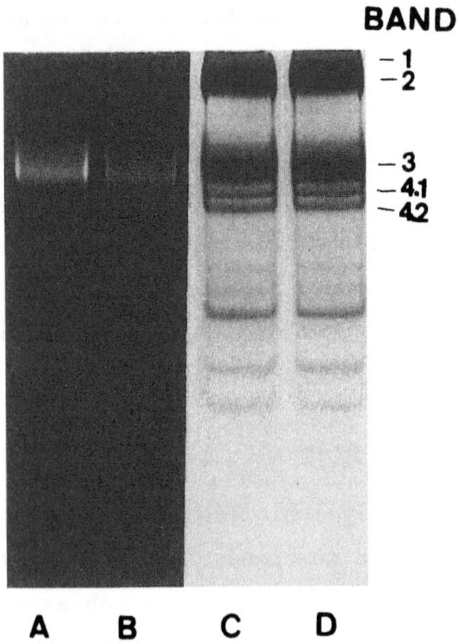

Fig.3. SDS-polyacrylamide gel electrophoresis patterns of membrane proteins isolated from erythrocytes with and without DIDS treatment before SBBr labeling (PO_4).
A, fluorescence (SBBr)
B, fluorescence (DIDS, SBBr)
C, dye complexes (SBBr)
D, dye complexes (DIDS, SBBr)

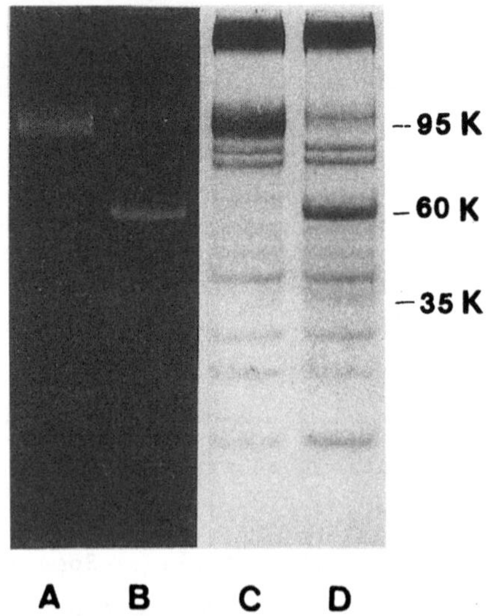

Fig.4A. SDS-polyacrylamide gel electrophoresis patterns of membrane proteins isolated from erythrocytes with and without chymotrypsin treatment after SBBr labeling (PO_4/NaCl).
A, fluorescence (SBBr)
B, fluorescence (SBBr, chymotrypsin)
C, dye complexes (SBBr)
D, dye complexes (SBBr, chymotrypsin)

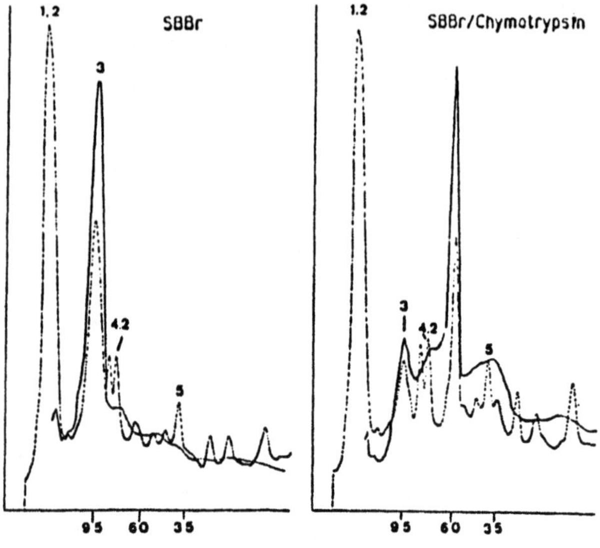

Fig.4B. Densitometric scans of gel patterns detected by either fluorescence or Coomassie blue absorption. (see caption, Fig.4A)
Fluorescence, ———
Dye absorption, -·-·-·-·-

high percentage of label found in the 6o kDa fragment, suggest that the SBBr-labeled sites are all or mostly within the 60kDa portion of Band 3.

Treatment of chymotrypsinized intact erythrocytes with DIDS at high pH prevents separation of the 60kDa and 35kDa fragments of Band 3, apparently due to the cross-linking of the fragments (Jennings & Passow, 1979). SBBr-labeled erythrocytes were treated with chymotrypsin, followed by incubation with DIDS and removal of peripheral proteins. As shown in Fig.5, DIDS was effective in cross-linking the 60kDa and 35kDa fragments after chymotrypsin treatment in both unlabeled and SBBr-labeled erythrocytes. Thus, SBBr did not prevent DIDS binding to the relevant sites under these conditions.

DIDS and related compounds bind to Band 3 and compete with the anionic substrates. After binding, DIDS forms a covalent link with one or two lysine residues and irreversibly inhibits anion transport. DIDS presumably is also linked to lysine residue(s) in the membrane part of the 35kDa domain, in addition to the identified lysines in the 17kDa hydrophobic membrane domain of the 60kDa fragment (Lodish,1988; Passow, 1986; Pimplikar et al, 1988). Since SBBr labels in the presence of DIDS, albeit more slowly, it appears that the labeling targets are groups other than the lysines attacked by DIDS.

Pretreatment of erythrocytes with NEM did not diminish significantly the SBBr-labeling. We tentatively propose that the labeled site(s) are within the membrane part of the anion exchange domain of Band 3, 'below' the sites exposed to DIDS binding. Thus, DIDS blocks the entry of SBBr into the anion exchange active region, whereas SBBr does not inhibit the reaction of DIDS. The SBBr-site(s) may be cysteine and/or lysine residues, the former inaccessible to NEM in membrane-bound Band 3. A cysteine residue in the membrane domain of the 17kDa fragment is known to be cryptic and unreactive to NEM (Ramjeesingh & Rothstein, 1983; Rao, 1979). Further studies are required for the identification of the amino acids involved in SBBr-binding.

Band 3 (anion exchange protein, AEP) is a major, transmembrane protein in the erythrocyte. AEP mediates chloride-bicarbonate exchange across the erythrocyte membrane and may also facilitate movement of water through the membrane (Lodish, 1988; Low, 1986; Passow, 1986). Band 3 also carries cer-

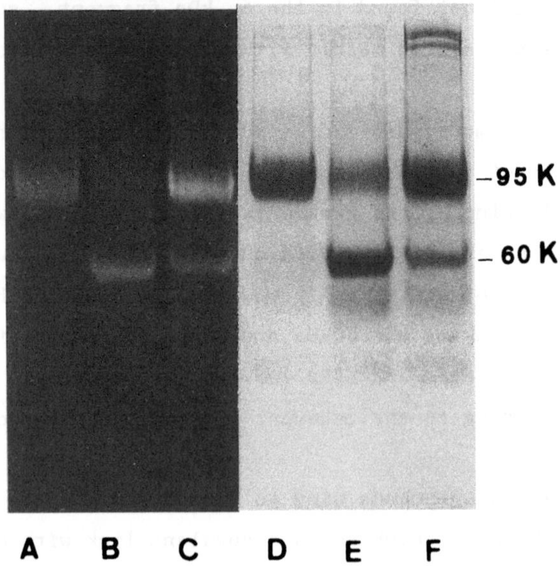

Fig.5. SDS-polyacrylamide gel electrophoresis of erythrocyte membranes solubilised after removal of peripheral proteins.
A, fluorescence (SBBr)
B, fluorescence (SBBr, chymotrypsin)
C, fluorescence (SBBr, chymotrypsin, DIDS)
D, dye complexes (SBBr)
E, dye complexes (SBBr, chymotrypsin)
F, dye complexes (SBBr, chymotrypsin, DIDS)

tain blood group specificities. Band 3 may be related to the appearance of the 'senescent' antigen on the aged erythrocyte (Kay, 1984). Band 3 is also important for the anchorage and binding of cytoskeletal and cytoplasmic proteins to the membrane (Low, 1986).

A number of anion exchange systems sensitive to DIDS have been described in various tissues. These AEP are responsible for Cl^-/HCO_3^- exchange, Na^+-dependent Cl^-/HCO_3^- exchange, Cl^- conductance, Na^+/HCO_3^- electrogenic symport, etc (Lodish, 1986). Proteins homologous to Band 3 have been found in other cells (Allen et al, 1987; Demuth et al, 1986; Lodish, 1986; Low,1986). Band 3 is a member of multigene family that encodes proteins of related structures and functions (Lodish, 1986; Pimplikar et al, 1988). Homologies in amino acid sequence which complement the similarities in anion exchange behavior have been noted for Band 3 and the $GABA_A$ receptor (Kosower,1988).

Various radioactive and fluorescent labeling agents have been used for Band 3 (Landry et al, 1987; Macara et al, 1983; May, 1987; Sheetz et al, 1980). Among the fluorescent reagents, SBBr has a number of advantages. First, the labeling moiety is relatively small, like DIDS and its derivatives. Second, the labeling reaction may be carried out on the intact cell at physiological pH and temperature. Other agents bearing isothiocyanato or dichlorotriazinyl linking groups require a pH near 9 for efficient covalent binding. Third, the bimane-labeled product is stable to high pH and light. Fourth, bimane-labeled derivatives are suitable for enzymatic degradation and peptide separation. (Kosower et al, 1979)

To conclude, the new bimane labeling agent, SBBr, offers a convenient means for fluorescent labeling of Band 3 and for studies on its behavior in anion transport, spatial relationship to other membrane constituents(Klugerman et al, 1984; Macara et al, 1983) and alterations in certain pathological conditions. SBBr may be of use in the study of Cl^- channels.

Allen DP, Low PS, Dola A, Maisel H (1987) Band 3 and ankyrin homologues are present in eye lens: Evidence for all major erythrocyte membrane components in same non-erythroid cell. Biochem Biophys Res Commun 149:266-275

Demuth DR, Showe LC, Ballantine M, Palumbo A, Fraser PJ, Cioe L, Rovera G, Curtis PJ (1986) Cloning and structural characterization of a human non-erythroid band 3-like protein. EMBO J 5:1205-1214

Fairbanks G, Steck TL, Wallach DFH (1971) Electrophoretic analysis of the major polypeptides of the human erythrocytic membrane. Biochemistry 10:2602-2617

Fahey RC, Newton GL (1987) Determination of low molecular weight thiols using monobromobimane fluorescent lableling and high-performance liquid chromatography. Meth in Enzymol. 143:85-96

Jennings ML, Passow H (1979) Anion transport across the erythrocyte membrane, in situ proteolysis of band 3 protein, and cross-linking of proteolytic fragments by 4,4'-diisothiocyano-dihydrostilbene-2,2'-disulfonate. Biochim Biophys Acta 554:498-519

Jessen F, Sjøholm C, Hoffmann EK (1986) Identification of the anion exchange protein of Ehrlich cells: A kinetic analysis of the inhibitory effects of 4,4'-diisothiocyano-dihydrostilbene-2,2'-disulfonate (DIDS) and labeling of membrane proteins with ^{3}H-DIDS. J Membrane Biol 92:195-205

Kay MMB (1984) Localization of senescent cell antigen on band 3. Proc Nat'l Acad Sci USA 81:5753-5757

Klugerman AH, Gaarn A, Parkes JG (1984) Effect of cholesterol upon the conformation of band 3 and its transmembrane fragment. Can J Biochem Cell Biol 62:1033-1040

Kosower EM (1988) HYPOTHESIS: A partial structure for the γ-aminobutyric acid ($GABA_A$) receptor is derived from the model for the nicotinic acetylcholine receptor. The anion exchange protein of cell membranes is related to the $GABA_A$ receptor. FEBS Lett 231:5-10

Kosower NS, Kosower EM (1987) Thiol labeling with bimanes. Methods Enzymol. 143:76-84

Kosower NS, Kosower EM, Newton GL, Ranney HM (1979) Bimane fluorescent Labels: Labeling of normal human ced cells Under physiological conditions. Proc Nat'l Acad Sci USA 76:3382-3386

Landry DW, Reitman M, Cragoe EJ,jr, Al-Awqati Q (1987) Epithelial chloride channel. Development of inhibitory ligands. J Gen Physiol 90:779-798

Lodish HF (1988) Anion-exchange and glucose transport proteins: Structure, function and distribution. Harvey Lec. 82:19-46

Low PS (1986) Structure and function of the cytoplasmic domain of band 3: center of erythrocyte membrane-peripheral protein interactions. Biochim Biophys Acta 864:145-167

Macara IG, Kuo S, Cantley LC (1983) Evidence that inhibitors of anion exchange induce a transmembrane conformational change in band 3. J Biol Chem 258:1785-1792

May JM (1987) Labeling of human erythrocyte band 3 with maltosylisothiocyanate. Interaction with the anion transporter. J Biol Chem 262:3140-3145

Passow H (1986) Molecular aspects of band 3 protein-mediated anion transport across the red blood cell membrane. Rev. Physiol. Biochem. Pharmacol. 103:61-223

Pimplikar SW, Reithmeier RAF (1988) Identification, purification and characterization of a stilbenedisulfonate binding glycoprotein from canine kidney brush border membranes. A candidate for a renal anion exchanger. J Biol Chem 263:4485-4493

Ramjeesingh M, Gaarn A, Rothstein A (1983) The locations of the three cysteine residues in the primary structure of the intrinsic segments of band 3 protein, and implications concerning the arrangement of band 3 protein in the bilayer. Biochim Biophys Acta 729:150-160

Rao A (1979) Disposition of the band 3 polypeptide in the human erythrocyte membrane. The reactive sulfhydryl groups. J Biol Chem 254:3503-3511

Steck TL, Kant JA (1974) Preparation of impermeable ghosts and inside-out vesicles from human erythrocyte membranes. Methods Enzymol. 31:172-180

Steck TL, Ramos B, Strapazon E (1976) Proteolytic dissection of band 3, the predominant transmembrane polypeptide of the human erythrocyte membrane. Biochemistry 15:1154-1161

Sheetz MP, Schindler M, Koppel DE (1980) Lateral mobility of integral membrane proteins is increased in spherocytic erythrocytes. Nature 285, 510-512

CHANGES IN MEMBRANE PHOSPHOPROTEINS ASSOCIATED WITH THE ATP MODULATION OF THE DELAYED RECTIFIER K^+ CHANNEL IN THE SQUID GIANT AXON.

Eduardo Perozo, William S. Agnew and Francisco Bezanilla
Department of Physiology, UCLA Medical School,
Los Angeles CA 90024 U.S.A. and Department of Physiology
Yale University School of Medicine, New Haven CT 06510,
U.S.A.

INTRODUCTION

Potassium channels are still among the most misterious of all membrane proteins, and practically no biochemistry is known of the most diverse and at the same time selective group of ion channels. This lack of information is due to the lack of a specific marker or toxin that could be used to identify and isolate most of the members of this family.

The present set of data tries to explore a novel approach to the identification and possible isolation of a K^+ channel. In squid axons, the K conductance is modulated by ATP (Bezanilla et al, 1986, Perozo et al., 1986, 1988), a property shared by several other K^+ and Ca^{++} channels (Kaczmarek and Levitan, 1987). However, this modulation occurs with relatively high affinity (10 uM, Perozo 1986), and the preparation itself allows experimental manipulations unthinkable in any other cell.

The rational is as follows: if a high enough amount of radiolabeled ATP is placed inside a giant axon, it should be possible to detect bands related to the phosphorylation of K channels. Since the phosphorylation of the channel has an electrophysiological equivalence, the potassium channel should display a positive correlation between the size of the electrophysiological parameter and the amount of radioactivity incorporated.

Fig. 1 shows the effect of ATP mediated phosphorylation on the I-V characteristic of the delayed rectifier current . There is a clear stimulation of the current density that can be up to 300 % at very depolarized potentials, and as has been demonstrated previously (Perozo et al. 1988), this effect is produced only when the phosphate transfer from ATP is allowed. This effect is very specific for ATP and no other nucleotide affects I_K when is placed intracellulary instead of ATP.

Several attempts have been made in the past to measure protein phosphorylation in nervous tissues, some of them incubating whole ganglia in ^{32}Pi and then analyzing phosphoproteins (Levitan and Barondes 1974, Paris et al, 1981). Another elegant approach consists in injecting a single cell under voltage clamp conditions, thus relating the phosphorylation with the electrical activity of the cell (Lemos et al., 1982, 1985); however, the interpretation was very complex because it was not possible to distinguish which protein was cytoplasmic and which was membrane bound. In the present report we approach the phosphorylation of membrane proteins in an isolated axon as a way to gain information about the molecular properties of a K channel.

NATO ASI Series, Vol. H 32
Molecular Biology of Neuroreceptors and
Ion Channels. Edited by A. Maelicke

METHODS

Electrophysiology

Experiments were performed on axons from the Atlantic squid *Loligo pealei.* Internally dialyzed axons (Brinley and Mullins, 1967) were voltage clamped using standard axial wire methods (Bezanilla et al, 1982). Experiments were controlled using a hardware system for waveform generation, acquisition and display on an 16-bit IBM AT type computer (Intel 80286 based), connected to a 12-bit A/D and a D/A converter. Ionic currents were recorded using the P/4 procedure, and both leakage and capacity transients subtracted using an analog circuit.

Intracellular Dialysis with ^{32}P-ATP

Freshly dissected axons were predialyzed for 1 hour with internal solution containing no ATP, in order to wash out small solutes (< 9 Kd) and to equilibrate the ionic composition of the axoplasm with the internal solution. One to two mCi of γ-^{32}P-ATP (New England Nuclear, specific activity of 3000 Ci/mmol) were concentrated in a speedvac, and dissolved in 250 ul of internal solution for a final concentration of 7 uM ^{32}P-ATP carrier free plus 50 uM carrier ATP. The labeling solution was applied internally after the initial predialysis period at a rate of 1 ul/min. The amplitude of the K current was measured throughout the experiment by pulsing to 0 mV from a holding potential of -50 mV. During the phosphorylation period, a system of guards and air gaps in the chamber prevents the contact of the internal solution with the extracellular side of the axon. The internal solution contained (in mM) 310 K, 30 PO_4, 110 Glycine, 4 Mg, 1 EGTA, 0 Na, 0 Ca. The external solution was artificial sea water plus 300 nM TTX and 1 mM NaCN.

Pore Gradient Electrophoresis and Autoradiography

Segments of axons from the region where the ionic current was recorded were taken out of the chamber and extruded of axoplasm using a rolling tube. Both, the membrane part and the axoplasm were placed in denaturing sample buffer containing 10 % SDS and 2-mercaptoethanol for five minutes, to be sonicated later 1 min at 90 C. Samples corresponding to each individual axon/experiment were run on a SDS polyacrylamide gradient of 4.5 to 15 %. The gels were stained with Comassie brilliant blue and dried for autoradiogram exposure of 3 to 11 days at - 70 C. Each band of the autoradiogram was densitometrically analyzed, and the intensity of the bands related to the magnitude of current stimulation measured electrophysiologically.

RESULTS AND DISCUSSION

Three different types of experiments were performed: 1) Direct labeling with ^{32}P-ATP and subsequent SDS-PAGE electrophoresis of both axolemma and axoplasm. 2) Direct labeling with ^{32}P-ATP, wash out of the label by dialyzing with a 0 ATP solution, and then SDS-PAGE. 3) Labeling with ^{32}P-ATP after the current has been previously stimulated with non-radioactive ATP-γ-S, then SDS-PAGE . In each of the experiments there is a simultaneous recording of the current density before and after the actual labeling.

The pattern of bands produced by the Comassie stain in each of the experiments was remarkably constant for all the axolemma lines as well as for the axoplasm, considering that each sample came from a different squid in different days and different experimental conditions. On the other hand, the level of phosphorylation was quite variable even for samples under the

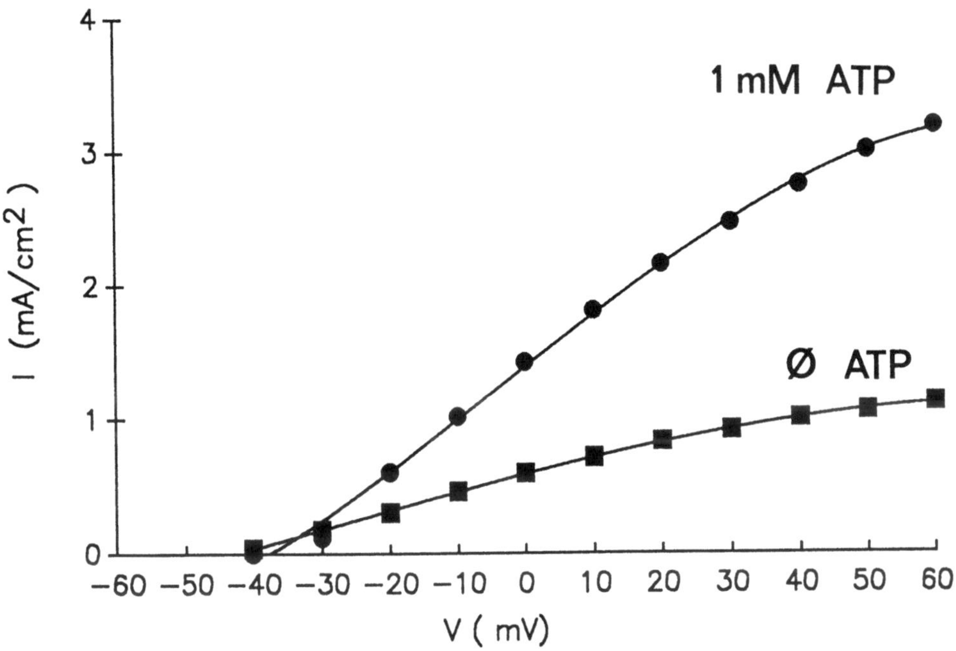

Figure 1. Current-voltage relationship for an axon dialyzed with 0 ATP and the increase in current density after dialysis with 1 mM ATP. Holding Potential, -50 mV, T= 14 C.

same experimental conditions. This fact should be considered when the correlation of electrical vs. labeling is made.

It is clear from an analysis of the autoradiograms that there are a high number of well defined bands; still, some of those bands appear consistently in every gel. However, as pointed out before, the actual availability of the nucleotide seems to differ from experiment to experiment. As a consequence of that, an internal normalization has to be performed in order to compare the level of phosphorylation in all the experiments. In every gel, there is a conspicuous band of about 100 Kd; this band, which is close to the reported MW of several ATPases (and will be called the putative "ATPase") was used to normalize the amplitude of the rest of the bands, regardless of its MW.

Fig. 2 shows densitograms for a selected gel in every condition. The amplitude of the peaks is relative to the "ATPase" band, but the level of radioactivity of the direct label experiment is much larger than either the washout experiment or the ATP-γ-S experiment.

The labeling pattern is apparently very complex, but on close examination it is evident that there are few bands which can be identified in all the conditions. Of those bands, we chose five of them to correlate with the relative stimulation in current density at the moment in which the experiment was stopped. The bands are at 25-26, 55-56, 100, 150-155 and 190 Kd. There are of course many more bands that do not appear in the present analysis, either because they can not be found consistently in all the gels, or their intensity is too dim to carry out a comparison. It is important to point out, that the labeling pattern of the isolated axoplasm does not show any of the bands under consideration. However, axoplasm and axolemma do have many common bands, indicating that most probably there was a thin layer of axoplasm remaining in the axolemma samples.

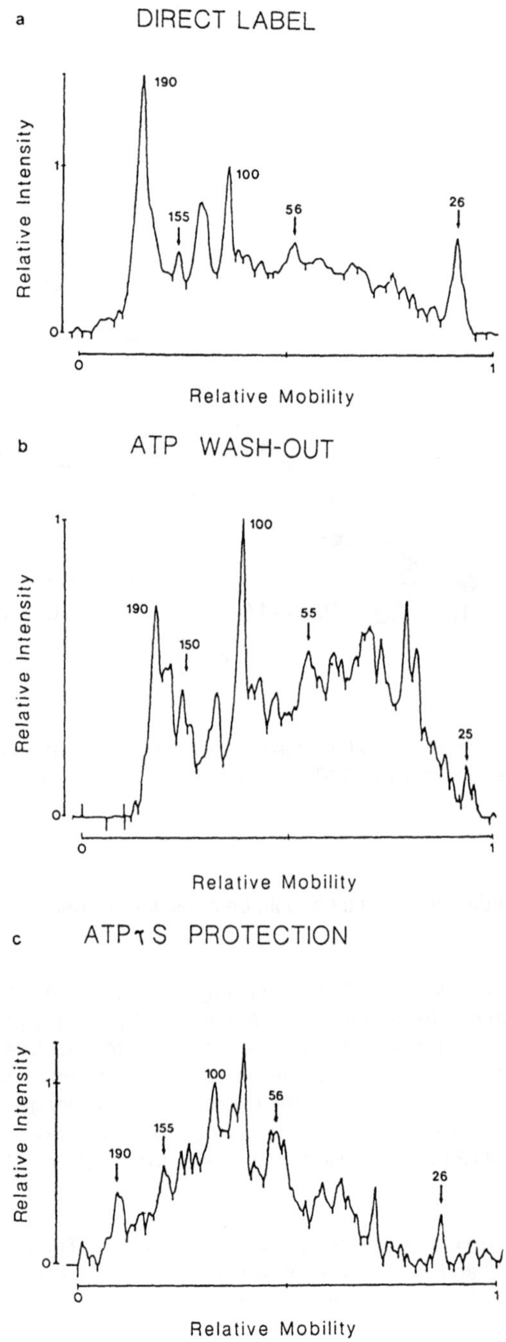

Figure 2. Densitometric scans of the autoradiograms for three experimental conditions. The scanner was centred on each gel line and the amplitude is relative to the 100 Kd band. In each scan, the numbers represent MW x 10^{-3}. a) Direct labeling with ^{32}P-ATP. b) Direct labeling with ^{32}P-ATP and wash out the current stimulation. c) Stimulation of the potassium current with ATP-γ-S and then labeling with ^{32}P-ATP.

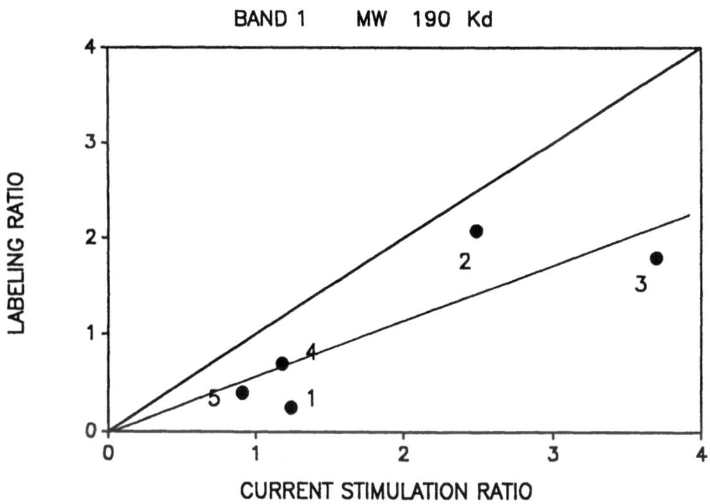

Figure 3. Correlation between the current stimulation at the end of the experiment with the relative labeling of band 1 (190 Kd). Each number near an experimental point represents a different type of experiment: axon 1 and 2 = direct labeling, axon 3 and 5 = ATP-γ-S protection, axon 4 = wash out.

A correlation between the labeling and the current simulation is shown in Fig. 3 and 4. For each band, the values for the labeling ratio (LR) were plotted against the values for the current stimulation ratio (CSR) in five different experiments. LR is defined as:

$$LR = A_x/A_{100\ MW}$$

where A_X is peak amplitude of band X, and $A_{100\ MW}$ is peak amplitude of the 100 Kd band. On the other hand, CSR is defined as:

$$CSR = I_f/I_{0\ ATP}$$

where I_f is the current density at the end of the experiment, and $I_{0\ ATP}$ is the current density at the end of the predialysis.

The high molecular weight band clearly shows a very good correlation between the labeling and the current stimulation (Fig. 3). This is specially true under direct labeling conditions (in which the size of the peak at 190 Kd is almost duble than the one at 100 Kd), and with ATP-γ-S protection, where the high molecular weight bands are very faint (Fig. 2a). Fig. 4 shows that the same kind of analysis performed for three other bands shows no or very little correlation, compared to the high MW band.

The question now arises, what is the nature of the bands showing positive correlation?. Using the present approach, we have the advantage of being able to sort out labeled bands of the axonal membrane from the ones of the axoplasm, thus eliminating the main source of uncertainty present in previous single-cell labeling experiments (Lemos et al., 1982, 1985, Volterra et al., 1988). Additionally, biochemical data from other preparations on the structural parameters of other ion channels and membrane proteins, make possible the putative

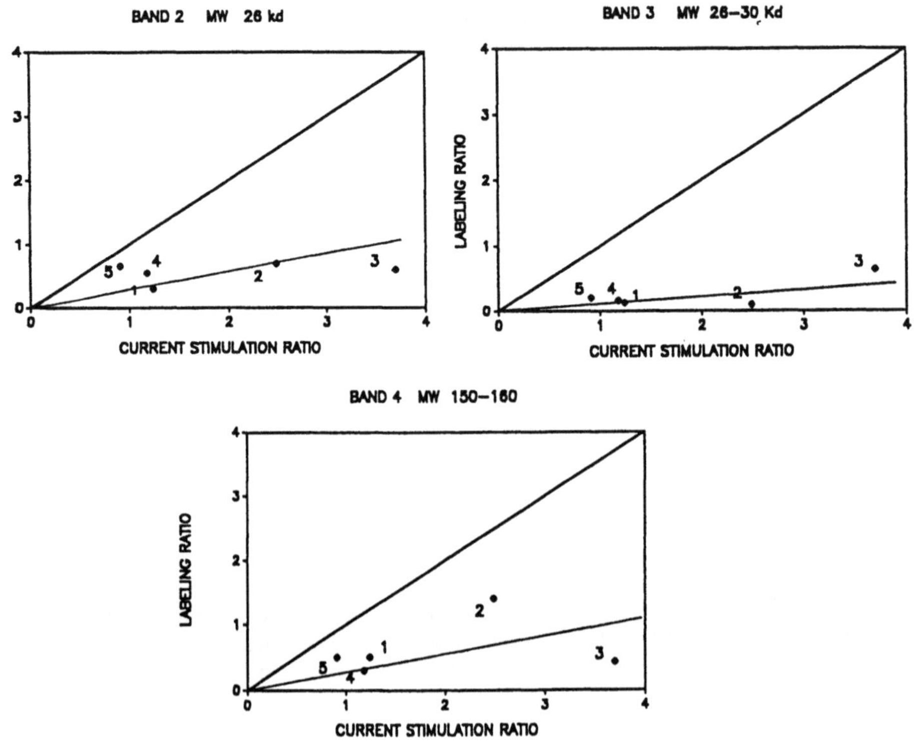

Figure 4. Correlation current stimulation-labeling for three other bands: 26-30 Kd, 56 Kd and 150-160 Kd. Conditions are the same than for Fig. 3.

identification of some of the bands like Na channel (Agnew, 1984, Barchi, 1988) or the alpha subunit of the Na-K ATPase (Jorgensen, 1982). Nevertheless, we are far from truly identifying any of the bands in our gels, and that will not be possible until a somewhat specific probe is made available.

This simplified view gets complicated by the fact that the delayed rectification current in produced by the ensemble of an inhomogeneous population of channels. Studies at the single channel level in the cut-pen axon preparation (Llano et al, 1988), have shown clearly that at least three well defined populations of kinetically and conductively distinct population coexist and form the delayed outward current. How closely related are they from a structural point of view? Probably to a large extent, a fact that would complicate the interpretation of the present results. Still, with a detailed electrophysiological characterization of each channel type, the study of the molecular properties of these proteins will show how subtle structural differences can have profound consequences on channel function.

CONCLUSION

We have shown that there is a strong correlation of the labeling of high molecular weight peptides and the stimulation of K currents in squid giant axon. However, it is clear that a rigorous quantitative study cannot be made using present tools; additionally, we have little evidence suggesting that indeed de 190 Kd band is a channel. This information nonetheless places some boundaries to determine the molecular identity of one or more of the K channels functioning during delayed rectification in squid axons.

REFERENCES

Agnew, W.S. (1984) Ann. Rev. Physiol. **46**: 517

Barchi R.L. (1988) Ann. Rev. Neurosci. **11**: 455

Bezanilla, F., C. Caputo, R. Dipolo and H. Rojas (1986) Proc. Nat. Acad. Sci. U.S.A. **83**: 2743

Bezanilla, F., J. Vergara and R.E. Taylor (1982) in: Methods of Experimental Physics vol. 20, pp 445 Academic Press

Brinley, F.J., and L.J. Mullins (1967) J. Gen. Physiol. **50**: 2303

Jorgensen, P.L. (1981) Biochim. Biophys. Acta 694: **27**

Kaczmarek, L.K. and I.B. Levitan (1987) Neuromodulation. The biochemical control of nuronal excitability. Oxford University Press. New York.

Lemos, J.R., I. Novak-Hofer and I.B. Levitan (1982) Nature 298: **64**

Lemos, J.R., I. Novak-Hofer and I.B. Levitan (1985) J. Biol. Chem. **260**: 3207

Levitan, I.B. and S.H. Barondes (1974) Proc. Nat. Acad. Sci. U.S.A. **71**: 1145

Llano I., C.K. Webb and F. Bezanilla (1988) J. Gen. Physiol. 92: 179

Paris, C., V. Castellucci, E. Kandel and J.H. Schwartz (1981) Cold Spring Harbor Conf. Cell Prolif. **8**: 1361

Perozo, E., R. Dipolo, C. Caputo, H. Rojas and F. Bezanilla (1986) Biophys. J. **49**: 215a

Perozo, E., F. Bezanilla, R. Dipolo (1989) J. Gen. Physiol. in press

Volterra, A., S.A. Siegelbaum, E.R. Kandel, J.D. Sweatt (1988) Soc. Neurosci. Abstracts pp 152.

CHANNEL MODELLING EXPERIMENTS IN PLANAR BILAYERS USING AMPHIPHILIC HELICAL POLYPEPTIDES

Günther Boheim, Inge Helbig and Sabine Meder
Dept. Cell Physiology, Ruhr-Universität Bochum,
Postfach 102148, D-4630 Bochum, F.R.G.
Brigitte Franz and Günther Jung
Inst. Organic Chemistry, Universität Tübingen,
Auf der Morgenstelle 18, D-74oo Tübingen, F.R.G.

Summary

An artificially designed and chemically synthesized polypeptide of 22 amino acid chain length and most likely helical structure forms well resolved ion channels in planar lipid bilayers. Data obtained with related polypeptides reveal the changes in stabilizing intermolecular forces which are caused by modifications at single amino acids. Results are discussed in relation to the putatively helix lining of biological protein channels.

Introduction

Hydropathy profiles (Kyte and Doolittle 1983) of integral channel proteins of biological membranes seem to indicate that its partial sequences, which putatively span the bilayer core, consist of mainly hydrophobic amino acids and adopt helical structure (Maelicke 1988). The amphipathic eicosapeptide alamethicin (AL) adopts helical structure in crystals and organic solvents (Fox and Richards 1982; Esposito et al. 1987). In membranes AL forms multi-state channels by long-lived stable aggregation of a varying number of helices (Boheim 1974). These channels show a strong voltage dependence in kinetic behaviour due to the intrinsic dipole moment of the helical rod (Boheim et al. 1983).

NATO ASI Series, Vol. H 32
Molecular Biology of Neuroreceptors and
Ion Channels. Edited by A. Maelicke

Recently we reported that completely hydrophobic polypeptides create transmembrane channel structures, too (Menestrina et al. 1986). However, channel states are non-stabilized, short-lived and experimentally not resolved. Here we report that by introducing hydrophilic end groups and a helix bend due to the insertion of one proline into the helix we find a different type of channel. This type is characterized by a single open state conductance, resolved lifetimes and weakly voltage dependent kinetics. Corresponding experiments with series of related synthetic polypeptides will give reliable data for computer simulation studies on channel structure.

Materials and Methods

Polypeptides were synthesized by fragment condensation. Starting from the pentapeptide Ala-Aib-Ala-Aib-Ala (P5) the polypeptides Boc-(Ala-Aib-Ala-Aib-Ala)$_n$-OMe with n = 2 (P10), 3 (P15) and 4 (P20) were obtained as described previously (Voges et al. 1987). The newly synthesized analogue "Dans-P21-Trp" is: Dansyl-(Ala-Aib-Ala-Aib-Ala)$_3$-Pro-(Ala-Aib-Ala-Aib-Ala)-Trp-OMe. For comparison we list the sequence of paracelsin A (PCA), a natural alamethicin (AL) analogue of similar ion channel forming activity (Brückner and Przybylski 1984): Ac-Aib-(Ala-Aib-Ala-Aib-Ala)-Gln-Aib-Val-Aib-Gly-Aib-Aib-Pro-Val-Aib-Aib-Gln-Gln-Pheol. Aib is α-aminoisobutyric acid and pheol phenylalaninol. Polypeptides are soluble in water at the given concentrations.

Virtually solvent-free planar bilayers were formed on a hole of 0.2 mm^2 diameter in a teflon sandwich septum (Menestrina et al. 1986) using the lipid mixture 1-palmitoyl-2-oleoyl-glycero-3-phosphatidylcholine (1,2-POPC): 1,2-dioleoyl-glycero-3-phosphatidylethanolamine (1,2-DOPE): cholesterol 85: 9:6 by weight dissolved in hexane. Lipids were purchased from Avanti (Pelham, Alabama).

Salt solution was identical for both compartments: 1.0 M NaCl, 10 mM Tris/HCl pH 7.0. Polypeptides were added to the cis compartment. Trans side was set to virtual ground. Temperature of 22 ^{o}C and filter setting at .3 kHz completes experimental conditions.

Results

If 20 µg/ml P20 are added to the cis side compartment, one observes a strongly voltage dependent current-voltage (I-V) curve. The steep current increase starts at about ± 200 mV (Menestrina et al. 1986). However, the nature of the elementary channel events is significantly different from that of AL-type antibiotics. After switching-on open channel current fluctuations by the application of an appropriate voltage discrete conductance levels cannot be discriminated up to an experimental time resolution of 0.3 ms. Apart from this apparent channel state destabilization the efficiency of P20 in channel formation is much less than that of AL and PCA. Whereas one observes strongly voltage dependent I-V curves similar to those with P20, the switching-on of channels at ca. 200 mV is already seen after addition of only 50 ng/ml PCA, i.e. at a concentration smaller by a factor of 400. Consequently there are two structural elements in the PCA sequence which on one hand reduce the critical concentration for aggregate formation and on the other stabilize the circular array of helices.

The addition of 100 µg/ml Dans-P21-Trp to the cis compartment and application of a voltage ramp leads to a similar steep increase in membrane current at high voltages as observed with P20. Single channel events under these conditions exhibit the same phenomena of conductance state destabilization. A surprising observation, however, is made in some of the experiments. Whilst membrane voltage is reduced and the "noisy" channels of large conductance (5 to 10 nS) are turned off, a particular different type of channel establishes and remains fluctuating. In fig. 1 single channel current fluctuations of this type of channel are shown. Channel activity appears in bursts (fig. 1a) which indicates at least 3 different channel states: one open state of fixed conductance and two closed states of zero conductance. In fig. 1b the discrete

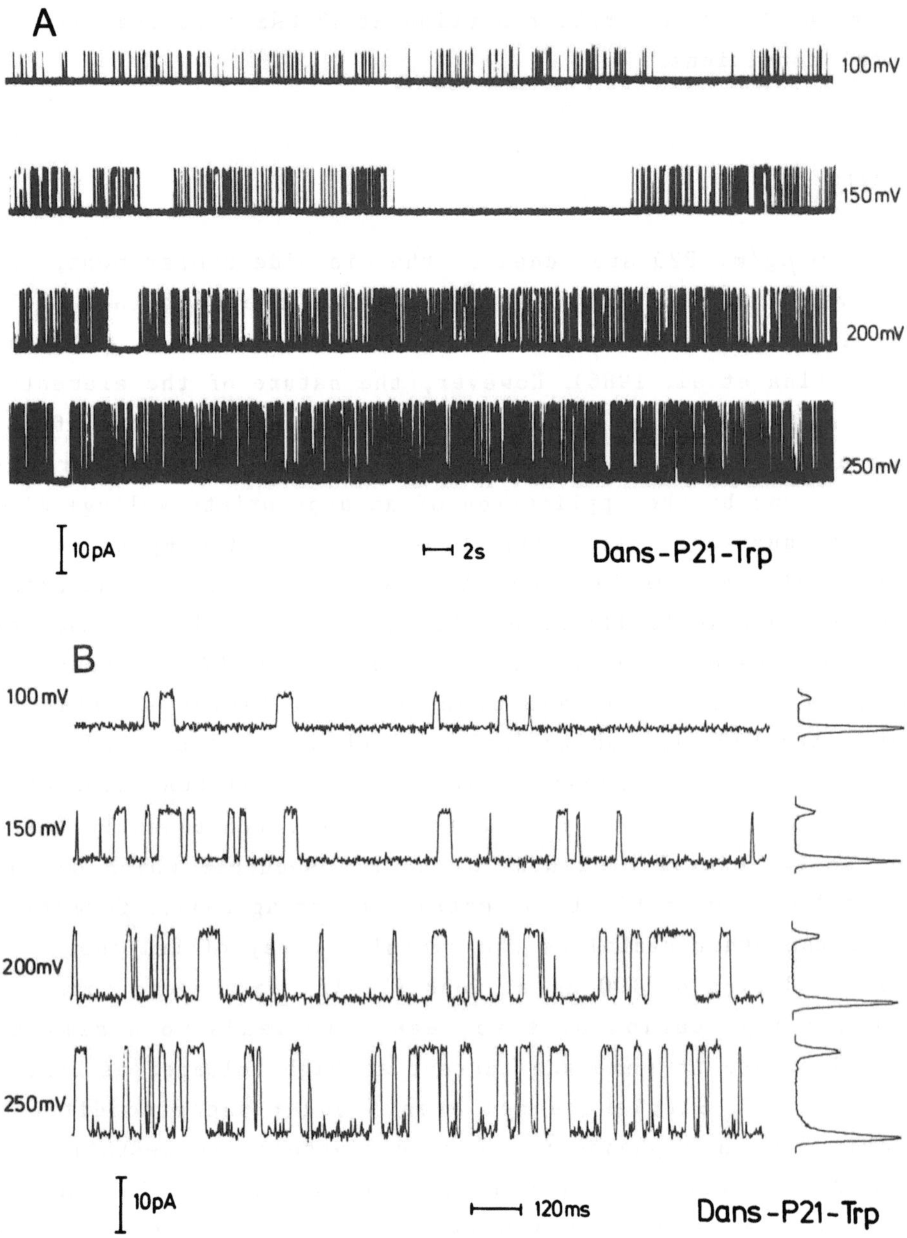

Fig. 1: Single channel current fluctuations of an ion channel formed by the synthetic polypeptide Dans-P21-Trp at various applied voltages a) at a condensed, b) at an expanded time scale. 50 µg Dans-P21-Trp were added to the cis side. For experimental conditions see Methods section.

current fluctuations are recorded at an expanded time scale. Conductance histograms of at least 100 single channel events at each voltage are inserted on the right. Open channel current increases linearly with the applied voltage as shown in fig. 2. The slope of the straight line fit corresponds to an open state conductance of 63 pS. This ohmic behaviour reflects a defined stabilized channel structure.

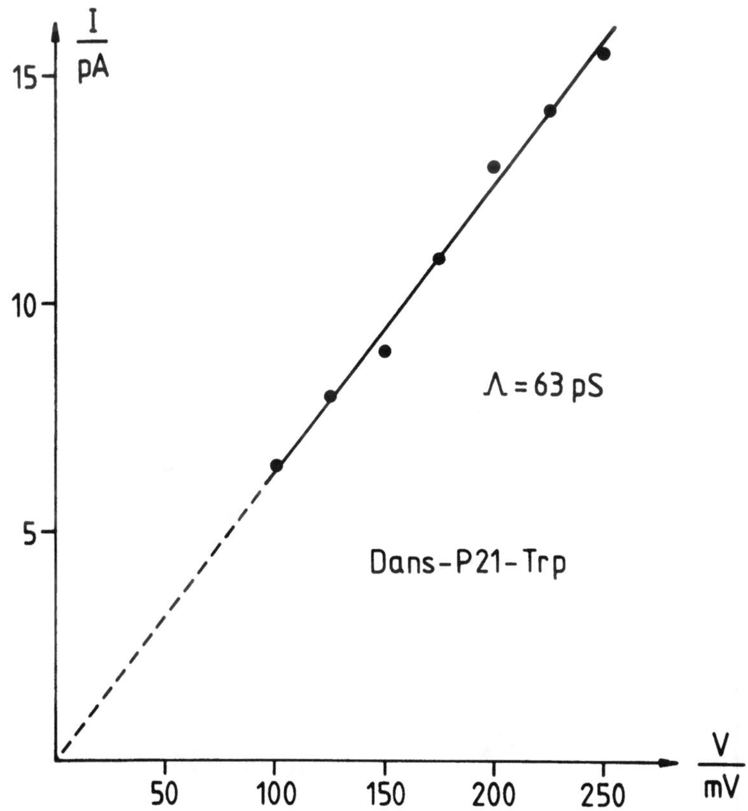

Fig. 2: Current-voltage characteristic of the open state of the Dans-P21-Trp channel. The voltage independent conductance value of 63 pS in 1.0 M NaCl reflects a fixed channel structure. Experimental conditions are the same as given for fig. 1.

Lifetime histograms of open and closed events were fitted by sums of exponential functions. Whereas the open event histogram could be described by a single exponential function, at least two exponential functions were needed to sufficiently fit the closed event histogram. The large majority of closed events are short-lived, i.e. they originate from adoption of the intraburst closed channel state. Mean lifetimes and distribution functions of the various states are voltage dependent. This is shown in fig. 3a for the open state and for the intraburst closed state. Long-lived interburst closed events

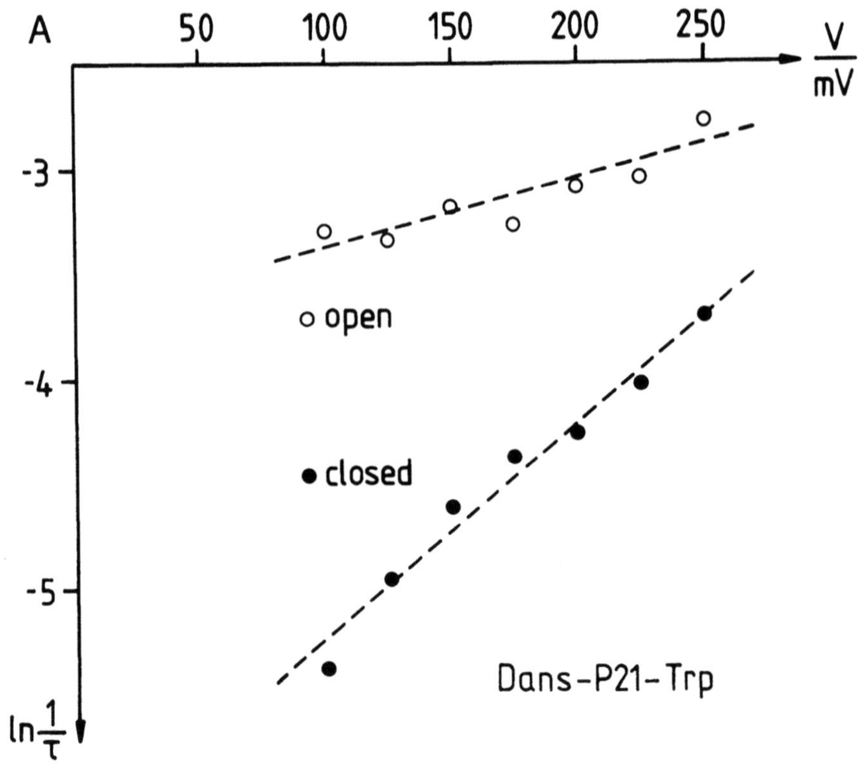

Fig. 3a: Voltage dependence of mean lifetimes of the open state and of the intraburst closed state for the Dans-P21-Trp channel. The opening rate of the channel is influenced by the applied voltage more strongly than the closing rate. Experimental conditions are the same as given for fig. 1.

are quite rare, and more experimental data are needed in order to get a reliable analysis. In fig. 3b the ratios of the mean open and corresponding mean intraburst closed state lifetimes are plotted versus the applied membrane voltage. This ratio is equal to the equilibrium constant of the two states. The straight line fit indicates a formal gating charge of +0.19 elementary charge. This value reflects a weak voltage dependence as compared to the voltage dependence of state distribution changes with AL and PCA which exhibits a formal gating charge of 0.9 (Boheim 1974, Gelfert-Peukert 1986). In addition

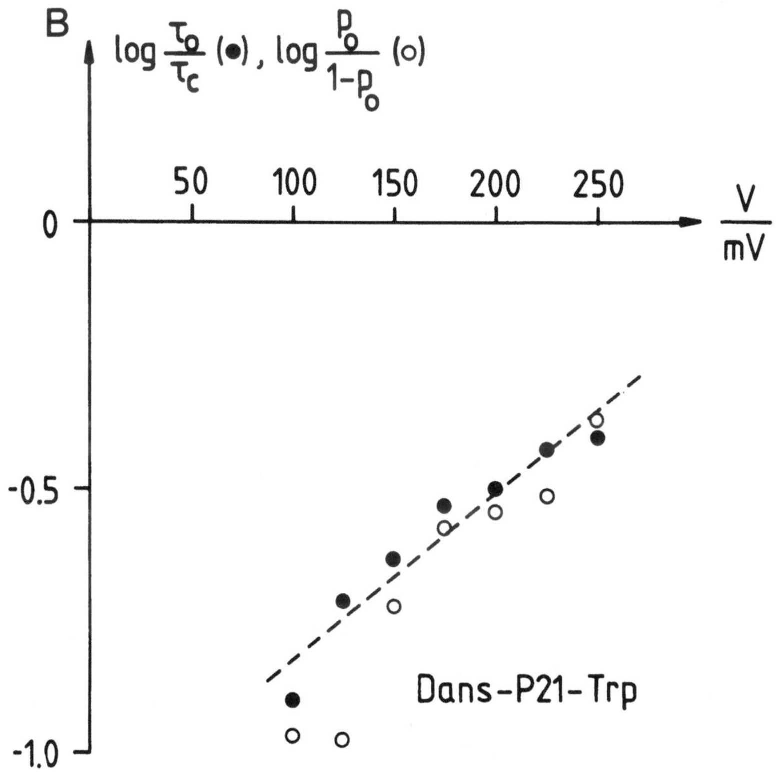

Fig. 3b: Ratios of the mean open state lifetimes to the mean closed state lifetimes from fig. 3a. This ratio is equal to the equilibrium constant of the two states. The weak voltage dependence is characterized by a formal gating charge of +0.19 elementary charges.

fig. 3b shows the voltage dependent ratios of the probability to find the channel in the open state divided by the probability to find it in either of the closed states. The small deviation of the two data sets indicates that the interburst closed state is adopted with low probability.

Discussion

It is quite impressive that small changes in polypeptide primary sequence, which seem not to alter its helical secondary structure, significantly affect its ability to form stabilized helix aggregates and ion permeable channels, respectively. With alamethicin the N-terminal exchange of acetyl(Ac)-Aib- to t-butyloxy-carbonyl(Boc)- causes complete destabilization of the channel states (Boheim et al. 1987). A weaker but still large reduction in stability is observed with the C-terminal exchange of phenylalaninol by leucinol (Gelfert-Peukert 1986). It appears to be difficult to predict those effects just from the actual modification in primary sequence.

In case of the polypeptides P20 and Dans-P21-Trp as shown here similar drastic changes occurred. The exchange of the end groups Boc-, -OMe by the more bulky groups Dansyl-, -Trp-OMe leads to a slight decrease in channel forming effectivity. However, the insertion of one proline in position 16 (total 22 amino acids) creates a stabilized channel aggregate. Similarly in AL and PCA the proline in position 14 (total 20 amino acids) seems to be an essential prerequisite for channel state stability (Boheim et al. 1987) and presumably for the particular mode of single channel current fluctuations.

In contrast to the subunits of biological channel proteins, which have a large molecular mass and several membrane spanning helices in antiparallel orientation (Guy 1984, Noda et al. 1984), the single helical rods are very flexible and respond to the aligning forces of the membrane electric field. This orientational flexibility had been taken into consideration for the proposal of the "dipole flip-flop gating model" of AL-type channel state fluctuations (Boheim et al. 1983).

The channel presented here, however, displays a qualitatively different fluctuation pattern, i.e. transitions between two conductance levels as compared to multi-level transitions in AL and PCA. The two level bursting mode of Dans-P21-Trp channel current fluctuations resembles very much those of the nicotinic acetylcholine receptor channel (Sakmann et al. 1980) and of the Ca^{++}-dependent K^{+}-channel (Methfessel and Boheim, 1982). This difference to AL and PCA is confirmed by a much smaller formal gating charge with Dans-P21-Trp.

The proposal of a feasible gating model is schematically drawn in fig. 4. At present we cannot determine the size of the stabilized channel. Does it consist of four or five heli-

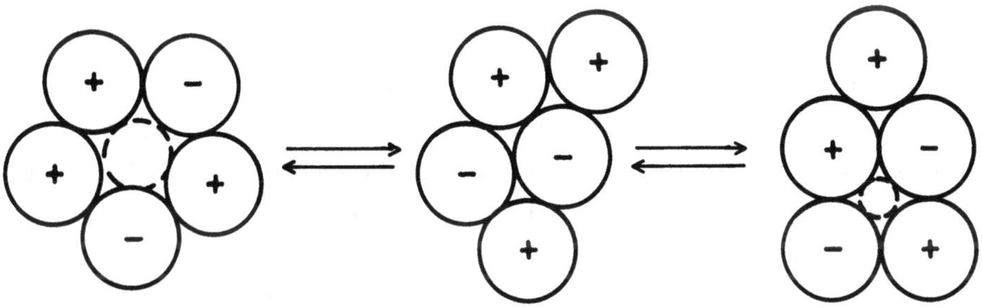

WEAKLY VOLTAGE DEPENDENT GATING
(SCHEMATIC MODEL)

Fig. 4: Schematic model for a weakly voltage dependent gating mechanism of an ion channel simply consisting of helical rods. State transitions occur by lateral sliding of single helix elements: the "helix lateral sliding model". Here helix dipoles might be oriented mainly antiparallel. This is in contrast to the "dipole flip-flop gating model" for the strongly voltage dependent channels formed by alamethicin (AL) and paracelsin (PCA), where helices adopt parallel orientation.

cal rods? The 63 pS conductance (in 1.0 M NaCl) corresponds to that of the acetylcholine receptor channel (95 pS in 1.0 M NaCl, Boheim et al. 1981)) as well as to the value of the Na^+-channel (25 pS in 0.14 M NaCl, Hanke et al. 1984). One open and two closed states would require at least three alternative helix arrangements. We propose the "helix lateral sliding model" with tightest helix packing in the interburst closed state and circular helix array in the open state. The proline induced bend in the helical rod creates a funnel-like structure of the open channel. Some significance on the role of proline residues in conformational changes of protein structures has been discussed (Freedman 1987). A simple cis-trans isomerization at a single proline induced bend might lead to the closure of the channel funnel. Then the bursting mode might be caused by statistical isomerization transitions and consequently bending fluctuations of the channel forming helices. In further experiments we plan to analyze the role of proline for channel gating in more detail by systematic variations of the amino acid sequence of synthetic polypeptides.

Are these experiments with synthetic model peptides relevant for elucidation of the behaviour of biological channel proteins? Structure and conformational fluctuations of channel proteins are not known. Therefore model experiments should give fundamental informations on feasible conformational fluctuations which are of electrophysiological relevance. Recently channel forming activity has been reported for a putative transmembrane helix segment of the Na^+-channel protein (Oiki et al. 1988). The peptide was synthesized using solid-phase methods. In view of our results presented here we conclude that experimental data obtained with channel lining helix segments will have only chance relations to those data recorded with the original protein (Imoto et al. 1988).

Acknowledgement

This work was supported by the Deutsche Forschungsgemeinschaft (Bo 468/2-1).

References

Boheim G (1974) Statistical analysis of alamethicin in black lipid membranes. J Membrane Biol 19:277-303

Boheim G, Hanke W, Barrantes FJ, Eibl H, Sakmann B, Fels G, Maelicke A (1981) Agonist-activated ionic channels in acetylcholine receptor reconstituted into planar lipid bilayers. Proc Natl Acad Sci USA 78:3586-3590

Boheim G, Hanke W, Jung G (1983) Alamethicin pore formation: voltage dependent flip-flop of α-helix dipoles. Biophys Struct Mech 9:181-191

Boheim G, Gelfert S, Jung G, Menestrina G (1987) α-Helical ion channels reconstituted into planar bilayers. In: Yagi K, Pullman B (eds) Ion Transport Through Membranes. Academic Press, Tokyo, pp. 131-145

Brückner H, Przybylski M (1984) Isolation and structural characterization of polypeptide antibiotics of the peptaibol class by high-performance liquid chromatography with field desorption and fast atom bombardment mass spectrometry. J Chromatography 296:263-275

Esposito G, Carver JA, Boyd J, Campbell ID (1987) High-resolution ^{1}H NMR study of the solution structure of alamethicin. Biochemistry 26:1043-1050

Fox RO, Richards FM (1982) A voltage-gated ion channel model inferred from the crystal structure of alamethicin at 1.5 Å resolution. Nature 300:325-330

Freedman R (1987) Folding into the right shape. Nature 329: 196-197

Gelfert-Peukert S (1986) Einfluss des Austauschs und der Modifikation einzelner Aminosäuren des Eicosapeptids Alamethicin auf die Eigenschaften des Ionenkanals. Thesis, Ruhr-Universität Bochum

Guy, HR (1984) A structural model of the acetylcholine receptor channel based on partition energy and helix packing calculations. Biophys J 45:249-261

Hanke W, Boheim G, Barhanin J, Pauron D, Lazdunski M (1984) Reconstitution of highly purified saxitoxin-sensitive Na^{+}-channels into planar lipid bilayers. EMBO J 3:509-515

Imoto K, Busch C, Sakmann B, ..., Numa (1988) Rings of negatively charged amino acids determine the acetylcholine receptor channel conductance. Nature 335:645-648

Kyte J, Doolittle RF (1983) A simple method for displaying the hydropathic character of a protein. J Mol Biol 157:105-132

Maelicke A (1988) Structural similarities between ion channel proteins. TIBS 13:199-202

Menestrina G, Voges KP, Jung G, Boheim G (1986) Voltage-dependent channel formation by rods of helical polypeptides. J Membrane Biol 93:111-132

Methfessel C, Boheim G (1982) The gating of single calcium-dependent potassium channels is described by an activation/ blockade mechanism. Biophys Struct Mech 9:35-60

Noda M, Shimizu S, ..., Numa S (1984) Primary structure of Electrophorus electricus sodium channel deduced from cDNA sequence. Nature 312:121-127

Oiki S, Danho W, Montal M (1988) Channel protein engineering: Synthetic 22-mer peptide from the primary structure of the

voltage sensitive sodium channel forms ionic channels in lipid bilayers. Proc Natl Acad Sci USA 85:2393-2397
Sakmann B, Patlak J, Neher E (1980) Single acetylcholine-activated channels show burst-kinetics in presence of desensitizing concentrations of agonist. Nature 286:71-73
Voges K-P, Jung G, Sawyer WH (1987) Depth-dependent fluorescent quenching of a tryptophan residue located at defined positions on a rigid 21-peptide helix in liposomes. Biochim Biophys Acta 896:64-76

STRUCTURAL AND DYNAMIC MOLECULAR MODELS FOR SODIUM CHANNELS

Edward M.Kosower
Biophysical Organic Chemistry Unit,
School of Chemistry, Sackler Faculty of Exact Sciences
Tel-Aviv University, Tel-Aviv 69978, ISRAEL

Abstract

Chemical logic (choice of hydrophobic and amphiphilic domains; electrostatic interactions), genetic homologies and single group rotation (SGR) theory (channel operation) are used to construct molecular models from the primary structures of eel and rat brain sodium channels. Channel structure and dynamics are accounted for with a detailed model for the bilayer portion and amino acid sequences for an outer control element (OCE) (tentatively, the Hodgkin-Huxley m-gate) and an inner control element (ICE) (tentatively, the Hodgkin-Huxley h-gate). The factors treated include gating current, sensitivity to changes in membrane potential (threshold for action potential), channel opening, a binding site for sodium, selectivity for sodium over potassium, capacity for rapid sodium flow, sensitivity to batrachotoxin and inactivation. Genetic homologies between "active" portions of the eel and three rat brain sodium channels support the models. Similar mechanisms but differences in charge distribution occur for the eel and rat channels. Local anesthetic (LA) action on the sodium channel is explained. BTX binds to the LA site. Portions of the ICE are identified as possible analogs to local anesthetics. The model is consistent with the neurophysiological results obtained with anti-Cl^+ peptide antibody. Molecular graphics studies on the central portion of the eel sodium channel model are described.

Introduction

Ion channels are proteins which allow ions to pass through biological membranes, either under ligand control (nicotinic acetylcholine receptor, nAChR) or under voltage control like the sodium channel. Voltage-gated channels are the primary active element in the transfer of depolarization within

NATO ASI Series, Vol. H 32
Molecular Biology of Neuroreceptors and
Ion Channels. Edited by A. Maelicke

neurons to the axon terminal. Channels are named for the ions which are most easily transferred, and include sodium channels, potassium channels, calcium channels and chloride channels, the latter being either ligand gated ($GABA_A$ receptor) or voltage gated.

A deep understanding of how channels produce action potentials (Hille, 1984) or the functioning of other components of the neuron (receptors and other biopolymers) cannot be achieved without knowledge about the molecular structure.(Catterall, 1986a) The sodium channel is a glycoprotein through which sodium enters a neuron down its concentration gradient after activation by a moderate depolarization of the normal membrane potential. The flow of sodium reaches a maximum in less than a millisecond and is usually turned off (inactivated) within several milliseconds. The local depolarization is lost in a somewhat longer time by the operation of an outward flow of potassium ions through a potassium channel. The minor imbalance in the usual intracellular high K^+/ low Na^+ ratio is corrected after a time by the operation of the metabolically driven "Na^+/K^+ pump" (Na^+/K^+ ATPase).

Sodium Channel Sequence

Numa and his coworkers (Noda et al, 1984) have elucidated the primary structures of the sodium channel of the electric eel, <u>Electrophorus electricus</u>, and three sodium channels from the rat brain (Noda et al, 1986a, Kayano et al, 1988). The eel channel is an 1820 amino acid sequence, calculated molecular weight 208,321. The α-subunits of the rat channels have molecular weights of 228,758 (I, 2009 amino acids), 227,840 (II, 2005 amino acids) and 221,375 (III, 1951 amino acids), respectively. The channel isolated from rat brain is accompanied by two subunits, noncovalently bound β_1 and disulfide-bound β_2 (Hartshorne & Catterall, 1984, Hartshorne et al, 1986, Talvenheimo, 1985). A functional rat brain sodium channel can be expressed in <u>Xenopus</u> oocytes using only α-subunit messenger RNA (Goldin et al, 1986, Hirono et al, 1985, Noda et al, 1986b), but channel properties are altered by the subunits (Messner et al, 1986, Krafte et al, 1987).

The eel channel protein was isolated in purified form (Agnew et al, 1978, Nakayama et al, 1982) and sequenced through cDNA. (Noda et al, 1986a)

The Kyoto group found genetic homologies between four sections of the eel protein, pointed out the four unusual positively charged sequences (and the possibility that these might be 3_{10}-helices) and noted the presence of unusual long sequences of negatively charged amino acids. A specific channel was not identified.

Current approach

Beginning with the principle that a charged amphiphilic segment was needed to form an ion channel (Kosower,1982, 1986, 1987), a model for the sodium channel was constructed. (Kosower, 1985) The present article is a revised formulation of the eel sodium channel model, and includes an extension to the rat brain channels I and II and partially, III.

Constructing the model: Folding proteins

Since the currently available rules for folding proteins are problematic (Kosower, 1987, Wallace et al, 1986, Chothia, 1984, Engelman et al, 1986), a variety of information on the shape and folding of the target protein must be used. Hydrophobic segments are usually easy to recognize, either by scanning the sequence or with the aid of a program which assigns "hydrophobicity" indices to individual side chains. Charged groups may be present in some hydrophobic helices, serving to stabilize the overall arrangement, as in the nAChR structural model (Kosower, 1987).

Search for ion channel

The eel sodium channel amino acid sequence does not have an ion channel like that of the nAChR model (Kosower, 1987), but the positively charged sequences noted by Numa have many hydrophobic groups interspersed with the charged groups, and thus are candidate ion channel elements. I placed the four sequences of this type in the bilayer and accepted the suggestion by Numa (Noda et al, 1984) that the repeat period of three amino acids for these sequences could correspond to a 3_{10}-helix. The 3_{10}-helix occurs with less frequency than α-helices in proteins (Schulz & Schirmer, 1979) but is common enough to be a reasonable choice.(Barlow & Thornton, 1988) Stretches of 3_{10}-helix with 20 a.a. have been found in a 30 kD "natural insectide" from Bacillus thuringiensis (Wilke et al, 1987) The 3_{10}-helix has been characterized in the crystal structure of an octapeptide derived from α-aminoisobutyric acid. (Toniolo et al, 1986) After placing all the 3_{10}-helices in the bilayer, it was then necessary to balance the large number of positive charges (20^{+}) now within the bilayer. The only compact sources of a large number of negatively charged groups are the highly unusual neighboring sequences present in the middle of the sodium channel sequence. I took the logical but surprising step of inserting two hydrophilic, negatively charged segments (net 22^{-}) into the bilayer, arranged as 3_{10}-helices on the basis of length (18 amino acids) and as a match for the positively charged helices. The functional part of the channel thus contains six 3_{10}-helices. The ion channel elements, $C1^{+}$, $C2^{+}$, $C3^{-}$, $C4^{-}$, $C5^{+}$ and $C6^{+}$, are listed in Fig.1.

210 arg^{+}	657 arg^{+}	1093 ala	1417 arg^{+}	910 asp^{-}	942 glu^{-}
211 thr	658 ser	1094 ile	1418 val	909 glu^{-}	943 glu^{-}
212 phe	659 leu	1095 lys^{+}	1419 ile	908 glu^{-}	944 glu^{-}
213 arg^{+}	660 arg^{+}	1096 asn	1420 arg^{+}	907 asp^{-}	945 glu^{-}
214 val	661 leu	1097 leu	1421 leu	906 ser	946 glu^{-}
215 leu	662 leu	1098 arg^{+}	1422 ala	905 ser	947 glu^{-}
216 arg^{+}	663 arg^{+}	1099 thr	1423 arg^{+}	904 asp^{-}	948 glu^{-}
217 ala	664 ile	1100 ile	1424 ile	903 val	949 glu^{-}
218 leu	665 phe	1101 arg^{+}	1425 ala	902 leu	950 glu^{-}
219 lys^{+}	666 lys^{+}	1102 ala	1426 arg^{+}	901 gly	951 pro
220 thr	667 leu	1103 leu	1427 val	900 glu^{-}	952 glu^{-}
221 ile	668 ala	1104 arg^{+}	1428 leu	899 glu^{-}	953 glu^{-}
222 thr	669 lys^{+}	1105 pro	1429 arg^{+}	898 glu^{-}	954 leu
223 ile	670 ser	1106 leu	1430 leu	897 ile	955 glu^{-}
224 phe	671 trp	1107 arg^{+}	1431 ile	896 glu^{-}	956 ser
225 pro	672 pro	1108 ala	1432 arg^{+}	895 ser	957 lys^{+}
226 gly	673 thr	1109 leu	1433 ala	894 glu^{-}	958 asp^{-}
227 leu	674 leu	1110 ser	1434 ala	893 gly	959 pro
$^{1}C^{+}$	$^{2}C^{+}$	$^{5}C^{+}$	$^{6}C^{+}$	$^{3}C^{-}$	$^{4}C^{-}$

Fig.1 Ion channel elements

The homology between the eel and rat (I/II/III) (+)-sequences is extraordinarily good (94.4-100%), counting both identity and functionally plausible substitutions (FPS, (+) --> (+), (-) --> (-), hydrophobic --> hydrophobic) (Noda et al, 1984). The eel and rat (I/II/III) (-)-sequences are more difficult to compare, but local homologies are at least equal to that of the overall channel homology.

Hydrophobic bilayer helices

A sufficiently long sequence of hydrophobic amino acids will be more stable within the hydrophobic phospholipid bilayer rather than the polar environments of the cytoplasm or cell exterior. The sequence will fold in a way that will minimize the interaction of the polar peptide links with the apolar milieu, with an α-helix a highly preferred choice. Such hydrophobic sequences are therefore presumed to form bilayer α-helices. Helix-end criteria (Richardson & Richardson, 1988) have not been tested.

The genetic homology discovered by Numa between four sections of the protein, and confirmed (Greenblatt et al, 1985) is taken into account, some polar groups perhaps stabilizing the hydrophobic helices (Gray and Matthews, 1984). Twenty hydrophobic helices are proposed, a number greater than originally suggested (Kosower, 1985). A role for salt and/or polar bridges in stabilizing membrane protein structure within the bilayer has been suggested for the nAChR; some charged groups are found in the bilayer region of rhodopsin.(Kosower,1988) The hydrophobic helices are similar to, but not the same as those of Numa(Noda et al,1984). The sequences assigned to the hydrophobic helices are summarized in the compact form shown later. The total number of bilayer helices (ion channel elements plus hydrophobic helices) in our model is 26. In the nAChR model, there are 25 bilayer helices for 2333 amino acids.

There is substantial genetic homology between the eel and rat(I/II) hydrophobic segments, with the sum of identity (I) and FPS amino acids being for most helices between 83 and 100%. The average I-homology is ca 62% (Noda et al, 1986a), although it is 87% between the rat I/II channels. A rough count of FPS in the amino acid sequences of the rat with respect to the eel suggests that the I + FPS homology for the eel/rat comparison might be 74%. (The homology between III and the others is high)

Arrangement of ion channel elements

The constraints on the local arrangements are clear. The (+)-elements are amphiphilic and therefore must be in contact with the more hydrophobic bilayer helices of the protein. The geometry of the (+)-elements is defined by the spacing of the (+)-charges and the proposition that all should be pointed in the same directions. A 3_{10}-helix satisfies this criterion. These criteria imply that the hydrophilic (-)-elements must lie within a region surrounded by (+)-elements. To match the charge spacing of the (+)-helices, it is presumed that the (-) elements are also 3_{10}-helices.

A square arrangement of the amphiphilic (+)-helices minimizes (+)-(+) repulsion and yields an internal space just large enough to admit the completely polar (-)-helices. The interaction of the the (+)-helices with the (-) helices is maximized if the charged (mostly arginine) groups are placed between the pairs of (-) groups. The 3_{10}-helices are oriented ("outer" portion) so that the least charge faces in the direction of the hydrophobic helices. The amino acids located on the outer and inner sides (A and B) of the (-)-helices as well as those on the inner side of the (+)-helices (the side oriented towards the center) are shown in Fig.2.

To minimize the number of charges in the region facing the hydrophobic portion of the channel protein, the inner sets of (-) groups are apposed to one another. The charge distributions for this arrangement are shown level by level on pentagons in Fig.2. The presentation in Fig.2 requires a word of explanation. The 3_{10}-helices (18 amino acids to span the bilayer) have been divided into 6 "layers". The charges in each "layer" have been projected onto a plane. Since there are 10 charge-carrying points per "layer", two apposed pentagons are a natural way to present the charge distribution in each layer. The average net charge on any side of the combination through the bilayer ranges from -2 to +1. A significant feature of the arrangement is that the net charge at a given level of the channel changes from negative (level 1) at the top of the bilayer to neutral at all other levels, the net charges being -2, 0,0,0,0,0.

The layer charges in the rat channels I and II are: Rat channel I, 0, +4, +1, +1, +2, 0 and -1; rat channel II, 0, +3, 0, +3, -1 and 0. A search was then made for mutations within the eel channel protein which replaced neutral amino acids with charged amino acids. There are 6 such mutations involving both rat I and rat II within or right next to segments assigned as bilayer helices. Rat III mutations correspond. [Eel numbering 121 N --> HHH

(H1), 146 A --> DDD (H2), 628 Q --> EEE (H8), 1075 G --> DDD (H13), and 1321 Q --> RKR]. Taking into account the probable disposition of the helices, and the change 1025 W --> DEE which occurs immediately under H11, there is an increase of between 2 and 4 negative charges in the rat I and II channels (and probably in III) within the bilayer near the + charges of the channel.

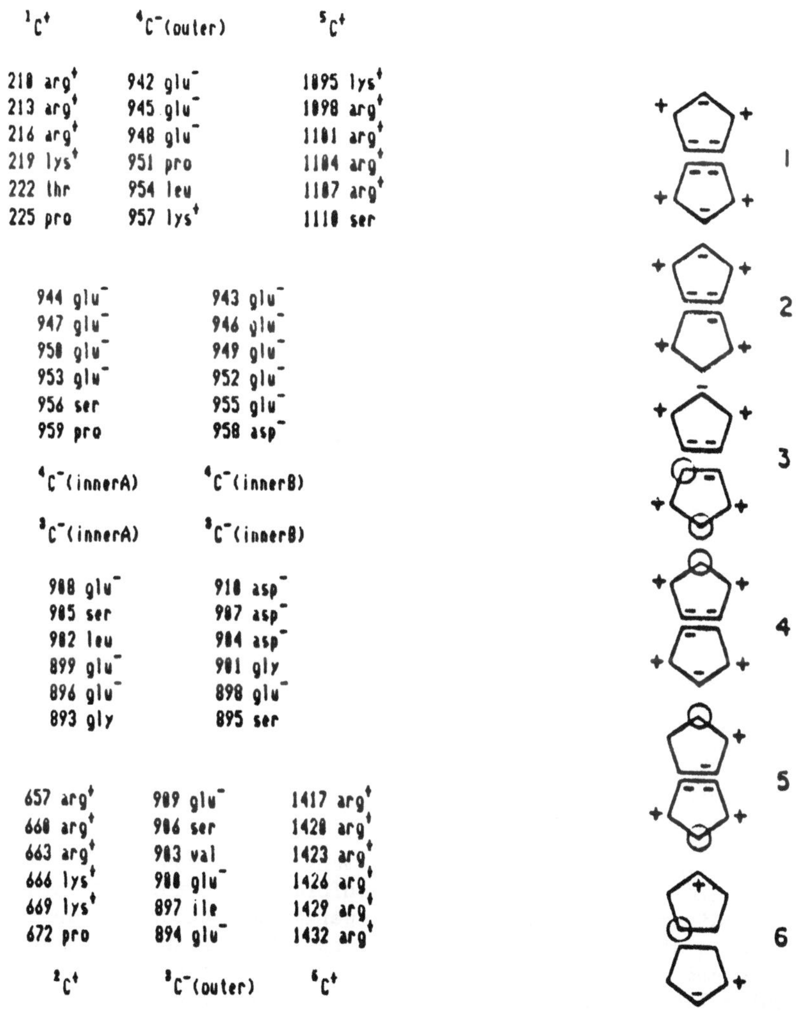

Fig.2 Ion Channel Active Amino Acids and Charge Distributions
Reprinted, with permission, from Kosower,E.M., The Molecular Basis of Learning and Memory, Princeton University Press, 1989
Sodium binding site

I applied SGR to the topmost set of groups on the channel elements, and found that, after single group rotation (SGR) of the (-) groups of the inner portion, a plausible binding site for sodium was created. (Fig.3) The site might discriminate (13/1?) in favor of the larger hydrated ion (Na^+) over the smaller (K^+) ion.

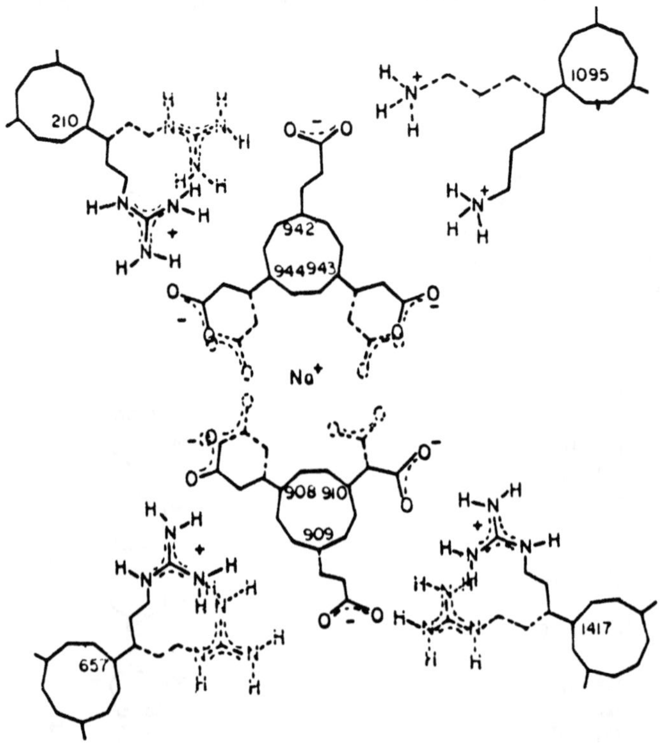

FIRST LEVEL
SODIUM CHANNEL

Reprinted, with permission, from Kosower,E.M., The Molecular Basis of Learning and Memory, Princeton University Press, 1989

Fig.3 A representation of the level 1 arrangement of the sodium channel. The two extreme single group rotation (SGR) conformations for arginine and glutamate are indicated with dashed lines superimposed on solid lines. Two remarkable features are readily discerned. First, SGR of the glutamate side chains away from positions near the guanidino groups of arginine leads to an arrangement which looks plausible for binding a cation. Second, the guanidino groups of the arginine may interact with either the inner or outer (see Fig.2) carboxylate groups according to the SGR conformation.

The bilayer helices may be arranged around the central channel elements, with the four (+) helices around the (-) helices according to (a) connectivity, with helices separated by short peptide links being near one another and (b) charge, with helices bearing negative charge being placed near the (+) helices. A schematic overall arrangement of the bilayer helices of the sodium channel is shown in Fig.4. A compact representation showing

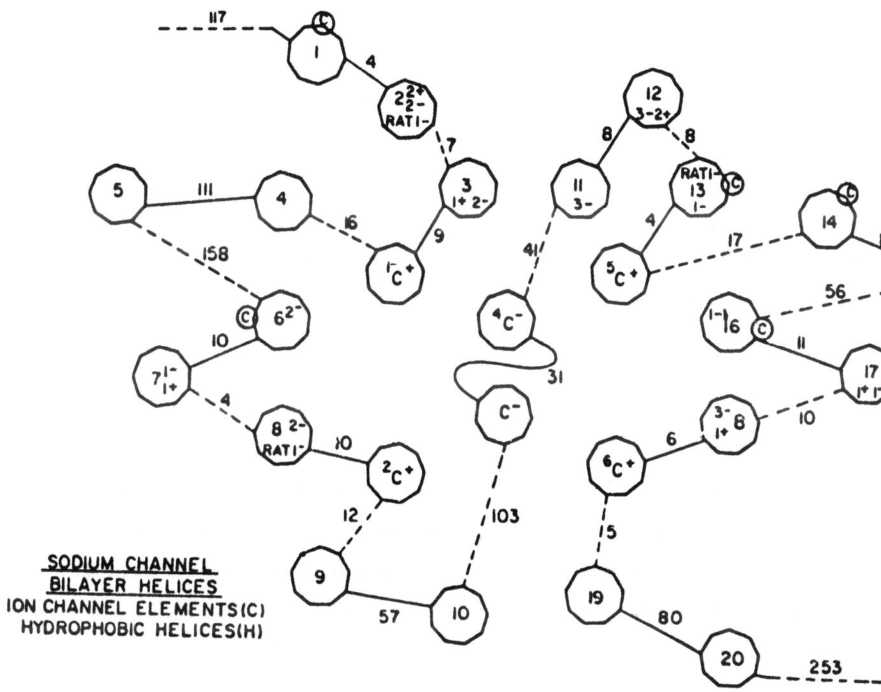

Fig.4 Schematic arrangement of the eel ion channel elements (Fig.1) and the hydrophobic helices (1-20). The solid lines are connections between elements or helices outside the bilayer while dashed lines represent connections on the cytoplasmic side. The connecting links contain the indicated number of amino acids. The signs on the channel elements indicate the predominant charge on a given 3_{10}-helix. It is interesting that the top of the third channel element is amino acid 910, exactly halfway through the 1820 amino acid sequence of the channel protein.
Reprinted, with permission, from Kosower, E.M., The Molecular Basis of Learning and Memory, Princeton University Press, 1989

the sequence numbers for the beginning and end of each bilayer helix is shown in Fig.5. Seven of the potential N-glycosylation sites [asn-X-ser (thr)] (X not pro) at 205, 278, 288, 317, 591, 1160 and 1174 are outside the bilayer. The molecular weight of the glycosyl portion of the channel is ca 60,800. (Messner & Catterall, 1985)

The additional negative charges present in the bilayer portions of the rat sodium channel as compared with the eel are in helices 2, 8 and 13, with another negative charge near helix 11. A positive charge has been added to helix 16. The extra charges are almost all in positions suitable for interaction with the central channel.

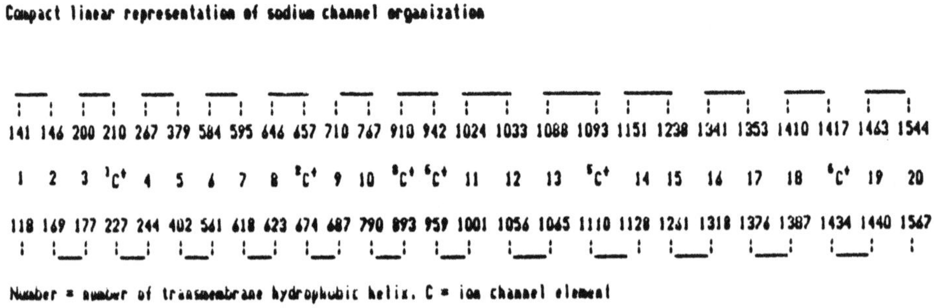

Fig.5
Reprinted, with permission, from Kosower,E.M., The Molecular Basis of Learning and Memory, Princeton University Press, 1989

The N- and C-termini of the sodium channel are almost certainly intracellular based on (a) intracellular side localization of the C-terminus by gold-labeled antibody to the eel 1783-1794 region (Gordon et al, 1987) (b) no leader peptide in cDNA sequence (Noda et al, 1984) and (c) probable four-fold symmetry of protein (Noda et al, 1984, Greenblatt et al, 1985).

Channel dynamics

Conductance of the sodium channel (3-9pS, 1 picoSiemen = 10^{-12} $ohms^{-1}$) is 1000 times greater than expected for ion carrier transport.(cf.Catterall, 1986a) The transport rate of Na^+ through the sodium channel, ca 10^7 ions/sec, is within an order of magnitude of that in water. At least two additional protein segments are required to control the ion flow, one outside the bilayer (outer control element or OCE), a second inside the cell (inner control element or ICE).

Outer control element (OCE)

The outer control element (OCE) closes an otherwise open binding site and channel, a logical choice being the 31-amino sequence connecting the (-) ion channel elements. (Fig.6) This sequence contains groups of a charge type and position (941 lys$^+$ and/or 913 lys$^+$, 915 lys$^+$) which can block the opening to the channel and maintain a closed state. It is possible that the OCE corresponds to the m-gate (Hodgkin and Huxley, 1952).

OCE

934 ser ----933 pro-----932 ser---- 931 tyr------930 asp$^-$

935 glu$^-$	917 ala--------918 leu		929 val
936 gln	916 his	919 asn	928 thr
937 asp$^-$	915 lys$^+$	920 asp$^-$	927 ser
938 pro	914 lys$^+$	921 glu$^-$	926 cys
939 leu	913 lys$^+$	922 asp$^-$	925 val
940 ala	912 asn	923 ser ---	924 ser
941 lys$^+$	911 thr		

ICE

791 ser	811 gln	831 thr	851 gly	871 glu$^-$	891 val
792 ser	812 val	832 val	852 glu$^-$	872 lys$^+$	892 lys$^+$
793 phe	813 ala	833 gln	853 glu$^-$	873 ile	
794 ser	814 ser	834 ala	854 gly	874 val	
795 ser	815 glu$^-$	835 leu	855 asp$^-$	875 asp$^-$	
796 asp$^-$	816 arg$^+$	836 val	856 asn	876 gly	
797 asn	817 ala	837 leu	857 glu$^-$	877 ile	
798 leu	818 ser	838 trp	858 gly	878 thr	
799 ser	819 arg$^+$	839 ile	859 lys$^+$	879 asn	
800 ser	820 ala	840 gln	860 lys$^+$	880 cys	
801 ile	821 lys$^+$	841 gly	861 asp$^-$	881 val	
802 glu$^-$	822 asn	842 lys$^+$	862 thr	882 glu$^-$	
803 glu$^-$	823 trp	843 lys$^+$	863 leu	883 ser	
804 asp$^-$	824 val	844 pro	864 pro	884 pro	
805 asp$^-$	825 lys$^+$	845 pro	865 leu	885 thr	
806 glu$^-$	826 ile	846 ser	866 asn	886 leu	
807 val	827 phe	847 asp$^-$	867 tyr	887 asn	
808 asn	828 ile	848 asp$^-$	868 leu	888 leu	
809 ser	829 thr	849 val	869 asp$^-$	889 pro	
810 leu	830 gly	850 val	870 gly	890 ile	

Fig.6 Outer control element at entrance to sodium channel (OCE) and sequence of inner control element (ICE). The OCE is the segment between C3$^-$ and C4$^-$. The ICE is contained in the segment between H10 and C4$^-$.

With a normal membrane potential, (+) groups (probably 2) from the OCE are associated with the level 1 groups of the sodium channel. Comparison of the eel and rat channels suggests that 913 and 915 lys$^+$ in the eel OCE correspond to (eel numbering) 915 and 917 lys$^+$ in the rat OCE. The rat OCE is somewhat shorter and carries much more negative charge than the eel OCE (Δ -5), possibly to counterbalance the decrease in negative charge present in the (-) channel elements (Δ +3).

Inner control element (ICE)

Once the channel is "activated", some closure mechanism is required. Since inactivation is affected by intracellular proteolysis (below), a protein segment, called the inner control element (ICE) is a suitable candidate for the function of activation. The ICE may correspond to the h-gate (Hodgkin and Huxley, 1952).

A decrease in membrane potential would allow (-) groups (e.g., 937 asp- and/or 920 asp-, 922 asp-) to compete for the (+) groups which block the channel. A likely choice for the ICE would be the sequence between 793 ser and 892 lys^+. The sequence includes a substantial number of (-) groups (18- versus 10^+), especially such sequences as those between 802 glu- and 806 glu-(5(-)/5 a.a.) and 847 asp- and 857 asp- (6(-)/11 a.a.) (Fig.6). A possible additional function of the ICE is to serve as a cationic blocking agent, like a local anesthetic.

Channel activation by depolarization

At a depolarization of membrane potential near the "threshold", the OCE moves away and opens up the binding site, allowing Na^+ to flow into the cell. The change in local ionic concentration arising from the flow of Na^+ within the cell (either from a neighboring sodium channel or from a ligand-gated channel), diminishes the attraction of the interior for the OCE. The attraction is mediated via (+) groups moving up and (-) groups moving down, with the outer channel region more strongly affected because of a lower dielectric constant.(Fig.7) The inactivation process involves interaction of the inner control element (ICE) with the exit region of the channel.

Gating current

The motion of the channel group charges can account for the gating current (estimated as equivalent to the transmembrane motion of at least 4 charges, or 144D for a 36A bilayer) (Armstrong & Bezanilla, 1974, Kuffler et al, 1984, Stimers et al, 1985, Bezanilla, 1985) which precedes channel open-

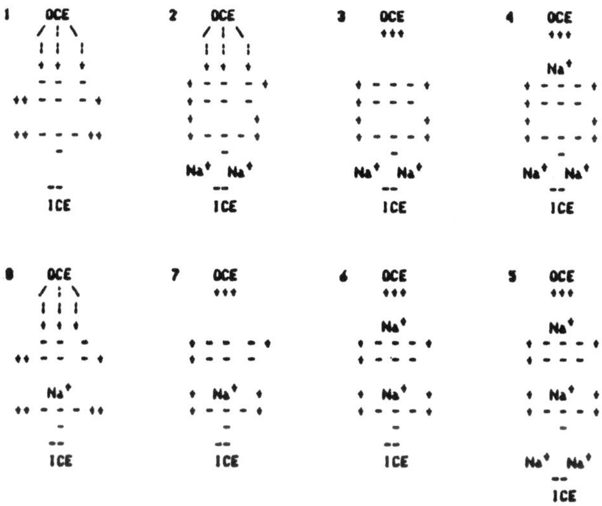

Fig.7 Schematic for channel operation: Dynamics of sodium channel as displayed schematically with the charges from the channel elements, the OCE and the ICE.

ing. (French & Horn, 1983) Charge motion in the eel model channel could involve SGRs of 20^+ over 4.5A (arg^+ or lys^+) and 22- over 2.5A (glu- or asp-), corresponding to 150D, in agreement with the estimate from gating current. The corresponding figures for the rat I and rat II model channels are 125D and 128D respectively, whereas 300D (8.5e-, 36A) is derived indirectly from rat brain channel II measurements on protein expressed in Xenopus laevis oocytes.(Stühmer et al, 1987) There is a substantial variation in the number of charges reported to be transferred across the membrane. The motion of other charges within the bilayer has not been included in the estimates for the charges transferred in the models. Some caution is thus required in making and using these comparisons. The release of the OCE leads to the exposure of a potential ion-binding site.

Channel states

At the binding site, arginine (+) charges can compete with Na^+ for the (-) charges of the ligand groups. If ligand binding interactions decrease, the Na^+ ion can move to the next lower level. Inspection of the charge distributions shown on the pentagons in Fig.2 show that levels 2, 3, 4 and 5 in the eel channel have no net charge. Motion of the ions would not be retarded through most of the channel. Level 6 has a net +1 charge, which may be influenced by (-) charges in the cytoplasmic portion of the channel.

A maximum of 3 Na^+ ions can occupy the channel at levels 1, 3 and 5. The fully occupied eel channel will have a net 1+ charge, and can attract the ICE which then reverses the charge displacements that led to opening of the channel via dissociation of the OCE. The ICE acts by helping to return the OCE to its initial state. An equation (1) summarizes the channel states. A plausible name for the intermediate state with ICE closed and OCE open is "anesthetized".

Channel states

```
                  OCE         OCE
      OCE                                 OCE
       C   -->     C   -->     C   -->     C
(1)                           ICE         ICE
                                           ||
      ICE         ICE                      ||
       ||__________________________________||
    resting      open    anesthetized  inactivated
    (closed)
```

OCE = outer control element; C = sodium carrying portion of channel; ICE = inner control element

Toxin effects and toxin classes

There are four types of sodium channel toxins, channel blockers [tetrodotoxin (TTX), saxitoxin (STX)], channel activators [batrachotoxin (BTX), grayanotoxin (GTX), veratridine, aconitine], channel inactivation blockers [α-scorpion toxin (North African Leiurus quinquestriatus), sea anemone toxins] and an activation enhancer [β-scorpion toxin, American Centuroides sculpturatus]. (Zell et al, 1985, Catterall, 1980)

Channel blockers: Tetrodotoxin and saxitoxin

A chemical analogy exists between the guanidino groups of tetrodotoxin

(TTX) and saxitoxin (STX) and the arginine groups of the channel. A plausible blocking site for STX and TTX is the binding site for Na^+. More information is needed on the other parts of the TTX binding site of the channel protein. The active form of the sodium channel with a high affinity for saxitoxin can be stabilized by intramolecular cross-linking with carbodiimides. (Tejedor et al, 1988)

Channel activators: Batrachotoxin

We have proposed that the compounds of the BTX- channel activator group have a constellation of oxygen atoms ("oxygen triad") suitable for complexation with the ϵ-ammonium ion of a lysine side chain(Kosower, 1983) The oxygen-oxygen distances in the triad vary within the group from 2.26 to 4.97A.

Externally added BTX does not activate the sodium channel immediately, implying that the BTX must act on the interior of the bilayer (Hille,1984). Binding to the local anesthetic site would prevent the binding of the ICE and thus prevent inactivation. Association of the oxygen triad with a (+)-group at the lower end of a (+)-charged helix of the channel (eel 1432 arg^+ becomes lys^+ in both rat channels I and II) could enhance the binding of the BTX, while the proximity of the BTX-binding site to the channel might lower the conductivity by partial occlusion. Keeping a lysine away from the channel could lower the rate of inactivation and might account for the shift of activation to more negative voltages with the BTX group of compounds (Leibowitz et al, 1986, Sutro, 1986) since one (+)-charge is already "neutralized" by the BTX. The LA site is occupied by the ICE or by BTX, so that gating charge is still immobilized. If the LA site is not occupied as happens as a result of pronase treatment, gating charge is no longer immobilized.(Tanguy et al, 1987) The cooperative effect of BTX-class compounds (BTX and veratridine) acting at the LA site and polypeptide toxins which complex with the OCE (Corbett et al, 1987, Garber & Miller, 1987) is readily explained since different sites are involved. The use of BTX in the study of sodium channels has been reviewed. (Khodorov, 1985) Another "oxygen triad" toxin, α-dihydrograyanotoxin II opens sodium channels from the inside of the squid axonal membranes. (Seyama et al, 1988)

Toxin effects on channel selectivity

Both equilibrium and kinetics play a role in determining the selectivity of the channel. The permeabilities to various ions in neuroblastoma cells and skeletal myotubes (Na^+: K^+: Rb^+: $(NH_2)_2C{=}NH_2{}^+$: $CH_3NH_3{}^+$, 1: 0.086: 0.012: 0.13: 0.007) change in the presence of channel-opening toxins (1: 0.4: 0.15: 0.65: 0.14). (Frelin, 1981) Using a simple kinetic model for the passage of ions, the amount of cation delivered via the channel to the interior of a cell would be determined by the ratio of the rate at which ions pass through the channel and the rate at which inactivation occurs times an equilibrium constant for ion binding. The more slowly the ion moves through the channel, the greater is the inactivation rate. In the presence of toxins, the OCE-toxin complex forms, inactivation does not occur, and the "intrinsic rates" of cation diffusion through the channel would control the amount of cation going through the membrane. (eq.1)

Effects of other agents

The removal of sodium channel inactivation by the internal perfusion of squid axons with N-bromoacetamide has been ascribed to cleavage of a peptide link next to tyrosine.(Oxford et al, 1978) Cleavage of the ICE at 867 tyr would interfere with the conformational change needed to initiate inactivation. Internal perfusion of squid axons with pronase removes sodium channel inactivation (Armstrong et al, 1973) as expected for the loss of an effective ICE. Pronase is an enzyme mixture in which the main activity is due to alkaline proteinase b, an endopeptidase similar in activity to trypsin, with cleavage expected at the peptide bond of arginine or lysine. The effect of pronase would be due to the removal of a cationic peptide, and was shown to be replaced by another cation (Rojas & Rudy, 1976). The reappearance of inactivation (lost after pronase treatment) through introduction of polyglycylarginine (Eaton et al, 1978) or octylguanidinium ion (Oxford & Yeh, 1982) suggests that a polypeptidic or hydrophobic guanidinium ion can simulate the action of the ICE. In the model, the 872 lys^+ is closest to the exit of the channel and is a candidate for the proposed positively charged amino acid, as is 825 lys^+. It is interesting that the 872 lys^+ is not far from the 867 tyr mentioned above. The effects of 1-propylguanidinium ion in decreasing peak Na^+ conductance have been adduced in support of a role for an arg^+ group in inactivation (Lo & Shrager, 1984a,b) There are two arg^+, 816 and 819, in the ICE.

Antibody effects: Antipeptide antibodies

Antibodies against solubilized sodium channels are species-specific and not physiologically active. Nevertheless, antibodies against native eel electroplax membranes which block mammalian nerve conduction have been generated. (Meiri et al, 1984, 1986)

Peptides selected from the eel channel sequence have been used to generate antibodies which are neurophysiologically effective against dissociated rat ganglion (DRG) cells. In particular, an anti-$C1^+$ antibody generated against the eel channel 210-223 sequence, RTFRVLRALKTITI, shifted the steady state sodium channel inactivation parameter, h_∞, to more negative potentials and changed the kinetics of activation (Meiri et al, 1987). That an anti-$C1^+$ antibody based on the eel channel 210-223 sequence was active against mammalian sodium channel was not surprising in view of the almost complete homology of these sequences between eel and rat.

In addition to the FRV (1416-1417-1418) triplet previously noted (Meiri et al, 1987), and the triplets of the $C1^+$ itself, two additional contiguous triplets might be epitopes to the polyclonal antibodies generated against $C1^+$-peptide. One is VLR, 655-656-657. The two first amino acids are outside the assigned region for $C2^+$. The second is LRA, 371-372-373. In the model, this connects bilayer helices 2 and 3. Several other triplets were found but are considered to be too buried in the bilayer for strong consideration.

Anesthetics

Anesthetics (Greek, <u>anaesthetos</u>, insensible) clearly decrease perception, either generally or locally, but their actions are difficult to define in physical terms.(Woodbury et al, 1975) The major target for local anesthetics appears to be the sodium channel (Hille, 1980, Strichartz, 1987), but the structural diversity of the compounds which can induce general anesthesia certainly precludes a single receptor as the target (Richards et al, 1980, Roth, 1979) Anesthesia is a reversible, drug-induced perturbation of neuronal behavior, arising from a complex combination of many effects.(Roth, 1979)

Local anesthetics

Local anesthetics inhibit the flow of ions through sodium channels more effectively in the activated than in the resting state. (Hille, 1977) The quaternary derivative (QX-314) of the local anesthetic, lidocaine, acts only on the inside of the squid axon and not when added on the outside. By inference, the site for the action of local anesthetics is at the intracellular exit of the sodium channel, a location which contains a reasonable binding place for local anesthetics.

Molecular graphics suggests that a binding site on the central part of the sodium channel of moderate specificity binds local anesthetics through a combination of ionic and hydrophobic interactions. An aryl group on the bound molecule physically blocks the main pathway for sodium ions to exit from the channel and pass into the cell. Proximity of the BTX and LA binding sites would lead to anesthetic-promoted dissociation of bound BTX.(Postma & Catterall, 1984, Creveling et al, 1986) The Na^+ channel is activated by BTX, but combination with an LA promotes the dissociation of BTX. The ability of LA to compete with BTX for binding to sodium channel is an index of the effectiveness of the LA (McNeal et al, 1985) The specific location implied for the LA is in accord with a variation in the effectiveness of use-dependent sodium channel block with LA structure.(Courtney, 1980, 1984) The competition of BTX with LA is consistent with the idea of a common binding site.

The ICE as the natural anesthetic

There is some resemblance between local anesthetics (most of which would be protonated at neutral pH) and the cationic molecules which replace the proteolyzed intracellular portions of the sodium channel responsible for inactivation. In analogy to the premise that opiate peptides are the natural analog of morphine, the ICE might be the natural analog of local anesthetics. The appropriate combination seems to be a lys or arg followed or preceded by a hydrophobic amino acid. These combinations are preserved from eel to rat for 816 arg^{+}-817 ile (rat, met), 825 lys^{+}-826 ile (824 val), 843 lys^{+}-844 pro and 872 lys^{+}-873 ile (rat, 871 leu). A relationship between blocking by quaternary LA and inactivation has been pointed out (Cahalan & Almers, 1979).

Prolines: Distribution

We note that two (eel) or three (rat I/II) intracellular regions of the eel sodium channel have unusually high proline contents. In the eel (54 - 63, 4 pro/10 a.a.), rat I/II (56-64, 3 pro/9 a.a.), rat (1300-1307, 4 pro/8 a.a.), eel (1619-1625, 3 pro/7 a.a.), rat I (1618-1625, 4 pro/8 a.a.), rat II (1618-1625, 3 pro/8 a.a.) and eel (1772 - 1788, 5 pro/17 a.a.). Each of the channels has a total of 74 prolines (less than 5% of the total number of amino acids). The identified proline-rich regions have less than 1.5% of the total a.a., but have 9/27 (eel), 11/25 (rat I) and 10/25 (rat II) prolines, and serve as anchors to collagen like proteins.(Kosower, 1987)

Other models

A "sliding helix model" (Catterall, 1986b) suggests that the positively charged α-helixes spiral upwards on depolarization by 5A and 60°. The (+)-charges are matched by (-) charges in other unspecified α-helixes, the latter remaining more or less in place. Each (+)-helix would move more or less independently. Although the proposal formally accounts for the gating current, it leaves unclear how and where the Na^{+} ions would traverse the channel. In addition, the large scale motions involved might not be consistent with the speed of sodium channel response to depolarization (20-50 microsecond range, with peak currents being reached within 1 msec).

Another model (Greenblatt et al, 1985) does not have a clear mechanism for channel operation and control. The SGRs of the charged groups are ignored, and the charge motion required to explain the gating current seems too large unless some substantial motions of the helices occur. A third model, either in the original or as modified (Guy & Seetharamulu, 1986, Guy, 1987, 1988) lacks suitable control mechanisms and requires large scale "screwlike" motion of an α-helix carrying positive charges. Both the second and third models rely very much on protein folding predictions.

Antibody experiments (Meiri et al, 1987, Rosenthal et al, unpublished) favor the present model as the best working hypothesis for further experiments. Antibody to eel 930-941 (corresponding to the OCE) produced labeling primarily on the cytoplasmic face in cryostat sections of eel electroplax (Gordon et al, 1987) As we have pointed out in connection with antibody labeling experiments on the nAChR (Kosower, 1987), hydrophilic peptides within the membrane can leave the membrane easily. In the case of our model, the LIFO ("last in, first out") principle would be particularly important. Release of natural constraints would expose the channel might allow the exit of the negatively charged segments into the cytoplasm together with the dissociation of the OCE. The reverse process would occur during assembly of the channel. The contrast between the neurophysiological results on intact nerve (in which a similar but monoclonal antibody against the eel 930-942 peptide was active against frog sciatic nerve when applied from the external surface of the membrane) (Rosenthal et al, unpublished) and the antibody experiments (Gordon et al, 1987) is striking and explicable.

Conclusions

The sodium channel model proposed here is a remarkably balanced and beautiful structure, which deserves the comment, "Si non e vero, e ben trovato". The structural features discovered with the model may well deserve incorporation into a synthetic model. In addition, these features are a chemical expression of a new physical model for channel gating(Finkelstein & Peskin, 1984) and define a molecular basis for local anesthetic blocking of the open channel (Starmer et al, 1984).

A Molecular Graphical Model of the Active Portion of the Eel Sodium Channel (collaboration with G.A.Petsko and R.L.Feldmann)

The manipulation of models is a required step in examining and developing the structural, spatial and dynamic relationships of molecules. Biologically important molecules like the sodium channel are so large that a physical model would be prohibitively large and expensive, and even if one were built, it would be too unwieldy for detailed analysis. In addition, looking at internal segments of large molecules would be difficult; moving the segments or small ligand molecules around next to impossible. The development of computer graphics for molecules, now commonly referred to as molecular graphics, has brought to fruition the treatment of large molecules like proteins. (Venkataraghavan & Feldmann, 1985)

We have examined only the central, ion-binding and ion-transporting portion of the eel sodium channel, checking the relationships between the channel side chain groups, the adequacy of the proposed sodium binding site and searching for a possible local anesthetic binding site.

The amino acids constituting the channel elements $C1^+$, $C2^+$, $C3^-$, $C4^-$, $C5^+$ and $C6^+$ were incorporated into 3_{10}-helices and arranged with the hydrophobic (+)-helices outside and the (-)-helices inside. A projection through the array in a direction parallel to the bilayer helices is shown in Fig.8, with the central + representing a sodium at the binding site. To test the transient neutrality of the channel groups, we rotated each positively charged side chain around its bond to the polypeptide so as to bring the charged end close to the oppositely charged end of a negatively charged group. (The local conformational change is exactly what we have referred to elsewhere as single group rotation or SGR) The distance over which the guanidino end of an arginine can move with an SGR is ca 4.5A; the corresponding distance for a glutamate side chain is ca 2.5A. It was found that every charged group in the channel could "communicate" with a countercharged group, consistent with the idea of the model.

The only local anesthetic for which a crystal structure was available was cocaine (Gabe & Barnes, 1963, Hyrnchuk & Barton, 1983), a classical local anesthetic (LA) introduced by Freud (Van Dyke & Byck, 1982) but

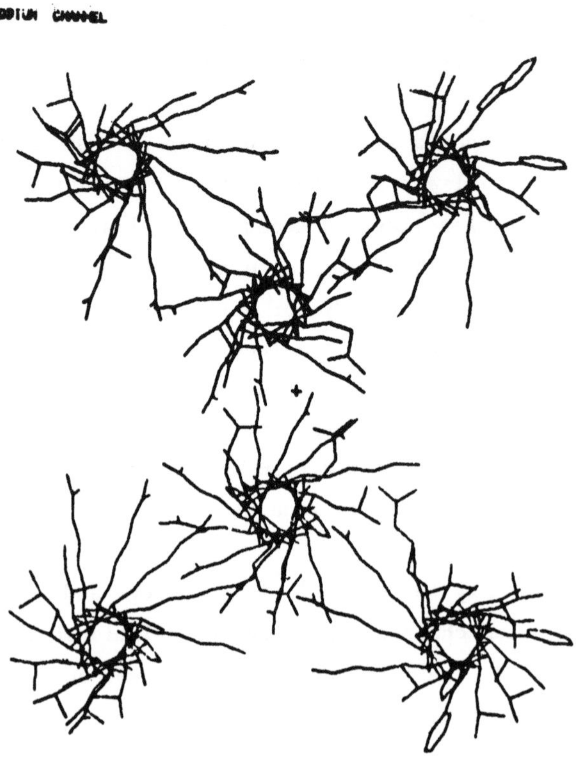

Fig.8 A "stick" model of the sodium channel,looking down through the central part of the sodium channel from the outside. Almost every side chain on the two central helices carries a carboxylic acid group (glutamate), while most of the chains which project towards the center are guanidino (arginine). The (+) sign in between the two central helices represents a Na+ ion. An E & S PS340-VAX 750 combination at M.I.T. was used to develop the model using the FRODO program (v6.0). Space-filled, stereo color models (not shown) were executed from the coordinates on the Apollo-Iris network at the N.I.H. (Collaboration with G.A.Petsko and R.J.Feldmann)
Reprinted, with permission, from Kosower,E.M., The Molecular Basis of Learning and Memory, Princeton University Press, 1989

abandoned because of addiction, the last presumably as a consequence from binding to other sites. (Banerjee et al, 1987) We brought the positively charged LA molecule to a readily available negatively charged group on the cytoplasmic side of the channel. With relatively little adjustment, the structure of a NaC.LA complex produced a surprise, that the phenyl group of the LA blocked the ion pathway. We thus explain local anesthesia. The mechanism would apply to all of the common LA each of which has a phenyl group in approximately the same position. Additional binding interactions are not excluded. We have inserted the LA, lidocaine, into a side view of the central part of the eel sodium channel to illustrate in a general way the position of the phenyl group in the ion pathway. (Fig.9)

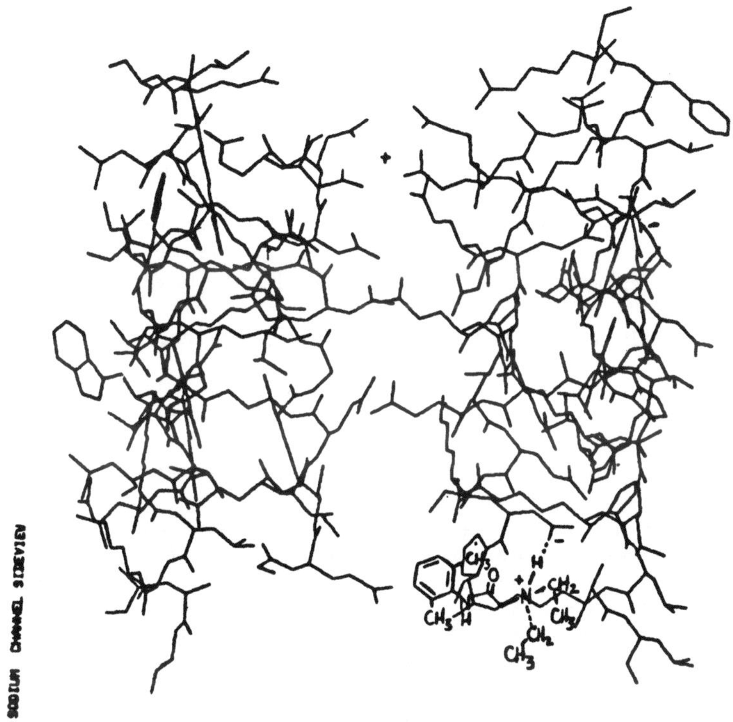

Fig.9 A side view of a "stick" model of the sodium channel, looking at the central part of the sodium channel. The (+) sign at the top between the two central helices represents a Na^+ ion within the sodium binding site to which has been added a molecule of lidocaine, a local anesthetic.

Reprinted, with permission, from Kosower,E.M., The Molecular Basis of Learning and Memory, Princeton University Press, 1989

References

Agnew WS, Levinson SR, Brabson JS, Raftery MA (1978) Purification of the tetrodotoxin-binding component associated with the voltage sensitive sodium channel from Electrophorus electricus electroplax membranes. Proc. Nat'l. Acad. Sci. USA 75:2606-2610

Armstrong CM, Bezanilla F (1974) Charge movement associated with opening and closing of the activation gates of the Na channels. J. Gen. Physiol. 63:533-552

Armstrong CM, Bezanilla F, Rojas E (1973) Destruction of sodium conductance inactivation in squid axons perfused with pronase. J. Gen. Physiol. 62:375-391

Banerjee DK, Lutz RA, Levine MA, Rodbard D, Pollard HB (1987). Uptake of norepinephrine and related catecholamines by cultured chromaffin cells: Characterization of cocaine-sensitive and -insensitive plasma membrane transport sites. Proc. Nat'l. Acad. Sci. USA 84:1749-1753

Barchi RL (1982) Biochemical studies of the excitable membrane sodium channel. Int'l. Rev. Neurobiol. 23:69-101

Barlow DJ, Thornton JM (1988) Helix geometry in proteins. J. Mol. Biol. 201: 601-619

Bar-Sagi D, Prives J (1985) Negative modulation of sodium channels in cultured chick muscle cells by the channel activator batrachotoxin. J. Biol. Chem. 260:4740-4744

Bezanilla F (1985) Gating of sodium and potassium channels. J.Membrane Biol. 88:97-111

Cahalan MD, Almers W (1979) Interaction between quaternary lidocaine, the sodium channel gates and tetrodotoxin. Biophys. J. 27:39-56

Catterall WA (1986a) Molecular properties of voltage sensitive sodium channels. Ann. Rev. Biochem. 55: 953-985

Catterall WA (1980) Neurotoxins that act on voltage-sensitive sodium channels in excitable membranes. Ann. Rev. Pharmacol. Toxicol. 20:15-43

Catterall WA (1986b) Voltage-dependent gating of sodium channels: correlating structure and function. Trends in Neurosci. 9:7-10

Chothia C (1984) Principles that determine the structure of proteins. Ann. Rev. Biochem. 53:537-572

Corbett AM, Zinkand WC, Krueger BK (1987) Activation of single neuronal sodium channels by veratridine and polypeptide neurotoxin in planar lipid bilayers. Biophys. J. 51:435a

Courtney KR (1980) Structure-activity relations for frequency-dependent sodium channel block in nerve by local anesthetics. J. Pharm. Exp. Therap. 213:114-119

Courtney KR (1984) Size-dependent kinetics associated with drug block of sodium current. Biophys. J. 45:42-44

Creveling CR, Lewandowsky GA, Daly JW (1986) Effect of local anesthetics on the dissociation rate of the batrachotoxin-sodium channel complex. Neurosci. Abstr. 16:1511

Daly JW, Myers CW, Warnick JE, Albuquerque EX (1980) Levels of batrachotoxin and lack of sensitivity to its action in poison-dart frogs (Phyllobates). Science 208, 1383-1385

Eaton DC, Brodwick MS, Oxford GS, Rudy B (1978) Arginine-specific reagents remove sodium channel inactivation. Nature 271:473-476

Engelman DM, Steitz TA, Goldman,A (1986) Identifying nonpolar transbilayer helices in amino acid sequences of membrane proteins. Ann. Rev. Biophys. Biophys. Chem. 15:321-333

Frelin C, Vigne P, Lazdunski M (1981) The specificity of the sodium channel for monovalent cations. Eur. J. Biochem. 119:437-442

French RJ, Horn R (1983) Sodium channel gating: Models, mimics and modifiers. Ann. Rev. Biophys. Bioeng. 12:319-356

Finkelstein A, Peskin CS (1984) Some unexpected consequences of a simple physical mechanism for voltage-dependent gating in biological membranes. Biophys. J. 46:549-558

Gabe EJ, Barnes WH (1963) The crystal and molecular structure of l-cocaine hydrochloride. Acta Cryst. 16:796-801

Garber SS, Miller C (1987) Single Na+ channels activated by veratridine and batrachotoxin. J. Gen. Physiol 89:459-480

Goldin AL, Snutch T, Lübbert H, Dowsett A, Marshall J, Auld V, Downey W, Fritz LC, Lester HA, Dunn R, Catterall WA, Davidson N (1986) Messenger RNA coding for only the α-subunit of the rat brain sodium channel is sufficient for expression of functional channels in Xenopus oocytes. Proc. Nat'l. Acad. Sci. USA 83:7503-7507

Gordon RD, Fieles WE, Schotland DL, Barchi RL (1987) Immunochemical testing of current molecular models of the voltage-dependent sodium channel from Electrophorus electricus. Biophys. J. 51:437a

Gordon RD, Fieles WE, Schotland DL, Hogue-Angeletti R, Barchi RL (1987) Topographical localization of the C-terminal region of the voltage-dependent sodium channel from Electrophorus electricus using antibodies raised against a synthetic peptide. Proc. Nat'l. Acad. Sci. USA 84:308-312

Gray TM, Matthews BW (1984) Intrahelical hydrogen bonding of serine, threonine and cysteine residues within α-helices and its relevance to membrane-bound proteins. J. Mol. Biol. 175:75-81

Greenblatt RE, Blatt Y, Montal M (1985) The structure of the voltage-sensitive sodium channel. Inferences derived from computer-aided analysis of the Electrophorus electricus channel primary structure. FEBS Lett. 193:125-134

Guy HR, Seetharamulu P (1986) Molecular model of the action potential sodium channel Proc. Nat'l. Acad. Sci. USA 83, 508-512

Guy HR (1987) Structural models of the sodium channel ion selectivity activation and inactivation. Biophys. J. 51:193a

Guy HR (1988) A model relating the sodium channel's structure to its function. In: Agnew WS, Sigworth F, Claudio T (eds) Molecular biology of ion channels: Current topics in membrane transport. Academic New York in press

Hartshorne RP, Catterall WA (1984) The sodium channel from rat brain. Purification and subunit composition. J. Biol. Chem. 259:1667-1675

Hartshorne R, Tamkun M, Montal M (1986) The reconstituted sodium channel from brain. In: Miller C (ed) Ion Channel Reconstitution. Plenum New York, Chap.5, pp.337-362

Hille B (1977) Local anesthetics: Hydrophilic and hydrophobic pathways for the drug-receptor reaction. J. Gen. Physiol. 69:497-515

Hille B (1980) Theories of anesthesia: general perturbations versus specific receptors. In: Molecular Mechanisms of Anesthesia. Vol.2 Fink BR (ed) Progress in Anesthesia. Raven New York pp.1-5

Hille B (1984) Ionic Channels of Excitable Membranes. Sinauer Sunderland Mass. USA

Hirono H, Yamagishi S, Ohara R, Hisanaga Y, Nakayama T, Sugiyama H (1985) Characterization of the mRNA responsible for induction of functional sodium channels in Xenopus oocytes. Brain Res. 359:57-64

Hodgkin AL, Huxley AF (1952) A quantitative description of membrane current and its application to conduction and excitation in nerve. J. Physiol. 117: 500-544

Hyrnchuk RJ, Barton RJ, Robertson BE (1983) The crystal structure of free base cocaine, $C_{17}H_{21}NO_4$. Can. J. Chem. 61:481-487

Kayano T, Noda M, Flockerzi V, Takahashi H, Numa S (1988) Primary structure of rat brain sodium channel III deduced from the cDNA sequence. FEBS Lett. 228: 187-194

Khodorov BI (1985) Batrachotoxin as a tool to study voltage-sensitive sodium channels of excitable membranes. Progr. Biophys. Mol. Biol. 45:57-148

Kosower EM (1982) Single Group rotation. In: Abstracts, Int'l Symposium, Structure and Dynamics of Nucleic Acids and Proteins. San Diego Calif. 5-9 Sept., pp.52-3

Kosower EM (1983) A hypothesis for the mechanism of action of sodium channel opening by batrachotoxin and related toxins. FEBS Lett. 163:161-164

Kosower EM (1985) A structural and dynamic molecular model for the sodium channel of Electrophorus electricus. FEBS Lett. 182:234-242

Kosower EM (1986) A Structural and Dynamic Model for the Nicotinic Acetylcholine Receptor. Maelicke A(ed) In: Nicotinic Acetylcholine Receptor: Structure and Function. NATO-ASI Series Vol H-3 Springer Berlin. pp.465-483

Kosower EM (1987) A structural and dynamic model for the acetylcholine receptor. Eur. J. Biochem. 168:431-449

Kosower EM (1988) Assignment of the groups responsible for the "opsin shift" and the light absorptions of rhodopsin and red, green and blue iodopsins (cone pigments). Proc. Nat'l. Acad. Sci. USA 85:1076-1080

Krafte D, Leonard J, Snutch T, Davidson N, Lester HA (1987) Slowly inactivating sodium currents from injection of high molecular weight rat brain RNA into Xenopus oocytes. Biophys. J. 51:194a

Kuffler SW, Nicholls JG, Martin AR (1984) From Neuron to Brain. 2nd edn. Sinauer Sunderland Mass., USA pp.157-160

Leibowitz MD, Sutro JB, Hille,B (1986) Voltage-dependent gating of veratridine-modified Na channels. J. Gen. Physiol. 87:25-46

Lo M-vC, Shrager P (1981a) Block and inactivation of sodium channels in nerve by amino acid derivatives I. Biophys. J. 35:31-43

Lo M-vC, Shrager P (1981b) Block and inactivation of sodium channels in nerve by amino acid derivatives II. Biophys. J. 35:45-57

McNeal ET, Lewandowski GA, Daly JW, Creveling CR (1985) [^{3}H]Batrachotoxinin A 20α-benzoate binding to voltage-sensitive sodium channels: A rapid and quantitative assay for local anesthetic activity in a variety of drugs. J. Med. Chem. 28:381-388

Meiri H, Goren E, Bergmann H, Zeitoun I, Rosenthal Y, Palti Y (1986) Specific modulation of sodium channels in mammalian nerve by monoclonal antibodies. Proc. Nat'l. Acad. Sci. USA 83:8385-8389

Meiri H, Spira G, Marei S, Namir M, Schwartz A, Komoriya A, Kosower EM, Palti Y (1987) Mapping a region associated with sodium channel inactivation using antibodies to a synthetic peptide corresponding to a part of the channel. Proc. Nat'l. Acad. Sci. USA 84:5058-5062

Meiri H, Zeitoun I, Grunhagen HH, Lev-Ram V, Eshhar Z, Schlessinger J (1984) Monoclonal antibodies associated with sodium channel block block nerve impulse and stain nodes of Ranvier. Brain Research 310:168-173

Messner DJ, Catterall WA (1985) The sodium channel from rat brain. Separation and characterization of subunits. J. Biol. Chem. 260:10597-10604

Messner DJ, Feller DJ, Scheuer T, Catterall WA (1986) Functional properties of rat brain sodium channels lacking the β_1- or β_2-subunit. J. Biol.

Chem. 261:14882-14890

Nakayama H, Withy RM, Raftery MA (1982) Use of a monoclonal antibody to purify the tetrodotoxin binding component from the electroplax of Electrophorus electricus. Proc. Nat'l. Acad. Sci. USA 79:7575-7579

Noda M, Ikeda T, Suzuki H, Takeshima H, Takahashi H, Kuno M, Numa S (1986b) Expression of functional sodium channels from cloned cDNA. Nature 322:826-828

Noda M, Shimizu S, Tanabe T, Takai T, Kayano T, Ikeda T, Takahashi H, Nakayama H, Kanaoka Y, Minamino N, Kangawa K, Matsuo H, Raftery MA, Hirose T, Notake M, Inayama S, Hayashida H, Miyata T, Numa S (1984) Primary structure of Electrophorus electricus sodium channel deduced from cDNA sequence. Nature 312:121-127

Noda M, Ikeda T, Kayano T, Suzuki H, Takeshima H, Kurasaki M, Takahashi H, Numa S (1986a) Existence of distinct sodium channel messenger RNAs in rat brain. Nature 320:188-192

Oxford GS, Wu CH, Narahashi T (1978) Removal of sodium channel inactivation in squid giant axons by N-bromoacetamide. J. Gen. Physiol. 71:227-247

Oxford GS, Yeh JZ (1982) Biophys.J. 37:104a

Postma SW, Catterall WA (1984) Inhibition of binding [^{3}H]Batrachotoxinin A 20α-benzoate to sodium channels by local anesthetics. Mol. Pharmacol. 25:219-227

Richards CD, Keightley CA, Hesketh TR, Metcalfe JC (1980) A critical evaluation of the lipid hypotheses of anesthetic action. In: Molecular Mechanisms of Anesthesia. Vol.2 Fink BR (ed) Progress in Anesthesia. Raven New York pp. 337-351

Richardson JS, Richardson DC (1988) Amino acid preferences for specific locations at the end of α-helices. Science 240:1648-1652

Rojas E, Rudy B (1976) Destruction of the sodium conductance inactivation by a specific protease in perfused nerve fibers from Loligo. J. Physiol. (Lond.) 262:501-531

Rosenberg RL, Tomiko SA, Agnew WS (1984) Reconstitution of neurotoxin-modulated ion transport by the voltage-regulated sodium channel isolated from the electroplax of Electrophorus electricus. Proc. Nat'l. Acad. Sci. USA 81:1239-1243

Rosenthal V, Meiri H, Spira G, Kraoz Z, Kosower EM, Palti Y. Interaction of the anti-eel antibody 5F3 with OCEIII and the frog sciatic nerve. unpublished results.

Roth SH (1979) Physical methods of anesthesia. Ann. Rev. Pharmacol. Toxicol. 19:159-178

Schulz GE, Schirmer RH (1979) Principles of Protein Structure. Springer Heidelberg pp.73-78

Seyama I, Yamada K, Kato R, Masutani T, Hamada M (1988) Grayanotoxin opens Na channels from inside the squid axonal membrane. Biophys. J. 53:271-274

Starmer CF, Grant AO, Strauss HC (1984) Mechanism of use-dependent block of sodium channels in excitable membranes by local anesthetics. Biophys. J. 46:15-27

Stimers JR, Bezanilla F, Taylor RE (1985) Sodium channel activation in the squid giant axon: Steady state properties. J. Gen. Physiol. 85:65-82

St hmer W, Methfessel C, Sakmann B, Noda M, Numa S (1987) Patch clamp characterization of sodium channels expressed from rat brain cDNA. Eur. Biophys. J. 14:131-138

Strichartz G (ed) (1987) Local Anesthetics. Vol.81 Handbk. Exp. Pharmacol. Springer Berlin 292 pp.

Sutro JB (1986) Kinetics of veratridine action on Na channels of skeletal muscle. J. Gen. Physiol. 87:1-24

Talvenheimo, A (1985) The purification of ion channels from excitable cells. J. Membrane Biol. 87:77-91

Tanguy J, Yeh JZ, Narahashi T (1987) Gating charge immobilization after removal of Na-channel inactivation. Biophys. J. 51:7a

Tejedor FJ, McHugh E, Catterall WA (1988) Stabilization of a sodium channel state with a high affinity for saxitoxin by intramolecular cross-linking. Evidence for allosteric effects of saxitoxin binding. Biochemistry 27:2389-2397

Toniolo C, Bonora GM, Bavoso A, Benedetti E, DiBlasio B, Pavone V, Pedone C (1986) A long, regular polypeptide 3_{10}-helix. Macromol. 19:472-479

Van Dyke C, Byck R (1982) Cocaine. Sci. Amer. 246:108-119

Venkataraghavan B, Feldmann RJ (eds) (1985) Macromolecular structure and specificity: Computer-assisted modeling and applications. Ann. N. Y. Acad. Sci. 439:1-209

Wallace B, Cascio M, Mielke DL (1986) Evaluation of methods for the prediction of membrane protein secondary structures. Proc. Nat'l. Acad. Sci. USA 83:9423-9427

Wilke ME, Mitra A, Yamamoto T, Stroud,R. (1987) Structural study of a natural insecticide. Biophys. J. 51:454a

Woodbury JW, D'Arrigo JS, Eyring H (1975) Physiological mechanism of general anesthesia: general blockade. In: Molecular Mechanisms of Anesthesia. Vol.1 Fink BR (ed) Progress in Anesthesia. Raven New York pp.53-87

Zell A, Ealick SE, Bugg CE (1985) Three-dimensional structures of scorpion neurotoxins. In: Molecular architecture of proteins and enzymes. Academic New York pp.65-97

PROCESSING, ASSEMBLY, AND LIGAND BINDING OF THE MOUSE NICOTINIC ACETYLCHOLINE RECEPTOR SUBUNITS IN STABLY TRANSFECTED FIBROBLASTS

Paul Blount and J. P. Merlie
Department of Pharmacology
Washington University School of Medicine
660 S. Euclid
St. Louis, MO. 63110

INTRODUCTION

In 1981 Sumikawa and his colleagues showed that messenger RNA isolated from Torpedo and injected into Xenopus oocytes produced an assembled form of the nicotinic acetylcholine receptor (AChR) within the injected oocyte. This system elegantly demonstrated that all the machinery for the biogenesis of the AChR, from message to assembled protein, was common among diverse species, and most probably common among tissues. Since then, the oocyte injection technique has been utilized to explore the properties of AChR subunit combinations (Kurolaki et al., 1987) as well as functional regions of the α subunit via site directed mutagenesis (Mishina et al., 1985). There are, however, disadvantages of this technique due to working with one large cell as opposed to a cell population: oocytes must be injected before every experiment, the long incubation times before receptor is expressed make pulse-chase experiments difficult, and finally, detailed binding and biochemical studies of the receptor are impractical because of the inadequate amount of protein produced. Transfection of eukaryotic cells is an alternative expression system that has recently been utilized to express functional AChR on the surface of mouse fibroblasts (Claudio et al., 1987). Because eukaryotic cells can be stably transfected with expression vectors then grown from clones to populations of cells, many of the disadvantages of the oocyte injection system are avoided. We therefore have chosen to use a eukaryotic expression system to study the processing of the α subunit and assembly of combinations of subunits of the mouse nicotinic AChR.

NATO ASI Series, Vol. H 32
Molecular Biology of Neuroreceptors and
Ion Channels. Edited by A. Maelicke

MATERIALS AND METHODS

Our expression vectors consist of a Rouse Sarcoma virus Long Terminal Repeat (RSV LTR) followed by the full length cDNA clone of the subunit to be expressed, inserted into the polylinker of an M13 RF. Expression vectors have been made for the α,β and δ subunits of the mouse AChR. Transfections into the quail fibroblast cell line, QT-6, (Moscovici et al., 1977) were performed by calcium phosphate precipitation (Graham and van der Eb, 1973). The fibroblast clones expressing AChR subunits were obtained by co-transfection of the subunit expression vector(s) and a plasmid vector that contained a neomycin resistance gene (Colbere-Garapin et al., 1981) under the transcriptional control of the RSV LTR. Subsequent to transfection, G-418 resistant colonies were selected, subcloned, and tested for subunit(s) expression by metabolic labeling with [^{35}S]methionine, immunoprecipitation, SDS-PAGE and fluorography. Clones described here were transfected with only the expression vector construct(s) for the subunit(s) they express and the neo resistant gene as outlined above. A more detailed materials and methods section including a description of the binding assay may be found in Blount and Merlie, 1988.

RESULTS AND DISCUSSION

The first clone isolated and characterized was the α subunit producing clone, Q-α5. In this system we could test directly whether the α subunit expressed independent of other subunits would acquire high affinity BTX binding. It had been previously shown in the muscle-like cell line, BC3H-1, that the α subunit of the AChR acquired α bungarotoxin (BTX) binding in a time dependent process prior to association with the β subunit (Merlie and Sebbane, 1981; Merlie and Lindstrom, 1983). The simplest interpretation of these experiments was that the α subunit could acquire high affinity BTX binding before assembly. Association of α with subunits other than β, however, could not be measured; also, the assay for association of the α and β subunits was co-immunoprecipitation, therefore a loose or transient association of α and β could not be ruled out. To directly address the question of whether high affinity BTX binding could be acquired independent of other subunits, Q-α5 cells were pulse-labeled for 5 min and chased for increasing amounts of time. The α subunit precipitated by mAb61 (Fig. 1A, α_{61}) showed no delay in its appearance. In contrast, the α subunit immunoprecipitated by toxin-antitoxin (Fig. 1B, α_{Tx}) showed a delay of 20 min before reaching its maximum value. Therefore, it appears that α subunits acquires high affinity BTX binding in a time dependent manor in the absence of other subunits of the AChR.

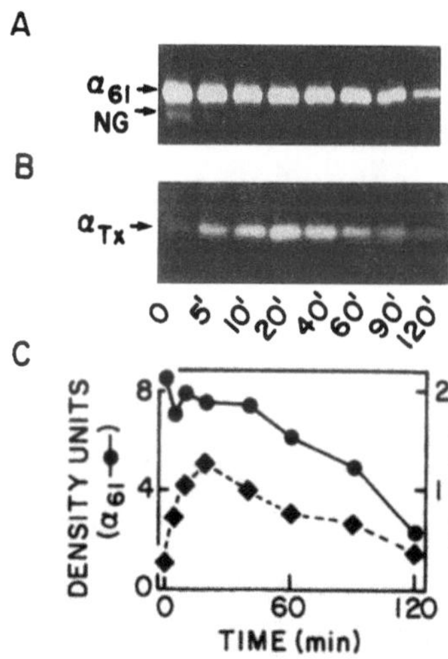

Fig. 1: **Kinetics of acquisition of α bungarotoxin binding of the transfected α subunit.** The transfected cell line, Q-α5, was pulse-labeled with [^{35}S]methionine for 5 min and chased for the times indicated. The cell extracts were immunoprecipitated with either mAb61, a monoclonal antibody made against SDS-denatured receptor (**Panel A**) (Tzartos et al., 1981), or, in **Panel B**, an affinity purified rabbit antiserum against BTX to precipitate BTX prebound to the α subunit (toxin-antitoxin) (Merlie and Sebbane, 1981; Merlie and Lindstrom, 1983). The initials 'NG' in panel A indicate the nonglycosylated form of the subunit. Appropriate exposures of α_{61} (**filled circles**) and and toxin-antitoxin (**filled diamonds**) were analyzed by quantitative densitometry and are shown in **Panel C** (from Blount and Merlie, 1988).

Having demonstrated that the α subunit could acquire a high affinity BTX binding site independent of other subunits, we next developed a membrane binding assay so a direct comparison between the binding of ^{125}I-BTX to the α subunit in Q-α5 and the binding of ^{125}I-BTX to the intact AChR of the BC3H-1 cell line could be made. We were surprised to find that the Q-α5 α subunit had a **higher** apparent affinity for BTX than the BC3H-1 intact receptor (!). Upon investigation of the binding kinetics, we found that the binding of ^{125}I-BTX was essentially irreversible when bound to either the Q-α5 or BC3H-1 α subunits; the half times of dissociation ($t_{1/2}$) were on the order of 100 hours or more in both cases. In contrast, the association rate of ^{125}I-BTX to the two forms of the α subunit were quite dissimilar. As seen in Fig. 2, the association rate of ^{125}I-BTX to the unassembled α subunit was several fold faster than to the intact AChR. The K_{obs} calculated from this data are 7.9×10^{-5} s^{-1} and 1.4×10^{-5} s^{-1} for the Q-α5 and BC3H-1 α subunits, respectively. Therefore, the higher apparent affinity of ^{125}I-BTX to the Q-α5 α subunit is due to a faster association rate constant for the toxin to the unassembled α subunit. One possible explanation for this data is that when the α subunit is in the intact receptor, its conformation is constrained by neighboring subunits, making BTX binding either less probable or more difficult to achieve.

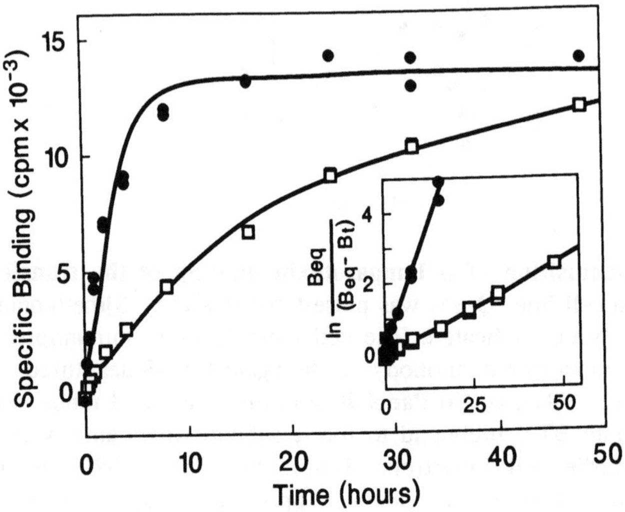

Fig. 2: **Time Dependence for Association of α Bungarotoxin with BC3H-1 and Q-α5 Membranes.** Membranes derived from BC3H-1 (**squares**) and Q-α5 (**circles**) cells were bound to 1.7nM ^{125}I-BTX. Duplicate samples were removed and filtered at the times indicated. Specific binding was determined using duplicate 'total' and 'nonspecific' samples at each time point. The inset shows a standard linearization of this data where B_{eq} is the calculated equilibrium of specific binding at this concentration of binding sites and BTX, and B_t is the specific binding at time t (from Blount and Merlie, 1988).

We next tested the Q-α5 membranes for small ligand agonist and antagonist binding sites using an 'inhibition of initial rates of BTX binding' assay. Of the five drugs tested, binding sites were found only for the two antagonists, D-tubocurarine and gallamine (Fig. 3). These drugs, however, were at least 10^3 times less potent for the Q-α5 than the BC3H-1 membranes. Carbamylcholine, decamethonium, and hexamethonium failed to inhibit initial rates of BTX binding to Q-α5 membranes up to 10 mM concentration but showed normal inhibition of BC3H-1 receptor binding (Blount and Merlie, 1988).

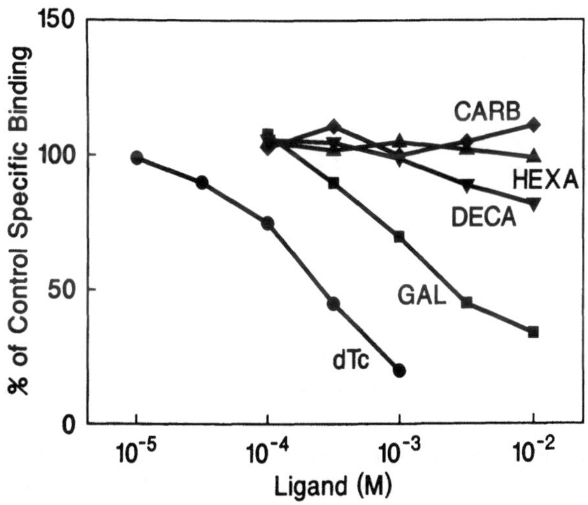

Fig. 3: **Analysis of Inhibition of Initial Rates of α Bungarotoxin Binding to Q-α5 Membranes Using AChR Agonists and Antagonists.** A membrane preparation of Q-α5 cells was preincubated with the indicated final concentration of ligand for 30 min in a total volume of 900 μl. After this time, 100 μl of 10nM ^{125}I-BTX was added and the sample was incubated for another 15 min. The sample was then filtered and counted. Nonspecific binding was defined by adding an excess of unlabeled BTX; 'control specific binding' was defined by addition of no ligand other than ^{125}I-BTX (total minus nonspecific binding). The abbreviations used are: **carb**, carbamylcholine; **hexa**, hexamethonium; **deca**, decamethonium; **gal**, gallamine; and **dTc**, d-tubocurarine (from Blount and Merlie, 1988).

More recently we have isolated a number of cell lines that have been transfected with and express more than one AChR subunit. Fig. 4 shows four G-418 resistant cell lines screened with a cocktail of monoclonal antibodies specific for the subunit proteins that should be expressed from the shuttle vector constructs. Lane A in each panel depicts a clone expressing the transfected subunits, while lane B is a G-418 resistant colony from the same transfection that does not express the transfected subunits.

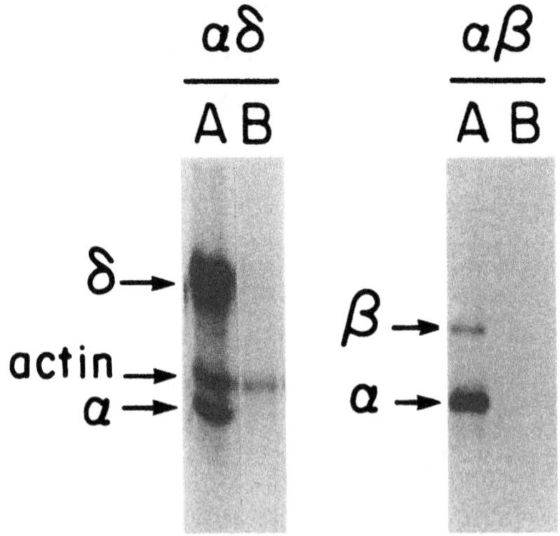

Fig. 4: **Specific immunoprecipitation of AChR subunits from stably transfected QT-6 cells.** The panels are labeled with the expression vector subunits transfected, α and β (αβ) or α and δ (αδ). Immunoprecipitations were carried out with a cocktail of monoclonal antibodies appropriate for the transfection. The α subunit was detected using mAb61 (Tzartos et al., 1981), the β subunit was detected using mAb148 (Gullick and Lindstrom, 1983), and the δ subunit was detected using mAb88B (Froehner et al., 1983). Lane A of each panel depicts a positive screening clone while the lane B is a G-418 resistant clone from the same transfection which does not express the transfected subunits. All clones incorporated roughly the same amount of [^{35}S]methionine and showed the same nonspecific bands (actin is shown in the αδ panel) upon long exposure.

We have begun to characterize one clone expressing both the α and δ subunits, Q-αδ2. Preliminary experiments have demonstrated that Q-αδ2 has an association rate constant for ^{125}I-BTX more like the BC3H-1 receptor than Q-α5 α subunit (not shown). It is difficult, however, to directly compare the association rate constant of Q-αδ2 to that of BC3H-1 receptor because of presence of free α subunit in Q-αδ2 that binds BTX with high affinity. The Q-αδ2 cell line also has a high affinity small ligand agonist and antagonist binding site as determined by inhibition of initial rate of ^{125}I-BTX binding. As demonstrated in Fig. 5, the binding of the partial agonist decamethonium to the high affinity site seen in the Q-αδ2 cell line is indistinguishable from the high affinity site seen in the BC3H-1 cell line, and at least 10^4 higher than

Q-α5 (compare Fig. 3). High affinity sites have also been observed in the Q-αδ2 cell line for the other small ligands carbamylcholine, hexamethonium, gallamine, and D-tubocurarine.

In contrast, clones expressing the α and β subunits appear to have similar binding properties to the Q-α5 cell line. Therefore, the δ subunit, but not the β subunit, in combination with the α subunit confers high affinity binding to small ligands.

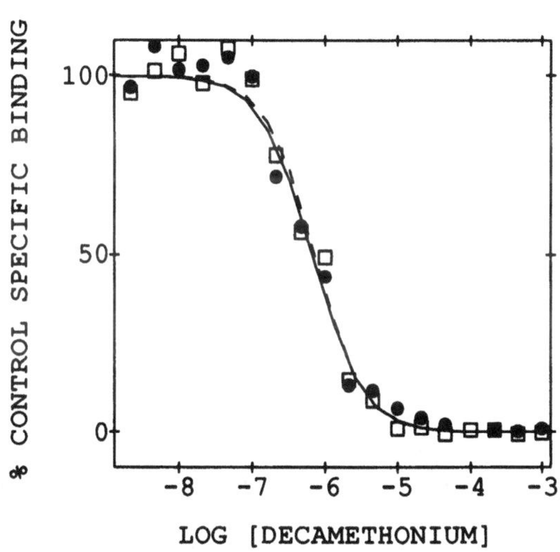

Fig. 5: **Analysis of Inhibition of Initial Rates of α Bungarotoxin Binding to Q-αδ2 and BC3H-1 Membranes Using Decamethonium.** As in Fig. 4, membrane preparations of Q-αδ2 **(filled circles)** and BC3H-1 **(squares)** cells were preincubated with the indicated final concentration of decamethonium for 30 min in a total volume of 900 μl. After this time, 100 μl of 10nM ^{125}I-BTX was added and the sample was incubated for another 2 hrs. The sample was then filtered and counted. Nonspecific binding for this high affinity binding site was defined by a series of four or more points at which addition of more decamethonium caused no more inhibition of ^{125}I-BTX binding; 'control specific binding' was defined by addition of no ligand other than ^{125}I-BTX (total minus nonspecific binding).

CONCLUSIONS

We have exploited a eukaryotic gene expression system to study the processing and ligand binding properties of combinations of AChR subunits. Previous studies suggested that the α subunit had the potential to bind BTX prior to assembly with the β subunit (Merlie and Sebbane, 1981; Merlie and Lindstrom, 1983). By expressing the α subunit in a fibroblast cell line we were able to confirm that the α subunit could acquire high affinity BTX binding. The time dependent nature of this phenomenon probably reflects a post-translational modification and/or conformational change in the α subunit not dependent upon other AChR subunits. We discovered that the unassembled α subunit that acquires high affinity BTX binding has a faster rate constant for association of this ligand than does the assembled receptor. Because of the essentially irreversible binding of this toxin, the faster association rate causes the unassembled α subunit to have a higher apparent affinity for BTX. In contrast, affinities of the unassembled α subunit for the small ligands, including D-tubocurarine and carbamylcholine, are either very weak or non existent. Finally, we found that the small ligand binding site can be conferred upon the α subunit by co-expressing the δ, but not the β subunit of the AChR.

REFERENCES

Blount, P., and Merlie, J. P. (1988). Native Folding of an acetylcholine receptor α subunit expressed in the absence of other subunits. J. Biol. Chem. **263,** 1072-1080.

Claudio, T., Green, W. N., Hartman, D. S., Hayden, D., Paulson, H. L., Sigworth, F. J., Sine, S. M., and Swedlund, A. (1987). Genetic reconstitution of functional acetylcholine receptor channels in mouse fibroblasts. Science **238,** 1688-1694.

Froehner, S. C., Douville, K., Klink, S., and Culp, W. J. (1983). Monoclonal antibodies to cytoplasmic domains of the acetylcholine receptor. J. Biol. Chem. **258,** 7112-7120.

Graham, F. L., and van der Eb, A. J. (1973). A new technique for the assay of infectivity of human adenovirus 5 DNA. Virology **52,** 456-567.

Gullick, W. J., and Lindstrom, J. M. (1983). Mapping the binding of monoclonal antibodies to the acetylcholine receptor. Biochem. **22,** 3312-3320.

Kurosaki, T., Fukuda, K., Konno, T., Mori, Y., Tanaka, K., Mishina, M., and Numa, S. (1987) Functional properties of nicotinic acetylcholine receptor subunits expressed in various combinations FEBS Letters **214,** 253-258

Merlie, J. P., and Sebbane, R. (1981). Acetylcholine receptor subunits transit a precursor pool before acquiring α-bungarotoxin binding activity. J. Biol. Chem. **256,** 3605-3608.

Merlie, J. P., and Lindstrom, J. (1983). Assembly in vivo of mouse muscle acetylcholine receptor: identification of an α subunit species that may be an assembly intermediate. Cell **34,** 747-757.

Mishina, M., Kurosaki, T., Tobimatsu, T., Morimoto, Y., Noda, M., Yamamoto, T., Terao, M., Lindstrom, J., Takahashi, T., Kuno, M., and Numa, S. (1984). Expression of functional acetylcholine receptor from cloned cDNAs. Nature **307,** 604-608.

Mishina, M., Tobimatsu, T., Imoto, K., Tanaka, K., Fujita, Y., Fukuda, K., Kurasaki, M., Takahashi, H., Morimoto, Y., Hirose, T., Inayama, S., Takahashi, T., Kuno, M., and Numa, S. (1985). Location of functional regions of acetylcholine receptor α-subunit by site-directed mutagenesis. Nature **313**, 364-369.

Moscovici, C., Moscovici, M. G., Jimenez, H., Lai, M. M. C., Hayman, M. J., and Vogt, P. K., (1977). Continuous tissue culture cell lines derived from chemically induced tumors of Japanese quail. Cell **11**, 95-103.

Sumikawa, K., Houghton, M., Emtage, J. S., Richards, B. M., and Barnard, E. A. (1981). Active multi-subunit Ach receptor assembled by translation of heterologous mRNA in **Xenopus** oocytes. Nature **292**, 862-864.

Tzartos, S. J., Rand, D. E., Einarson, B. L., and Lindstrom, J.M. (1981). Mapping of surface structures of electrophorous acetylcholine receptor using monoclonal antibodies. J. Biol. Chem. **256**, 8635-8645.

ASSEMBLY AND INSERTION OF A MULTI-SUBUNIT NICOTINIC ACETYLCHOLINE RECEPTOR INTO PLASMA MEMBRANES.

Katumi Sumikawa and Ricardo Miledi
Laboratory of Cellular and Molecular Neurobiology
Department of Psychobiology
University of California
Irvine, California 92717 USA

INTRODUCTION

The nicotinic acetylcholine receptor (AChR) is one of a family of ion channel receptors, which are gated directly by neurotransmitters. Other receptors of this type include the GABA and glycine receptors. An important characteristic of these receptors is that they are composed of multiple distinct subunits all of which have a similar transmembrane topology (for a review see Barnard et al., 1987). The AChR is an oligomeric glycoprotein consisting of five subunits (α_2, β, γ, and δ) of four different types. Five subunits span the plasma membrane and are arranged in a rosette to form an ion channel. The binding of acetylcholine (ACh) to the α subunits causes the membrane channel to open (for reviews see Karlin, 1980; Conti-Tronconi & Raftery,1982; Barrantes,1983; Anholt et al.,1984, and Popot & Changeux,1984). Each of the four subunits is coded by separate mRNAs. Thus, the production of a complete functional AChR molecule involves complex processes including not only the synthesis of the individual subunits but also their assembly in a particular order, and their insertion into the plasma membrane with the correct orientation (for a review see Merlie & Smith, 1986). In addition to the structural and functional information available about this receptor, the availability of tools such

NATO ASI Series, Vol. H 32
Molecular Biology of Neuroreceptors and Ion Channels. Edited by A. Maelicke

as antibodies and highly specific antagonists such as α-bungarotoxin (α-BuTX), as well as the cloned AChR genes, make this an attractive and important model system for studying the assembly and membrane insertion of a functional receptor of this family.

<u>Expression of incomplete Torpedo AChR molecules in Xenopus oocytes by using antisense RNAs</u>.

<u>In vitro</u> transcription systems provide a large amount of specific mRNAs and their antisense counterparts, which are known to block specifically the translation of the corresponding mRNA (for a review see Green et al., 1986).

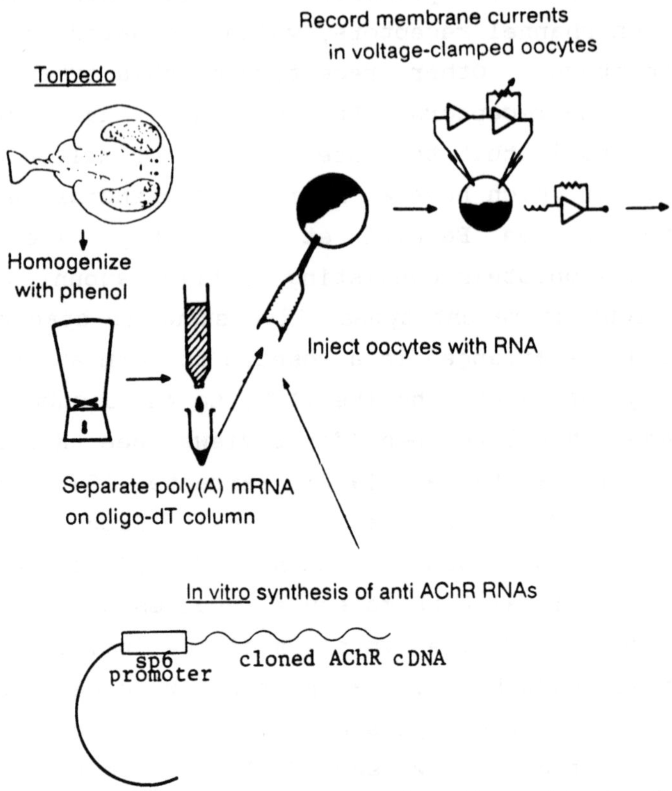

Fig. 1. Procedures for studying the effect of antisense RNAs on the expression of <u>Torpedo</u> AChRs by electrophysiological techniques.

A common mechanism of this form of inhibition involves hybridization between the mRNA and its counterpart antisense RNA (Melton,1985; and Harland & Weintraub,1985). We examined the effect of antisense RNAs on the expression of AChR in Xenopus oocytes. Antisense RNAs were synthesized from cDNA clones coding for the four AChR subunits and their effect on the expression of AChR molecules was studied using biochemical and electrophysiological techniques (Fig. 1). For biochemical studies, oocytes injected with a mixture of four synthetic AChR subunit mRNAs (instead of the natural Torpedo electric organ mRNA) alone, or plus one of the antisense RNAs, were cultured in the presence of ^{35}S methionine. The translation products were then subjected to immunoprecipitation with AChR antibodies. SDS gel electrophoretic analysis of the immunoprecipitates demonstrated that each antisense RNA inhibited the translation of its counterpart mRNA specifically and very efficiently, if not completely (Fig. 2A). We next examined the oocytes, injected with Torpedo mRNA alone or with one of the subunit antisense RNAs for the presence of functional AChR, by measuring the response to ACh (see Fig. 1). We found (Sumikawa & Miledi, 1988a) that oocytes injected with Torpedo mRNA alone gave large responses to ACh, while the amplitude of the currents elicited by ACh, applied to the oocytes injected with the α antisense RNA, was almost completely abolished. However, in oocytes injected with the other antisense RNAs there was always some residual activity (Fig. 2B), suggesting that AChR molecules consisting of only three subunits, but including the α subunit, are able to form functional molecules (cf. Kurosaki et al., 1987 and White et al., 1985) Oocytes injected with Torpedo mRNA also acquire a voltage-activated Cl^- channel (Sumikawa et al, 1984). We found (Sumikawa & Miledi, 1988a) that none of AChR subunit antisense RNAs had an effect on the expression of Cl^- channels (Fig.2C). This indicates that blocking of the expression of functional AChR by the antisense RNA occurs in a specific manner.

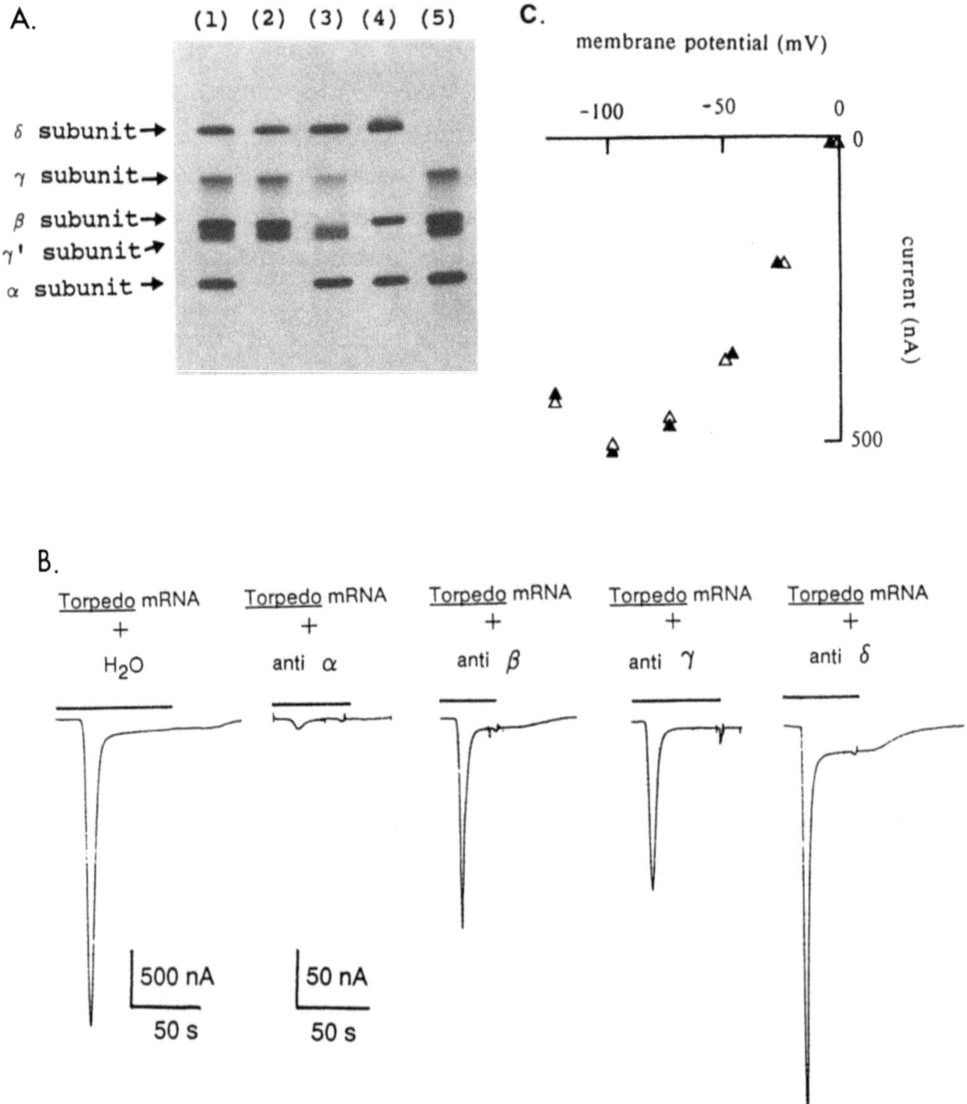

Fig. 2. (A). SDS gel electrophoresis of AChR molecules synthesized in oocytes. A mixture of the four subunit mRNAs was injected into 10 oocytes with (1) H_2O, (2) antisense α RNA, (3) antisense β RNA, (4) antisense γ RNA, or (5) antisense δ RNA for immunoprecipitation assay. Full-length AChR cDNA clones (Claudio et al., 1983,1987) were kindly provided by Dr. T. Claudio (α,β,γ and δ cDNA clones) and Dr. S. Heinemann (γ subunit cDNA clone) and used for the synthesis of capped AChR subunit mRNAs or AChR subunit antisense RNAs after insertion into plasmids pSP65 (β and δ), pGEM3 (γ) or pGEM4 (α).

Fig. 2. cont. (B). ACh-activated currents recorded in Xenopus oocytes injected with Torpedo mRNA alone or together with antisense RNAs. ACh (10^{-4}M) was applied by bath perfusion for the durations indicated by the bars. (C). Current/voltage relationship of Cl^-current induced in the oocytes by Torpedo mRNA in the presence (△) or absence (▲) of antisense RNA from the α subunit of the AChR.

Assembly of all ACh receptor subunits is a pre-requisite for their efficient insertion into plasma membranes.

AChR molecules devoid of either the β, γ or δ subunits were still capable of responding to ACh, although the current elicited were much smaller than those generated by the complete AChR molecules (Fig. 2B and Kurosaki et al., 1987). The smaller responses could be explained in several ways. For example, incomplete AChRs might be assembled less efficiently, or they might be inserted poorly into the plasma membranes. Alternatively, incomplete AChRs might have different channel characteristics so as to give a smaller overall current, with receptor numbers remaining unchanged. Therefore, we examined whether incomplete AChRs are assembled and inserted into the plasma membrane as efficiently as the complete AChRs. For this purpose, we injected oocytes with various combinations of subunit mRNAs always including the α subunit mRNA and incubated them in the presence of ^{35}S methionine. Initially, we examined whether incomplete AChRs are assembled in the absence of one or two AChR subunits by co-immunoprecipitation of the other subunits with α-subunit specific antibody, followed by SDS gel electrophoresis. We found (Sumikawa & Miledi, 1988b; also see Fig. 3A) that the α-subunit is capable of associating with any of the other subunits, although in some cases the efficiency of assembly may be slightly reduced. The results clearly indicated that incomplete AChR molecules are assembled in the oocyte.

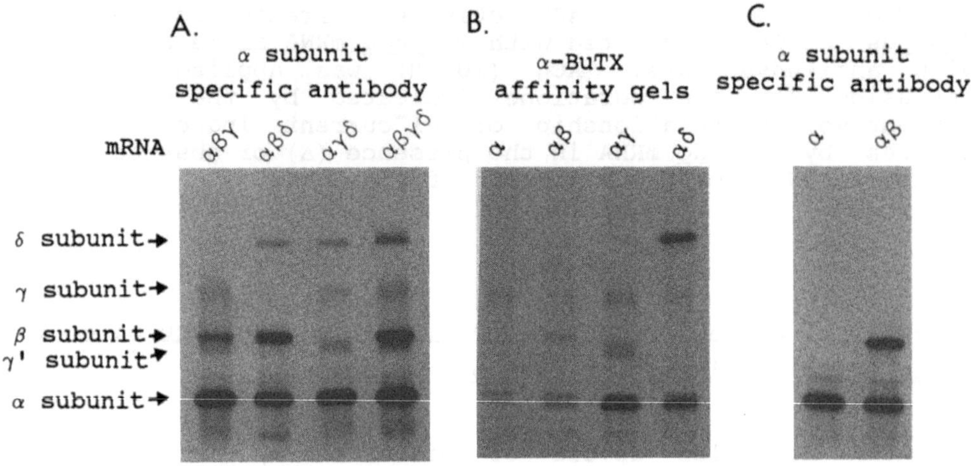

Fig. 3. Electrophoretic analysis of the assembly of AChR subunits. (A) and (C). Detection of subunit assembly by co-immunoprecipitation with the α subunit specific antibody. Oocytes were injected with subunit mRNAs in various combinations as indicated and incubated with ^{35}S methionine. (B). Acquisition of high affinity toxin binding by subunit assembly. Oocytes were injected with the α subunit, singly or in combination with other subunits as indicated and incubated with ^{35}S methionine. To identify the subunits with high affinity toxin binding, each of the oocyte extracts was incubated with toxin affinity gels.

We also used α-bungarotoxin-Sepharose gels (α-BuTX gels) to study receptor assembly (Sumikawa & Miledi, 1988b). It is now well established that α-BuTX can bind to the isolated α subunit (Haggerty & Froehner, 1981; Gershoni et al.,1983; Oblas et al., 1983; and Tzartos & Changeux, 1983) and even to a synthetic peptide consisting of 32 amino acid residues of the α subunit (Wilson et al., 1985) with low affinity. The α subunit synthesized in a variety of cells, by expressing the cloned α subunit cDNA, binds to α-BuTX with different affinity (Fujita et al., 1986; Kurosaki et al., 1987; Claudio et al, 1987; Blount & Merlie, 1988). These results suggest that the α subunit alone is required for the formation of α-BuTX binding sites and that assembly with the other subunits is not required. However, the α subunit synthesized in a cell free system is incapable of binding to α-BuTX (Anderson & Blobel,1981), and in muscle cells the α subunit does not acquire the ability to bind to α-BuTX with high affinity for several minutes after translation (Merlie & Lindstrom,1983). Interestingly α-BuTX binding to this α subunit species was not blocked by curare or decamethonium (Carlin et al.,1986).

To see if incomplete AChRs synthesized in oocytes acquire α-BuTX binding capacity with high affinity, oocytes were injected with the α subunit mRNA alone or together with one of the other subunit mRNAs and incubated in the presence of ^{35}S methionine. ^{35}S methionine labeled translation products were solubilized from membranes and incubated with α-BuTX affinity gels, and then the affinity purified proteins were analyzed on SDS gels. When the α subunit mRNA was injected alone, a protein was always purified with the same mobility as the Torpedo AChR α subunit on SDS gels, confirming that the α subunit synthesized in oocytes is capable of binding to α-BuTX (Fig. 3B; also see Sumikawa & Miledi, 1988b). However, the intensity of the band obtained with affinity gels was always weaker than that of α subunit obtained with immunoprecipitation with α subunit specific antibody (Fig. 3C), suggesting that only some portion of the α subunit population acquires α-BuTX binding capacity with high affinity. Since α-BuTX gels were washed extensively before elution, the procedure will detect only α subunits which bind to α-BuTX with high affinity. The α subunits with low affinity for α-BuTX are most likely lost during washing of the affinity gels. Thus, the assembly of the other subunits to the α subunit (with high affinity binding capacity for α-BuTX) can be monitored using α-BuTX affinity gels. Using this approach we also found that the amount of α subunit recovered from oocytes injected either with the α subunit mRNA alone, or in combination with the β subunit mRNA, was less than that obtained from oocytes injected with α subunit mRNA in combination with γ or δ subunit mRNAs (Fig. 3B). Since the association of the α subunit with the β subunit occurs efficiently (see Fig. 3C), the results suggest that many of the αβ subunits failed to bind to α-BuTX with high affinity. These results also imply that association of the α subunit with either the γ or δ subunit is important for the formation of α-BuTX binding sites which resemble those of mature AChR molecules. Kurosaki et al (1987) have also found, although without studying the assembly, that the injection of α subunit mRNA with either the γ or δ subunit mRNA increased ^{125}I-BuTX binding activity in oocytes, while injection with the β subunit mRNA did not. In addition they observed that the ^{125}I-α-BuTX binding activities of the $\alpha\gamma$ and the $\alpha\delta$ molecules were suppressed by carbamylcholine, while the binding activities of the α and the $\alpha\beta$ were not, suggesting that association of the α subunit with either the γ or δ subunit is a prerequisite for formation of the agonist binding site.

Incomplete AChR molecules appear to be assembled about as efficiently as complete AChR molecules. Therefore, we next examined whether incomplete AChR molecules were inserted in the membrane as efficiently as the complete receptors (Sumikawa & Miledi, 1988b). The surface AChRs in intact oocytes can be specifically labeled with α-BuTX, and after detergent solubilization the α-BuTX-AChR complexes can be immunoprecipitated with anti-α-BuTX antibody for further characterization. Antibodies which recognize the extracellular domains of the receptor can also be used for a similar purpose. However, in some cases where only three (or less) of the AChR subunits are expressed, isolation of the surface membranes may be necessary to study the structure of the AChR molecules inserted. Since it has been demonstrated that the Torpedo α subunit expressed in yeast cells was inserted in the outside membrane with abnormal orientation (Fujita et al.,1986), it could be that translation of an incomplete set of subunit mRNAs in the oocyte may also result in insertion of AChR subunits with unusual conformation. Therefore, α-BuTX and/or some antibodies, which normally recognize the extracellular surface of mature AChRs, may not recognize the incomplete AChR molecules in the surface membranes of the intact oocytes. For this reason, it is necessary to isolate the surface membranes prior to characterization. Once AChR molecules are solubilized from the isolated membranes, α-BuTX and/or any antibodies which recognize any part of the AChR molecules may be used. The oocyte is an ideal system for this, since the large size of the oocyte allows us to dissect out the surface membrane manually (Sadler & Maller, 1981) to characterize the surface AChRs (Sumikawa & Miledi, 1988b).

Various combinations of the subunit mRNAs were injected into oocytes and the surface membranes manually dissected. After detergent solubilization, the compositions of the surface AChR complexes were studied by immunoprecipitation with the α subunit specific antibody and SDS gel electrophoresis. In all cases only a low level of incomplete molecules appeared to be inserted into the surface membranes, but most of them remained in intracellular compartments (Fig. 4; also see Sumikawa & Miledi, 1988b). In contrast, completely assembled AChR molecules were inserted into the surface membranes much more efficiently than incompletely assembled molecules (Fig. 4). Thus, this result indicates that assembly of receptors comprised of all four subunits is required for their efficient insertion into the surface membrane, and that the small membrane current responses observed with incomplete receptors can be explained, at least partially, by inefficient membrane insertion of these molecules.

α subunit specific antibody
surface membranes

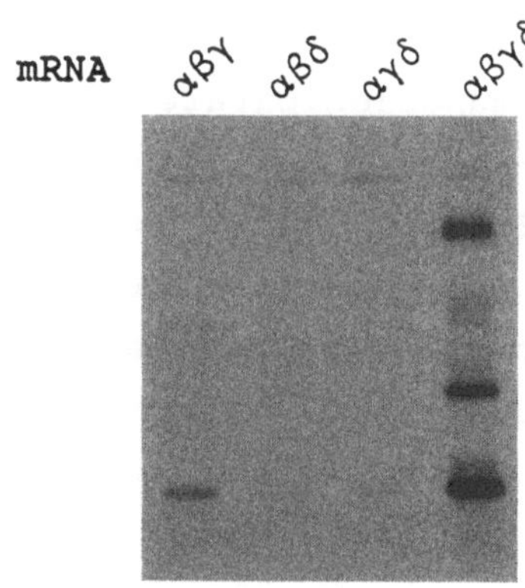

Fig. 4. Assembly and surface membrane insertion of AChR subunits. Oocytes were injected with subunit mRNAs as indicated and cultured with ^{35}S methionine. The surface membranes (with surrounding connective tissues and cells) of each batch of five oocytes were separated manually and processed as in Fig. 3A.

What kinds of post-translational modification are required for assembly and membrane insertion?

Using tunicamycin to block protein N-glycosylation, it was shown that N-linked carbohydrate attachment is required for functional expression of the AChR in cultured muscle cells. In these studies tunicamycin has been reported either to block the assembly of AChR subunits (Merlie et al., 1982) or to increase the rate of degradation of the AChR molecules (Prives & Olden,1980). The importance of N-glycosylation in the expression of functional AChR was confirmed by site directed mutagenesis. When a single potential N-glycosylation site in the α subunit was eliminated by this technique, the expression of the functional AChR in the oocytes was completely abolished (Mishina et al., 1985). However, in that study it was not determined whether the mutated α subunit was capable of being assembled.

In attempting to understand the role of N-glycosylation in the assembly of AChR subunits, we injected the synthetic AChR subunit mRNAs into Xenopus oocytes, cultured them in the presence of tunicamycin and characterized the synthesized AChR molecules by immunoprecipitation and SDS gel electrophoresis (Sumikawa & Miledi, 1988c). We first examined whether the assembly of the α and β subunits occurred in the presence of tunicamycin. For this purpose the two mRNAs were injected into the oocytes and the assembly was assayed by immunoprecipitation with the α subunit specific antibody, followed by SDS gel electrophoresis. As shown in Fig. 5, the α subunit specific antibody co-immunoprecipitated the non-N-glycosylated β subunit, suggesting the non-N-glycosylated β subunit was associated with the α subunit. However, since tunicamycin failed to block N-glycosylation completely, the association might occur only between the glycosylated α subunit and the non-N-glycosylated β subunit. In order to exclude this possibility the glycosylated proteins were adsorbed on concanavalin A-Sepharose gel and non-N-

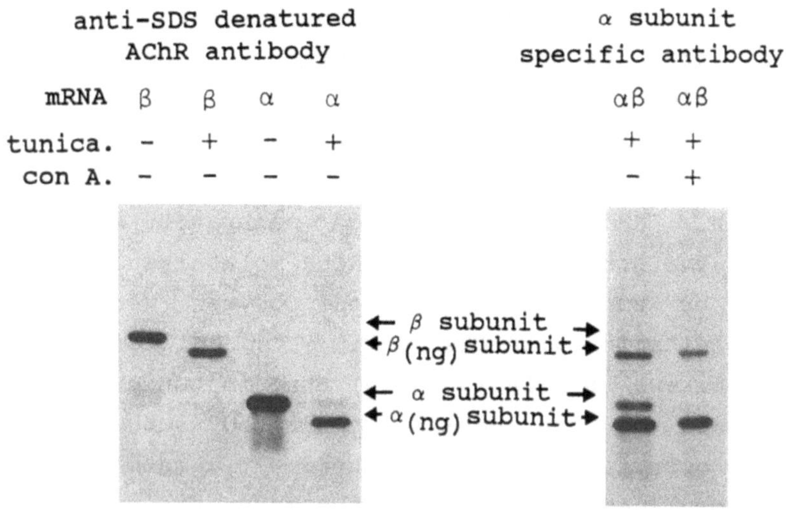

Fig. 5. Oocytes were injected with AChR subunit mRNAs as indicated and cultured in the absence or presence of tunicamycin. The synthesized AChR subunits were immunoprecipitated with antibody against SDS denatured AChR or the α subunit specific antibody as indicated.

glycosylated proteins were subjected to immunoprecipitation with the α subunit specific antibody. The result (Fig.5) clearly indicates that association of the non-N-glycosylated α and β subunits occurred. In the similar studies, the mixture of all four subunit mRNAs was injected into the oocytes. Although at the concentration used tunicamycin failed to block N-glycosylation completely, the results obtained again clearly suggest that assembly of the non-N-glycosylated subunits occurred. However, the association of the non-N-glycosylated δ subunit appeared to be less efficient (Fig.6; also see Sumikawa & Miledi, 1988c). In most cases tunicamycin treated oocytes appeared slightly unhealthy, which might explain why the intensity of AChR subunit bands obtained from tunicamycin treated oocytes were always weaker than those from control oocytes (Sumikawa & Miledi, 1988c). The possibility , that non-N-glycosylated AChR subunits degraded more rapidly than glycosylated subunits, remains to be investigated using site-directed mutant AChR subunits.

Since N-glycosylation of the subunits was apparently not required for assembly, we next examined the role of N-glycosylation in the insertion of AChR subunits into plasma membranes (Sumikawa & Miledi, 1988c). Oocytes injected with all four AChR subunit mRNAs were cultured in the presence of tunicamycin, and their surface membranes were isolated manually. The structures of the AChR molecules in the plasma membranes of tunicamycin treated oocytes were studied by immunoprecipitation with the anti α subunit antibody. We found (Sumikawa & Miledi, 1988c) that in tunicamycin treated oocytes most of the non-N-glycosylated subunits remained inside the oocytes, even though there was assembly of all four non-N-glycosylated subunits. Furthermore, the small number of non-N-glycosylated α and β subunits inserted in the plasma membranes could be adsorbed on concanavalin A-Sepharose gels (Fig. 6), suggesting that these non-N-glycosylated subunits were associated with other glycosylated subunits and, hence had been inserted into the plasma membranes. Thus, it appears that assembly of the four different non-N-glycosylated subunits occurs in tunicamycin treated oocytes, but that this non-N-glycosylated complex is not inserted in the plasma membranes (Fig. 6). The results suggest that N-glycosylation plays an important role in the insertion of AChR molecules into plasma membranes. The main problem with using tunicamycin is that it does not block N-glycosylation completely. The presence of both non-N-glycosylated and N-glycosylated subunits makes it difficult to study the efficiency of the non-N-glycosylated subunit assembly, and membrane insertion. This point will be studied at a later time using site directed mutant subunits.

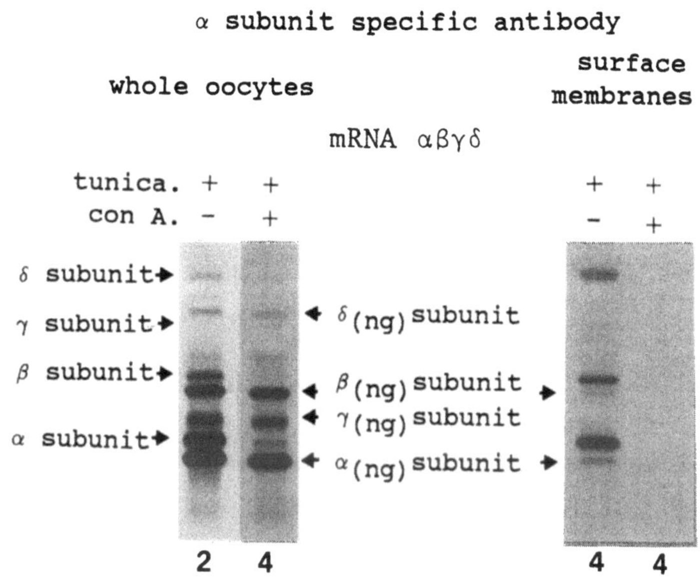

Fig. 6. Assembly and surface membrane insertion of AChR subunits in tunicamycin treated oocytes. All four AChR subunit mRNAs were injected into tunicamycin treated oocytes and cultured as in Fig.5. The synthesized AChR subunits were immunoprecipitated with the α subunit specific antibody. The non-N-glycosylated γ and δ subunits were identified by experiments similar to Fig. 5. The surface membranes (with enveloping cells) of each batch of 5-10 oocytes were separated manually. The surface membranes were then washed carefully to reduce contamination by the intracellular material. Detergent extract from surface membranes was immunoprecipitated with the α subunit specific antibody as in Fig. 5. Note the presence of both non-glycosylated and glycosylated subunits in tunicamycin treated oocytes. Numbers along bottom of gels indicate exposure times (days) to x-ray films; ng = non-N-glycosylated.

CONCLUSIONS

The use of Xenopus oocytes to translate functional receptor/channels from foreign mRNA has become widely popular following our original demonstration that functional ACh receptors could be expressed by mRNAs from Torpedo electric organ (Sumikawa et al, 1981; Barnard et al., 1982) and cat muscle (Miledi & Sumikawa, 1982; Miledi et al., 1982). Most studies using this system have employed electrophysiological methods to examine the properties of the expressed receptors (for reviews see Dascal, 1987; Miledi et al., 1988). However, when mutant receptors are expressed to study structure-function relationships, it is necessary to characterize the receptors biochemically, as well as electrophysiologically. This is because some mutations may affect the biogenesis of receptor-channel complexes through alterations through alterations at stages including subunit assembly and membrane insertion. Thus, conclusions drawn from electrophysiological data alone may be incomplete. We demonstrate here the utility of the Xenopus oocyte system for study of the biogenesis of receptor and ion channel molecules. Using this system we described two factors which are important for the insertion of AChRs into plasma membranes, and which are likely to be of general importance for the insertion of many other types of receptor molecule.

ACKNOWLEDGEMENTS

We are very grateful to Drs. T. Claudio (Yale University) and S. Heineman (Salk Institute) for kindly providing Torpedo AChR cDNA clones. This work was supported by the NIH and the Academic Senate Committee on Research, U.C.I.

REFERENCES

Anderson, D. J. & Blobel, G. (1981) In vitro synthesis, glycosylation, and membrane insertion of the four subunit of Torpedo acetylcholine receptor. Proc. Natl. Acad. Sci. USA 78, 5598-5602.

Anholt, R., Lindstrom, J. & Montal. M. (1984) The molecular basis of neurotransmission: structure and function of the nicotinic acetylcholine receptor. In: The Enzymes of Biological Membranes, Ed. Martonosi, A. Vol. 3, pp 335-401, Plenum Press, New York.

Barnard, E.A., Miledi, R. & Sumikawa, K. (1982) Translation of exogenous messenger RNA coding for nicotinic acetylcholine receptors produces functional receptors in Xenopus oocytes. Proc. R. Soc. Lond. B215, 241-246.

Barnard, E.A., Darlison, M.G. & Seeburg, P. (1987) Molecular biology of the GABA receptor: the receptor/channel superfamily. Trends Neurosci. 10, 502-509.

Barrantes, F.J. (1983) Recent developments in the structure and function of acetylcholine receptor. Int. Rev. Neurobiol. 24, 259-341.

Blount, P. & Merlie, J. P. (1988) Native folding of an acetylcholine receptor α subunit expressed in the absence of other receptor subunits. J. Biol. Chem. 263, 1072-1080.

Carlin, B.E., Lawrence, J. C., Lindstrom, J. M. & Merlie, J. P. (1986) An acetylcholine receptor precursor α subunit that binds α-bungarotoxin but not d-tubocurarine. Proc. Natl. Acad. Sci. USA 83, 498-502.

Claudio, T., Ballivet, M., Patrick, J. & Heinemann, S. (1983) Nucleotide and deduced amino acid sequences of Torpedocalifornica acetylcholine receptor γ subunit. Proc. Natl. Acad. Sci. USA 80, 1111-1115.

Claudio, T., Green, W. N., Hartman, D. S., Hayden, D., Paulson, H. L., Sigworth, F. J., Sine, S. M., & Swedlund, A. (1987) Genetic reconstitution of functional acetylcholine receptor channels in mouse fibroblasts. Science, 238, 1688-1694.

Conti-Tronconi, B.M. & Raftery, M.A. (1982) The nicotinic cholinergic receptor: correlation of molecular structure with functional properties. Ann. Rev. Biochem. 51, 491-530.

Dascal, N. (1987) The use of Xenopus oocytes for the study of ion channels. CRC Crit. Rev. Biochem.22, 317-387.

Fujita, N., Nelson, N., Fox, T., Claudio, T., Lindstrom, J., Reizman, H. & Hess, G. (1986) Biosynthesis of the Torpedocalifornica receptor α subunit in yeast. Science, 231, 1284-1287.

Gershoni, J. M., Hawrot, E. & Lentz, T. L. (1983) Binding of α-bungarotoxin to isolated α subunit of the acetylcholine receptor of Torpedo californica: quantitative analysis with protein blots. Proc. Natl. Acad. Sci. USA 80, 4973-4977.

Green, P. J., Pines, O. & Inouye, M. (1986) The role of antisense RNA in gene regulation. Annu. Rev. Biochem. 55, 569-597.

Haggerty, J. G. & Froehner, S. C. (1981) Restoration of ^{125}I-α-bungarotoxin binding activity to the α-subunit of Torpedo acetylcholine receptor isolated by gel electrophoresis in sodium dodecyl sulfate. J. Biol. Chem. 256, 8294-8297.

Harland, R. & Weintraub, H. (1985) Translation of mRNA injected into Xenopus oocytes is specifically inhibited by antisense RNA. J. Cell. Biol. 101, 1094-1099.

Karlin, A. (1980) Molecular properties of nicotinic acetylcholine receptors. In: Cell Surface and Neuronal Function. Eds. Cotman, C.W., Poste, G. & Nicolson, G.L. Vol. 6, pp. 192-260, Elsevier Biomedical, Amsterdam, New York.

Kurosaki, T., Fukuda, K., Konno, T., Mori, Y., Tanaka, K., Mishina, M. & Numa, S. (1987) Functional properties of nicotinic acetylcholine receptor subunits expressed in various combinations. FEBS Letts. 214, 253-258.

Melton, D. (1985) Injected anti-sense RNAs specifically block messenger RNA translation in vivo. Proc. Natl. Acad. Sci. U.S.A. 82, 144-148.

Merlie, J.P., Sebbane, R., Tzartos, S. & Lindstrom, J. (1982) Inhibition of glycosylation with tunicamycin blocks assembly of newly synthesized acetylcholine receptor subunits in muscle cells. J. Biol. Chem. 257, 2694-2701.

Merlie, J.P. & Lindstrom, J. (1983) Assembly in vivo of mouse muscle acetylcholine receptor: identification of an α subunit species that may be an assembly intermediate. Cell, 34, 747-757.

Merlie, J. P. & Smith, M. M. (1986) Synthesis and assembly of acetylcholine receptor, a multisubunit membrane glycoprotein. J. Memb. Biol. 91, 1-10.

Miledi, R. & Sumikawa, K. (1982) Synthesis of cat muscle acetylcholine receptors by Xenopus oocytes. Biomed. Res. 3, 390-399.

Miledi, R., Parker, I. & Sumikawa, K. (1982) Properties of acetylcholine receptors translated by cat muscle mRNA in Xenopus oocytes. EMBO J. 1, 1307-1312.

Miledi, R., Parker, I. & Sumikawa, K (1988) Transplanting receptors from brains into oocytes. In: Fidia Award Lecture Series. Ed. Smith, J., Raven Press (in press).

Mishina, M., Tobimatsu, T., Imoto, K., Tanaka, K., Fujita, Y., Fukuda, K., Kurasaki, M., Takahashi, H., Morimoto, Y., Hirose, T., Inazama, S., Takahashi, T., Kuno, M. & Numa, S. (1985) Location of functional regions of acetylcholine receptor α-subunit by site-directed mutagenesis. Nature, 313, 364-369.

Oblas, B., Boyd, N. D. & Singer, R. H. (1983) Analysis of receptor-ligand interactions using nitrocellulose gel transfer: application to Torpedo acetylcholine receptor and alpha-bungarotoxin. Anal. Biochem. 130, 1-8.

Popot, J-L. & Changeux, J-P. (1984) Nicotinic receptor of acetylcholine: structure of an oligomeric integral membrane protein. Physiol. Rev. 64, 1162-1239.

Prives, J.M. & Olden, K. (1980) Carbohydrate requirement for expression and stability of acetylcholine receptor on the surface of embryonic muscle cells in culture. Proc. Natl. Acad. Sci. USA 77, 5263-5267.

Sadler, S.E. & Maller, J.L. (1981) Identification of a steroid receptor on the surface of Xenopus oocytes by photoaffinity labeling. J. Biol. Chem. 256, 6368-6373.

Sumikawa, K., Houghton, M., Emtage, J.S., Richards, B.M. & Barnard, E.A. (1981) Active multi-subunit ACh receptor assembled by translation of heterologous mRNA in Xenopus oocytes. Nature, 292, 862-864.

Sumikawa, K., Parker, I. & Miledi, R. (1984) Separate fractions of mRNA from Torpedo electric organ induce chloride channels and acetylcholine receptors in Xenopus oocytes. EMBO J. 3, 2291-2294.

Sumikawa, K. & Miledi, R. (1988a) Repression of nicotinic acetylcholine receptor expression by antisense RNAs and an oligonucleotide. Proc. Natl. Acad. Sci. USA. 85, 1302-1306.

Sumikawa, K., & Miledi, R. (1988b) Assembly of all ACh receptor subunits is a prerequisite for their efficient insertion into plasma membranes. (submitted).

Sumikawa, K., & Miledi, R. (1988c) N-glycosylation of ACh receptor is required for insertion into plasma membranes, but not for subunit assembly. (submitted).

Tzartos, S. J., & Changeux, J.-P. (1983) High affinity binding of α-bungarotoxin to the purified α-subunit and to its 27,000-dalton proteolytic peptide from Torpedo marmorata. Requirement for sodium dodecyl sulfate. EMBO J. 2, 381-387.

White, M.M., Mayne, K.M., Lester, H.A. and Davidson, N. (1985) Mouse-Torpedo hybrid acetylcholine receptors: Functional homology does not equal sequence homology. Proc. Natl. Acad. Sci. USA, 82, 4852-4856.

Wilson, P. T., Lentz, T. L. & Hawrot, E. (1985) Determination of the primary amino acid sequence specifying the α bungarotoxin binding site on the α-subunit of the acetylcholine receptor from Torpedo californica. Proc. Natl. Acad. Sci. 82, 8790-8794.

STABLE EXPRESSION OF MULTISUBUNIT PROTEIN COMPLEXES IN MAMMALIAN CELLS

Toni Claudio, Deborah S. Hartman, William N. Green, Anthony F. Ross, Henry L. Paulson, and Deborah Hayden

Department of Cellular & Molecular Physiology
Yale University School of Medicine
333 Cedar Street
New Haven CT 06510
USA

INTRODUCTION

There are advantages and disadvantages to each of the different protein expression systems available. Whereas ease, speed, and expense are important factors in determining which type of expression system to use, the complexity of the protein to be expressed and the kinds of questions addressed will often be the factors that ultimately determine which system is used. Our interest has been in expressing the nicotinic acetylcholine receptor (AChR). This protein is a heterologous multisubunit complex composed of four different subunits in the stoichiometry $\alpha_2\beta\gamma\delta$. In addition to the complex subunit composition, each subunit undergoes several posttranslational modifications and spans the membrane a multiple number of times. In order to reconstitute all of the functional properties of the AChR, the four subunits must be present and at least some of the posttranslational modifications must be executed correctly. Thus, whichever gene transfer technique one uses to express AChRs, the ability to introduce all four of the subunits into the same cell and to introduce them into a cell type that is capable of performing the necessary modifications, is of considerable importance. A mechanical technique such as microinjection would clearly be the easiest solution to the problem of multiple subunits. Microinjection into *Xenopus* oocytes of the the mRNA isolated from *Torpedo* electric organ (Sumikawa et al., 1981) or cRNA prepared *in vitro* (Mishina et al., 1984; White et al., 1985; Claudio, 1987) has proven to be a very effective method for expressing functional AChRs. One of the goals of our laboratory is to be able to study the process of nerve-induced AChR clustering. For this and other studies, we require an expression system in which large numbers of identical cells express receptors on their surface for long periods of

NATO ASI Series, Vol. H 32
Molecular Biology of Neuroreceptors and Ion Channels. Edited by A. Maelicke

time. These requirements are not readily attained with transient systems, and thus we have chosen to establish a stable expression system in cultured cell lines.

COTRANSFECTION

A problem unique to heterologous multisubunit protein complexes is how to introduce each of the different cDNAs into the same cell. Using a calcium phosphate precipitation method (Graham & van der Eb, 1973; Wigler et al., 1977) we recently showed that the cotransfection efficiency of introducing four different cDNAs plus a selectable marker gene into the same cell was 80% (Claudio, 1987). In that study, we described the introduction of the four *Torpedo* subunit cDNAs plus the adenine phosphoribosyltransferase gene (aprt) into mouse fibroblast L cells deficient in thymidine kinase (tk) and aprt (Ltk^-aprt^-). We found (by genome blotting) that nine of eleven clonal isolates had incorporated the four cDNAs plus aprt into their genomes and that the majority of the material had been integrated correctly (see Fig. 2 in Claudio, 1987). In addition, of the nine positive cell lines, half had picked up approximately equal copy numbers (10 to 20) of each of the four cDNAs. These results strongly suggested that this method of gene transfer could be used effectively to introduce the DNAs of multisubunit protein complexes into cells. In that particular study, the cDNAs were not engineered into vectors that would express efficiently in mammalian cells and thus no information was available on the efficiency of protein expression.

In a subsequent study (Claudio et al., 1987), the four AChR subunit cDNAs engineered into simian virus 40 (SV40) expression vectors (pSV2) (Subramani et al., 1981) and the tk gene were cotransfected into Ltk^-aprt^- cells. Analysis of two clonal isolates revealed that one line (all-11) had integrated copies of all four subunit cDNAs (4:2:2:8 copies per cell for α, β, γ, δ, respectively) and the other line (all-6) had integrated three of the four subunit cDNAs (0:1:8:4 for α, β, γ, δ, respectively). The all-11 cell line was analyzed biochemically, pharmacologically, and electrophysiologically and it was shown to express fully functional cell surface *Torpedo* AChRs (Claudio et al., 1987). The conclusion from these studies was that stable expression in cell lines, established by cotransfection, was a method that could be used to study a multisubunit protein.

Both of the previous studies used a selectable marker gene that complimented a cellular mutation (tk^- or $aprt^-$). In a more recent study, we cotransfected the four AChR subunit cDNAs plus a dominant selectable marker gene into cultured cells. The marker gene was the neomycin resistance gene (neo^r) which encodes G418 (a neomycin analog) resistance in mammalian cells (Davies & Jimenez, 1980). The recipient cell line used was the mouse fibroblast NIH3T3 cell line. As in the other studies, the DNAs were introduced into cells using a calcium phosphate precipitation procedure, the cells were then put into selective medium, individual colonies were isolated using cloning cylinders, and they were grown into stable cell lines. Of twenty-three clonal isolates, seventeen were analyzed at the protein level to determine which ones were expressing which subunits. Four lines were identified that expressed all four subunits (one of these, all-15, is shown in Fig. 1, 4, 5), four lines expressed three subunits, five lines expressed two subunits, two lines expressed one subunit, and two lines expressed no subunits. Approximately one-fourth of the transfected cells that had incorporated the selectable marker gene, expressed all four subunits. Protein analysis on some of these cell lines is shown in Fig. 1. We have not yet performed genome blots on these lines and thus we do not know whether our cotransfection efficiency was lower than in previous studies, or whether it remained high (~80%) but the efficiency of protein expression of all four subunits was low (25%).

VIRAL INFECTION

Another method we used to transfer DNA into cells was to make retroviral recombinants and virally infect cells. This method was employed when we wished to introduce single copy cDNAs into genomes, or when we needed to introduce DNA into cell types that were resistant to taking up DNA by calcium phosphate precipitation methods. The cell lines of interest included several muscle cell lines such as rat L6 cells (Yaffe, 1968), mouse C2 cells (Yaffe & Saxel, 1977; Blau et al., 1983), and mouse BC_3H-1 cells (Schubert et al., 1974). We constructed our α cDNA into retroviral vectors kindly provided by R.C.

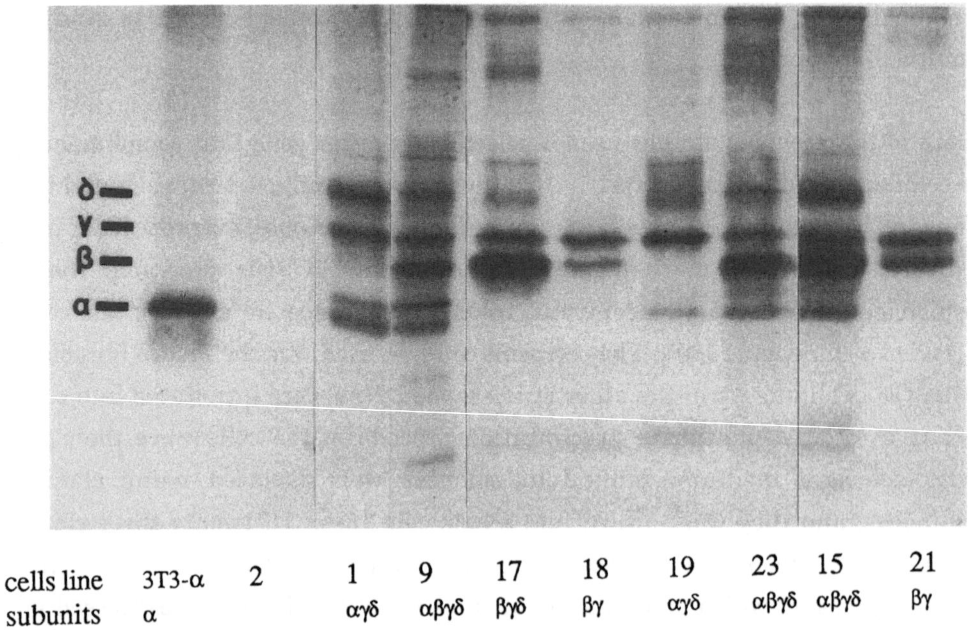

Fig. 1. Autoradiograph of metabolically labeled and immunoprecipitated fibroblast cell lines stably expressing different *Torpedo* AChR subunit combinations. **Methods:** A 100 mm dish of cells was metabolically labeled with 400 μCi of [^{3}H]leucine (lanes 1, 2, 5-10) or 400 μCi of [^{35}S]methionine (lanes 3, 4), the cells were solubilized, incubated with subunit-specific antiserum, and precipitated with Protein A-sepharose. The precipitates were electrophoresed on SDS gels, the gels were treated for fluorography, and X-OMAT film was exposed at -70°C with an intensifying screen as previously described (Claudio et al., 1989a). Typically, films in which cells had been labelled with [^{35}S]methionine were exposed for 10 to 18 hours and films in which cells had been labelled with [^{3}H]leucine were exposed for 24 to 72 hours. The band migrating between α and β in lanes 3 and 4 is actin. An actin band is usually visible when cells are labeled with [^{35}S]methionine but not visible when cells are labeled with [^{3}H]leucine.

Mulligan (Korman et al., 1987) or J. Morgenstern (J. Morgenstern and H. Land, unpublished). In both vectors, the gag, pol, and env genes of murine leukemia virus (MLV) were deleted and a cDNA can be introduced at a site between the two long terminal repeats (LTR). Downstream from the cDNA are a SV40 early promoter, a neo^r gene, and a pBR322 origin of replication. Transcription of the cDNA is driven by the LTR while an independent transcriptional unit is driven by the internal SV40 promoter which provides expression of the G418^r gene. These retroviral α-recombinants were transfected into the NIH3T3 packaging cell line φ2 (Mann et al., 1983), helper-free infectious viral particles containing the recombinant were harvested, and recipient cell lines were infected (Claudio et al., 1989b). Cells were put into selective medium, individual colonies were isolated using cloning cylinders, and grown into stable cell lines. Cell lines were analyzed for the expression of α polypeptide by metabolic labeling, followed by immunoprecipitation, SDS gel electrophoresis, and autoradiography. NIH3T3, L6, and C2 cells infected with the packaged retroviral α-recombinant are shown in Fig. 2.

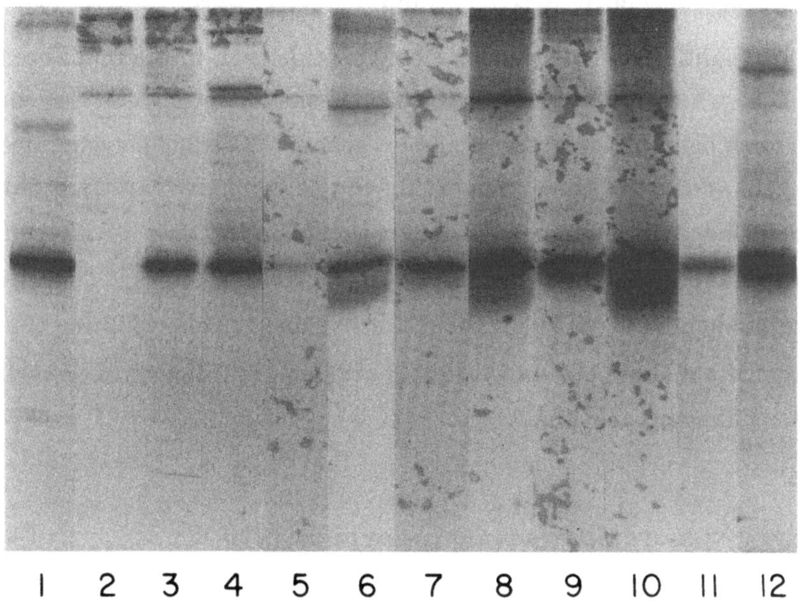

Fig. 2. Autoradiograph of metabolically labeled and immunoprecipitated fibroblast and muscle cell lines stably expressing the *Torpedo* AChR α subunit. In lane 1, confluent (~10^7 cells) NIH3T3-α cells; lane 2, 75% confluent, undifferentiated L6 cells; lane 3, 75% confluent, undifferentiated L6-α cells; lane 4, confluent, differentiated (expressing endogenous AChRs) L6-α cells; lane 5, 75% confluent, undifferentiated C2 cells; lane 6, 75% confluent, undifferentiated C2-α cells; lane 7, confluent, undifferentiated C2 cells; lane 8, confluent, undifferentiated C2-α cells; lane 9, confluent, differentiated C2 cells; lane 10, confluent, differentiated C2-α cells; lane 11, 75% confluent, undifferentiated BC_3H-1 cells; lane 12, confluent, undifferentiated BC_3H-1 cells. **Methods:** A 100 mm dish of cells was metabolically labeled with 400 μCi of [^{3}H]leucine, the cells were solubilized, incubated with anti-α antiserum, and precipitated with Protein A-sepharose. The precipitates were electrophoresed on SDS gels, the gels were treated for fluorography, and X-OMAT film was exposed at -70°C with an intensifying screen as previously described (Claudio et al., 1989a).

As shown in Fig. 2, *Torpedo* α polypeptide can easily be detected in all of those cell lines infected with the α-recombinant (lanes 1, 3, 4, 6, 8, 10). The particular anti-α antiserum used for these immunoprecipitations was a polyclonal serum prepared in rabbits and directed against gel purified and gel isolated *Torpedo californica* α subunit (Claudio & Raftery, 1977). Although prepared against *Torpedo* α, this antiserum cross-reacted quite strongly with rat and mouse α polypeptides, as well. This can be seen in lanes 7, 9, 11, and 12, the non-infected parent muscle cell lines. It is interesting to note that C2 (lane 7) and especially BC_3H-1 (lane12) cells were clearly expressing significant quantities of endogenous α polylpeptide even in the undifferentiated state, a time when no cell surface AChR expression is ever detected. In order to distinquish between endogenous and exogenous *Torpedo* α polypeptides in L6 and C2 cells, one strategy was to assay the cells when they were only 75% confluent. Under these conditions, expression of endogenous α was barely (L6, lane 2) or slightly (C2, lane 5) detectable. In contrast, significant quantities of α were detectable in 75% confluent BC_3H-1 cells (lane 11) necessitating the use of species-specific monoclonal antibodies in order to distinguish exogenous from endogenous α in this cell line.

THE EFFECTS OF TEMPERATURE ON EXPRESSION

We have observed that functional *Torpedo* AChRs are not formed in mammalian cells at 37°C. At temperatures lower than 37°C (~26°C), functional AChR complexes are formed. Such complexes can be detected both internally and properly inserted into the plasma membrane. Typical expression in all-11 cells is ~25,000 α-bungarotoxin (BuTx) binding AChR molecules on the surface per cell. This quantity is sufficient for biochemical and pharmacological analyses, as well as electrophysiological analyses using single channel recording techniques (Claudio et al., 1987).

It is likely that most, if not all, of the cellular functions occuring in normal mammalian cells (ie. transcription, translation, protein degradation, etc.) are retarded at 26°C compared with 37°C. As shown in Fig. 3, the peak surface expression of AChRs in all-11 cells is not achieved until about day 12 at 26°C.

We have shown that transcription is greatly retarded in these cells at 28°C. There is about a 100-fold increase in the level of transcription at 37°C compared with 28°C (see Fig. 2 lanes 1 vs. lanes 3 in Claudio et al., 1987). Significant levels of transcripts are not seen until days four or five. Thus, the slow rate of surface expression shown below in Fig. 3 can be attributed, at least in part, to the reduced rate of transcription at this temperature. Not shown in this figure, but shown in Fig. 3A of Claudio et al., 1987, there is absolutely no surface (or internal) expression of AChR complexes at 37°C. If the temperature of the incubator remains constant at 26°C, the level of expression also remains constant for approximately 10 to 20 days, after which time, the cells begin to die. If the temperature rises more than a few degrees above 26°C, a sharp decline in the level of expression of receptor occurs. We routinely observe stable levels of expression at 26°C and now believe that the sharp decline shown in Fig 3A of Claudio et al., 1987 was probably due to a rise in the temperature of the incubator.

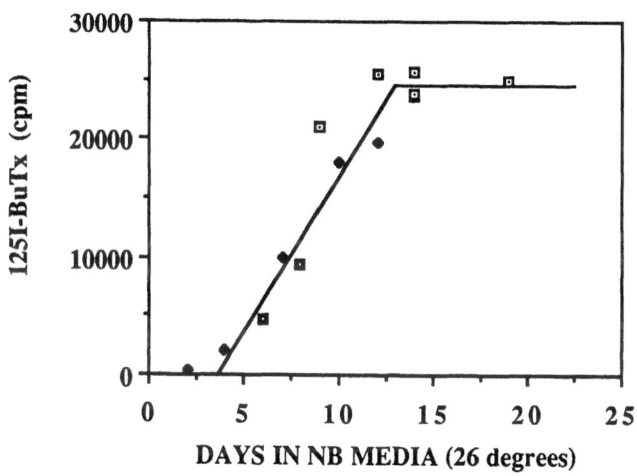

Fig. 3. Time course of surface AChR expression in all-11 cells at 26°C. **Methods:** All-11 cells were grown in a 37°C, 5% CO_2 incubator until confluent. Once confluent, the medium was supplemented with 10 mM sodium butyrate (NB) and dishes were placed in a 5% CO_2 incubator that was maintained at 26°C (time 0). On the days indicated, dishes of cells were surface labeled with 4-10 nM [^{125}I]BuTx for 2 1/2 hours at room temperature. Unbound toxin was washed away, cells were solubilized in a buffer containing 1% Triton X-100, and then counted in a gamma counter. The different symbols represent three different experiments.

In addition to the profound effect on the formation of functional AChR complexes, temperature had differential effects on the individual subunits. We tested the stability of each of the subunits in cell lines expressing only one of the subunits. When cells were metabolically labeled with [^{3}H]leucine or [^{35}S]methionine for 10 minutes, then chased for 0, 10, 20, 40, 90, or 180 minutes, we found that the α, γ, and δ subunits had virtually the same short half life of about 43 minutes at 37°C. In contrast, the β subunit had an extremely short half life of only about 12 minutes. When half life studies were performed at 26°C, each subunit half life increased. Again, the α, γ, and δ subunits displayed similar half lives of about 70 minutes whereas the β subunit half life was about 50 minutes. The observations that the β subunit has a half life different from the other three subunits, and that it is significantly shorter than the others, is suggestive that it might play some key regulatory role in the assembly process. If functional AChR complexes are allowed to form in cells grown at 26°C, the receptors are very stable and display a surface half life of approximately 60 to 100 hours when maintained at 26°C.

CELL SURFACE DISTRIBUTION

The results we had obtained thus far suggested that this expression system was capable of producing large quantities of identical cells which expressed fully functional, cell surface AChRs for long periods of time. In order to begin clustering studies, it was necessary to first determine the cell surface distribution of the receptors. This was achieved by incubating AChR-expressing cells with rhodamine-BuTx. As shown in Fig. 4 panels A' and B', the cells expressed quantities of cell surface receptors that were sufficient to be visualized by rhodamine-BuTx labeling. If cells were incubated with rhodamine-BuTx in the presence of 10 mM carbamylcholine, no fluorescence was detected (panel C'). The lack of BuTx binding in the presence of carbamylcholine indicates that the binding observed in panels A' and B' was specific for AChRs. In addition, the surface distribution of AChRs appeared to be fairly uniform. The question now was, were these AChRs capable of forming clusters on the surface of fibroblasts?

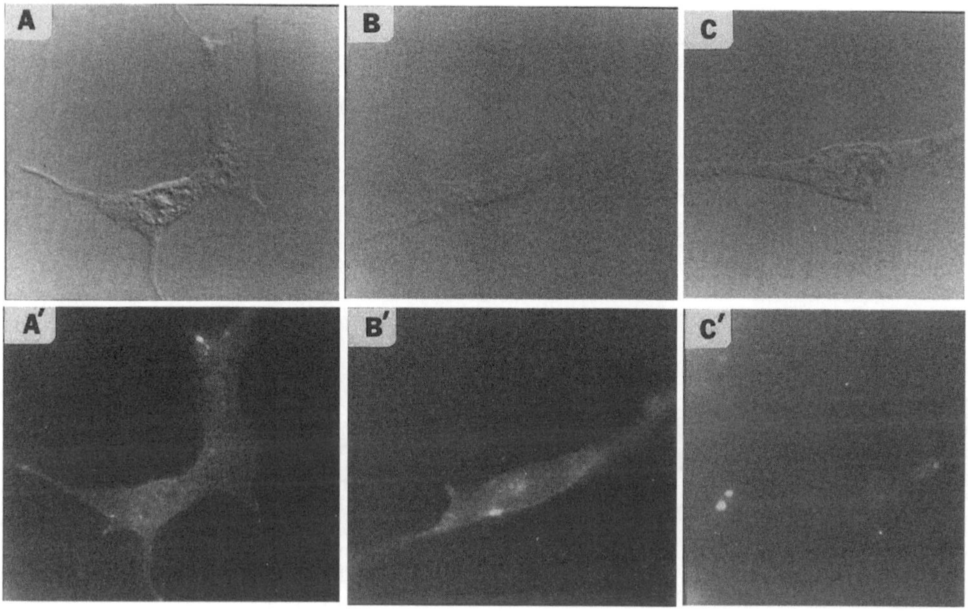

Fig. 4. Cell surface distribution of *Torpedo* AChRs expressed in mouse fibroblasts. AChR-fibroblast (all-15) cells were labeled with rhodamine-BuTx in the absence (panels A, A' and B, B') or presence (panels C, C') of carbamylcholine. Panels A, B, and C are cells visualized with Zeiss differential interference contrast (DIC) optics. Panels A', B', and C' are the same cells visualized with rhodamine fluorescence. **Methods:** Cells were grown on glass cover slips at 26°C in the presence of 10 mM sodium butyrate for six days. They were then labelled for 60 minutes at room temperature with tetra-rhodamine-BuTx (Molecular Probes Inc., Eugene, OR), washed, fixed in 20% acetone, and visualized in a Zeiss IM35 microscope with a Plan-NeoFluar 63x objective. In panel C (C'), 10 mM carbamylcholine was included during the labeling.

Coculturing muscle cells with nerve cells leads to the appearance of AChR clusters on the muscle cell surface. Additionally, several agents have also been reported to induce clusters in various muscle cell preparations (reviewed in Rubin & McMahan, 1982). Some of these agents include extracellular matrix (ECM), brain extract, and conditioned medium from nerve cells. AChR-expressing fibroblasts were incubated with *Torpedo* ECM for one or two days and then labelled with rhodamine-BuTx. As shown in Fig. 5, there seemed to be a real difference in the cell surface distribution of AChRs in these cells after treatment with ECM. Several smaller clusters of AChRs were apparent in the cells treated for one day with ECM and one large cluster was visible in the cell treated for two days with ECM.

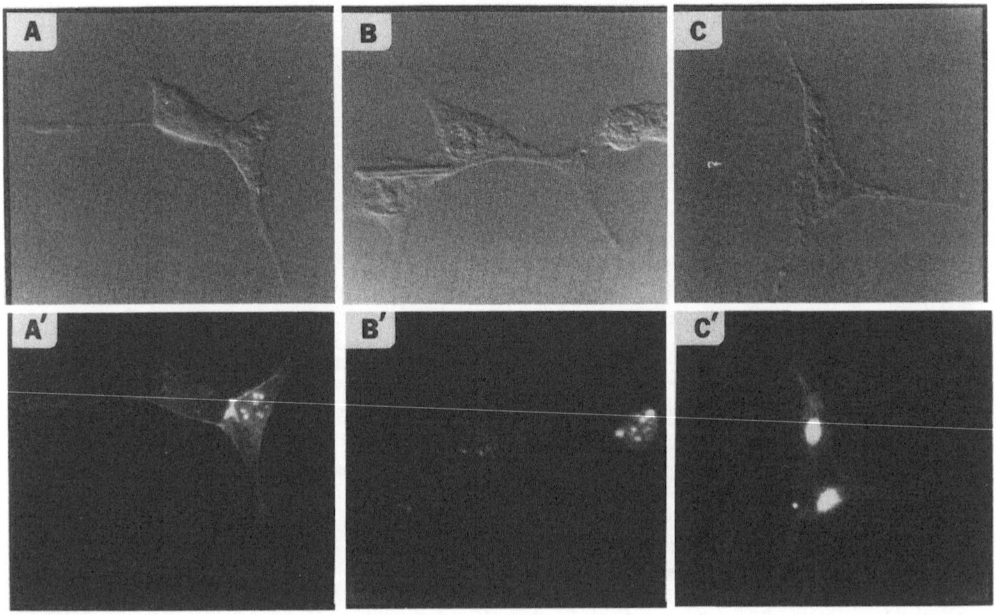

Fig. 5. Cell surface distribution of *Torpedo* AChRs expressed in mouse fibroblasts after incubation with *Torpedo* extracellular matrix. AChR-fibroblast (all-15) cells were labeled with rhodamine-BuTx after incubation with ECM for one day (panels A, A' and B, B') or two days (panels C, C'). Panels A, B, and C are cells visualized with Zeiss DIC optics. Panels A', B', and C' are the same cells visualized with rhodamine fluorescence. **Methods:** Cells were grown on glass cover slips at 26°C in the presence of 10 mM sodium butyrate for five days, then 10μl of *Torpedo* ECM (final concentration, 33 μg/ml) was added, and the cells were returned to the 26°C incubator for another one (panels A & B) or two days (panel C). Cells were then labelled for 60 minutes at room temperature with tetra-rhodamine-BuTx and processed as described in the legend to Fig. 4.

CONCLUSIONS

Using either cotransfection or viral infection techniques we have established cell lines containing one, two, three, or four different AChR subunits. In those cell lines expressing all four subunits, fully functional, cell surface AChRs were expressed. Although this system is much more time-consuming to establish than several transient expression systems, it has several advantages over other systems for the types of experiments our laboratory is conducting. In order to try to unravel the process of nerve-induced AChR clustering that one observes at the neuromuscular junction, we needed a system in which large quantities of identical cells were expressing cell surface AChRs for long periods of time. We have shown that the AChRs expressed in these fibroblasts contain all of the proper pharmacological and physiological properties expected of *Torpedo* AChRs. In addition, the results we have obtained thus far with one clustering agent, extracellular matrix, are suggestive that we will be able to study clustering in this artificial muscle system.

ACKNOWLEDGEMENTS

This work was supported by National Institutes of Health: grants NS21714 and HL38156 to TC, and postdoctoral training grant NS07102 to WNG and AFR.

REFERENCES

Blau, H.M., Chiu, C.-P., and C. Webster. 1983. Cytoplasmic activation of human nuclear genes in stable heterocaryons. *Cell* **32**, 1171-1180.

Claudio, T. (1987) Stable expression of transfected *Torpedo* acetylcholine receptor α-subunits in mouse fibroblast L cells. *Proc. Natl. Acad. Sci. USA* **84**, 5967-5971

Claudio, T. and Raftery, M.A. (1977) Immunological comparison of acetylcholine receptors and their subunits from species of electric ray. *Arch Biochem Biophys* **181**, 484-489

Claudio, T., Green, W.N., Hartman, D., Hayden, D., Paulson, H. L., Sigworth, F.J., Sine, S.M., and Swedlund, A. (1987) Genetic reconstitution of functional acetylcholine receptor-channels in mouse fibroblasts. *Science* **238**, 1688-1694.

Claudio, T., Paulson, H.L., Hartman, D., Sine, S., and Sigworth, F. (1989a) Establishing a stable expression system for studies of acetylcholine receptors. In: *Current Topics in Membranes and Transport*, "Molecular Biology of Ion Channels", Academic Press, New York, Vol. **33** pp 219-247.

Claudio, T., Paulson, H. L., Green , W.N., Ross, A.F., Hartman, D.S., and Hayden, D. (1989b) Fibroblasts transfected with *Torpedo* acetylcholine receptor β, γ, and δ subunit cDNAs express functional AChRs when infected with a retroviral-α recombinant. (submitted)

Davies, J. and A. Jimenez. 1980. A new selective agent for eukaryotic cloning vectors. *Am. J. Trop. Med. Hyg.* (Suppl.) **29**, 1089-1092.

Graham, R. and A. van der Eb. 1973. A new technique for the assay of infectivity of human adenovirus 5 virus. *Virol.* **52**, 456-467.

Korman, A.J., Frantz, J.D., Strominger, J.L., and R.C. Mulligan. 1987. Expression of human class II major histocompatibility complex antigens using retrovirus vectors. *Proc. Natl. Acad. Sci. USA* **84**, 2150-2154.

Mann, R., Mulligan, R.C., and D. Baltimore. 1983. Construction of a retrovirus packaging mutant and its use to produce helper-free defective retrovirus. *Cell* **33**, 153-159.

Schubert, D., Harris, A.J., Devine, C.E., and S. Heinemann. 1974. Characterization of a unique muscle cell line. *J. Cell Biol.* **61**, 398-413.

Subramani, S., Mulligan, R.C., and P. Berg. 1981. Expression of the mouse dihydrofolate reductase cDNA in simian virus 40 vectors. *Mol Cell Biol* **1**, 854-864.

Sumikawa, K., Houghton, M., Emtage, J.S., Richards, B.M. and Barnard, E.A. (1981)

Active multi-subunit ACh receptor assembled by translation of heterologous mRNA in *Xenopus* oocytes. *Nature* **292**, 862-864.

Mishina, M., Kurosaki, T., Tobimatsu, T., Morimoto, Y., Noda, M., Yamamoto, T., Terao, M., Lindstom, J., Takahashi, T., Kuno, M. and Numa, S. (1984) Expression of functional acetylcholine receptor from cloned cDNAs. *Nature* **307**, 604-608.

Rubin, L.L. & McMahan, U.J. (1982) Regeneration of the neuromuscular junction: steps toward defining the molecular basis of the interaction between nerve and muscle. In: Schotland, D.L. (ed) *Disorders of the Motor Unit*, John Wiley & Sons, pp. 187-196.

White, M.M., Mixter-Mayne, K., Lester, H.A. and Davidson, N. (1985) Mouse-*Torpedo* hybrid acetylcholine receptors: functional homology does not equal sequence homology. *Proc Natl Acad Sci* **82**, 4852-4856.

Wigler, M., Silverstein, S., Lee, L.-S., Pellicer, A., Cheng, Y.-c. and Axel, R. 1977. Transfer of purified herpes virus thymidine kinase gene to cultured mouse cells. *Cell* **11**, 223-232.

Yaffe, D. 1968. Retention of differentiation potentialities during prolonged cultivation of myogenic cells. *Proc. Natl. Acad. Sci USA* **61**, 477-483.

Yaffe, D. and O. Saxel. 1977. Serial passaging and differentiation of myogenic cells isolated from dystrophic mouse muscle. *Nature* **270**, 725-727.

REGULATION OF ACETYLCHOLINE RECEPTOR GENE EXPRESSION DURING DEVELOPMENT OF THE NEUROMUSCULAR JUNCTION

Jean-Pierre Changeux, Jean Cartaud[1], Anne Devillers-Thiéry, Bertrand Fontaine, Hoang Nghiêm, André Klarsfeld, Ralph Laufer, Marten Osterlund[2], and Jacques Piette
Molecular Neurobiology Laboratory, Pasteur Institute
25 rue du Dr Roux
75724 Paris Cedex 15, France

Introduction

The nicotinic acetylcholine receptor (AChR) is a membrane-bound allosteric protein which operates, in the millisecond timescale, as an acetylcholine-gated cationic channel. The functional architecture of this heterologous α_2 β γ δ pentamer is actively investigated (Changeux, 1981 ; Popot and Changeux, 1984 ; Hucho, 1986 ; Changeux et al., 1987 ; Lindstrom et al., 1987 and several papers in this Volume) and the regions of the molecule which form its two main functional domains : the acetylcholine binding site (Kao et al., 1984 ; Dennis et al., 1988) and the agonist gated ion channel (Giraudat et al., 1986, 1988 ; Oberthür et al., 1986 ; Hucho et al., 1986 ; Imoto et al., 1986, 1988 ; Leonard et al. in press ; Oiki et al. in press) have recently been defined at the amino acid level.

The AChR protein is a dominant component of the postsynaptic membrane of the motor endplate. The subcellular distribution of AChR is therefore an important feature relevant to its involvment in signal transduction. In the adult motor endplate the density of AChR reaches 10.000 molecules per μm^2 and drops more than a 1.000 fold a few microns away from the endplate. In other words, the AChR is strictly localized at the level of the subneural domain of the muscle fiber. Moreover, in the adult, this distribution can be modified by section of the motor nerve or by chronic pharmacological blocking of synaptic

1 Institut Jacques Monod
Université Paris VII
2 Place Jussieu,
75251 Paris Cédex 05, France
2 Present address : Karo Bio,
Huddinge University Hospital, R64
S-14186 Huddinge, Sweden.

NATO ASI Series, Vol. H 32
Molecular Biology of Neuroreceptors and Ion Channels. Edited by A. Maelicke

transmission. The muscle fiber becomes hypersensitive to ACh and AChR molecules appear outside the endplate. A long-term regulation of AChR number and distribution takes place (review Fambrough, 1979 ; Salpeter and Loring, 1985). Here, we will report recent studies from this laboratory the molecular mechanisms engaged in this spatio-temporal evolution.

Spatio-temporal evolution of AChR distribution and properties during endplate formation

The focal distribution of the AChR in adult muscle is the outcome of a complex set of regulatory processes that affect AChR biosynthesis during embryogenesis and early postnatal development (for reviews, see Fambrough, 1979 ; Changeux, 1981 ; Salpeter and Loring, 1985 ; Merlie and Smith, 1986 ; Schuetze and Role, 1987 ; Changeux et al., 1985, 1987 ; Laufer and Changeux, 1989). Four main aspects of these regulations may be tentatively distinguished :

1) Upon differentiation of myoblasts and their fusion into myotubes, the biosynthesis of AChR increases AChR molecules become diffusely distributed all along the myotubes at a density of about 100-500 molecules per μm^2, exhibit significant rotational and lateral motion, undergo rapid metabolic turnover (half-life 17-22 hours) and possess a long mean channel open time (3-10 msec).
2) As soon as the growth cone of an exploratory motor axon contacts the surface of the myotube, the AChR molecules start to cluster under the nerve ending, become immobile and electrical signals are transmitted via release of ACh from the nerve terminals.
3) When the first neuromuscular contacts become functional, the surface density of extrajunctional AChR sharply declines as a consequence of a repression of its biosynthesis by the neurally evoked electrical firing of the muscle fiber.
4) Meanwhile, around birth in the rat (or more than 3 weeks after hatching in the chick), the metabolic half-life of the AChR molecule increases up to a value of about 11 days in adult muscle. Then, about a week after birth in some species (frog, rat calf), but not in the chick, the mean duration of channel openings becomes 3-5

times shorter (review Schuetze and Role 1987). Throughout this phase, the overall number of AChR molecules in the postsynaptic domain steadily augments and after birth this increase can be as large as 15 to 20-fold.

In the past years, the genes coding for AChR subunits in vertebrate muscles have been cloned and sequenced (Ballivet et al., 1983 ; Noda et al., 1983 ; La Polla et al., 1984 ; Nef et al., 1984 ; Boulter et al., 1985 ; Klarsfeld and Changeux, 1985) allowing an analysis of several of these mechanisms at the nucleic acid level.

Differential expression of AChR α-subunit gene in muscle nuclei under the endplate and outside the endplate during development (Fontaine et al., 1988 ; Fontaine and Changeux, 1989).

The simplest hypothesis which accounts for the uneven distribution of the AChR protein under the motor nerve ending and outside the endplate takes into account the presence of multiple nuclei in the adult muscle fiber. Accordingly, the biosynthesis of AChR would take place, in a privileged manner, at the level of the nuclei which underlie the endplate ; these nuclei which were referred to as "fundamental" by Ranvier (1875), because of their distinct morphology, and tentatively related to the biosynthesis of synapse-specific proteins, already, by Couteaux in 1978.

Merlie and Sanes (1985) have compared steady-state levels of AChR α-and δ-subunit mRNAs in hand-dissected endplate-rich and extra-junctional regions of mouse diaphragm and noticed that the levels of α and δ-subunit mRNA were 3-and 9- to 14-fold higher, respectively, in the subsynaptic than extrasynaptic regions. They considered that such distribution could arise as a result of : 1) a directed transport of mRNA toward the synaptic regions, 2) an enhanced stability of AChR mRNA near the endplates, 3) an increased transcription of AChR subunit genes by the nuclei underlying the synapse.

To decide between these possibilities, in situ hybridization to mature and unspliced mRNA transcripts was carried out using fragments of chicken genomic DNA coding for the α-subunit, respectively with

(Fontaine et al., 1988 ; Fontaine and Changeux, 1989) or without (Fontaine and Changeux, 1989) exonic sequences as templates for the synthesis of antisense ^{35}S labelled transcripts.

In two different muscles, the slow, multi-innervated, anterior latissimus dorsi (ALD) and the fast, singly innervated, posterior latissimus dorsi (PLD), from 15-day old chick, AChR α-subunit mRNA was detected in situ at discrete regions on myofibers. These domains colocalize (80 % correspondence) with motor endplates identified by histochemical staining for acetylcholinesterase. At this level, the autoradiographic grains accumulated on and around the "fundamental" nuclei.

Denervation of ALD and PLD has been reported to increase ACh sensitivity in extra-junctional areas (Bennett et al., 1973) as in other vertebrate skeletal muscles (review Salpeter and Loring, 1985). Northern blot analysis of AChR α-subunit mRNA reveals important increases in the steady-state levels of mature mRNA in PLD and ALD 4 days after denervation of 11-day-old chicks (Fontaine et al., 1988) as observed in other systems (Merlie et al., 1984 ; Klarsfeld and Changeux, 1985 ; Goldman et al., 1985 ; Evans et al., 1987 ; Moss et al., 1987 ; Shieh et al., 1988). In situ, clusters of grains appear at the level of about 10 % of the nuclei distributed at random all along the muscle fibers. A similar picture was observed with the strictly intronic probe supporting the view (see also Shieh et al., 1987) that the transcriptional activity of individual nuclei varied under these conditions. In contrast, mRNAs for actin were found diffusely distributed all over muscle fibers during development and did not significantly change after denervation (Fontaine et al., 1988 ; Fontaine and Changeux, 1989).

To analyse the development of the adult pattern, in situ hybridization studies were carried out with developing embryos, first, at the level of the somites where the myoblast lineage arises. Consistent with the observations of Meiniel and Bourgeois (1982) with labeled α-bungarotoxin, AChR α-subunit mRNA was found to accumulate at high levels in the mono-nucleated myotomal cells of the somites and in the early stages of myotomes differentiation, but not in the presomitic mesoderm. Baldwin et al. (1988) have recently reported that AChR subunits mRNAs accumulate in the somites of Xenopus, yet, in this species, myotomal muscle cells do not fuse into myotubes making the comparison difficult. These high levels of α-subunit message persist when the myoblasts fuse into myotubes. In ED (embryonic day) 11 PLD, both mature and precursor mRNAs are detected all over the developing muscle fiber. Subsequently, at ED 16, the total number of grains distributed along the muscle fiber decreases but several clusters of grains persist. At ED 19, only one cluster is observed per muscle fiber. The distribution of AChR α-subunit mRNAs in PLD becomes restricted to the newly formed endplate (Fontaine and Changeux, 1989).

The evolution of the number and distribution of AChR molecules in the muscle fiber during endplate formation thus results, at least in part : 1) from the occurrence of nuclei in different transcriptional states for AChR genes within the cytoplasm of the muscle fiber, and 2) from their differential regulation under the nerve endings and outside the endplates.

The activity-dependent repression of extra-junctional AChR biosynthesis (Klarsfeld and Changeux, 1985 ; Fontaine and Changeux, 1989 ; Klarsfeld et al., submitted for publication)

Chronic paralysis of the chick embryo in ovo by snake venom α-toxins or curare-like agents does not affect the initial onset of AChR biosynthesis (Giacobini et al., 1973) but prevents the elimination of extra-junctional AChR (step 2) (Giacobini-Robecchi et al., 1975 ; Burden, 1977 ; Bourgeois et al., 1978b ; Betz et al., 1980). This elimination occurs without change of the metabolic stability of the AChR protein and therefore represents an activity-dependent repression of AChR biosynthesis. Interestingly, a similar regulation can be reproduced with primary cultures of myotubes which exhibit spontaneous (non-neurogenic) firing (Cohen and Fischbach, 1973 ; Shainberg and Burstein, 1976 ; Betz and Changeux, 1979). Blocking this spontaneous electrical activity by tetrodotoxin, an inhibitor of the voltage-dependent Na^+ channel, results after 24 and 48 hours of treatment, in a 1.5 to 2 fold increase in surface AChR, accompanied by a respective 4.5 and 13 fold increase in α-subunit mRNA levels (Klarsfeld and Changeux, 1985) although differences in the amplitude of the effect were observed (Shieh et al., 1988, Harris et al., 1988). To test for an eventual regulation at the transcriptional level, Harris et al (1988) attempted to measure the steady-state levels of partially spliced precursor forms of the α-subunit mRNA. No significant changes of these levels were observed after exposure to TTX. However, recent experiments (Klarsfeld et al., submitted) with two probes which cover different parts of the α-subunit precursor mRNA (including either exon 7 or exons 2 and 3) demonstrate that TTX does significantly increase both precursor and mature mRNA levels in parallel. These results do not favour the hypothesis of a differential regulation of a particular step in precursor splicing. Upon denervation of adult muscle, partially spliced forms of α-subunit mRNA were also shown to vary in parallel with its mature form (Shieh et al., 1987). Although examples exist of precursor stabilization without increased transcription (Leys et al., 1984 ; Narayan and Towle, 1985), this parallelism suggests that electrical activity represses AChR biosynthesis at the level of transcription.

The same conclusion was reached by in situ hybridization on chick cultured myotubes with strictly intronic probes derived from the α subunit gene (Fontaine and Changeux, 1989) (Figure 1). After 6 days in culture, the spontaneously contracting myotubes displayed little labeling ; on the other hand, after 48 h treatment with TTX, hybridization was strikingly enhanced. Yet a significant heterogeneity of labeling was noticed. In a number of myotubes, some nuclei were found significantly labeled, while others were not. The reasons for the coexistence of nuclei with different states of gene expression within the same cytoplasm are not known. Tentatively one may speculate that such switch of gene expression results from intrinsic regulatory circuits, including feedback loops, able to create, at the level of a single nucleus, discrete steady-states of gene transcription (see Monod and Jacob, 1961 ; Changeux et al., 1985).

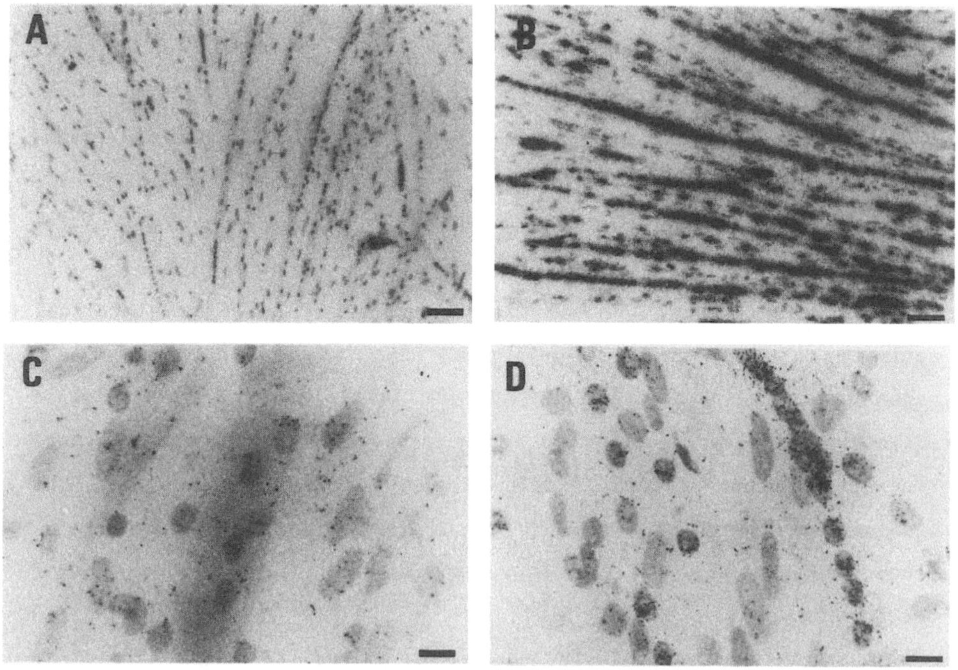

Figure 1. Control and TTX-treated muscle cells in primary culture of embryonic chick myotubes. **A, B** : Cultured muscle cells not treated and treated by TTX (0.2 μg/ml) respectively hybridized with the 2.3 kb probe ; bar = 150 μM. **C, D** : Muscle cells in culture not treated and treated by TTX (0.2 μg/ml) respectively hybridized with the AChR α-subunit intron-VII probe ; bar = 20 μM. After 4 days in culture (**A,B,C,D**), TTX was added directly into the culture medium. After 48 hours of treatment, cells were fixed and processed for the in situ hybridization. Autoradiograms were exposed 6 days in **A** and **B**, 15 days in **C** and **D**. Sections were stained with hematoxyline (Pictures taken from Fontaine and Changeux (1989). J. Cell Biol. In Press).

As mentioned above, this regulation can also be studied by following the increase of AChR levels caused by denervating the adult muscle. In this system, the direct assay of the α-subunit primary (unspliced) gene transcript suggests that this regulation (which can be reversed by electrical stimulation of the muscle) takes place at the level of transcription (Shieh et al., 1987 ; Goldman et al., 1988). Thus, the spontaneously active primary cultures of chick myotubes appear as a convenient model system to analyse the activity-dependent repression of AChR biosynthesis which occurs in vivo in extrajunctional areas.

Intracellular signals linking electrical activity to gene expression (Fontaine et al., 1987 ; Laufer and Changeux, 1989a ; Klarsfeld et al., submitted for publication)

The intracellular "second messengers" which link the electrical activity of the muscle cell to the transcription of AChR genes are not yet securely identified. Muscle cells from the mouse mutant muscular dysgenesis fire action potentials but do not contract. Yet, the mutant fibers still display a derepression of AChR biosynthesis by TTX, which is therefore due to a block of the electrical activity of the myotubes and not of their mechanical contraction.

One of the second messengers postulated to link electrical activity to the expression of AChR genes is Ca^{2+} (review Laufer and Changeux, 1989b). Upon depolarization of the cell membrane, the cytoplasmic Ca^{2+} concentration transiently rises. Moreover, several lines of evidence, mostly pharmacological, support the notion that a rise of intracellular Ca^{2+} causes a decrease of surface AChR numbers (Shainberg et al., 1976 ; Birnbaum et al., 1980 ; Pezzementi and Schmidt, 1981 ; McManaman et al., 1982 ; Smilowitz et al., 1988).

This question has been reexamined with chick muscle primary cultures (Klarsfeld et al., submitted). Treatment of the cultures with the Ca^{2+} channel blocker verapamil (10 µM) was found to increase α-subunit mature and precursor mRNA levels by an average factor of 11.5, close to the 14.9-fold increase observed with TTX. The latter effect was completely abolished in the presence of the Ca^{2+} ionophore A23187 (0.25 µM). Dantrolene, which blocks Ca^{2+} efflux from the sarcoplasmic reticulum, only caused a 1.7-fold increase in α-subunit mRNA levels, which was additive with that elicited by TTX. These results tend to exclude a major regulatory role of Ca^{2+} ions released from the sarcoplasmic reticulum, but support the view that the entry of Ca^{2+} through the sarcolemmal membrane during electrical activity contributes to the repression of α-subunit gene transcription.

The Ca^{2+}-activated and phospholipid-dependent protein kinase C is an intracellular regulatory enzyme that may be activated by muscle depolarization (Vergara et al., 1985 ; Richter et al., 1987). Short-term exposure (up to 40 h) of chick muscle cultures to tetradecanoyl-12, 13-phorbol acetate (TPA), an activator of protein kinase C, causes an approx. 40 % decrease of surface AChR number without significant change in the steady-state level of α-subunit mRNA but blocks the increase in AChR α-subunit mRNA and surface AChR levels caused by TTX in spontaneously active myotubes (Fontaine et al., 1987). This observation raises the possibility that protein kinase C contributes to the repression of AChR biosynthesis. To investigate further the role of this enzyme, exposure to TPA was prolonged for an extra day, probably leading to a "down-regulation" of the kinase, as described in many systems (reviewed by Niedel and Blackshear, 1986). Indeed, it was found that surface AChR levels recover from an initial 40 % decrease to reach values almost 2-fold above control, while α-subunit mRNA levels increase by a factor of 7.8 ± 2.9, when compared to matched controls (Klarsfeld et al., submitted).

The protein kinase C inhibitor, staurosporine (Takaomi et al., 1986) caused a concentration-dependent increase in surface AChR and α-subunit mRNA levels (Klarsfeld et al., submitted) respectively, up to 5- and 10-fold above control at staurosporine concentration of 10 ng/ml. At concentrations where it only marginally affects myotube morphology and does not prevent their spontaneous contractions, staurosporine also reversed the short-term decrease of surface AChR number elicited by TPA. Moreover, its effect on AChR expression was not found additive with that of TTX, suggesting that staurosporine acts upon a downstream step in the signalling pathway(s) that link(s) membrane electrical activity to AChR biosynthesis. These results thus further support the suggestion (Fontaine et al., 1987 ; Changeux et al , 1987) that protein kinase C contributes to the regulation of AChR α-subunit gene expression by the electrical activity of the sarcolemmal membrane. The precise relationships that may exist between the depolarisation of the muscle membrane, Ca^{2+} entry and protein kinase C activation, remain to be elucidated.

Subneural regulation of AChR gene expression (Fontaine et al., 1986, 1987 ; Laufer and Changeux, 1987, 1989a,b ; Osterlund et al., submitted for publication)

Even after the transcriptional activity of the extrajunctional nuclei has been switched off, the biosynthesis of the AChR by the subneural nuclei persists. Soluble factors from neural tissues which increase AChR number on cultured myotubes in vitro have been postulated to play a role in this maintenance (review Salpeter and Loring, 1986 ; Schuetze and Role, 1987) such as a trypsin-sensitive polypeptide of about 42 KD purified from chick brain (Usdin and Fischbach, 1986) and, among lower molecular weight factors, ascorbic acid (Knaack et al., 1986).

Other potentially important compounds are the neuropeptides known to coexist (Hökfelt et al., 1986 ; Villar et al., 1988) with ACh in spinal cord motor neurons. These include vasoactive-intestinal-peptide (VIP), somatostatin and calcitonin gene-related peptide (CGRP) (Villar et al., 1988). CGRP (Amara et al., 1982 ; Rosenfeld et al., 1983 ; Fischer and Born, 1987) is a peptide of 37 amino acids, which has been localized in spinal cord motor neurons (Rosenfeld et al., 1983, Gibson et al., 1984, Fontaine et al., 1986 ; New and Mudge, 1986) and at the level of the motor nerve endings (Takami et al., 1985a,b), where it is stored within dense-core vesicles (Matteoli et al., 1988). This peptide has been reported to potentiate contraction of electrically stimulated skeletal muscle fibers (Takami et al., 1985).

To test for an eventual role of CGRP as an anterograde "trophic" factor of neural origin, primary cultures of chick myotubes were exposed to CGRP (10^{-8} - 10^{-6}M). As a result both surface and total AChR levels increased (30-50 %) without change of AChR degradation rate (Fontaine et al., 1986 ; New and Mudge, 1986). Moreover, CGRP elevated the steady-state level of α-subunit mRNA by about 3-fold (Fontaine et al., 1987). Under these conditions, high molecular weight unspliced mRNA increased in parallel supporting the view - with the limitations mentioned above - that CGRP stimulates the transcription of the α-subunit gene (Osterlund et al., submitted).

The response to CGRP occurs independently of the rise of AChR numbers caused by TTX and is thus not influenced by the electrical activity of the muscle fiber. Moreover, under the conditions where the phorbol ester TPA abolishes the effect of TTX on AChR biosynthesis, the response to CGRP persists (Fontaine et al., 1987). Also, in differentiating primary cultures, the response to CGRP develops earlier than the response to TTX (Osterlund et al , submitted). Different intracellular signalling mechanisms are thus involved in the regulation of AChR gene expression by CGRP and by electrical activity (Fontaine et al., 1986, 1987).

CGRP elevates intracellular levels of cAMP and activates adenylate cyclase in skeletal muscle (Laufer and Changeux, 1987, Kobayashi et al., 1987). Moreover, AChR biosynthesis is enhanced by membrane permeant analogs of cAMP and by compounds that activate adenylate cyclase (prostaglandin, β-adrenergic agonists, cholera toxin, forskolin) (Betz and Changeux, 1979 ; Blosser and Appel, 1980, Fontaine et al., 1987 ; Harris et al., 1988). Thus, cAMP may serve as one of the second messengers that mediate the action of CGRP.

Recent evidence suggested an implication of phosphoinositide turnover in muscle cell differentiation (Wakelam, 1985) and excitation-contraction coupling (Volpe et al., 1986). The possibility that CGRP may also modulate this intracellular pathway was thus considered (Laufer and Changeux, 1989a). In cultured chick myotubes whose phosphoinositides have been labelled with [^{3}H]inositol, rat CGRP elicits a transient increase in [^{3}H]inositol 1,4,5 - trisphosphate, and a more sustained elevation of [^{3}H]inositol 1,3,4 - triphosphate levels. In the presence of Li^+, rCGRP evokes an approximately 3-fold increase of [^{3}H]inositol monophosphate levels, which persist for up to one hour.

Since rCGRP also accelerates the rate of synthesis of [^{3}H]inositol-containing lipids, the peptide appears to act by stimulating phosphoinositide turnover in chick myotubes. Agents that either mimick or elevate intracellular cyclic AMP also enhance the synthesis of [^{3}H]inositol-containing lipids, and the accumulation of inositol phosphates, suggesting that the effects of rCGRP are mediated, at least in part, via the activation of adenylate cyclase. These results provide an example of positive interaction between cAMP and the phosphoinositide system and point to a pleitropic action of CGRP upon its target cell. Furthermore, they raise an interesting problem, since cAMP and protein kinase C (which constitutes the second branch of the phosphoinositide signalling system) exert opposite effects on AChR biosynthesis in chick muscle cultures. A possibility, which remains to be explored, is that the various components of the two signalling pathways have a different topological distribution in junctional vs extrajunctional areas of the innervated adult muscle fiber. Accordingly, the cAMP-stimulated turnover of phosphoinositides elicited by CGRP might act in conjunction with depolarisation and Ca^{2+}-dependent signal to repress AChR synthesis in extrajunctional areas, while, under the nerve endings, the observed "net" positive effect of CGRP would maintain the expression of AChR genes (Laufer and Changeux, 1989a,b).

It should be borne in mind, however, that we ignore at present whether CGRP is physiologically involved in endplate morphogenesis and/or maintenance. It certainly is a plausible candidate, along with other neural "anterograde" trophic factors, which may act synergistically and/or sequentially during development. Also, intracellular signalling systems different from cAMP may mediate the increase of subsynaptic AChR numbers. These second messengers may, for instance, simply reverse the activity-dependent repression of AChR gene expression by antagonizing one of the signalling steps activated by membrane depolarisation (A. Duclert, unpublished results, see also Harris et al., 1988).

Molecular mechanisms of differential AChR gene expression during endplate formation : a model (Changeux et al., 1985, 1987 ; Laufer and Changeux, 1989b).

The results are consistent with a model (Changeux et al., 1985,

1987, Laufer and Changeux, 1989b) according to which : 1) in the mature innervated muscle the nuclei which underlie the nerve endings are in a state of AChR gene expression different from those located in extrajunctional areas and 2) the development and maintenance of this pattern is compartmentalized and involves different intracellular signalling pathways. For instance, in extrajunctional areas, the electrical activity of the muscle membrane may repress the transcription of the AChR genes by an intracellular pathway involving calcium and/or the activation of protein kinase C. On the other hand, in subneural areas, CGRP or another neural factor coreleased with ACh by the motor nerve ending would maintain their expression via a distinct intracellular pathway involving cyclic AMP or any mechanism which would overcome the second messenger system linked with electrical activity.

Accordingly, a set of distinct DNA regulatory sequences and their corresponding DNA-binding regulatory proteins, sensitive to the second messengers mentioned above, would be involved in the differential regulation of AChR genes in subneural vs extrajunctional nuclei (Changeux et al., 1985, 1987). An important outcome of this model thus becomes the identification of these regulatory elements.

DNA elements involved in the cis-regulation of α-subunit gene (Klarsfeld et al., 1987 ; Piette et al., submitted for publication)

To look for such regulatory DNA sequences the upstream and downstream flanking regions of the AChR α-subunit gene were cloned and sequenced. The upstream (850 bp) flanking region was inserted in front of the bacterial chloramphenicol acetyltransferase (CAT) gene (Klarsfeld et al., 1987). This "chimeric" construction was introduced into cultured living cells and the expression of the reporter gene under the command of the AChR regulatory sequences was investigated by assaying CAT activity in cell extracts prepared a few days later. High CAT expression was obtained in myotubes, but not in cultured myoblasts, nor in fibroblasts (Klarsfeld et al., 1987). Hence, this 5'-flanking region of the α-subunit gene contains DNA elements which selectively control the expression of AChR α-subunit in muscle cells during myotube differentiation. Subsequently, analysis of the γ-subunit gene in the mouse revealed that the 5'-flanking sequence together with the 5' end of the gene, including the first intron, regulates transcription (Gardner et al., 1987). More recently, Wang et al. (1988) have reported that the -116 to -41 region of the chicken α-subunit gene behaves as a cis-acting enhancer in the mouse C2C12 muscle cell line.

Consistent with the latter finding, deletion mapping of the 5' flanking sequence of the chicken AChR α-subunit gene following transient expression of chimeric CAT plasmids in homologous primary cultures of chicken myotubes shows that the -110 to -45 segment confers developmental control of expression (Piette et al., submitted) (Figure 2).

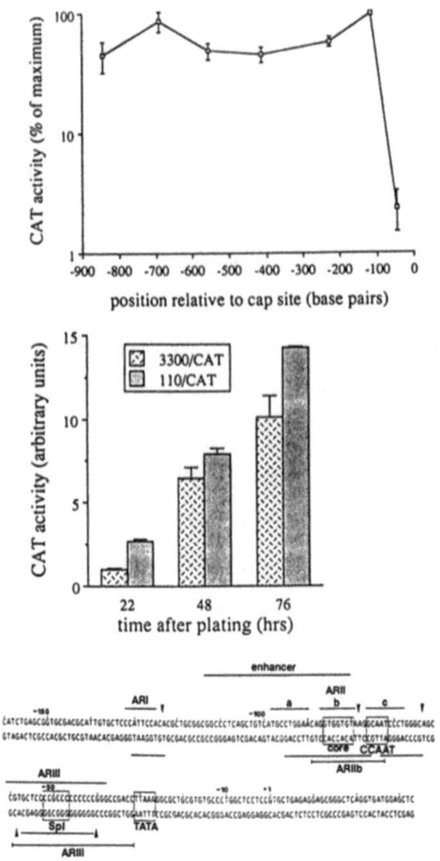

Figure 2. Effect of progressive 5'-deletions on CAT expression driven by an upstream flanking region of the chicken AChR α-subunit

Top : primary cultures of chicken muscle cells were transfected either at the myotube stage (after 3-4 days in culture), or at the myoblast stage and were then left to differentiate for at least 3 days. CAT values are means of 6 independent experiments, 2 of which were performed by electroporation. Bars represent standard errors.
Middle : increases in CAT activity during the course of myoblast differentiation. Transfections were performed immediately after plating, and cells were harvested after the indicated times in culture. CAT activities were normalized to the protein content of the extracts, and then to the activity obtained with pSVECAT transfected in parallel culture dishes. These normalized activities are expressed in arbitrary units relative to equimolar amounts of plasmid DNA, taking 1 as the activity obtained at the first time point with pαACh3300/CAT. Bars represent the variation between duplicate dishes.
Bottom : nucleotide sequence of the upstream region of the α-subunit gene of the acetylcholine receptor with the results of the DNase I footprinting experiments (see Fig. 3).
The sequence is numbered with respect to the startpoint of transcription (Klarsfeld et al., 1987) and was verified by the Maxam-Gilbert method. Homologies to consensus sequences are boxed. Protected regions are indicated by horizontal lines, hypersensitive sites by vertical arrows. Also indicated is the muscle-specific enhancer sequence mapped by Wang et al. (1988), and the position of oligonucleotides ARIIb and ARIII in the sequence (Pictures taken from Piette, Klarsfeld and Changeux, submitted for publication).

In agreement with these functional assays, Piette et al. (submitted) have found by DNase I footprinting and gel retardation assays, that among the 800 nt upstream sequence of the α-subunit gene, only the most proximal 140 nt sequence displays significant interactions with nuclear proteins prepared from cultured myotubes or embryonic muscle. Three domains of interactions referred to as ARI, II and III have been identified in this region, adjacent to the transcription start point (Figure 3). The level of several of the factors interacting with these DNA elements factors was found to change during fusion of myoblasts into myotubes (ARIIb and III) and, also, as a consequence of denervation (ARIIb and III). The identification of these DNA-binding factors is in progress (Piette et al., submitted). Accordingly, a rather short segment of DNA would be involved in the regulation of AChR gene transcription. Yet, it is still not known whether this segment is sufficient to control the localisation and stabilisation of the AChR at the adult endplate.

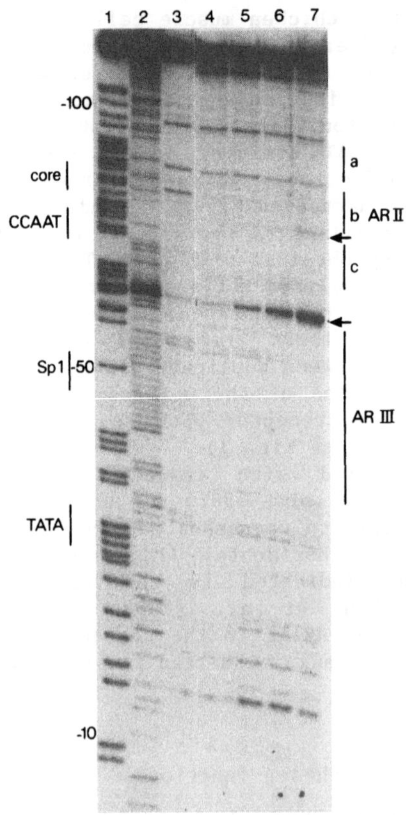

Figure 3. Footprint of nt -100 to -5 of the coding strand. Lane 1 : G+A sequence ladder, lane 2 : C+T sequence ladder, lane 3 : DNase I digestion in the absence of nuclear extract. Subsequently : DNase I digestion in the presence of 2 µg (lane 4), 4 µg (lane 5), 8 µg (lane 6) and 16 µg (lane 7) of a nuclear extract of myotubes. Lines indicate protected regions, arrows indicate hypersensitive sites to DNase I. (Picture taken from Piette, Klarsfeld and Changeux, submitted for publication).

Discoordinate regulation of AChR subunit genes (O. Heidmann et al., 1986, Osterlund et al. submitted for publication)

In the mouse, the genes coding for the four subunits of the AChR have been located on three different chromosomes by analyzing restriction fragment length polymorphisms between the domestic mouse and Mus spretus (Robert et al., 1985). The γ- and δ-subunit genes cosegregate in this species, as in the chick (Nef et al., 1984) and have been allocated to chromosome 1 (Heidmann et al., 1986). The β-subunit gene is located on chromosome 11 (Heidmann et al., 1986). The α-subunit gene was found closely linked to the α-cardiac actin gene (Heidmann et al., 1986) but the latter was incorrectly assigned to chromosome 17 by Czosnek et al. (1983). Recent unpublished work by Siracusa et al (in preparation) demonstrates linkage of the α-subunit gene with the Abl gene and thus definitively assigns this gene to chromosome 2. Such partial dispersion of the subunit genes rules out a coordinate regulation mediated by a common cis-acting regulatory sequence. Consistent with this view, denervation causes a differential increase of some subunit transcripts over the others in rat (Evans et al., 1987) and chick (Shieh et al., 1988 ; Moss et al., 1987).

In primary cultures, a similar discoordination has been reported to occur (Harris et al., 1988). Finally, at late stages of synapse maturation in the rat (Witzemann et al., 1987), an additional subunit, referred to as ε (Mishina et al., 1986) probably replaces the γ subunit and may account for some of the changes of channel properties recorded after birth in mammalian species and in the frog (see Witzemann et al. this Symposium).

In this laboratory, Osterlund et al. (submitted) have compared the steady-state levels of the mRNAs for α,γ and δ subunits in chick muscle primary cultures during differentiation and in the presence of TTX and CGRP. It was found that the levels of α-subunit message vary to a much larger extent than those of γ-and δ-subunit mRNAs (see also Harris et al., 1988). The levels of all these mRNAs first increase as a consequence of differentiation. Then α-subunit mRNA levels decrease drastically at the onset of spontaneous electrical activity, to rise again in the presence of TTX or CGRP, whereas γ- and δ-subunit mRNA levels remain much more stable throughout this evolution.

Posttranscriptional regulations (Kordeli et al., 1989)

Regulation of transcription is only one of the multiple processes which contribute to the localisation and stabilisation of the AChR in the postsynaptic membrane. These processes have been studied primarily in cell cultures (for review see Merlie and Smith, 1986 ; Peng and Poo, 1986). Yet, in the adult, the AChR is integrated into a supramolecular assembly which includes, on the extracellular side, components of the basal lamina (review Burden, 1987) and, on the intracellular side, elements of the cytoskeleton (review Kordeli et al., 1986 ; Laufer and Changeux, 1989b). Denervation of the adult electric organ of Electrophorus electricus (Bourgeois et al., 1973, 1978) and of the neuromuscular junction in mammals (Frank et al., 1975) showed that the subsynaptic agregate of AChR persists, several weeks in the absence of nerve terminals (review Salpeter and Loring, 1985). Thus, neither the membrane of the nerve ending, nor the neurotransmitter (or coexisting neuro-peptides) are required for the maintenance of the postsynaptic AChR agregate in the adult endplate. As discussed (Bourgeois et al., 1978), the fluidity of the lipid bilayer and the molecular weight of the AChR are such that in the absence of interaction with extrinsic components, the AChR agregate would disperse in a few hours (review Rousselet et al., 1981).

Among the various structural elements which may contribute to this stability, one candidate is an extrinsic protein of apparent M.W. 43 000 daltons, referred to as 43 kD protein. Initially identified in this laboratory (Sobel et al., 1977, 1978) as a major component of purified postsynaptic membranes from T. marmorata electric tissue, its cDNAs have been recently cloned and sequenced from T. californica (Frail et al., 1987).

Immunocytochemical studies showed that a protein immunologically related to the 43kD is present at mammalian motor end plates (Froehner et al., 1981 ; Froehner, 1984). Similarly in rat and Xenopus myotubes in culture, AChR clusters arising spontaneously in the presence or in the absence of nerve cells (Bloch and Froehner, 1987 ; Burden, 1985) or experimentally elicited by various factors such as coated latex beads (Peng and Froehner, 1985) have been found associated with the 43kD protein. It was further reported that clustering of the AChR and the association with the 43kD protein were taking place at the same moment within the time-limits of the technique (reviewed in Peng, 1987).

The developing electrocyte from T. marmorata eleotric tissue represents a useful model system to study the early stages of AChR-43kD protein interaction because of the relative abundance of these two components of the cholinergic synapse and of the unique architecture of the electrocyte that displays an asymmetric distribution of the AChR across the whole cell (see Witzemann et al., 1983). The attempt was thus made to follow, in situ, the assembly of the AChR and of the 43kD protein and their intracellular biosynthetic pathways in the developing electrocyte (Kordeli et al., 1989). Immunochemical studies were carried out at three stages of embryonic development : 1) the 45 mm (body length) stage, before the entry of the nerve endings, 2) the 80-85 mm stage where the innervated electrocyte displays a flattened asymmetric structure and 3) the adult. At the 45 mm stage, in agreement with the earlier work of Witzemann et al. (1983), AChR visualized by α-bungarotoxin (with, or without, amplification with a rabbit-antitoxin antibody) was found accumulated at the ventral pole of the developing electrocyte. Double fluorescence experiments, however, never revealed 43KD protein immunoreactivity associated with the AChR but faint and diffuse fluorescence mostly with cytoplasmic distribution. On the other hand, a transient asymmetric distribution of the extracellular protein, laminin which paralleled that from the AChR was noticed. At the subsequent 80-85 mm stage investigated, the AChR and 43KD protein were found codistributed on the ventral face of the cell, as in the adult electrocyte. These studies show that, in the electrocyte, the AChR may cluster independently of the 43 KD protein and before the entry of the nerve terminals. The presence of the extracellular matrix protein, laminin, at the moment AChR molecules agregate is consistent with the observation of Burden et al. (1979) and Sanes et al. (1978) that, at the motor endplate, the synaptic basement membrane directs both pre- and postsynaptic specialisations in the course of in vivo regeneration experiments. It also offers one explanation for our initial observation that AChR agregates persist with a high surface density several weeks after denervation (Bourgeois et al., 1973, 1978).

Perspectives

The analysis of the molecular mechanisms involved in the genesis of the motor endplate is at its beginnings. Yet, the data presently available show that this analysis is feasible both at the transcriptional and post-transcriptional levels. Several of the processes involved are regulated by the state of activity of the muscle cell and may serve as model for activity-dependent changes of synapse properties. Extension of the conclusions reached with the motor endplate to interneuronal-synapses looks, to a large extent, legitimate. Nevertheless, important differences may arise since, at variance with the muscle fibers, nerve cells are mononucleated (for discussion, see Laufer and Changeux, 1989b). In any case, because of its simplicity and of the knowledge of its basic molecular components, the motor endplate still offers many advantages to further explore the mechanisms of synapse biogenesis and plasticity at the molecular level.

Acknowledgements

This work was supported by grants from the Muscular Dystrophy Association of America, the Collège de France, the Ministère de la Recherche, INSERM, (contract n° 872 004), and DRET (contract n° 87/211).

References

Amara SG, Jonas V, Rosenfeld MG, Ong ES, Evans RM (1982) Alternative RNA processing in calcitonin gene expression generates mRNAs encoding different polypeptide products. Nature 298:240-244

Ballivet M, Nef P, Stalder R, Fulpius B (1983) Genomic sequences encoding the alpha-subunit of acetylcholine receptor are conserved in evolution. Cold Spring Harb Symp Quant Biol 48:83-87

Baldwin TJ, Yoshihara CM, Blackmer K, Kintner, CR, Burden, SJ (1988) Regulation of acetylcholine receptor transcript expression during development in Xenopus laevis. J Cell Biol 106:469-478

Bennett MR, Pettigrew AG (1974) The formation of synapses in striated muscle during development. J Physiol (London) 241:515-545

Betz H, Changeux, JP (1979) Regulation of muscle acetylcholine receptor synthesis in vitro by cyclic nucleotide derivatives. Nature 278:749-752

Betz H, Bourgeois JP, Changeux, JP (1980) Evolution of cholinergic proteins in developing slow and fast skeletal muscles from chick embryo. J Physiol 302:197-218

Birnbaum M, Reiss M, Shainberg A (1980) Role of calcium in the regulation of acetylcholine receptor synthesis in cultured muscle cell. Pflueger's Archiv 385:37-43

Bloch RJ, Froehner, SC (1987) The relationship of the postsynaptic 43 K protein to acetylcholine receptors in receptor clusters isolated from cultured rat myotube. J Cell Biol 104:645-654

Blosser JC, Appel, SH (1980) Regulation of acetylcholine receptor by cyclic AMP. J Biol Chem 253:3088-3093

Boulter J, Luyten W, Evans K, Mason P, Ballivet M, Goldman D, Stengelin S, Martin G, Heinemann S, Patrick J (1985) Isolation of a clone coding for the alpha-subunit of a mouse acetylcholine receptor. J Neurosci 5:2545-2552

Bourgeois JP, Popot JL, Ryter A, Changeux, JP (1973) Consequences of denervation on the distribution of the cholinergic (nicotinic) receptor sites from Electrophorus Electricus revealed by high resolution autoradiography. Brain Res 62:557-563

Bourgeois JP, Popot JL, Ryter A, Changeux JP (1978) Quantitative studies on the localization of the cholinergic receptor protein in the normal and denervated electroplaque from Electrophorus Electricus. J Cell Biol 79:200-216

Burden S (1977) Development of the neuromuscular junction in the chick embryo : the number, distribution and stability of the acetylcholine receptors. Dev Biol 57:317-329

Burden SJ (1985) The subsynaptic 43 KD protein is concentrated at developing nerve-muscle in vitro. Proc Natl Acad Sci USA 82:7805-7809

Burden SJ (1987) The extracellular matrix and subsynaptic sarcoplasm at nerve-muscle synapses. In : "The Vertebrate Neuromuscular junction", Alan R Liss, Inc pp 163-186

Burden SJ, Sargent PB, McMahan UJM (1979) Acetylcholine receptors in regenerating muscle accumulate at original synaptic sites in the absence of the nerve. J Cell Biol 82:412-425

Changeux JP (1981) The acetylcholine receptor : an "allosteric" membrane protein. In "Harvey Lectures". Acad Press Inc 75:85-254

Changeux JP, Klarsfeld A, Heidmann T (1985) The acetylcholine receptor and molecular models for short and long term learning. Dahlem Konferenzen. In :"The cellular and molecular bases of learning", published in 1987, J.P. Changeux and M. Konishi, ed. Wiley London, pp 31-83

Changeux JP, Devillers-Thiéry A., Giraudat J, Dennis M, Heidmann T, Revah F, Mulle C, Heidmann O, Klarsfeld A, Fontaine B, Laufer R, Nghiêm H.O, Kordeli E, Cartaud J (1987) The acetylcholine receptor : functional organisation and evolution during synapse formation. In : "Strategy and prospects in Neuroscience". Taniguchi Symposia on Brain Sciences n° 10. O. Hayaishi ed., Japan Scientific Societies Press Tokyo VNU Science Press BV Utrecht, pp 29-76

Cohen SA, Fischbach GD (1973) Regulation of muscle acetylcholine sensitivity by muscle activity in cell culture. Science 181:76-78

Couteaux R (1978) Recherches morphologiques et cytochimiques sur l'organisation des tissus excitables, 225 pp. Robin et Marenge Paris

Czosnek H, Nudel U, Mayer Y, Barker PE, Pravtcheva DD, Ruddle FH, Yaffe D (1983) The genes coding for the cardiac muscle actin, the skeletal muscle actin and the cytoplasmic β-actin are located on three different mouse chromosomes. EMBO J 2:1977-1979

Dennis M, Giraudat J, Kotzyba-Hibert F, Goeldner M, Hirth C, Chang JY, Lazure C, Chrétien M, Changeux JP (1988) Amino acids of the Torpedo marmorata acetylcholine receptor α subunit labeled by a photoaffinity ligand for the acetylcholine binding site. Biochemistry 27:2346-2357

Evans S, Goldman D, Heinemann S, Patrick J (1987) Muscle acetylcholine receptor biosynthesis. J Biol Chem 262:4911-4916

Fambrough D (1979) Control of acetylcholine receptors in skeletal muscle. Physiol Rev 59:165-227

Fischer JA, Born W (1987) Calcitonin gene products : evolution, expression and biological targets. Bone and Mineral 2:347-359

Fontaine B, Klarsfeld A, Hökfelt T, Changeux, JP (1986) Calcitonin gene-related peptide, a peptide present in spinal cord motoneurons, increases the number of acetylcholine receptors in primary cultures of chick embryo myotubes. Neurosci Lett 71:59-65

Fontaine B, Klarsfeld A, Changeux JP (1987) Calcitonin-gene related peptide and muscle activity regulate acetylcholine receptor α-subunit mRNA levels by distinct intracellular pathways. J Cell Biol 105:1337-1342

Fontaine B, Sassoon D, Buckingham M, Changeux JP (1988) Detection of the nicotinic acetylcholine receptor α-subunit mRNA by in situ hybridization at neuromuscular junctions of 15-day old chick striated muscles. EMBO J 7:603-609

Fontaine B, Changeux JP (1989) Localization of nicotinic acetylcholine receptor α-subunit transcripts during myogenesis and motor endplate in the chick. J Cell Biol (in press)

Frail DE, Mudd J, Shah V, Carr C, Cohen JB, Merlie JP. (1987) cDNAS for the postsynaptic 43-kDa protein of Torpedo electric organ encode two proteins with different carboxyl termini. Proc Natl Acad Sci USA 84:6302-6306

Frank E, Gautvik K, Sommer-Schild H (1975) Cholinergic receptors at denervated mammalian endplates. Acta Physiol Scand 95:66-76

Froehner SC, Gulbrandsen V, Hyman C, Jeng AY, Neubig RR, Cohen JB (1981) Immunofluorescence localization at the mammalian neuromuscular junction of the M_r 43,000 protein of Torpedo postsynaptic Membranes. Proc Natl Acad Sci USA 78:5230-5234

Froehner SC, Douville K, Klink S, WJ (1983) Monoclonal antibodies to cytoplasmic domains of the acetylcholine receptor. J Biol Chem 258:7112-7120

Gardner PD, Heinemann S, Patrick, J (1987) Transcriptional regulation of nicotinic acetylcholine receptor genes : identification of control elements of a γ-subunit gene. Molecular Brain Res 3:69-76

Giacobini-Robecchi MG, Giacobini G, Filogamo G, Changeux, JP (1975) Effect of the type A toxin from C. Botulinum on the development of skeletal muscles and of their innervation in chick embryo. Brain Res 83:107-121

Gibson SJ, Polak JM, Bloom SR, Sabate IM, Mulderry PM. Ghatel MA, McGregor GP, Morrison, JFB, Kelly JS, Evans RM, Rosenfeld MG (1984) Calcitonin gene-related peptide immunoreactivity in the spinal cord of man and of eight other species. J. Neurosci 4:3101-3111

Giraudat J, Dennis M, Heidmann T, Chang JY, Changeux JP (1986) Structure of the high affinity binding site for noncompetitive blockers of the acetylcholine receptor : Serine-262 of the δ subunit is labeled by ^{3}H chlorpromazine. Proc Natl Acad Sci USA 83:2719-2723

Giraudat J, Dennis M, Heidmann T, Haumont PT, Lederer F, Changeux JP (1987) Structure of the high-affinity binding site for noncompetitive blockers of the acetylcholine receptor : ^3H chlorpromazine labels homologous residues in the beta and delta chains. Biochemistry 26: 2410-2418

Goldmann D, Boulter J, Heinemann S, Patrick, J (1985) Muscle denervation increases the levels of two mRNAs coding for the acetylcholine receptor α-subunit. J Neurosci 5:2553-2558

Goldmann D, Brenner HR, Heinemann, S (1988) Acetylcholine receptor α-,β-, γ-, and δ-subunit mRNA levels are regulated by muscle activity. Neuron 1:329-333.

Harris DA, Falls DL, Dill-Devor RM, Fischbach GD. (1988) Acetylcholine receptor-inducing factor from chicken brain increases the level of mRNA encoding the receptor α-subunit. Proc Natl Acad Sci USA 85:1983-1987

Heidmann O, Buonanno A, Goeffroy B, Robert B, Guénet JL, Merlie JP, Changeux JP (1986) Chromosomal localization of the nicotinic acetylcholine receptor genes in the mouse. Science 234:866-868

Hökfelt T, Holets VR, Staines W, Meister B, Melander T, Schalling M, Schultzberg M, Freedman J, Björklund H, Olson L, Lindk B, Elfvin LG, Lundberg J, Lindgren JA, Samuelsson B, Terenius L, Post C, Everitt B, Goldstein M (1986) Coexistence of neuronal messengers - an overview. Progr Brain Res 68:33-70

Hucho F (1986) The nicotinic acetylcholine receptor and its ion channel Eur J Biochem 158:211-226

Hucho FL, Oberthür W, Lottspeich F (1986) The ion channel of the nicotinic acetylcholine receptor is formed by the homologous helices M II of the receptor subunits. FEBS Lett 205:137-142

Imoto K, Busch C, Sakmann B, Mishina M, Konno T, Nakai J, Bujo H, Mori Y, Fukuda K, Numa S (1988) Rings of negatively charged amino acids determine the acetylcholine receptor channel conductance. Nature 335:645-648.

Imoto K, Methfessel C, Sakmann B, Mishina M, Mori Y, Konno T, Fukuda K, Kurasaki M, Bujo H, Fujita Y, Numa S (1986) Location of a δ-subunit region determining ion transport through the acetylcholine receptor channel. Nature 324:670-674

Kao P, Dwork A, Kaldany R, Silver M, Wideman J, Stein S, Karlin A (1984) Identification of the α-subunit half-cystine specifically labeled by an affinity reagent for the acetylcholine receptor binding site. J Biol Chem 259:11662-11665

Klarsfeld A, Changeux JP (1985) Activity regulates the level of acetylcholine receptor alpha-subunit mRNA in cultured chick myotubes. Proc Natl Acad Sci USA 82:4558-4562

Klarsfeld A, Daubas P, Bourachot B, Changeux JP (1987) A 5' flanking region of the chicken acetylcholine receptor alpha-subunit gene confers tissue-specificity and developmental control of expression in transfected cells. Mol Cell Biol 7:951-955

Klarsfeld A, Laufer R, Fontaine B, Devillers-Thiéry A, Dubreuil C, Changeux JP (submitted) Regulation of muscle AChR α-subunit gene expression by electrical activity : involvement of protein kinase C and Ca^{++}

Knaack D, Shen I, Salpeter MM, Podleski TR (1986) Selective effects of ascorbic acid on acetylcholine receptor number and distribution. J. Cell Biol 102:795-802

Kobayashi H, Hashimoto K, Uchida S, Sakuma J, Takami K, Tohyama M, Izumi F, Yoshida H (1987) Calcitonin gene related peptide stimulates adenylate cyclase activity in rat striated muscle. Experientia 43: 314-316

Kordeli E, Cartaud J, Nghiêm H.O, Pradel LA, Dubreuil C, Paulin D, Changeux JP (1986) Evidence for a polarity in the distribution of proteins from the cytoskeleton in Torpedo marmorata electrocytes. J. Cell Biol 102:748-761

Kordeli E, Cartaud J, Nghiêm HO, Devillers-Thiéry A, Changeux, JP (1989) Asynchronous assembly of the acetylcholine receptor and of the 43KD-ν_1 protein in the postsynaptic membrane of developing Torpedo marmorata electrocyte. J Cell Biol (in press)

LaPolla RJ, Mixter Mayne K, Davidson N (1984) Isolation and characterization of a cDNA clone for the complete protein coding region of the δ subunit of the mouse acetylcholine receptor. Proc. Natl. Acad. Sci. USA 81:7970-7974

Laufer R, Changeux JP (1987) Calcitonin gene-related peptide elevates cyclic AMP levels in chick skeletal muscle : possible neurotrophic role for a coexisting neuronal messenger. EMBO J 6:901-906

Laufer R, Changeux, JPa (1989) Calcitonin gene-related peptide and cyclic AMP stimulate phosphoinositide turnover in skeletal muscle cells : interaction between two second messenger systems. J Biol Chem (in press)

Laufer R, Changeux, JPb (1989) Activity dependent regulation of gene expression in muscle and neuronal cells. Mol. Neurobiol. (in press)

Leonard RJ, Labarca CG, Charnet P, Davidson N, Lester HA (in press) Evidence that the M2 membrane spanning region lines the ion channel pore of the nicotinic receptor. Science

Leys EJ, Crouse GF, Kellems RE (1984) Dihydrofolate reductase gene expression in cultured mouse cells is regulated by transcript stabilization in the nucleus. J Cell Biol 99:180-187

Lindstrom J, Schoepfer R, Whiting P (1987) Molecular studies of the neuronal nicotinic acetylcholine receptor family. Mol Neurobiol 1: 281-337

MacManaman JL, Blosser JC, Appel SH. (1982) Inhibitors of membrane depolarization regulate acetylcholine receptor synthesis by a calcium-dependent, cyclic nucleotide independent mechanism. Biochem Biophys Acta 720:28-35

Matteoli M, Haimann C, Torri-Tarelli F, Polak JM, Ceccarelli B, De Camilli P (1988) Differential effect of α-latrotoxin on exocytosis from small synaptic vesicles and from large dense-core vesicles containing calcitonin gene-related peptide at the frog neuromuscular junction. Proc. Natl. Acad. Sci. USA 85:7366-7370

Meiniel R, Bourgeois, JP (1982) Appearance and distribution in situ of nicotinic acetylcholine receptors in cervical myotomes of young chick embryos. Anat Embryol 164:349-368

Merlie J, Sanes JR (1985) Concentration of acetylcholine receptor mRNA in synaptic regions of adult muscle fibers. Nature 317:66-68

Merlie JP, Isenberg KE, Russell SD, Sanes JR (1984) Denervation supersensitivity in skeletal muscle : analysis with a cloned cDNA probe. J Cell Biol 99:332-335

Merlie JP, Smith MM. (1986) Synthesis and assembly of acetylcholine receptor, a multisubunit membrane glycoprotein. J Memb Biol 91:1-10

Mishina M, Takai T, Imoto K, Noda M, Takahashi T, Numa S, Methfessel, C, Sakmann B. (1986) Molecular distinction between fetal and adult forms of muscle acetylcholine receptor. Nature 321:406-411

Monod J, Jacob F (1961) General conclusions : teleonomic mechanisms in cellular metabolism, growth and differentiation. Cold Spring Harbor Symp. Quant. Biol. 26:389-401

Moss SJ, Beeson DM, Jackson JF, Darlison MG, Barnard EA. (1987) Differential expression of nicotinic acetylcholine receptor genes in innervated and denervated chicken muscle. EMBO J 6:3917-3921

Narayan P, Towle HC (1985) Stabilization o a specific nuclear mRNA precursor by thyroid hormone. Mol Cell 5:2642-2646
Nef P, Mauron A, Stalder R, Alliod C, Ballivet M (1984) Structure, linkage and sequence of the two genes encoding the delta and gamma subunits of the nicotinic acetylcholine receptor. Proc Natl Acad Sci USA 81:7975-7979
New HV, Mudge AW (1986) Calcitonin gene-related peptide regulates muscle acetylcholine receptor synthesis. Nature 323:809-811
Noda M, Takahashi H, Tanabe T, Toyosato M, Kikyotani S, Hirose T, Asai M, Takashima H, Inayama S, Miyata T, Numa S (1983) Primary structures of beta and delta-subunit precursors of Torpedo californica acetylcholine receptor deduced from cDNA sequences. Nature 301:251-255
Oberthür W, Muhn P, Baumann H, Lottspeich F, Wittmann-Liebold B, Hucho F (1986) The reaction site of a noncompetitive antagonist in the δ-subunit of the nicotinic acetylcholine receptor EMBO J 5:1815-1819
Oiki S, Danho W, Madison V, Montal M (in press) M2 δ, a candidate for the structure lining the ionic channel of the nicotinic cholinergic receptor. Proc Nat Acad Sci USA
Osterlund M, Fontaine B, Devillers-Thiéry A, Geoffroy B, Changeux JP (submitted) Acetylcholine receptor biosynthesis in primary cultures of embryonic chyick myotubes : I. Discoordinate regulation of the α-, γ- and δ-subunit gene expression by the calcitonin-gene related peptide and by muscle electrical activity
Peng HB (1987) Development of the neuromuscular junction in tissue culture. CRC Critical Reviews in Anat Sci 1:91-131
Peng HB, Froehner SC (1985) Association of the postsynaptic 43K protein with newly formed acetylcholine receptor clusters in cultured muscle cells. J Cell Biol 100:1698-1705
Peng HB, Poo MM (1986) Formation and dispersal of acetylcholine receptor clusters in muscle cells. Trends in Neurosci 9:125-129
Pezzementi L, Schmidt J (1981) Rapid modulation of acetylcholine receptor synthesis. FEBS Lett 135:103-106
Piette J, Klarsfeld A, Changeux JP (1989) Interaction of nuclear factors with the upstream of the α-subunit gene of chicken muscle acetylcholine receptor : variations with muscle differentiation and denervation (submitted)
Popot JL, Changeux JP (1984) The nicotinic receptor of acetylcholine : structure of an oligomeric integral membrane protein. Physiol Rev 64:1162-1184
Ranvier L (1875) Traité technique d'histologie. Librairie F. Savy, Paris
Richter EA, Cleland PJF, Rattigan S, Clark MG (1987) Contraction-associated translocation of protein kinase C in rat skeletal muscle. FEBS Lett 217:232-236
Rosenfeld MG, Mermod JJ, Amara SG, Swanson LW, Sawchenko PE, Rivier J, Vale WW, Evans RM (1983) Production of a novel neuropeptide encoded by the calcitonin gene via tissue-specific RNA processing. Nature 304: 129-135
Rousselet, A., Cartaud, J. and Devaux, P. F. (1981) Effects of temperature, lipid modification and pH on the mobility of the major proteins of the receptor-rich membranes from Torpedo marmorata. Biochim Biophys Acta 648:169-185
Salpeter M, Loring RH (1985) Nicotinic acetylcholine receptors in vertebrate muscle : properties, distribution and neural control. Progr Neurobiol 25:297-325

Sanes JR, Marshall LM, McMahan UJ (1978) Reinnervation of muscle fiber basal lamina after removal of myofibers. Differentiation of regenerating axons at original synaptic sites. J Cell Biol 78:176-198

Schuetze SM, Role LW (1987) Developmental regulation of nicotinic acetylcholine receptor. Ann Rev Neurosci 10:403-457

Shainberg A, Burstein M (1976) Decrease of acetylcholine receptor synthesis in muscle cultures by electrical stimulation. Nature (London) 264:368-349

Shainberg A, Cohen S.A, Nelson PG (1976) Induction of acetylcholine receptors in muscle cultures. Pflügers Archiv 361:255-261

Shieh BH, Ballivet M, Schmidt J (1987) Quantitation of an alpha subunit splicing intermediate : evidence for transcriptional activation in the control of acetylcholine recpetor expression in denervated chick skeletal muscle. J Cell Biol 104:1337-1341

Shieh BH, Ballivet M, Schmidt J (1988) Acetylcholine receptor synthesis rate and levels of receptor subunit messenger RNAs in chick muscle. Neuroscience 24:175-187

Siracusa LD, Silan CM, Justice MJ, Jenkins NA, Copeland NG (in preparation) A molecular genetic map of the proximal portion of mouse chromosome 2 - Linkage of **Abl** and **Achra**

Smilowitz H, Smart E, Bowik C, Chang RJ (1988) Regulation of the number of α-bungarotoxin binding sites in cultured chick myotubes by a 1.4 dihydropyridine calcium channel antagonist. J Neurosci Res 19:321-325

Sobel A, Weber M, Changeux JP (1977) Large scale purificdation of the acetylcholine receptor protein in its membrane-bound and detergent-extracted forms from Torpedo marmorata electric organ. Eur J Biochem 80:215-224

Sobel A, Heidmann T, Hofler J, Changeux JP (1978) Distinct protein components from Torpedo marmorata membranes carry the acetylcholine receptor site and the binding site for local anesthetics and histrionicotoxin. Proc Natl Acad Sci USA 75:510-514

Takami K, Hashimoto K, Uchida S, Tohyama M, Yoshida H (1986) Effect of calcitonin gene-related peptide on the cyclic AMP level of isolated mouse diaphragm. Japan J Pharmacol 42:345-350

Takami K, Kawai Y, Uchida S, Tokyama M, Shiotani Y, Hoshida H, Emson, PC, Girgis SH, Hillyard CJ, MacIntyre I (1985) Effect of calcitonin gene-related peptide on contraction of striated muscle in the mouse. Neurosci Lett 60:227-230

Usdin TB, Fischbach GD (1986) Purification and characterization of a polypeptide from chick brain that promotes the accumulation of acetylcholine receptors in chick myotubes. J Cell Biol 103:493-507

Vergara J, Tsien RY, Delay M (1985) Inositol 1,4,5- triphosphate ; a possible chemical link in excitation-contraction coupling in muscle. Proc Natl Acad Sci USA 82:6352-6356

Villar MJ, Huchet M, Hökfelt T, Changeux JP, Fahrenkrug J, Brown JC (1988) Existence and coexistence of calcitonin gene-related peptide, vasoactive intestinal polypeptide-and somatostatin-like imminoreactivities in spinal cord motoneurons of developing embryos and post-hatch chicks. Neurosci Lett 86:114-118

Volpe P, Di Virgilio F, Pozzan T, Salviati G (1986) Role of inositol 1,4,5-trisphosphate in excitation-contraction coupling in skeletal muscle. FEBS 197:1-4

Witzemann V, Richardson G, Boustead C (1983) Characterization and distribution of acetylcholine receptors and acetylcholinesterase during electric organ development in Torpedo marmorata. Neuroscience 8:333-349

Witzemann V, Barg B, Nishikawa Y, Sakmann B, Numa, S. (1987) Differential regulation of muscle acetylcholine receptor gamma and epsilon subunits mRNAs. FEBS Lett 223:104-112

Wakelam MJO (1985) Review article : the fusion of myoblasts. Biochem J 228:1-12

Wang Y, Xu HP, Wang XM, Ballivet M, Schmidt, J (1988) A cell type-specific enhancer drives expression of the chick muscle acetylcholine receptor α-subunit gene. Neuron 1:527-534

DEVELOPMENTAL REGULATION OF ACETYLCHOLINE RECEPTOR-SUBUNIT mRNA AT THE RAT NEUROMUSCULAR JUNCTION

V. Witzemann
Abteilung Zellphysiologie
Max-Planck-Institut for biophysikalische Chemie
Postfach 2841
D-3400 Göttingen, FRG

INTRODUCTION

The acetylcholine receptors (AChR) of mammalian muscle undergo drastic changes in their cellular distribution, metabolic stability and functional properties during the development of the neuromuscular synapse (for review see Schuetze and Role, 1987). AChR-channels in fetal muscle, when activated by acetylcholine, remain open 3-5 times longer and have a 30% lower conductance than the channels in the adult end-plate. The difference in channel properties is most probably the result of structural changes of the AChR complex (Mishina et al., 1986). While low conductance AChRs are comprised of α-, β-, γ- and δ-subunits, the high conductance AChRs contain α-, β-, ϵ- and δ-subunits (Mishina et al., 1986; Witzemann et al., 1987). Little is known about the mechanisms regulating the development, cellular distribution and functional properties of the AChRs. Both neurotrophic substances as well as nerve induced muscle activity have been postulated to control the expression of the two AChR complexes. In an attempt to obtain cues on the developmental regulation of AChR properties, we have analyzed the levels of all five AChR-subunit mRNAs in developing rat muscle. Denervation and reinnervation experiments were conducted to investigate the mechanisms by which the motor nerve changes the expression of the different forms of AChR channels. To discriminate between neurotrophically and muscle activity mediated control, denervated muscles were chronically stimulated. The results support the

NATO ASI Series, Vol. H 32
Molecular Biology of Neuroreceptors and
Ion Channels. Edited by A. Maelicke

notion that all five AChR-subunit mRNAs are probably regulated differentially by as yet unknown nerve dependent signals and by signals mediated by muscle activity.

RESULTS

Differential expression of AChR-subunit specific mRNAs during postnatal development of rat muscle

DNA probes specific for the α-, β-, γ-, δ- and ϵ-subunit mRNAs coding for the five subunit types of the rat nicotinic AChR were used for blot hybridization analysis of total RNA isolated from lower leg muscle of rat. The levels of the five AChR-subunit mRNA species exhibit characteristic changes during postnatal development. While the γ-subunit mRNA is present only at early postnatal stages and becomes undetectable in the adult muscle the other four subunit mRNAs, α, β, δ and ϵ, are present throughout development and in the adult muscle. The autoradiographs of Fig. 1 show blot hybridization experiments of total RNA isolated from muscle at various stages of development and of adult muscle. Largely varying amounts of the mRNA levels indicate that the five AChR-subunit mRNAs are regulated differentially during development.

Analysis of RNA isolated at additional times of postnatal development and densitometric evaluation of the resulting autoradiographs was performed to follow the time course of the individual subunit mRNA levels at increased time resolution. The α-subunit mRNA is most abundant during the first 10 days after birth to reach significantly lower levels of adult muscle within 4 to 6 weeks. The β-subunit mRNA is present at low levels at day 1, reaches maximal amounts at the end of the second postnatal week and then decreases gradually within 6 weeks after birth to the adult levels. The ϵ-subunit mRNA shows a similar profile. Only very low amounts are detected at day 1. Within the first weeks a strong increase is measured and maximal levels are expressed at the end of the second week.

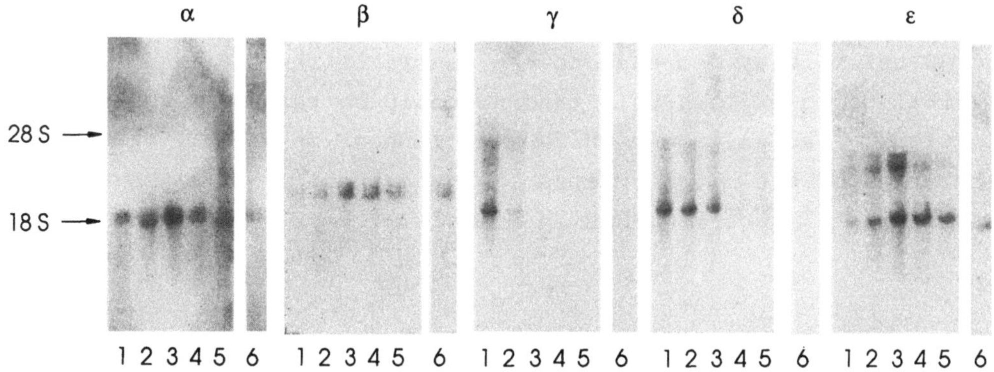

Fig. 1: AChR-subunit mRNA levels during development of rat muscle. Total RNA (10μg each) from lower leg muscle was analyzed by blot hybridization analysis using α-, β-, γ-, δ- and ϵ-subunit specific DNA probes. The corresponding subunit-spcific mRNA species was visualized upon autoradiography at -70°C using an intensifying screen. The numbers 1-6 below the lanes correspond to the developmental stage of the muscle used for RNA extraction. Number 1-5 correspond to postnatal days 4, 8, 13, 23 and 42, number 6 corresponds to adult muscle. Distinct differences in transcript size demonstrate the specificity of the probes used. Characteristic changes in mRNA levels support the view that all five AChR-subunit mRNAs are regulated differentially during development.

Quite different changes occur in the abundance of γ- and δ-subunit specific mRNAs. At early postnatal stages γ-subunit levels are high but decrease rapidly to low levels within the first 10 days after birth. Like the γ-subunit mRNA the δ-subunit mRNA is most abundant at early postnatal stages. Relatively high levels are present during the first 10 days, followed by a sharp decrease between day 10 and 20. The δ- and γ-subunit specific DNAs of rat (unpublished observation) are located as in other species on the same gene (Nef et al., 1984; Shibahara et al., 1985). Both subunit genes could therefore underly a common transcriptional regulation during early muscle development. At later postnatal stages, however, different regulatory elements become involved. While the γ-subunit mRNAs are no longer detectable in late postnatal stages as in the adult muscle, the δ-

subunit mRNAs reveal even slightly elevated levels between day 20 and 30 before they decrease to the levels of the adult. Another possibility to explain this finding could be the existence of an additional δ-subunit gene which would appear only after complete maturation of the motor synapse. To date there is no evidence for two distinct δ-subunit mRNAs in mammals.

Correlation of functional and structural changes of the AChR complex at the developing rat neuromuscular junction.

During the postnatal development of rat muscle the most dramatic changes are observed for the γ- and ϵ-subunit mRNAs. The reciprocal change in the abundance of these two subunit mRNAs (Fig. 2) indicates as in bovine muscle that the structural composition of the AChR-complex is under developmental control. Assuming that the changing mRNA levels reflect properly on the AChRs expressed these findings suggest that at early postnatal stages the receptors consist predominantly of α-, β-, γ- and δ-subunits. Within 10 days after birth however this receptor type decreases and is replaced by the adult type AChR being comprised of α-, β-, ϵ- and δ-subunits. Due to the fine time resolution it has been possible to demonstrate that the γ-/ϵ-subunit mRNA switch occurs during the first and second postnatal week.

A functional conversion of channel behaviour in developing end-plates has been demonstrated electrophysiologically by several authors (reviewed by Schuetze and Role, 1987). Analysis of the decay of miniature end-plate currents and of acetylcholine induced membrane current fluctuations recorded at different developmental stages suggests that the early postnatal AChR channels have longer mean open times and a lower conductance than the adult type channels with short mean open times and a higher conductance (Michler, A. and Sakmann, B., 1980; Vicini, S. and Schuetze, S.M., 1985; Hamill, O.P. and Sakmann, B., 1981). Reciprocal changes of γ- and ϵ-subunit mRNA levels are observed immediately after birth, suggesting that the γ-/ϵ-subunit mRNA switch preceeds the conversion of end-plate channel properties as schematically shown in Fig. 2. Additional experiments will be needed to show directly the appearance of ϵ-subunit containing AChRs in the end-plate membrane before one can explain the observed lag phase between mRNA switch and channel conversion.The results support the notion that the changing amounts of γ- and ϵ-subunit mRNAs being developmentally regulated result in the expression of fetal and adult type AChR.

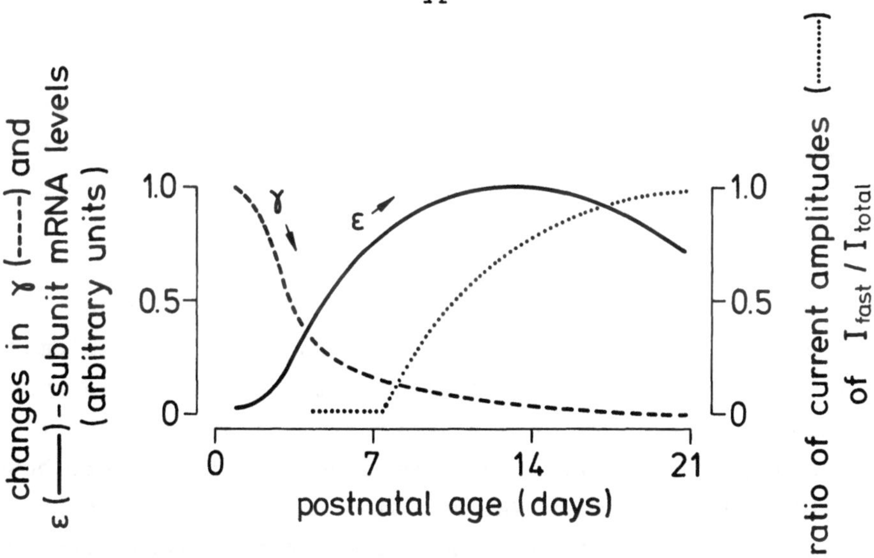

Fig. 2: Schematic representation of developmental changes occuring with respect to the γ- and ϵ-subunit mRNA levels and the AChR channel properties.
Changing levels of ϵ-subunit (solid line) and γ-subunit (broken line) mRNA during postnatal development are indicated. The dotted line (modified from Michler and Sakmann, 1980) represents the increase in the relative amplitude (I_{fast}/I_{total}) of the fast relaxation component of miniature end-plate currents recorded from rat soleus neuromuscular junctions as a function of postnatal age.

Nerve dependent alterations in the expression of subunit specific mRNA levels.

An important question is in how far during development the innervating nerve can affect the expression of the various subunit mRNAs. Is the observed γ-/ϵ-subunit mRNA switch and the conversion of AChR channel properties solely the result of a muscle specific developmental program or does it also depend on additional signals, e.g. on nerve induced muscle activity or on factors originating from the presynaptic nerve terminal? To investigate mechanisms by which the motor nerve controls the expression of the different forms of AChR channels we examined the effects of denervation and reinnervation on the AChR-subunit mRNA levels and the AChR-channel properties in the muscle fibre membrane.

Denervation of adult muscle results in a large increase of newly synthesized AChRs which become mostly incorporated into extrasynaptic regions. The majority of these AChRs have distinct biophysical properties resembling those of fetal AChRs and appear to be biochemically different from adult AChRs at the synapse (Schuetze and Role, 1987). The large increase of AChR synthesis is correlated with the increase of α- and δ-subunit mRNAs (Merlie et al., 1984; Klarsfeld and Changeux, 1985; Goldmann et al., 1985, Covault et al., 1986; Boulter et al., 1986) whereas the β-subunit mRNA shows only a moderate increase (Evans et al., 1987; unpublished observations). Of special interest are however the relative contents of mRNAs encoding the γ- and ϵ-subunits of the AChR. The γ-subunit mRNA, not detectable in innervated adult muscle, shows a strong increase upon denervation (Evans et al., 1987; Witzemann et al., 1987). On the other hand the level of the ϵ-subunit mRNA is relatively independent of innervation and increases only moderately. Motor nerve crush yields a similar result. Upon reinnervation, the γ-subunit mRNA levels decrease again to undetectable amounts within 20-25 days. These results indicate that the γ-subunit mRNA level is under tight control of the motor nerve whereas the ϵ-subunit mRNA level is relatively independent of denervation or reinnervation of adult muscle (Witzemann et al., 1987).

Single channel recordings on innervated and denervated muscles demonstrate the presence of different types of AChR. In innervated muscle a single class of high conductance (~60pS) channels with short mean open times is found. Upon denervation the vast majority of the channels, now being no longer restricted to the end-plate region has a low conductance (~40pS) and longer mean open channel times and resemble the fetal type channels (Witzemann et al., 1987). Again the switch in γ- and ϵ-subunit mRNA levels and the changing channel properties indicate that the relative abundance of the γ- and ϵ-subunit mRNAs determines the expression of two different classes of AChR channels. In innervated muscle where only the ϵ-subunit mRNA is found (besides the α-, β- and δ-subunit mRNAs) the high-conductance channel is expressed. In denervated muscle, both ϵ- and γ-subunit mRNAs are present, but the γ-subunit mRNA is much more abundant. This is reflected by the finding that the low-conductance channels predominate in denervated muscle.

<u>Effect of muscle activity on the expression of the various AChR-subunit mRNAs.</u>

The changing levels of AChR-subunit mRNAs and the concomitantly occuring channel conversion observed during innervation and upon denervation could be controlled by neurotrophic factors and/or by the nerve evoked contractile activity of the muscle fibre. Both regulatory signals are eliminated upon denervation. Chronic muscle stimulation via implanted electrodes enables one then to examine the effect of muscle activity alone on the expression of the AChR-subunit mRNAs. A high frequency stimulation pattern of denervated muscles previously found to suppress the accumulation of newly synthesized AChRs (Lomo and Westgard, 1975) was therefore applied to denervated soleus muscles in adult animals and the levels of AChRs mRNAs determined. In agreement with results described by Goldmann et al., 1988, it could be shown that muscle activity prevents the strong increase of α-, γ- and δ-subunits mRNAs which would occur otherwise in denervated muscle. Little if any change is observed for the β-subunit mRNA. Stimulation did not suppress, however, the moderate increase in ϵ-subunit mRNA, being induced upon denervation.

These results suggest that the five AChR-subunit mRNAs respond differently to nerve dependent electrical activity of the muscle. The difference is most conspicuous comparing the γ-and ϵ-subunit mRNAs. The γ-subunit mRNA level remains under tight neuronal control troughout development and can be repressed by muscle activity. The mechanism by which electrical activity of the muscle can regulate the γ-subunit mRNA (and other AChR-subunit mRNAs) remains to be elucidated.

The ϵ-subunit mRNA in denervated muscle is not affected by stimulation. It seems that once the expression of this subunit mRNA is induced during muscle development it is not regulated by nerve evoked activity. As for the other AChR-subunit mRNAs nothing is known about signals leading to the respective gene activation. Whether this is controlled by muscle specific differentiation programs or whether it depends also on neuronal factors remains to be examined.

CONCLUSIONS

The AChR and its corresponding subunit genes represent an ideal model system to investigate in how far neuronal signals contribute to the expression of a multisubunit receptor/channel complex in muscle cells. Recombinant DNA techniques combined with electrophysiological measurements have shown that two types of AChR channels are expressed in mammalian muscle, and it is possible to correlate distinct structural alterations of the receptor complex with two functional subclasses.

The change in synaptic transmission during development is preceeded by the reciprocal change of γ- and ϵ-subunit mRNA levels. Both subunits respond in a different manner to nerve evoked muscle activity. The precise nature of the regulatory signals involved is not known yet.

The observed exchange of distinct subunits of a multisubunit receptor complex could represent a general example of how functional properties of receptor and channel proteins are modulated to fulfill their specific physiological functions in the cell.

REFERENCES

Boulter, J., Evans, K., Martin, G., Mason, P., Stengelin, S., Goldmann, D., Heinemann, S. and Patrick, J. (1986). Isolation and Sequence of cDNA clones coding for the precursor to the γ-subunit of mouse muscle nicotinic acetylcholine re - ceptor.J. J. Neurosci. Res. 16, 37-49

Brenner, H.R., Lomo, T. and Williamson, R. (1987). Control of end-plate channel properties by neurotrophic effects and by muscle activity in rat. J. Physiol., 388, 367-381

Covault, J., Merlie, J.P., Goridis, C. and Sanes, J.R. (1986). Molecular forms of N-CAM and its RNA in developing and denervated skeletal muscle. J. Cell. Biol. 102, 731-739

Evans, S., Goldmann, D., Heinemann, S. and Patrick, J. (1987). Muscle acetylcholine receptor biosynthesis. J. Biol. Chem., 262, 4911-4916

Goldmann, D., Boulter, J., Heinemann, S. and Patrick, J. (1985). Muscle Denervation increases the levels of two mRNAs coding for the acetylcholine receptor α-subunit. J. Neurosci., 5, 2553-2558

Goldmann, D., Brenner, H.R., and Heinemann, S. (1988) Acetyl - choline receptor α-, β-, γ-, and δ-subunit mRNA levels are regulated by muscle activity. Neuron, 1, 329-333

Hamill, O.P. and Sakmann, B. (1981). Multiple conductance states of single acetylcholine receptor channels in embryonic muscle cells. Nature, 294, 462-464

Klarsfeld, A. and Changeux, J.-P. (1985). Activity regulates the levels of acetylcholine receptor α-subunit mRNA in cultured chicken myotubes. Proc. Natl. Acad. Sci. USA, 82, 4558-4562

Lomo, T. and Westgaard, R. (1975). Further studies on the control of ACh sensitivity by muscle activity in the rat. J. Physiol.252, 603-626

Merlie, J.P., Isenberg, K.E., Russel, S.D. and Sanes, J.R. (1984). Denervation Supersensitivity in skeletal muscle: Analysis with a cloned cDNA probe.J. Cell. Biol. 99, 332-335

Michler, A. and Sakmann, B. (1980). Receptor stability and channel conversion in the subsynaptic membrane of the de - veloping mammalian neuromuscular junction. Develop. Biol. 80, 1-1

Mishina, M., Takai, T., Imoto, K., Noda, M., Takahashi, T., Numa, S., Methfessel, C. and Sakmann, B. (1986). Molecular di - stinction between fetal and adult forms of muscle acetyl choline receptor. Nature, 321, 406-411

Nef, P., Mauron, A., Stadler, R., Alliod, C. and Ballivet, M. (1984). Structure, linkage, and sequence of the two genes encoding the γ and δ subunits of the nicotinic acetylcholine receptor. Proc. Natl. Acad. Sci. USA, 81, 7975-7979

Schuetze, S.M. and Role, L. (1987). Developmental regulation of nicotinic acetylcholine receptors. Ann. Rev. Neurosci. 10, 403-457

Shibahara, S., Kubo, T., Perski, H.J., Takahashi, H., Noda, M. and Numa, S. (1985). Cloning and sequence analysis of human genomic DNA encoding γ subunit precursor of muscle acetylcholine receptor. Eur. J. Biochem. 146, 15-22

Witzemann, V., Barg, B., Nishikawa, Y., Sakmann, B. and Numa, S. (1987). Differential regulation of muscle acetylcholine receptor γ- and ϵ-subunit mRNAs. FEBS Lett. 223, 104-112

Vicini, S. and Schuetze, S.M. (1985). Gating properties of acetylcholine receptors at developing rat endplates. J. Neurosci. 5, 2212-2224

MYOTUBE SPECIFIC EXPRESSION OF THE MOUSE ACHR DELTA SUBUNIT GENE IS CONFERRED BY 148 BP OF 5′ FLANKING DNA

T.J. Baldwin and S.J. Burden
Biology Department, 16-820
Massachusetts Institute of Technology
Cambridge, Massachusetts 02139
USA

The skeletal muscle AChR is a ligand-gated channel composed of four structurally related subunits. The four subunits assemble into a heteropentamer ($alpha_2$, beta, gamma and delta) which binds acetylcholine and thus initiates a permeability change in the myofiber membrane that ultimately couples release of neurotransmitter to myofiber contraction (Fambrough, 1979; Karlin, 1980; Anderson, 1987).

The availability of probes to identify and purify the skeletal muscle AChR has resulted in a rather thorough description of developmental and physiological events that control AChR expression (Lomo and Westgaard, 1975; Fambrough, 1979; Salpeter, 1987). These studies demonstrate that synaptic factors and electrical activity can regulate expression of AChRs and provide a framework for more detailed understanding of the mechanisms that synaptic factors and electrical activity use to regulate AChR expression. Moreover, the wealth of information regarding biosynthesis, assembly and regulation of the skeletal muscle AChR and the magnitude of the changes in AChR expression that are elicited by changes in myofiber electrical activity make analysis of AChR expression a favorable system to elucidate the steps and mechanisms that couple synaptic factors and electrical activity to expression of neuronal products.

The four AChR subunits are encoded by separate genes (Nef et al., 1984; Heidmann et al., 1986) and these genes are activated coordinately during embryonic development (Baldwin et al., 1988). AChRs are expressed during muscle differentiation as myoblasts withdraw from the cell cycle and fuse to form multinucleated myotubes (Patterson and Prives, 1973). Subsequent to this

NATO ASI Series, Vol. H 32
Molecular Biology of Neuroreceptors and Ion Channels. Edited by A. Maelicke

initial expression in primary myotubes, the quantity of AChR and AChR subunit transcripts decreases as myofibers are innervated and acquire an appropriate pattern and intensity of electrical activity (Burden, 1977; Klarsfeld et al., 1985).

In innervated, adult muscle the quantity of AChR and AChR transcripts is low and increases 50-100 fold in myofibers that have been deprived of electrical activity by denervation or by pharmacological antagonists of synaptic transmission or nerve/muscle action potentials (Miledi and Potter, 1971; Berg et al., 1972; Lomo and Westgaard, 1975; Merlie et al., 1984). Thus, regulation of AChR expression during development and by electrical activity is controlled largely by the availability of transcript encoding each of the AChR subunits: during myogenesis and following denervation of adult muscle an increase in the level of AChR mRNA is accompanied by a similar increase in the level of AChR protein (Merlie et al., 1984; Evans et al., 1987).

The wealth of information concerning physiological parameters that regulate AChR expression and the magnitude of the changes in AChR expression that are elicited by changes in myofiber electrical activity make studies of AChR regulation a particularly attractive system in which to study mechanisms involved in use-dependent changes in gene expression in electrically excitable cells. As a first step to understand how expression of AChR is controlled both during development and by electrical activity we have isolated the gene encoding the delta subunit of the AChR and have identified a cis-acting regulatory region that is sufficient to control appropriate expression of the delta subunit gene during muscle differentiation.

METHODS

Isolation of the mouse AChR delta subunit gene

400,000 phage from a genomic library constructed from mouse (AJ) liver DNA partially digested with MboI and cloned into EMBL3 phage DNA (Frischauf et al., 1983) were screened with a mouse AChR delta subunit cDNA clone kindly provided by Drs. Mixter-Mayne and Davidson (LaPolla et al., 1984). DNA from three positive phage was isolated and the mouse DNA was characterized by Southern blotting and DNA sequencing (Sanger et al., 1977). One phage contains the gene encoding the AChR delta subunit and the other two phage contain the gene encoding the AChR gamma subunit.

Construction of Delta gene-CAT gene fusions

A 2.0 kb PvuII fragment that includes 1.8 kb of 5′ flanking DNA, the first exon and part of the first intron of the mouse delta subunit gene was subcloned in the sense orientation into the SmaI site of SP65 DNA. This DNA was linearized with HindIII and digested with Bal31 to remove the methionine initiator codon of the delta subunit gene. The Bal31 digested DNA was digested further with EcoRI and the fragment containing the delta subunit gene was subcloned into EcoRI and HincII digested SP65 DNA (SP-PvuII,MD). A HindIII fragment which contains the CAT coding sequence and the small intron and polyadenylation site of SV40 t antigen (Gorman et al., 1982; Laimins et al., 1982) was subcloned into the HindIII site of SP-PvuII,MD (SP-PvuII,MD-CAT). 5′ deletions of SP-PvuII,MD-CAT DNA were generated by digestion with appropriate restriction endonucleases or by Bal31 digestion. An additional 1330 bp of 5′ flanking DNA was added by digestion of SP-PvuII,MD-CAT with SacI, removal of the 433 bp 5′ PvuII to SacI fragment and ligation of a 1762 bp SacI fragment (figure 1). The constructs were analyzed by mapping with restriction endonucleases and by DNA sequencing. Cells were also transfected with a plasmid containing the SV40 early promoter fused to the the coding sequence of the CAT gene (SV40-CAT) (Gorman et al., 1982) and with a plasmid (SP65) containing the CAT coding sequence on a HindIII fragment but without a promoter (SP65-CAT).

Transfection of mouse cell lines

DNA was transfected into mouse cell lines as a calcium phosphate precipitate (Wigler et al., 1979). For transient transfection assays we transfected a skeletal muscle cell line (C_2C_{12} cells), a fibroblast cell line (NIH 3T3 cells) and a multipotential mesodermal stem cell line (C3H 10T1/2 cells) with a mixture of SP-MD-CAT plasmid DNA and plasmid DNA containing the human growth hormone (HGH) gene (Selden et al., 1986). C2 cells were grown in DME supplemented with 15% fetal calf serum and 50 ug/ml gentamycin; 3T3 cells were grown in DME supplemented with 10% fetal calf serum and 50 ug/ml of penicillin/streptomycin; 10T1/2 cells were grown in DME supplemented with 15% fetal calf serum and 50 ug/ml of penicllin/streptomycin. Each 10 cm plate was transfected with 10 ug of SP-MD-CAT DNA and 2 ug of HGH DNA. Since differentiation and fusion of semi-confluent C2 cells is initiated by a shift from 15% fetal calf serum to 2-10% serum, all cell lines were maintained for two-three days in DME containing 5% adult bovine serum following transfection. CAT expression in 3T3 cells and 10T1/2 cells could not be induced by growing cells in 10-15% fetal calf serum following transfection. Cells were harvested 2-3 days following transfection.

Transfection of cells with the HGH gene was performed to monitor the efficiency of transfection. We measured the level of secreted HGH by radioimmunoassay as described by the supplier of an HGH immunoassay kit (Nichols Institute, San Juan Capistrano, Calif.) and corrected for variations in tranfection efficiency by expressing CAT activity relative to HGH activity (see legend to Table I). HGH activity varied less than two fold among the three different cell lines, and there was less than two fold variation in HGH activity in different experiments in any one cell line.

Stabily transformed C2 cells were isolated by co-transfecting cells with SP-MD-CAT and pSV2-neo DNA (10:1 molar ratio of SP-MD-CAT:pSV2-neo DNA) and selecting for resistance to G418 (700 ug/ml) (Southern and Berg, 1982). CAT assays were performed on pooled, stabily transformed cells. Expression in stabily transformed cells was normalized to total cellular protein and relative CAT specific activities in myoblasts and myotubes are presented in Table II.

CAT assay

Cells were washed twice in ice-cold PBS, scraped-off the plate in PBS with a rubber policeman, collected by centrifugation and resuspended in 150 ul of ice-cold 0.25 M Tris, pH 7.9. Crude cell extracts were prepared by 3 cycles of freezing/thawing (dry ice:ethanol/60°C). Cell debris was removed by centrifugation and deacylation activity minimized by heating the supernatant for 10 minutes at 60°C. CAT assays were performed with 50-80% of crude extract from transiently transfected cells and 10-20% of extract from stabily transformed cells as described (Gorman et al., 1982). The amount of extract in each assay from transient transfections was normalized to levels of secreted HGH (160 ng of HGH per assay). Acetylated chloramphenicol was resolved from unacetylated chloramphenicol on a TLC plate developed with 20:1 chloroform:methanol. Radioactive chloramphenicol spots were identified by autoradiography, excised from the TLC plate and radioactivity determined by scintillation counting.

RNase protection assay

Total RNA was isolated from cells (Chirgwin et al., 1979) and used in RNase protection assays (Melton et al., 1984; Baldwin et al., 1988). The probe for CAT mRNA was synthesized from SP65-CAT DNA (250 bp HindIII-EcoRI fragment from the 5′ end of the CAT gene subcloned into EcoRI and HindIII digested SP65 DNA) linearized with HindIII; the probe is 260 nucleotides long and protects 250 nucleotides of CAT RNA. The probe for neomycin phosphotransferase (neo) mRNA was synthesized from SP65-neo DNA (1.26 kb HindIII-SmaI fragment from pSV2-neo subcloned into SmaI and HindIII digested

SP65 DNA) linearized with PvuII; the probe is 720 nucleotides long and protects 695 nucleotides of neo RNA. The probe for mapping the transcription initiation site was synthesized from SP65-MD DNA (PvuII-PvuII fragment extending from nucleotide -1821 to nucleotide +234 subcloned into SmaI digested SP65 DNA) linearized with XhoII; the probe is 400 nucleotides long. The probe for mapping the transcription start site in delta-CAT gene fusions was synthesized from SP65 DNA containing the 786 bp EcoRI fragment from SP65-XmnI,MD-CAT subcloned into the EcoRI site of SP65 DNA. This EcoRI fragment contains 499 bp of 5′ flanking region and 24 bp of 5′ untranslated region from the delta subunit gene, 13 bp from the polylinker of SP65 DNA and 250 bp from the CAT gene. The DNA was linearized with BamHI. The probe is 831 nucleotides long and protects 286-287 nucleotides of RNA (250 nucleotides of CAT RNA, 13 nucleotides encoded by the polylinker and 23-24 nucleotides from the delta subunit gene) (figure 6). Preparation of ^{32}P-labelled probes, hybridization, digestion and analysis of protected labelled products was as described (Melton et al., 1984; Baldwin et al., 1988).

Primer-extension

A 42-mer synthethic oligodeoxynucleotide which is complementary to nucleotides 3 to 44 in the mouse delta subunit cDNA (lambda58 in LaPolla et al., 1984) was synthesized on an DNA synthesizer (Applied Biosystems, Foster City, Calif.). The 5′ end of the oligonucleotide was labelled with ^{32}P (Maniatis et al., 1982) and the labelled primer was hybridized to RNA (in 0.4 M sodium chloride, 0.1% SDS, 10 mM Pipes, pH 6.4) at 70°C for 8 hours. Addition of nucleotides to the annealed primer was catalyzed by reverse transcriptase (Hu et al., 1986); the reaction products were resolved in a DNA sequencing gel.

RESULTS

Isolation and characterization of the delta subunit gene

The gene encoding the delta subunit of the murine AChR was isolated from a murine genomic library (Methods). The phage DNA contains approximately 18 kb of murine DNA including 5.5 kb encoding the entire protein coding sequence of the mouse delta subunit, ~7 kb 5′ to the transcription start site and ~5.5 kb 3′ to the translation termination codon. A restriction map of the murine DNA was determined by Southern blotting and a partial restriction map is illustrated in figure 1. Approximately 2.5 kb of DNA that contains 2 kb of 5′ flanking DNA, the first two exons and two introns of the mouse delta subunit

a. **Mouse AChR Delta Subunit Gene: Restriction Map of 5' Flanking Region**

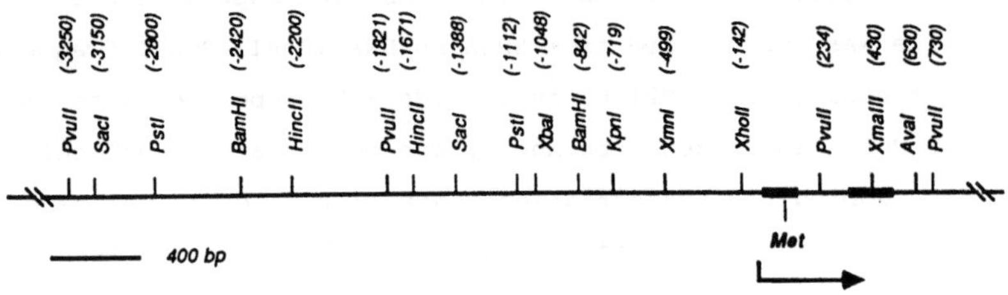

b. **Sequence of the mouse AChR delta subunit gene**

```
GAGCTCTTTGTACTTCATGCTGGACCTTGTCAGCTTCCTGATGGTGTCTTTGGAGCAGACATTGA
^-1392                                    ^
TCATTTTGCTCATATGTGAGGGTGCAGGGGCTCTCTTAGAGACCAAGAGATGAACCTCAGGGCAT
                           ^-1300
CCTTCTTGCTGTCATTGGTCTCATAGGTTGCATCATTGAGGGCAGAGTGGCAGCCTGGTCTGGCA
            ^                                                  ^-1200
GGGTCTTGTCATAGGTTTGGGCCACCTCCAAACTTCTGCTTAGAACAAAGAATAGGCTCAGACTA
                                                ^
TGGCTGCTAGTGCATGCTGCAGTCGGCTCCTCTCTTGTCAGACCCTATGGGCATCCCTTCACCCC
                                ^-1100
ATTGCTAGTCTCCCCGACATCTAGATGGCCTCCTCCCTCCATGCCTGACGGTTCTGGCTTTCTGA
                 ^
TTCCTATTAGATAGAAGCAGATCACGGTGCCCAGGCCATTTGAGGGTTGTATTCTCTTAAGCATA
  ^-1000                                             ^
CTCTTAGTGCCATCTATCCCGACCACTGACTTCTAGCCACAGTCATTACAACGGACAGGCCTGAC
                                     ^-900
TCTGTGTGGAAAGACCCTCCTTGCCCCAGTGGATCCCCTTTTCTTTCCCTTTGGTTTTGGCTACT
                       ^
CATAGATGCAGGAACGGTGAGGAGACTTCAGTTCAAATTAAAGAAAGAAAAGAAATGGGCCAGTA
       ^-800                                              ^
AGGTGGCTTAGCAAATAAAGGTACCTGCTGCCGAGCCATGTGAAAGGAAAAAAAAATTTAACTCC
                                          ^-700
TTCAAGTTTCCTCTGACTTCCACACATGCCACAAACTGCGCAGAAACACACTGAGTGGATGGGTG
                            ^
GATGGATGGATGGATGGATGGATTTTATTTATTTACTTATTTATTTATTTATAAATCTTTGGGGT
            ^-600                                          ^
TTTTTGGAGACATGGCTTCTCTGTGTAACAACCTGGCAGTCCTGGAACTCATTCTGTAGACCAGG
                                               ^-500
CTGGCCTTGAACTCAGAGATCCTTCCTCTCTTCCCAAGTGCTGGGATTAAAAGCATACACTGCCA
                                 ^
CACCCAGCTTAGAGGTAATATTTTTAAAGATGGGGGAAAAAATACACAAGTACAGGTGGAGCACT
                 ^-400
AGCAAGATGGCTCAACAGGTAACAGTTGCCTGCTGTACAAGAAGTTTCACCTCAGAACCCTCTGT
   ^                                                 ^-300
AGAAAGAGAAGCAATTGTGACAGGTGTACACACACACACACACACACACACACACACACACACAC
                                      ^
ACAGAGGATTCGCCCCCCACCCTCTGCCACCCTACCCCACCAGGCCCCACCCTACCAACAATCTG
                      ^-200
TGACAATCCTGCCTGGGATCTTTTCGTTCTGCCCTTGGCTCCTGCCCTAACTGGCAAACCCCACC
        ^ [-148                                           ^
CCCTCATCACCAGCTTTCAAGTATCAGATTGCGTTTCCGGCCTCTTCTTTCCAAACCCCTAAACC
                                           ^                 ]-32
ACCAGCACCTGTCCCCTTGCTTGCCTCATTCCACAGCAACAGGCTGAAGGGAAGACAAACCCTAG
                            **                     ]+24
TCAGTCAGAGGTGGGGGATGGCAGGGCCTGTGCTCACACTGGGGCTGCTGGCTGCCCTTGTTGTG

TGTGgtaaggatgacataacccttcctgggggccttctactgcacctggccaccgcctccctaac

attatacctaattcccactcccactctgccctttctaaccctaggcatcccacctcagggagcag
```

c. **Homology between the mouse AChR delta subunit gene and the chicken AChR alpha subunit gene**

```
                 .-128          .-113
M.D.    TTTTCGTTCTGCCCTTGGCTCCTGCCCTAACTGGCAAACCCCACCCCCTCATCACCAGCTTTCAAGTAT
                 TGCCCT-GGCTCCT-C
C.A.             TGCCCT-GGCTCCTCC
                 ^-15           ^-1
                                                           .-17       .-7
M.D.    CAGATTGCGTTTCCGGCCTCTTCTTTCCAAACCCCTAAACCACCAGCACCTGTCCCCTTGCTTGCCTC
                                                           GT-CCCT-GCT
C.A.                                                       GTGCCCTGGCT
                                                           ^-16       ^-6
```

```
        .-128              .-108                     .-100      .-90
M.D.    TGCCCTTGCTCCTGCCCTAA              C.A.       CATGCC-TGGAA
        TGCCCT-G    TGCCCT--                           TGCCCTGG
C.A.    TGCCCTGG    TGCCCTGG              C.A.         TGCCCTGG
        ^-15   ^-8  ^-15   ^-8                         ^-15   ^-8
```

Figure 1. Restriction map and sequence of the delta subunit gene. (a) Restriction map of the 5′ end of the delta subunit gene. The first two exons are indicated by bold lines; two introns follow these exons. Sites for restriction enzymes and the position of the methionine initiator codon are indicated. The direction of transcription is illustrated by an arrow; the transcription start site is number 1 and nucleotides that preceed this nucleotide are indicated by negative numbers. (b) Sequence of the delta subunit gene from the SacI site (-1388) to the PvuII site (+234). The cap sites (+1 and +2) are underlined and marked with an asterix. The methionine initiator codon (+56 to +58) is underlined; the first intron is written in small letters. The 5′ nucleotide (-148) of the shortest delta-CAT gene fusion that results in myotube specific expression is underlined and marked with a bracket ([). The 3′ nucleotide (+24) of the delta subunit gene in CAT gene fusions is underlined and marked with a bracket (]); 3′ deletion to nucleotide -32 (underlined and marked with a]) results in loss of CAT expression in myotubes. A tract of alternating AC residues is underlined. The carrots are spaced at 50 nucleotide intervals. (c) Homology with the chicken AChR alpha subunit gene. Two regions of the mouse delta subunit gene (M.D.) have homology with a region of the chicken alpha subunit gene (C.A.) that is immediately 5′ to the cap site (Klarsfeld et al., 1987). A shorter portion of this sequence is repeated in the chicken alpha subunit gene and occurs for a third time in the mouse delta subunit gene. The sequences are aligned and mismatches in the sequences are indicated with a dash (-). The first transcribed nucleotide is designated as +1 and nucleotides that are 5′ to the cap site are indicated with negative numbers.

gene was sequenced (figure 1). The DNA sequence within the two exons is identical to the corresponding region of the delta subunit cDNA isolated from the BC3H1 mouse muscle cell line (LaPolla et al., 1984) with the exception of an arginine residue at amino acid number 25 encoded by CGU rather than an alanine residue encoded by GCU. Moreover, the positions of the exon/intron boundaries in the mouse delta subunit gene are similar to those in the delta subunit gene from chicken (Nef et al., 1984): the first intron in the mouse delta subunit gene occurs between the first and second nucleotides encoding amino acid -7 of the signal sequence and the second intron occurs between nucleotides encoding amino acids 45 and 46 of the mature protein (figure 1).

The transcription initiation sites were assigned by primer extension and RNase protection analysis (figure 2). Two transcription initiation sites are detected: one transcription start site is at an A residue assigned as nucleotide +1 and a second start site is at a T residue at nucleotide +2. Thus, the delta subunit gene encodes a 54-55 bp long 5′ untranslated region. Although a canonical TATA element is not present, an A/T rich region that is bounded by C rich regions is present 30-33 bp upstream from the transcription initiation site.

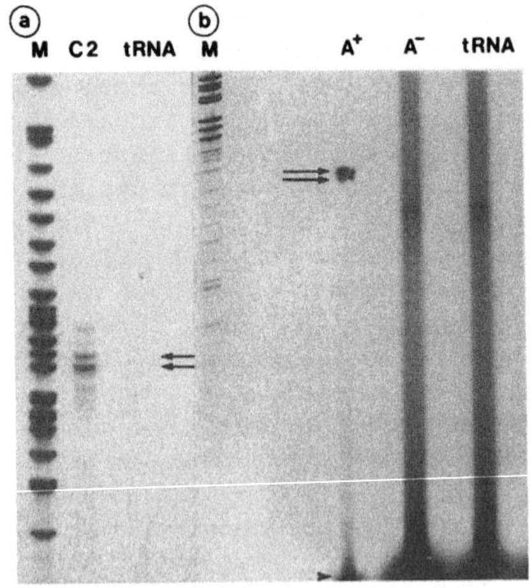

Figure 2. Identification of the transcription initiation site in the mouse AChR delta subunit gene. (a) RNase protection analysis of the transcription initiation site. 10 ug of total RNA from C2 myotubes (C2) was hybridized to a ^{32}P-labelled RNA probe extending from nucleotide -142 to nucleotide +234 in the mouse AChR delta subunit gene. 107 and 105 nucleotide ^{32}P-labelled RNA fragments (arrows) are protected from digestion by RNase (Methods); the 3′ ends of the protected fragments correspond to nucleotides +1 and +3 in the mouse AChR delta gene. Protected fragments are not detected in hybridizations with tRNA (tRNA). The gel was exposed to x-ray film with an intensifying screen at -70°C for 16 hours. The size of labelled products was determined by comparison to a DNA sequencing ladder (M). (b) Primer-extension analysis of the transcription initiation site. A ^{32}P-labelled 42-mer oligodeoxynucleotide primer (Methods) was hybridized to 8 ug of poly $(A)^+$ RNA from C2 myotubes, 20 ug of poly $(A)^-$ RNA from C2 myotubes or 20 ug of tRNA, extended and analyzed as described in Methods. The extended products from C2 myotube poly $(A)^+$ RNA are 52 and 53 nucleotides long (arrows); the 3′ end of the longer extended product corresponds to nucleotide +1 and the 3′ end of the shorter product corresponds to nucleotide +2 in the mouse AChR delta subunit gene (figure 1). Extended products are not detected with C2 myotube poly $(A)^-$ RNA or with tRNA. The position of the primer is indicated with an arrowhead. The gel was exposed to x-ray film with an intensifying screen at -70°C for 48 hours. The size of labelled products was determined by comparison to a DNA sequencing ladder (M).

221-260 bp upstream from the transcription initiation site is a tract of alternating purine-pyrimidine residues composed of 20 AC repeats. Such purine-pyrimidine repeats may be involved in recombination (Bullock et al., 1986; Treco and Arnheim, 1986) and may reflect duplication of AChR subunit genes during evolution (Noda et al., 1983). In this regard, it is noteworthy that the 5′ flanking region of the mouse AChR gamma subunit gene contains a purine-pyrimidine tract composed of 8 AC repeats approximately 800 bp upstream from the methionine initiator codon (unpublished data).

TABLE I. Quantitation of CAT expression in transient transfections

CAT Gene Fusion	CAT Activity		
	C2 Myotubes	3T3	10T1/2
Delta (-3150/+24)	100%	1.4%	2.0%
Delta (-1821/+24)	100%	0.6%	1.9%
Delta (-1388/+24)	95%	2.6%	1.2%
Delta (-1048/+24)	91%	0.7%	2.3%
Delta (-842/+24)	99%	1.3%	1.2%
Delta (-499/+24)	95%	0.3%	0.5%
Delta (-256/+24)	88%	0.3%	0.3%
Delta (-180/+24)	120%	0.5%	0.5%
Delta (-148/+24)	100%	0.6%	0.5%
Delta (-128/+24)	7.0%	0.3%	0.3%
Delta (-107/+24)	2.0%	0.3%	0.3%
Delta (-1821/-32)	0.1%	ND	ND
SV40 early promoter	120%	98%	106%
No promoter (SP65-CAT)	0.8%	0.5%	0.3%

CAT activity was measured as described in Methods. The 5′and 3′ nucleotides of each delta-CAT construct are indicated in parenthesis. CAT activity present in C2 myotubes transfected with the gene fusion Delta (-3150/+24) was assigned as 100% (0.33 - 0.36 pmol diacetylated chloramphenicol/min/ng HGH); all other values for CAT activity are expressed relative to this value. All assays were done in duplicate; duplicate values varied by less than 10%. Several independent assays were performed with each construct; variation of CAT activity in independent assays was ~10% for C2 myotubes transfected with SV40-CAT or delta-CAT and ~three fold for 3T3 and 10T1/2 cells transfected with delta-CAT.

A sequence homology with the chicken AChR alpha subunit gene is present in the delta subunit gene: 7-17 bp upstream from the cap site of the delta subunit gene is a 9/11 nucleotide match with a sequence immediately upstream from the cap site in the chicken alpha subunit gene (Klarsfeld et al., 1987). Moreover, this homology with the chicken alpha subunit gene is repeated in the mouse delta subunit gene 111 bp further upstream (-113 to -128) where the match is longer and consists of 14/16 residues (figure 1). A portion of this sequence (-8 to -15) is repeated in the chicken alpha subunit gene (-92 to -98) and occurs for a third time in the mouse delta subunit gene (-108 to -115) (figure 1).

Identification of upstream region that confers cell-type specific expression of the delta subunit gene

Gene fusions between the 5′ flanking region of the mouse delta subunit gene and the CAT gene were constructed and transfected into mouse cell lines (Methods). The methionine initiator codon was removed by Bal31 exonuclease digestion to ensure that translation initiated at the methionine codon of the CAT gene. Initially, we transfected two delta-CAT gene fusions containing 3′ deletions in the delta subunit gene into C2 cells and assayed for CAT activity in myotubes to determine whether promoter activity was retained in these 3′ deletion constructs. The two gene fusions had identical 5′ ends from the delta subunit gene (PvuII site at -1821), but differed in the amount of delta subunit DNA removed by Bal31 exonuclease and thus had different 3′ ends (figure 1). The construct with a 3′ end at nucleotide -32 produced no CAT activity in C2 myotubes, whereas the construct with a 3′ end at nucleotide +24 produced CAT activity in C2 myotubes that was comparable to that produced with SV40 early promoter (Table I). Therefore, further analysis was restricted to the construct with a 3′ end at nucleotide +24. Thus, all delta-CAT gene fusions include 24 bp of the 5′ untranslated region and terminate 31 bp prior to the methionine initiator codon of the delta subunit gene.

Figure 3 illustrates the amount of DNA from the delta subunit gene that is included in each construct. The amount of 5′ flanking DNA from the delta subunit gene included in different gene fusions ranged from ~3.2 kb to 107 bp.

The level of CAT activity produced with each construct in different cell lines is presented in Table 1. CAT expression in myotubes is relatively constant with constructs containing as much as 3.2 kb of 5′ flanking DNA and as little as 148 bp of 5′ flanking DNA. These same constructs produce ~100 fold less CAT activity in 3T3 cells and 10T1/2 cells (figure 4, Table I). Similar low level of CAT activity is detectable in cells transfected with a vector that contains the CAT protein coding region but lacks a promoter (figure 4, Table I) and in non-transfected cells. Thus, removal of 3 kb of 5′ flanking DNA from the delta subunit gene did not result in loss of expression in myotubes nor in appearance of expression in fibroblasts or mesodermal stem cells. In contrast to the cell-type specific expression conferred by the delta subunit upstream region, the SV40 early promoter confers equal expression of CAT in myotubes, 3T3 and 10T1/2 cells (figure 4, Table 1).

We also transfected several CAT gene fusions (Delta, -1821; Delta, -499; SV40 early promoter; figure 3) into primary cultures of embryonic chick skeletal muscle and assayed for CAT activity in myotubes (2-3 days after transfection). The mouse delta subunit gene promoter was active in chick myotubes and produced CAT activity comparable to the SV40 early promoter.

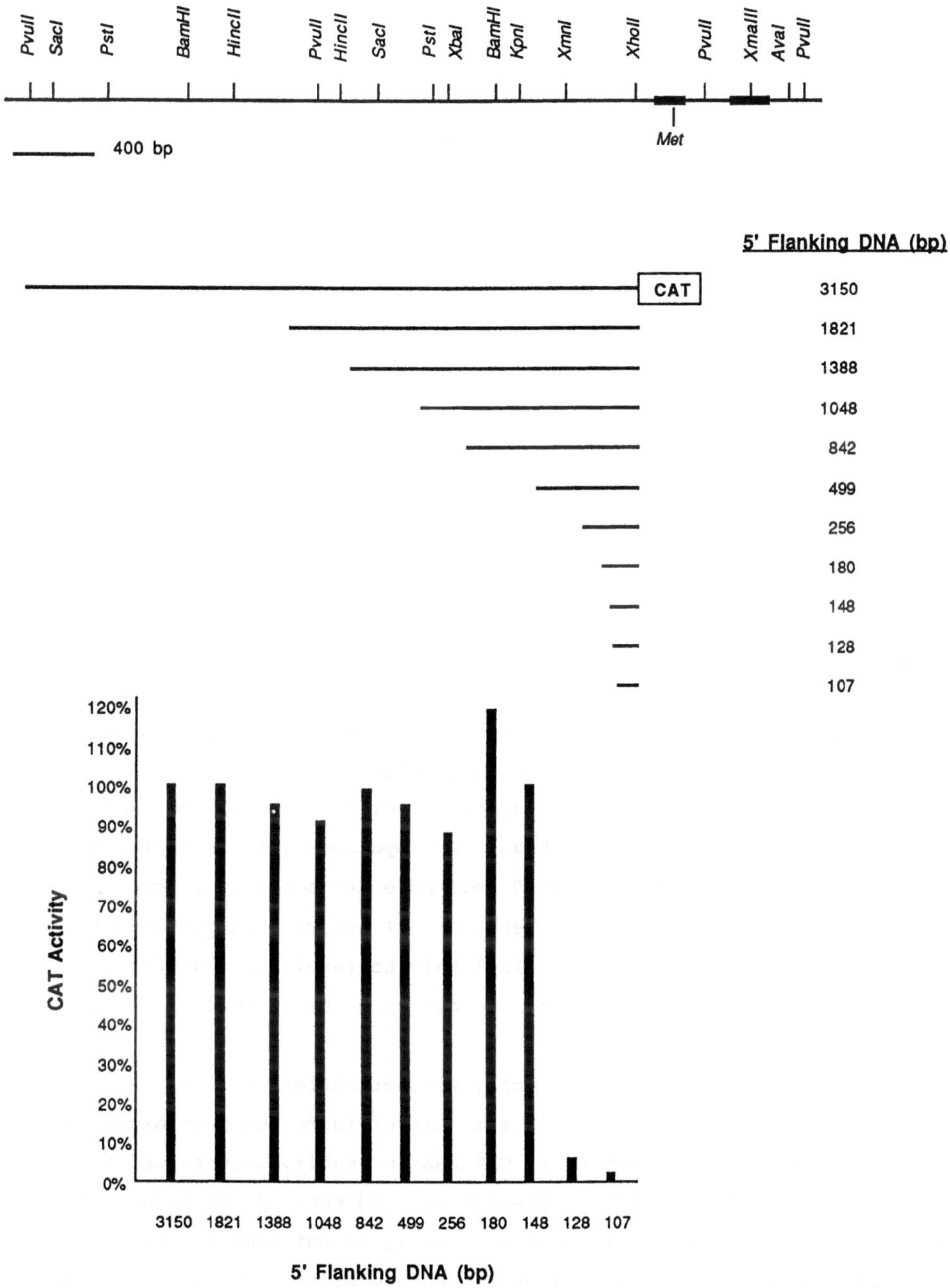

Figure 3. Delta subunit-CAT gene fusions. The amount of DNA from the delta subunit gene that is included in each construct is illustrated. The number of nucleotides that are 5′ to the cap site is indicated for each construct. The bar graph illustrates the amount of CAT activity present in C2 myotubes transfected with each construct; CAT activities present in C2 myotubes, C2 myoblasts, 3T3 cells and 10T1/2 cells transfected with each construct are presented in Table 1.

Nucleotides -148 through -108 are required for expression of the delta subunit gene in C2 myotubes

Further deletion of the 5′ flanking region of the delta subunit gene resulted in a reduction in CAT activity in C2 myotubes: deletion of nucleotides -148 through -129 resulted in a 16-fold decrease in CAT activity in C2 myotubes and further deletion of nucleotides -128 through -108 resulted in an additional 3.5-fold decrease in CAT activity. This reduction in CAT activity in C2 myotubes was not accompanied by appearance of CAT activity in C2 myoblasts, 3T3 and 10T1/2 cells. Thus, the DNA sequence from nucleotides -148 through -108 is important for expression of the delta subunit gene in C2 myotubes.

148 bp of 5′ flanking DNA confers differentiation-dependent expression of the delta subunit gene

C2 myotubes appear in culture 1-3 days after semi-confluent C2 myoblasts are grown in medium containing a low concentration of serum. Thus, transient transfection assays that result in differential expression of CAT activity in C2 myoblasts (1 day after transfection) and C2 myotubes (2-3 days after transfection) could arise as a result of accumulation of CAT during this differentiation period. To distinguish between time-dependent accumulation of CAT activity and differential rates of transcription in myotubes and myoblasts, we isolated stabily transformed C2 cells (Methods) and measured CAT activity in stabily transformed myoblasts and myotubes (figure 4, Table II). Stabily transformed cells express ~100 fold more CAT activity in myotubes than myoblasts (Table II). Thus, the presence of CAT activity in myotubes and the absence of CAT activity in myoblasts is likely to result from differential rates of transcription from the delta subunit promoter in myoblasts and myotubes.

In order to determine more directly whether different levels of CAT activity in myotubes and myoblasts is a result of increased gene expression in myotubes, we measured the quantity of CAT RNA in stabily transformed myoblasts and myotubes. CAT RNA is readily detected in tranfected C2 myotubes, and not detectable in transfected C2 myoblasts: stabily transformed C2 cells express at least 35 fold more CAT mRNA in myotubes than myoblasts (figure 5). Thus, differential CAT activity in myotubes and myoblasts is a result of increased delta subunit gene expression in myotubes. In contrast to the delta subunit promoter, the SV40 early promoter is active in myoblasts and myotubes and confers similar expression of mRNA encoding neomycin phosophotransferase in myoblasts and myotubes (figure 5).

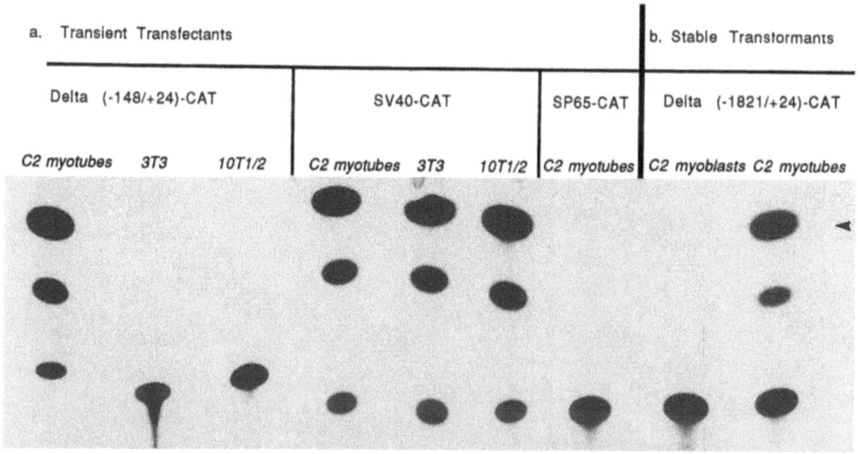

Figure 4. The mouse delta subunit promoter is active in myotubes, but not in myoblasts, 3T3 cells or 10T1/2 cells. (a) A representative autoradiogram of TLC fractionation of the radioactive reaction products in a CAT assay from transiently transfected cells demonstrates that CAT activity is present in myotubes, but not in 3T3 cells or 10T1/2 cells transfected with delta (-148/+24)-CAT. Similar CAT activity is present in all cell types transfected with SV40-CAT, whereas no CAT activity is evident in C2 myotubes transfected with SP65-CAT. (b) A representative autoradiogram of TLC fractionation of the radioactive reaction products in a CAT assay from stabily transformed C2 cells demonstrates that CAT activity is present in myotubes, but not in myoblasts. These C2 cells were stabily transfected with delta (-1821/+24)-CAT and with pSV2-neo DNA. The position of diacetylated chloramphenicol, resolved from monoacetylated chloramphenicol and chloramphenicol, is indicated with an arrowhead. The TLC plates were exposed to x-ray films at room temperature overnight.

TABLE II. Quantitation of CAT expression in stabily transformed C2 cells

CAT Gene Fusion	CAT Specific Activity	
	C2 Myoblasts	C2 Myotubes
Delta (-1821/+24)	0.6%	100%
Delta (-499/+24)	0.8%	100%
Delta (-180/+24)	0.8%	100%
Delta (-148/+24)	0.9%	100%

CAT specific activity was determined as described in Methods. Stabily transformed C2 cells were grown to semi-confluence (4 days) and either harvested as myoblasts or allowed to differentiate for two days and harvested as myotubes (Methods). Since we did not measure the number of integrated copies in cell lines transfected with different constructs, no attempt was made to compare CAT activity in cell lines transformed with different gene fusions. Rather, CAT specific activity in myotubes was assigned as 100% for each construct and CAT specific activity in myoblasts was expressed relative to the value for myotubes.

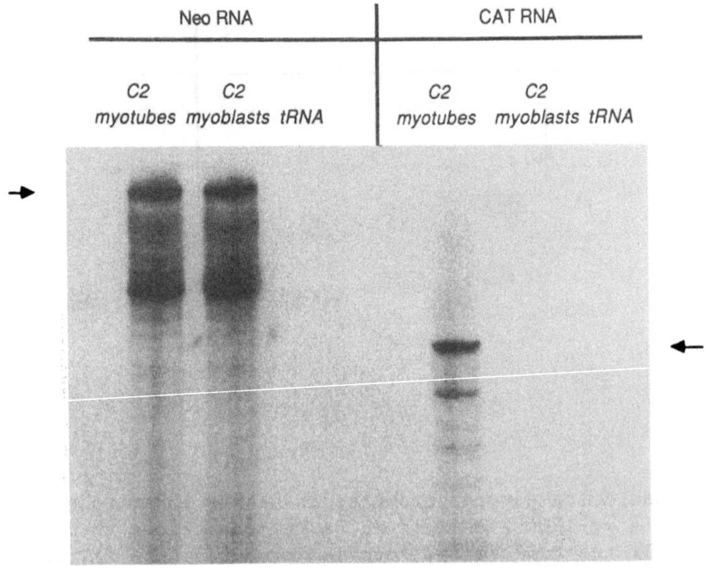

Figure 5. The delta subunit promoter is active in myotubes but not myoblasts, whereas the SV40 early promoter is equally active in myoblasts and myotubes. 35 fold more CAT mRNA is expressed in myotubes than myoblasts whereas neomycin phosphotransferase mRNA is 1.2 fold more abundant in myotubes than myoblasts. These C2 cells were stabily transformed with delta (-148/+24)-CAT and with pSV2-neo DNA. 10 ug of total cellular RNA from myoblasts and myotubes were included in each hybridization. The gel was exposed to x-ray film with an intensifying screen at -70°C for 16 hrs.

To determine whether transcription was initiated correctly in delta-CAT gene fusions, the transcription initiation site was determined for delta (-1821/+24)-CAT and for delta (-148/+24)-CAT. Figure 6 demonstrates that transcription was initiated at nucleotides +1 and +2 in the transfected delta subunit gene. Therefore, transcription was initiated at the same two nucleotides in the endogenous and transfected genes. Thus, 148 bp of 5′ flanking DNA from the delta subunit gene is sufficient to confer expression that is accurate and myotube-specific.

DISCUSSION

This study demonstrates that a 148 bp region of the AChR delta subunit gene contains all the information required for cell-type specific and differentiation-dependent expression of the AChR delta subunit.

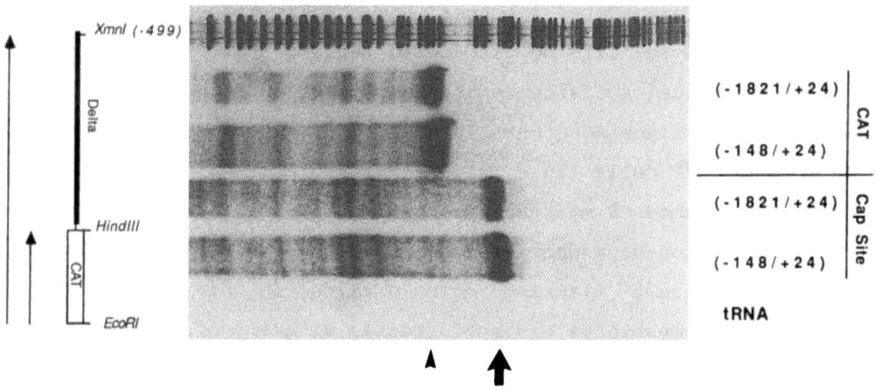

Figure 6. Transcription initiates at the same nucleotides in the transfected and endogenous delta subunit genes. The diagram illustrates the DNA template used to synthesize RNA probes. CAT RNA levels were measured by RNase protection as described (Methods and Figure 5); a 250 nucleotide ^{32}P-labelled RNA fragment is protected from digestion (arrowhead). The probe for mapping the cap site in delta-CAT gene fusions was synthesized from SP65 DNA containing the 786 bp EcoRI fragment from SP65-XmnI,MD-CAT subcloned into the EcoRI site of SP65 DNA (Methods and diagram). ^{32}P-labelled RNA fragments (arrow) at 286 and 287 nucleotides are protected from digestion by RNase (Methods); the 3′ ends of the protected fragments correspond to nucleotides +1 and +2 in the mouse AChR delta gene. 10 ug of RNA from C2 myotubes transfected with the indicated constructs were included in hybridizations with ^{32}P-labelled RNA probes. The size of the protected fragments was determined by comparison to a DNA sequencing ladder. The gels were exposed to x-ray film with an intensifying screen at -70° C for 16 hours.

Although this study demonstrates that equivalent expression is obtained in C2 myotubes containing either 148 bp or 3.2 kb of of 5′ flanking DNA from the delta subunit gene, it is possible that additional cis-acting elements can enhance or diminish expression in myotubes. Thus, cis-acting regulatory regions which are responsive to hormones and signals produced by electrical activity may not be contained within this 148 bp. In this regard, since stabily transformed C2 cells allow one to study regulation of the delta subunit gene in differentiated myotubes, the stabily transformed cell lines should be helpful in identifying cis-acting regions that are responsive to hormonal and electrical activity dependent regulators.

Nevertheless, since AChR mRNA levels increase 50 fold during differentiation of C2 cells (Yu et al., 1986) and CAT RNA is ~35 fold more abundant in transformed C2 myotubes than C2 myoblasts, it is unlikely that a potent differentiation-dependent enhancer regulates delta subunit expression and is absent from the 5′ flanking region examined in this study.

Expression in myoblasts was not detected with any of the delta subunit gene constructs we tested. Thus, our data is consistent with activation of delta subunit gene expression during the transition from myoblast to myotube rather than loss of repression in myotubes. Further studies are necessary, however, to distinguish among the possibilities.

The 5′ flanking region of the delta subunit gene contains a tract of alternating purine-pyrimidine residues composed of 20 AC repeats. In addition to a role in recombination (Bullock et al., 1986; Treco and Arnheim, 1986), it has been suggested that such tracts of purine-pyrimidine repeats may be transcriptional regulatory elements (Hamada et al., 1984). Since deletion of this purine-pyrimidine repeat in the delta subunit gene has no effect on CAT expression, this AC tract neither enhances nor is required for expression of the delta subunit gene in C2 myotubes.

Four DNase I hypersensitive sites are detected near the delta subunit gene in DNA isolated from muscle cells but not fibroblasts (Crowder and Merlie, 1986). One of these hypersensitive sites is within 1 kb of the delta subunit gene 5′ flanking region, whereas the other three hypersensitive sites are 3′ to the translation termination codon of the delta subunit gene (Crowder and Merlie, 1986). Since 148 bp of upstream sequence from the delta subunit gene is sufficient to confer muscle-specific expression, the 3′ DNase hypersensitive sites are not required for expression of the delta subunit gene in myotubes.

Since the chicken AChR alpha subunit gene is the only other AChR subunit gene in which gene expression has been analyzed at present (Klarsfeld et al., 1987), it is difficult to make meaningful sequence comparisons with other AChR subunit genes. Moreover, the region of the chicken alpha subunit gene that confers myotube-specific expression contains 850 bp of 5′ flanking DNA and the essential elements within this region have not yet been delineated further. Thus, although we have identified three sequences within the 148 bp cis-acting element of the mouse delta subunit gene that have homology to a sequence in the chicken alpha subunit gene, we have no evidence that these sequences are important for expression. Nevertheless, it is interesting that this conserved sequence is repeated twice within a region of the mouse delta subunit gene (nucleotides -148 through -108) that is essential for expression in C2 myotubes.

Several other genes that are expressed selectively in vertebrate skeletal myofibers have been characterized and cis-acting regulatory elements that confer myotube specific expression identified. This list includes the genes encoding creatine kinase (Jaynes et al., 1986; 1988), skeletal muscle actin (Grichnik et al., 1986; Bergsma 1986), myosin light chains 1/3 (Billeter et al., 1988; Shirakata et al., 1988), troponin I (Konieczny and Emerson, 1987) and AChR alpha subunit (Klarsfeld et al., 1987). Nevertheless, the skeletal muscle actin gene is the only skeletal muscle gene in which this regulatory region been delineated to less than several hundred bp upstream from the transcription start site (Bergsma, 1986). Although attempts have been made to identify a common sequence motif that is shared by the myotube-specific regulatory element(s) of these genes, no such sequence has been identified. Thus, it is not clear whether these genes are activated during muscle differentiation by identical factors that interact with different DNA sequences, or by unique DNA binding proteins that interact with cis-acting regulatory elements of the different genes. Further analysis of the cis-acting region of these genes to delineate the essential sequences and identification of the proteins that interact with these sequences is required to distinguish among the possibilities.

Nevertheless, this study demonstrates that a small region of the delta subunit gene is sufficient to confer myotube-specific expression and suggests that a similar analysis of gene fusions in electrically active and electrically quiescent stabily transformed cells may identify the region(s) of the delta subunit gene that confers electrical activity dependent regulation.

ACKNOWLEDGEMENTS

We thank Ms. Katharine Alexander for technical assistance and Ms. Jeanne Prisbylla for assistance with the manuscript. We thank Dr. Ken Walsh for helpful discussions and comments on the manuscript. We are grateful to Drs. LaPolla, Mixter-Mayne and Davidson for providing us with a cDNA clone encoding the mouse AChR delta subunit. This work was supported by grants from the Muscular Dystrophy Association and NIH (NS21579). Much of the data and several of the figures presented in this paper are currently in press (Baldwin, T.J. and S.J. Burden. December, 1988. *Journal of Cell Biology*); previously published data is printed here with the kind permission of the *Journal of Cell Biology*.

REFERENCES

Anderson, D.J. 1987. Molecular biology of the acetylcholine receptor: structure and regulation of biogenesis. In: The Vertebrate Neuromuscular Junction. M.M. Salpeter, ed. Alan R. Liss, Inc. N.Y., N.Y. 285-315.

Baldwin, T.J., C.M. Yoshihara, K. Blackmer, C.R. Kintner and S.J. Burden. 1988. Regulation of acetylcholine receptor transcript expression during development in *Xenopus laevis*. *J. Cell Biol.* 106:469-478.

Berg, D.K., R.B. Kelly, P.B. Sargent, P. Williamson and Z.W. Hall. 1972. Binding of alpha-bungarotoxin to acetylcholine receptors in mammalian muscle. *Proc. Natl. Acad. Sci. USA*. 69:147-151.

Bergsma, D.J., J.M. Grichnik, L.M. Gossett and R.J. Schwartz. 1986. Delimitation and characterization of *cis*-acting DNA sequences required for the regulated expression and transcriptional control of the chicken skeletal alpha-actin gene. *Mol. Cell. Biol.* 6:2462-2475.

Billeter, R., W. Quitschke and B.M. Patterson. 1988. Approximately 1 kilobase of sequence 5′ to the two myosin light-chain $1_f/3_f$ gene cap sites is sufficient for differentiation-dependent expression. *Mol. Cell. Biol.* 8:1361-1365.

Bullock, P., J. Miller and M. Botchan. 1986. Effects of poly [d(pGpT)-d(pApC)] and poly [d(pCpG)-d(pCpG)] repeats on homologous recombination in somatic cells. *Mol. Cell. Biol.* 6:3948-3953.

Burden, S.J. 1977. Development of the neuromuscular junction in the chick embryo: the number, distribution and stability of acetylcholine receptors. *Dev. Biol.* 57:317-329.

Chirgwin, J.M., A.E. Przybyla, R.J. MacDonald and W.J. Rutter. 1979. Isolation of biologically active ribonucleic acid from sources enriched in ribonuclease. *Biochemistry*. 18:5294-5299.

Crowder, C.M. and J. P. Merlie. 1986. DNase I-hypersensitive sites surround the mouse acetylcholine receptor delta-subunit gene. *Proc. Natl. Acad. Sci. USA*. 83:8405-8409.

Evans, S., Goldman, D., S. Heinemann and J. Patrick. 1987. Muscle acetylcholine receptor biosynthesis: regulation by transcript availability. *J. Biol. Chem*. 262:4911-4916.

Fambrough, D.M. 1979. Control of acetylcholine receptors in skeletal muscle. *Physiol. Rev*. 59:165-227.

Frischauf, A.M., H. Lehrach, A. Poustka and N. Murray. 1983. Lambda replacement vectors carrying polylinker sequences. *J. Mol. Biol*. 170:827-842.

Gorman, C.M., L.F. Moffat and B.H. Howard. 1982. Recombinant genomes which express chloramphenicol acetyltransferase in mammalian cells. *Mol. Cell. Biol*. 2: 1044-1051.

Grichnik, J.M., D.J. Bergsma and R.J. Schwartz. 1986. Tissue restricted and stage specific transcription is maintained within 411 nucleotides flanking the 5′ end of the chicken alpha-skeletal actin gene. *Nucleic. Acids Res*. 14:1683-1701.

Hamada, N., M. Seidman, B. H. Howard and C.M. Gorman. 1984. Enhanced gene expression by the poly (dT-dG)-poly (dC-dA) sequence. *Mol. Cell. Biol*. 4:2622-2630.

Heidmann, O., A. Buonanno, B. Geoffroy, B. Robert, J.L. Guenet, J.P. Merlie and J.P. Changeux. 1986. Chromosomal localization of muscle nicotinic acetylcholine receptor genes in the mouse. *Science*. 234:866-868.

Hu, M.C.T. S.B. Sharp and N. Davidson. 1986. The complete sequence of the mouse skeletal alpha-actin gene reveals several conserved inverted repeat sequences outside the protein coding region. *Mol. Cell. Biol*. 6:15-25.

Karlin, A. 1980. Molecular properties of nicotinic acetylcholine receptors. In: The Cell Surface and Neuronal Function. C. Cotman, G. Post, and G. Nicolson, eds. Elsevier/North Holland. 191-260.

Klarsfeld, A. and J.P. Changeux. 1985. Activity regulates the levels of acetylcholine receptor alpha-subunit mRNA in cultured chicken myotubes. *Proc. Natl. Acad. Sci. USA*. 82:4558-4562.

Klarsfeld, A., P. Daubas, B. Bourachot and J.P. Changeux. 1987. A 5′-flanking region of the chicken acetylcholine receptor alpha-subunit gene confers tissue specificity and developmental control of expression in transfected cells. *Mol. Cell. Biol*. 7:951-955.

Konieczny, S.F. and C.P. Emerson. 1987. Complex regulation of the muscle-specific contractile protein (troponin I) gene. *Mol. Cell. Biol*. 7:3065-3075.

Laimins, L.A., G. Khoury, C. Gorman, B. Howard and P. Gruss. 1982. Host-specific activation of gene expression of 72 base pair repeats of Simian virus 40 and Moloney murine leukemia virus. *Proc. Natl. Acad. Sci. USA*. 79:6453-6457.

LaPolla, R.J., K. Mixter-Mayne and N. Davidson. 1984. Isolation and characterization of a cDNA clone for the complete protein coding region of the delta subunit of the mouse acetylcholine receptor. *Proc. Natl. Acad. Sci. USA*. 81:7970-7974.

Lomo, T. and R.H. Westgaard, 1975. Control of ACh senstivity in rat muscle fibers. Cold Spring Harbor Symp. Quant. Biol. 40:263-274.

Maniatis, T., E.F. Fritsch and J. Sambrook. 1982. Molecular cloning. Cold Spring Harbor Laboratory, Cold Spring Harbor, N.Y.

Melton, D.A., P.A. Krieg, M.R. Rebagliati, K. Zinn and M. Green. 1984. Efficient *in vitro* synthesis of biologically active RNA and RNA hybridization probes from plasmids containing a bacteriophage SP6 promoter. *Nucleic Acids Res*. 12:7035-7056.

Merlie, J.P. K.E. Isenberg, S.D. Russell and J.R. Sanes. 1984. Denervation supersensitivity in skeletal muscle: analysis with a cloned cDNA probe. *J. Cell Biol*. 99:332-335.

Miledi, R. and L.T. Potter. 1971. Acetylcholine receptors in muscle fibers. *Nature*. 233:599-603.

Nef, P., A. Mauron, R. Stalder, C. Alliod and M. Ballivet. 1984. Structure, linkage, and sequence of the two genes encoding the delta and gamma subunits of the nicotinic acetylcholine receptor. *Proc. Natl. Acad. Sci. USA*. 81:7975-7979.

Noda, M., H. Takahashi, T. Tanabe, M. Toyosato, S. Kikyotani, T. Hirose, M. Asai, H. Takashima, S. Inayama, T. Miyata and S. Numa. 1983. Structural homology of *Torpedo californica* acetylcholine receptor subunits. *Nature*. 302:528-532.

Patterson, B. and J. Prives. 1973. Appearance of acetylcholine receptor in differentiating cultures of embryonic chick breast muscle. *J. Cell Biol.* 59:241-245.

Salpeter, M.M. 1987. Development and neural control of the neuromuscular junction and of the junctional acetylcholine receptor. In: The Vertebrate Neuromuscular Junction. M.M. Salpeter, ed. Alan R. Liss, Inc., N.Y. 55-115.

Sanger, F., S. Nicklen and A.R. Coulson. 1977. DNA sequencing with chain termination inhibitors. *Proc. Natl. Acad. Sci. USA*. 74:5463-5467.

Selden, R.F., K.B. Howie, M.E. Rowe, H.M. Goodman and D.D. Moore. 1986. Human growth hormone as a reporter gene in regulation studies employing transient gene expression. *Mol. Cell. Biol*. 6:3173-3179.

Shirakata, M., Y. I. Nabeshima, K. Konishi, and Y. Fujii-Kuriyama. 1988. Upstream regulatory region for inducible expression of the chicken skeletal myosin alkali light-chain gene. *Mol. Cell. Biol*. 8:2581-2588.

Southern, P.J. and P. Berg. 1982. Transformation of mammalian cells to antibiotic resistance with a bacterial gene under control of the SV40 early region promoter. *J. Mol. Appl. Gen*. 1:327-341.

Treco, D. and N. Arnheim. 1986. The evolutionarily conserved repetitive sequence d(TG-AC)$_n$ promotes reciprocal exchange and generates unusual recombinant tetrads during yeast meiosis. *Mol. Cell. Biol*. 6:3934-3947.

Wigler, M., A. Pellicer, S. Silverstein, R. Axel, G. Urlaub and L. Chasin. 1979. DNA mediated transfer of the adenine phospho-ribosyltransferase locus into mammalian cells. *Proc. Natl. Acad. Sci. USA*. 76:1373-1376.

Yu, L., R. J. LaPolla and N. Davidson. 1986. Mouse muscle nicotinic acetylcholine receptor gamma subunit: cDNA sequence and gene expression. *Nucleic Acids Res*. 14:3539-3555.

TRANSCRIPTIONAL REGULATION OF ACETYLCHOLINE RECEPTOR GENES IN TRANSFECTED MUSCLE CELLS AND TRANSGENIC MICE.

Andres Buonanno[1], Lucia Casabo, Jon Kornhauser, C. Michael Crowder, and John P. Merlie.
Dept. of Pharmacology, Box 8103
Washington University Medical School
660 S. Euclid Ave.
St. Louis, MO 63110

During early development of skeletal muscle, single cell myoblasts fuse to form multinucleated myotubes. The levels of nicotinic acetylcholine receptors (nAChR), which are undetectable in mouse myoblasts, increase dramatically in embryonic myotubes. At this early developmental stage receptors are distributed throughout the length of the muscle fiber surface. After motor neuron innervation, nAChR levels decrease significantly and their expression is confined to the neuromuscular junction. The repression of receptor expression is thought to be a consequence of innervation because when skeletal muscle is denervated the nAChR levels once again increases by 50 to 100-fold. Interestingly, if denervated muscle is electrically stimulated with electrodes in absence of the nerve, the expression of nAChRs is repressed (reviewed in Fambrough, 1979). This phenomena has been referred to as "activity regulation". Our laboratory has been interested in understanding the complex molecular mechanisms underlying development, localization, and activity regulation of nAChR expression. Herewith, we present evidence that transcription plays a major role in nAChR regulation, and show how we have used gene transfer into cultured muscle cells and transgenic mice to identify the DNA sequences that can account for the complex regulation of nAChR gene transcription.

[1]Present address:
Unit on Molecular Neurobiology
LDN, NICHD, NIH
Bldg. 36, Room 2A-21
Bethesda, MD 20892

NATO ASI Series, Vol. H 32
Molecular Biology of Neuroreceptors and Ion Channels. Edited by A. Maelicke

Coordinate regulation of nAChR subunits during myogenesis, innervation, and denervation of mouse skeletal muscle.

The skeletal muscle cell lines C2 and C2i have similar properties to muscle primary cultures; single cell myoblast can be induced to stop dividing, differentiate, and fuse to form myotubes which express high levels of nAChRs. Utilizing mouse cDNA clones for the nAChR α (Isenberg et al,1986), β (Buonanno et al, 1986), γ (Buonanno et al, 1986) and δ (La Polla et al, 1984) subunits, we measured the steady-state levels of the corresponding mRNAs on Northern blots. As shown in Figure 1A, the 4 subunit mRNAs were almost undetectable in myoblasts but increased by 10 (β,δ) to 40 (α,γ) fold during differentiation. The transcripts accumulated coordinately in the C2 myotubes.

Poly A+ RNA from newborn, 3-week-old, and 4 week-old mice was isolated to determine if nAChR mRNA levels are regulated by muscle innervation. As shown in Fig. 1B, the steady-state mRNA levels for the 4 subunits diminish during the time functional synapses mature, but the kinetics of decrease may differ among subunits. A similar decrease for the 4 calf nAChR mRNAs was observed by Mishina *et al* (1986). In addition, they reported an increase in the ε subunit mRNA levels during post-natal calf development; ε is a γ subunit isoform (Takai et al, 1985). A γ/ε switch during muscle development could account for the electrophysiological and biochemical differences between embryonic and adult nAChRs (Mishina et al, 1986). Although the levels of nAChRs located in extra-junctional regions dramatically decrease with innervation, receptors continue to be expressed at the neuromuscular junction. In this regard it is interesting, that the steady-state levels of nAChR mRNAs in innervated muscle fibers is highest at the endplate region (Merlie and Sanes, 1985; Fontaine et al, 1988; Witzeman et al, 1987). The localized accumulation of receptor mRNAs to synaptic regions could account for the restricted localization of receptors at the endplate of adult muscle. To determine the effects of denervation on nAChR subunit transcripts, blots containing RNA isolated from the hind-leg extensor muscles from denervated mice were probed with the 4 subunit cDNAs (Fig. 1C). The mRNAs for the four subunits accumulated coordinately after denervation. The changes in nAChR levels observed during myogenesis, innervation, and after denervation are always accompanied by parallel changes in the steady-state levels of the four subunit mRNAs. Thus, regulation of mRNA levels may play a major role in controlling the amounts of receptors expressed on the cell surface.

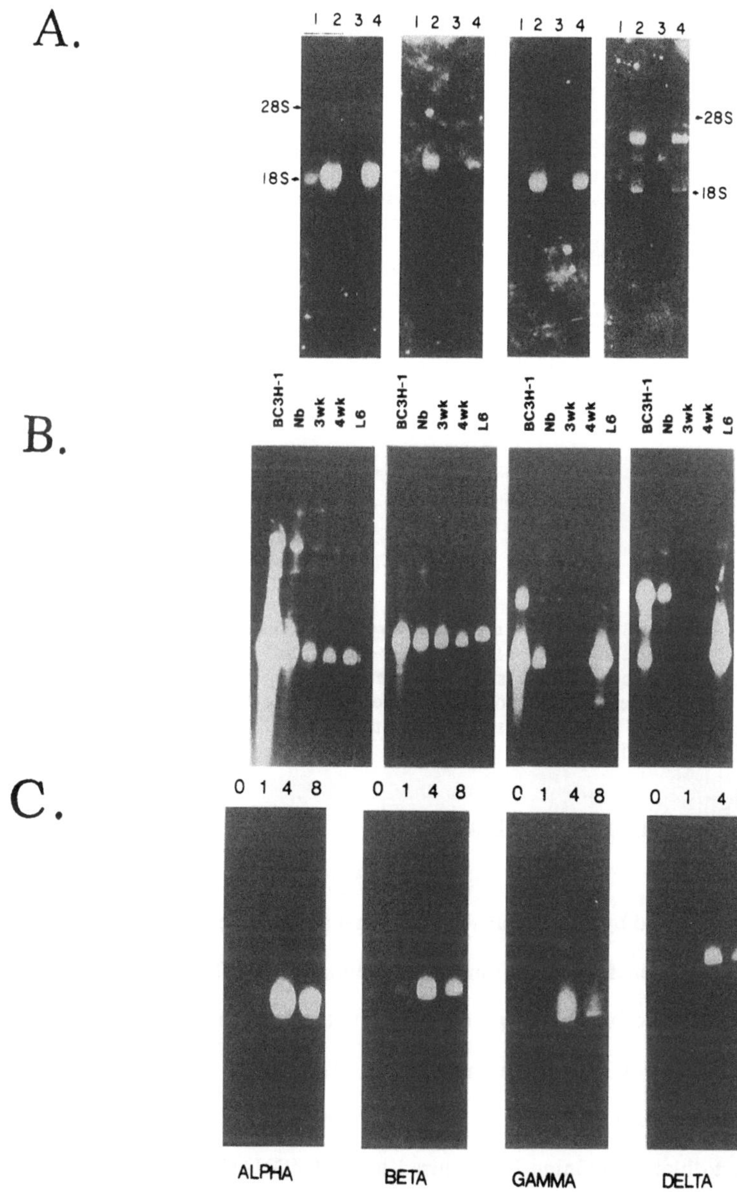

Figure 1. Steady-state levels of the 4 mouse nAChR mRNAs during myogenesis, innervation, and after denervation. Northern blots were hybridized with the inserts of cDNA clones coding for the α, β,γ, and δ receptor subunits. *Panel A.* Lanes 1 and 2 contain 5 μg of total RNA from undifferentiated and differentiated C2 cells, respectively. Lanes 3 and 4 contain 5 μg of total RNA from C2i cells, respectively. *Panel B.* The three center lanes contain 10μg of poly-A RNA isolated from newborn (Nb), 3-week-old and 4-week-old mouse lower extensor muscle. RNA from the muscle cell lines L6 and BC3H-1 were run as controls. *Panel C.* All lanes contain 10 μg of total RNA isolated from the lower extensor muscle of mice that had been denervated for 0, 1, 4 and 8 days. The probes were labeled by nick-translation to a specific activity of 1 to 3 X 10^8 cpm/μg DNA.

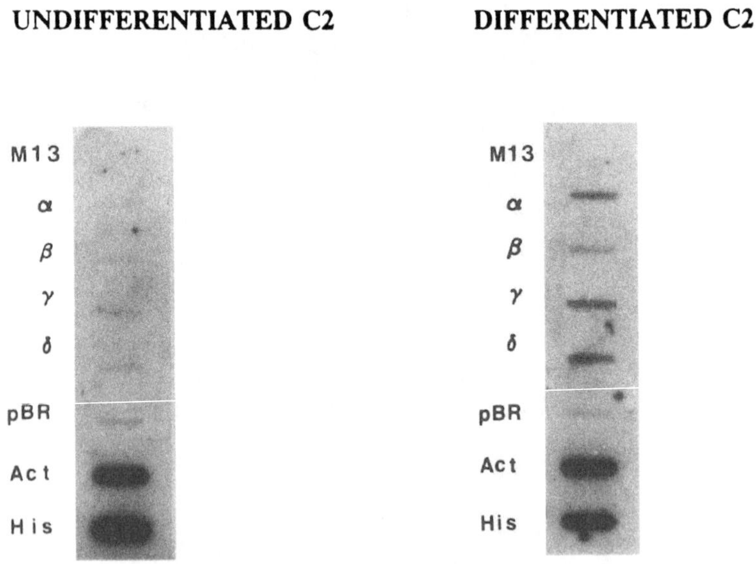

Figure 2. Transcription rates of nAChR genes measured by nuclear run-on assays. [α-^{32}P]UTP labeled transcripts, made in isolated nuclei from 3-day-old undifferentiated or 7-day-old differentiated C2 cells, were hybridized to slots containing either 5 μg of single-stranded M13mp18 vector, or single-stranded cDNAs coding for the 4 receptor subunits cloned in M13. Positive controls were cDNAs for β actin and histone 3.2 cloned in pBR322; both genes are known to be expressed in undifferentiated cells.

Coordinate transcriptional regulation of nAChR genes.

Previously, we determined by nuclear run-on assays that the relative rates of transcription of the nAChR α and δ subunit genes increased during differentiation of C2 cells (Buonanno and Merlie, 1986). We were interested in extending these studies to the 4 nAChR genes to determine if transcription is regulated coordinately during myogenesis. To measure relative transcription rates, nascent transcripts were labeled in nuclei isolated from undifferentiated and differentiated C2 cells and hybridized to blots containing immobilized nAChR cDNAs (Fig. 2). The signals obtained from blots hybridized with the RNA from differentiated cell nuclei were 5- to 8-times higher than background. In contrast, the signals on blots hybridized with the RNA from C2 myoblasts were indistinguishable from background. The lack of nAChR signals in the undifferentiated blot is specific because the genes coding for β-actin and histone, which are actively transcribed in myoblasts, gave strong hybridizing signals. Thus, the accumulation of nAChR subunit mRNAs during myoblast differentiation is associated with a coordinate activation of receptor genes.

Nuclear run-on experiments have not been done in muscle, but recently, α subunit splicing intermediates were quantified in innervated and denervated muscle as an indirect measure of transcriptional activity (Shieh et al, 1987). The results suggested that the increase of receptor steady-state mRNA levels after denervation result from higher transcription rates. In conclusion, transcriptional mechanisms have been shown to play a major role in defining the complex regulation of nAChR expression.

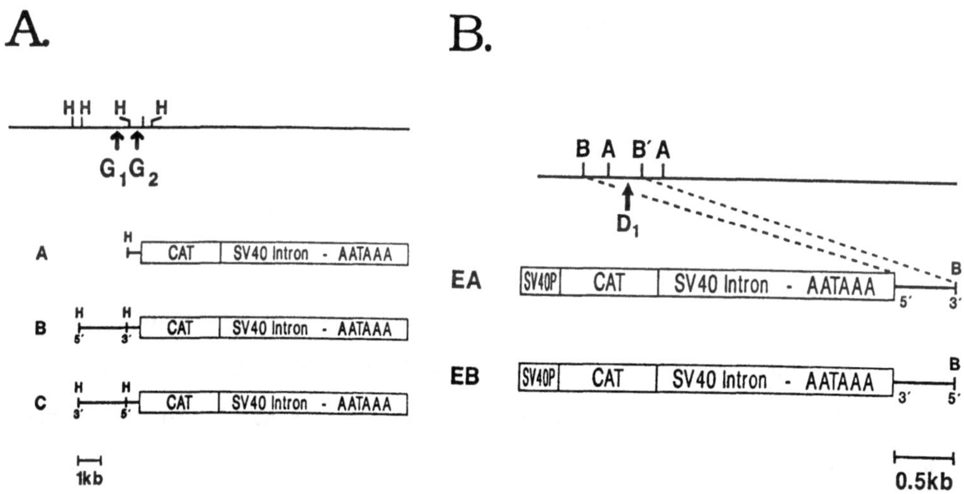

Figure 3. Constructs of the upstream sequences of the mouse γ and δ subunit genes cloned into CAT expression vectors. The upper lines represent simple maps of the nAChR γ and δ subunit genes. *Panel A*. The Hind III (H) and Ban I (stitch) restriction sites, and two DNase I hypersensitive sites (G_1 and G_2) located 5' of the γ gene are shown. The three constructs differ in the fragments or orientation of fragments cloned upstream of the CAT gene. The chimera contain: the 0.75 kb H-B fragment (γA), the 0.75 kb H-B fragment + the 2.3 kb H-H fragment in the proper orientation (γB), or the 0.75 kb H-B fragment + the 2.3 kb H-H fragment in the opposite orientation (γC). *Panel B*. A BamH 1 (B), an artificially synthesized BamH 1 (B') and Acc 1 (A) restriction sites, as well as, an upstream DNase I hypersensitive site (D_1) are shown. The constructs were made by cloning the 0.53 kb B-B' fragment in the sense and non-sense orientation 3' to the CAT gene, in an expression vector containing a SV40 early-promoter.

Upstream region of the mouse nAChR gamma and delta subunit genes confers developmental and tissue-specific expression of CAT in transfected cells.

The mouse α, β, γ, and δ nAChR genes are coordinately regulated and reside on three chromosomes (Heidmann et al, 1986), suggesting that these genes may share common cis-elements that play a role in regulating their expression. We cloned the α (Merlie, unpublished), γ, δ

(Crowder and Merlie, 1988), β and ϵ (Buonanno et al, to be published) nAChR genes hoping to characterize and define the transcription regulatory elements associated with these genes. Initially, fragments surrounding the translation start-site of both the γ and δ genes were subcloned and sequence. The transcription start-site of the δ gene was determined from RNase protection and primer extension experiments; definite mapping of the γ gene start-site has not yet been obtained. The transcription start-site for the δ gene was mapped to an adenosine located 55 bp upstream of the translation initiation site (Crowder and Merlie, 1988).

In order to identify potential transcription regulatory elements, we carried out chloramphenicol acyltransferase (CAT) assays (Gorman et al, 1982) using small fragments of DNA located 5' to the translation initiation site of the γ and δ genes. DNA fragments from the γ subunit gene were subcloned upstream of a CAT gene construct made in a pUC9 expression vector (kindly provided by N. Rosenthal). A simple map of the upstream region of the γ subunit gene, and constructs made in the CAT expression vector are shown in Fig. 3A. The constructs contained: a 0.75kb fragment extending from a Ban I site located 4 bp upstream of the initiator methionine to a Hind III site (γA), the same fragment plus an adjacent upstream 2.3kb Hind III fragment in the sense orientation (γB), and the 0.75kb fragment plus the 2.3kb Hind III fragment in the non-sense orientation (γC). We were interested in testing both fragments because they contain muscle-specific (0.75 kb fragment) and developmentally regulated (2.3kb fragment) DNase I hypersensitive sites (Crowder and Merlie, 1988); transcriptional enhancers frequently are hypersensitive sites (reviewed in Thomas et al, 1985). These recombinants, in addition to a vector containing the β-galactosidase gene used for normalization, were co-precipitated by calcium phosphate and transfected into C2 myoblasts or L929 fibroblasts. Cell extracts were made from myoblast 36 after transfection, and from myotubes and fibroblasts 72 hours after transfection. Extracts containing equal units of β-galactosidase were assayed for CAT activity. As shown in Table I, myoblasts and L929 fibroblast transfected with the γ constructs had low levels of CAT activity. In contrast, differentiated C2 cells transfected with the different constructs had 20 to 40-fold higher CAT activity than L929 fibroblasts. The upstream region of the γ subunit gene also conferred developmental regulation; C2 myotubes had 8 to 15-fold more CAT activity than myoblasts (Table I). Cells transfected with the expression vector lacking the γ subunit fragments had minimal levels of CAT activity. We used as positive controls CAT constructs containing the promoter and enhancer for either, a rat myosin light chain gene (Donoghue et al, to be published) or a chick nAChR α subunit gene (Klarsfeld et al, 1987). These chimera were properly regulated in the L929 fibroblasts and C2 cells (Table I). Thus, the 750 bases located upstream of the nAChR γ subunit gene can confer proper tissue and developmental expression when transfected into mouse muscle and fibroblast cells in culture. It is not clear yet if the DNase I hypersensitive site located in the 2.3kb Hind III fragment is involved in the transcriptional regulation of the gamma/delta gene locus.

TABLE I. Percentage of nonacetylated chloramphenicol converted to the monoacetylated form in cell extracts assayed for CAT activity[b].

	%CONVERSION			
	Experiment 1[a]		Experiment 2[a]	
CONSTRUCT	Fibroblasts	Myotubes	Myoblasts	Myotubes
Vector	1.3	0.4	0.6	0.8
Mouse γA	0.9	17.6	0.8	7.8
Mouse γB	0.4	24.2	1.0	14.4
Mouse γC	0.2	7.2	ND	ND
Chick α	2.4	61.7	ND	ND
Rat $MLCf_1$	5.2	24.7	1.6	30.4

[a] Experiment 1; mouse L929 fibroblasts or C2 cell myoblasts were cotransfected with 10 μg of the constructs shown above (described in the text), and 5 μg of a SV40-β-galactosidase expression vector used for normalization of transfection efficiencies. Cell extracts were made 72 h after transfection. Experiment 2; C2 myoblasts were transfected as discussed above. Cells were collected to make extracts 36h after transfection (myoblasts), or allowed to differentiate and collected 72 h after transfection (myotubes).

[b] Extracts containing equal units of galactosidase were assayed for CAT activity. Reaction products were separated by thin layer chromatography, and the un- and monoacetylated forms of chloramphenicol quantitated by measuring ^{14}C decay in a liquid scintillation counter.

Next, a 527 bp fragment located upstream of the δ subunit gene translation initiation site was tested for enhancer function, by assaying if it could stimulate transcription when cloned in both orientations 3' of the CAT gene. A simple map of the upstream region of the δ gene and the constructs is shown in Fig. 3B. The chimeric genes, as well as the controls, were cotransfected with β-galactosidase into fibroblasts and C2 muscle cells. Both δ gene constructs had undetectable or low levels of expression when transfected into fibroblasts or C2 myoblasts, respectively. In contrast, the δ subunit gene fragment enhanced transcription by approximately 7 to 10-fold in differentiated myotubes (Fig. 4). The construct containing the chick α subunit gene promoter and enhancer (discussed above) was properly regulated. The vector, which contained a SV40 early-gene promoter upstream of the CAT gene but not an enhancer, did not drive the expression of

CAT in fibroblast or muscle cells. Thus, the 0.5 kb fragment of the δ subunit gene behaves as a cellular enhancer; it confers proper tissue and developmental expression in transfected C2 cells irrespective of orientation and distance.

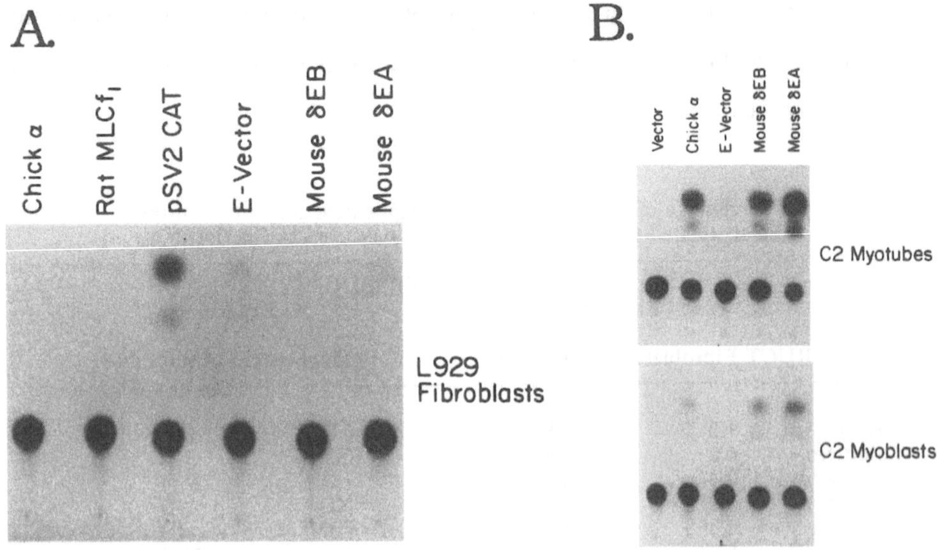

Figure 4. Expression of CAT driven by the δ subunit enhancer in transfected C2 cells and L929 fibroblasts. *Panel A.* Mouse L929 fibroblasts were transfected with CAT expression vectors that contained: the 0.85 kb 5'-flanking region of the chick α gene (chick α), the rat myosin light chain enhancer (Rat $MLCf_1$), the SV40 early-gene enhancer plus promoter, the expression vector containing the SV40 promoter but no other insert (E-vector), or the 0.5 kb mouse δ gene upstream region cloned in both orientations 3' to CAT (Mouse δEA and δEB). *Panel B.* The same constructs as in *panel A* transfected into C2 myoblasts or C2 myotubes.

Transcription of nAChR subunits genes is coordinately regulated during myogenesis and may also be coregulated during innervation and denervation. Because the genes are localized on different chromosomes (Heidmann et al, 1986), they could share common cis-acting sequences that serve to coordinately regulate transcription when recognized by shared trans-acting factors. In this regard, we have reported that a homologous sequence is found in the chick α and mouse β, γ, and δ subunit genes which lies upstream of the translation initiation site (Crowder and Merlie, 1988). The nAChR subunit homologous upstream element was given the acronym "SHUE-box". The SHUE-box described by Crowder and Merlie mapped 5 bases downstream of a 36 bp region located 5' of the chick α subunit gene shown to stimulate transcription of a reporter gene in a distance and orientation-independent manner (Wang et al, 1988). Interestingly, a sequence very homologous to the SHUE-box is centered within the 36 bp chick α subunit enhancer. As shown above,the upstream sequences of the mouse γ and δ genes can also direct tissue-specific and developmental regulation of a reporter gene when transfected into mouse cells; these regions also contain the SHUE-box. Further experiments will be necessary to assess if the SHUE-box plays a role in coordinately regulating transcription of receptor genes during development and after denervation.

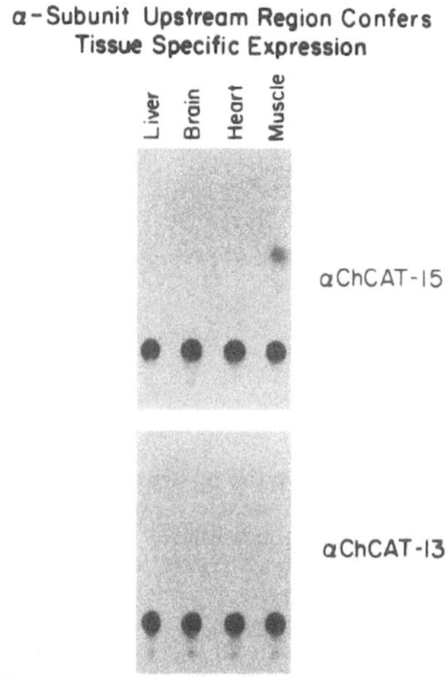

Figure 5. The upstream region of the chick α subunit gene confers skeletal-muscle specific expression of CAT in transgenic mice. A fragment containing the 0.85 kb 5'-flanking region of the chick α subunit gene, joined to the CAT gene plus the SV40 small t-antigen intron and polyadenylation signal, was used to inject mouse embryos and obtain transgenic mice. Tissue extracts were made from liver, brain, heart, and skeletal muscle of mature mice that had incorporated the CAT gene DNA into their genome (mouse lines # 13 and # 15). Extracts containing equal amounts of protein were assayed for CAT activity; the chromatograms are shown above.

Upstream sequences of the chick nAChR alpha-subunit confer proper muscle-specific and activity regulation in transgenic mice.

Klarsfeld et.al. (Klarsfeld et al, 1987) showed that an 850-bp-long 5'-flanking region of the chick α subunit gene conferred developmental and tissue-specific expression when transfected into mouse cells. These results suggested that transcription regulatory sequences and factors are conserved in the chicken and mouse. To test if the chick α upstream region restricts expression to skeletal muscle, and confers proper developmental and neuronal-activity regulation in vivo, transgenic mice containing the chick α-CAT chimeric construct were generated. CAT activity was

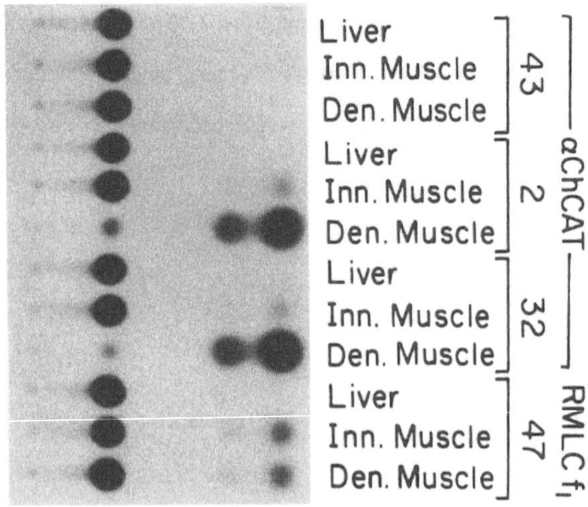

Figure 6. The expression of CAT driven by the upstream region of the chick α subunit gene is regulated by innervation and denervation of muscle. Transgenic mouse lines were generated using the fragment from the chick α construct (mouse lines # 43, 2, and 32), or the fragment from the Rat $MLCf_1$ construct (line # 47); see previous figure legends and text for description. Mature mice that had stably incorporated the CAT gene into their genome were denervated by transection of the sciatic nerve at mid-thigh. Four days later, the plantar extensor muscle groups were taken from the denervated and innervated contralateral legs. Muscle extracts containing equal amounts of protein were tested for CAT activity; the chromatogram is shown above.

assayed in extracts made from tissues of mice that had stably incorporated the construct into their genome. The results showed that CAT activity was restricted to skeletal muscle of adult mice from transgenic line # 15 (also lines # 2 and # 32, see below), it was not detectable in either liver, brain, or heart (Fig. 5). The amount of CAT activity in skeletal muscle from mature transgenic mice is low but specific. These results are consistent with the observation that receptor mRNA accumulation, and possibly transcription rates of these genes, is low in innervated muscle. The mouse line # 13 stably integrated the chimeric gene but did not express it; presumably the lack of expression is due to positional effects. Next, we investigated if the expression of the chimeric gene was under neuronal control. The plantar extensor hind-leg muscle group of adult mice (lines # 43, 2, 32, and 47) was denervated by transection of the sciatic nerve; 4 days later, cell extracts were made from the denervated and innervated (contralateral leg) muscles. The levels of CAT activity in muscle extracts from mice originating from lines # 2 and # 32 were approximately 100-fold higher in denervated than innervated muscle (Fig. 6). Mice from line #

43 integrated the construct but did not express CAT. A transgenic mouse line (#47) expressing CAT, driven by the myosin light chain enhancer, exhibited CAT expression specifically in skeletal muscle (data not shown) but did not show a response to denervation. Therefore, the response to denervation of the 850-bp-long 5'-flanking region of the nAChR alpha subunit is specific. In recent experiments using mice from transgenic line #15, we have observed that the levels of CAT activity increase in the hind-leg muscles during early postnatal development but dramatically start falling at approximately 2 weeks after birth (data not shown). The decrease in levels of transcription during this post-natal period could account for the decreasing steady-state mRNA levels seen during development (see Fig. 1B). In conclusion, the 5'-flanking sequence of the chick nAChR α subunit gene, when put in transgenic mice, can: restrict expression of the CAT reporter gene to skeletal muscle, confer proper developmental regulation, and mediate the proper responses to innervation and denervation. Presumably, the cis-acting sequences that interact with cellular factors, whose activity may be regulated during development and by electrical activity, should be present in this DNA fragment. Eventually, we hope to dissect and identify the molecular components involved in the intricate regulation of nAChR genes.

ACKNOWLEDGEMENTS

The authors wish to thank N. Davidson for providing δ clone p6H, N. Rosenthal for providing the CAT expression vectors and the rat $MLCf_1$ construct, and A. Klarsfeld and J.-P. Changeux for providing pαAChCAT+. We are also grateful to Despina Ghement for her technical assistance in cell culture. This research was supported by research grants to JPM from the NIH, Muscular Dystrophy Associations of America, Inc. and the Monsanto Company.

Buonanno A, Merlie JP (1986a) Transcriptional regulation of nicotinic acetylcholine receptor genes during muscle development. J Biol Chem 261:11452-11455

Buonanno A, Mudd J, Shah V, Merlie JP (1986b) A universal oligonucleotide probe for acetylcholine receptor genes. J. Biol Chem 261:16451-16458

Buonanno A, Mudd J, Merlie JP (to be published) Isolation and characterization of the beta and epsilon subunit genes of the mouse acetylcholine receptor

Crowder CM, Merlie JP (to be published) Studies of acetylcholine receptor subunit gene expression: chromatin structural changes during myogenesis. In: Ciba Fdn Symp

Crowder CM, Merlie JP (1988) Stepwide activation of the mouse acetylcholine receptor delta and gamma subunit genes in clonal cell lines. Mol Cell Biol 8:5257-5267

Donoghue M, Ernst H, Wentworth B, Nadal-Ginard B, Rosenthal N (to be published) A muscle-specific enhancer in the myosin light chain 1/3 gene locus. Genes and Development

Fambrough DM (1979) Control of acetylcholine receptors in skeletal muscle. Physiol Rev 59:165-227

Fontaine B, Sassoon D, Buckingham M, Changeux, JP (1988) Detection of the nicotinic acetylcholine receptor alpha subunit mRNA by in situ hybridization at neuromuscular junctions of 15-day-old chick striated muscles. EMBO J 7:603-609

Gorman CM, Moffat LF, Howard BH (1982) Recombinant genomes which express chloramphenicol acetyltransferase in mammalian cells. Mol Cell Biol 2:1049-1051

Heidmann O, Buonanno A, Geoffrey B, Robert B, Guenet JL, Merlie JP, Changeux JP (1986) Chromosomal localization of muscle nicotinic acetylcholine receptor genes in mouse. Science 234:866-868

Isenberg K, Mudd J, Shah V, Merlie JP (1986) Nucleotide sequence of the mouse muscle nicotinic acetylcholine receptor alpha subunit. Nucleic Acids Res 14:5111

Klarsfeld A, Daubas P, Bourachot B, Changeux JP (1987) A 5'-flanking region of the chicken acetylcholine receptor alpha subunit gene confers tissue specificity and developmental control of expression in transfected cells. Mol Cell Biol 7:951-955

La Polla RJ, Mixter-Mayne K, Davidson N (1984) Isolation and characterization of a cDNA clone for the complete protein coding region of the delta subunit of the mouse acetylcholine receptor. Proc Natl Acad Sci USA 81:7970-7974

Merlie JP, Sanes JR (1985) Concentration of acetylcholine receptor mRNA in synaptic regions of adult muscle fibres. Nature 317:6032-6033

Mishina M, Takai T, Imoto K, Noda M, Takahashi T, Numa S, Methfessel C, Sakmann B (1986) Molecular distinction between fetal and adult forms of muscle acetylcholine receptor. Nature 321:406-411

Shieh BH, Ballivet M, Schmidt J (1987) Quantitation of an alpha subunit splicing intermediate: evidence for transcriptional activation in the control of acetylcholine receptor expression in denervated chick skeletal muscle. J Cell Biol 104:1337-1341

Takai T, Noda M, Mishina M, Shimizu S, Furutani Y, Kayano T, Ikeda T, Kubo T, Takahashi H, Takahashi T, Kuno M, Numa S (1985) Cloning, sequencing and expression of cDNA for a novel subunit of acetylcholine receptor from calf muscle. Nature 315:761-764

Thomas GH, Siegfried E, Elgin SCR (1985) DNase I hypersensitive sites: A structural feature of chromatin associated with gene expression. In: Reeck G et al. (eds) Chromosomal proteins and gene expression. Plenum Press, New York, p 77-101

Wang Y, Xu H-P, Wang X-M, Ballivet M, Schmidt J (1988) A cell type-specific enhancer drives expression of the chick muscle acetylcholine receptor alpha subunit gene. Neuron 1:527-534

Witzemann V, Barg B, Nishikawa Y, Sakmann B and Numa S (1987) Differential regulation of muscle acetylcholine receptor gamma and epsilon-subunit mRNAs. FEBS Lett 223:104-112

THE INTERMEDIATE LOBE OF THE PITUITARY : A MODEL FOR STUDYING MECHANISMS OF TRANS-SYNAPTIC REGULATION OF SPECIFIC GENE EXPRESSION

J.P LOEFFLER

Laboratoire de Physiologie Générale

UA 309 CNRS - Université Louis Pasteur

21, rue René Descartes

67084 Strasbourg Cedex (France)

ABSTRACT

The role of the second messenger Ca^{++} on POMC gene expression was analyzed in primary intermediate lobe (IL) cultures. A single stranded POMC complementary DNA was used to quantitate POMC mRNA levels. We show that Ca^{++} entry through voltage dependent Ca^{++} channels is needed for efficient POMC gene expression and that the initial signal generated by changes in intracellular Ca^{++} is mediated by two Ca^{++} dependent enzymes : calmodulin and protein kinase C. Further, this Ca^{++} pathway can be stimulated by thyrotropin releasing hormone (TRH) and probably inhibited by dopamine.

NATO ASI Series, Vol. H 32
Molecular Biology of Neuroreceptors and
Ion Channels. Edited by A. Maelicke

INTRODUCTION

The intermediate lobe of the pituitary is a homogeneous endocrine population of melanotrope cells which can be easily maintained in primary cultures. Thus it provide an ideal model for studying secretory processes, gene expression and development. In addition, these cells display a spontaneous electrical activity, as certain derivatives of the neural crest do, eg sympathetic neurons or chromaffin cells. In situ, these cells are directly innervated by dopaminergic and GABAergic pathways, so they can be used to analyse transynaptic regulations, of electrical activity, secretory processes and of specific gene expression at a given time. Furthermore, since the precise pattern of innervation during ontogenesis is known, they constitute a model for transynaptic control of development of a given phenotype ; here an endocrine phenotype.

Many membrane located receptors have been described on these cells : GABAergic, dopaminergic, beta-adrenergic, corticotropin releasing hormone receptors, and in this paper we will provide the first evidence that these cells are also responsive to thyrotropin releasing hormone (TRH). These different receptors can be grouped into two classes, those transducing immediate membrane effects (ion channels) or those receptors coupled to second messenger systems (cAMP or Ca^{++}) via G proteins.

The two second messenger pathways triggered by cAMP or Ca^{++} play a pivotal role in development and cell differentiation (see for exemple Demeneix and Kley, this volume) which is the result of the expression of a set of genes (coding for receptors or secretory products) at a given time. In this paper, we focus our attention on the gene expression of the main secretory precursor product of the melanotrope cell, namely proopiomelanocortin (POMC) (Eipper and Mains, 1980). In previous reports, we have shown that cAMP and Ca^{++} are implicated in the regulation of POMC gene expression (Loeffler et al., 1986 ; Eberwine et al., 1987 ; Loeffler et al., 1988). We shall limit this analysis to the Ca^{++} pathway, its contribution to the stimulatory effects of TRH and its possible involvement in the inhibitory effects of dopamine (DA) on POMC gene expression at the level of POMC mRNA levels.

METHODS

Cell culture : Porcine melanotrope cells were prepared as previously described (Demeneix et al., 1986). Cells were plated in 24 multiwell plates at a density of 10^5 cells/well, and allowed to attach for two days in the presence of 5 % foetal calf serum. From then on, the cells were grown in the same medium (75 % DMEM/25 % HAM) without serum and appropriate drugs were added.

Preparation of the single-stranded POMC probe : The porcine POMC plasmid PJA 26 (Boileau et al., 1983) which contains the full length insert coding for POMC (a generous gift from Dr J. Drouin, Montreal) was used as a source of complementary DNA (cDNA). The insert was subcloned into the Pst I site of M13 (MP18). A M13 recombinant containing the message sense insert was used as a template for the synthesis of the single stranded cDNA as previously described (Pittius et al., 1985). This probe recognizes a single species of mRNA of about 1100 bases in length. POMC mRNA measurements have been reported in detail previously (Loeffler et al., 1986).

Second messengers measurements : cAMP measurement : Cells were allowed to attach for two days and then switched to preequilibrated serum free medium containing the phosphodiesterase inhibitor RO 20-1724 (100 uM) and after 10 min., drugs were added. After another 30 min., medium was removed and cells were stopped with ice cold HCl (0,1 N). Acetylated cAMP was measured by radioimmunoassay as reported. (Harper and Brooker, 1975).

G proteins : Ribosylation of IL cells membranes and identification of Go by "western" blot analysis with a antibody directed against a Go specific synthetic peptide was performed as previously described (Lang et Costa, 1987).

RESULTS AND DISCUSSION

Thyrotropin releasing hormone and POMC mRNA levels : In order to characterise a receptor which activates specifically the Ca^{++} pathway, the effect of TRH on IL cells was analysed. Indeed biochemical experiments showed that TRH increased inositol phosphate breakdown in these cells (manuscript in preparation) but did not change cAMP levels either alone or in combination with other neurotransmitters acting on the adenylate cyclase (eg : corticotropin releasing factor or drugs acting on the beta-adrenergic receptor).

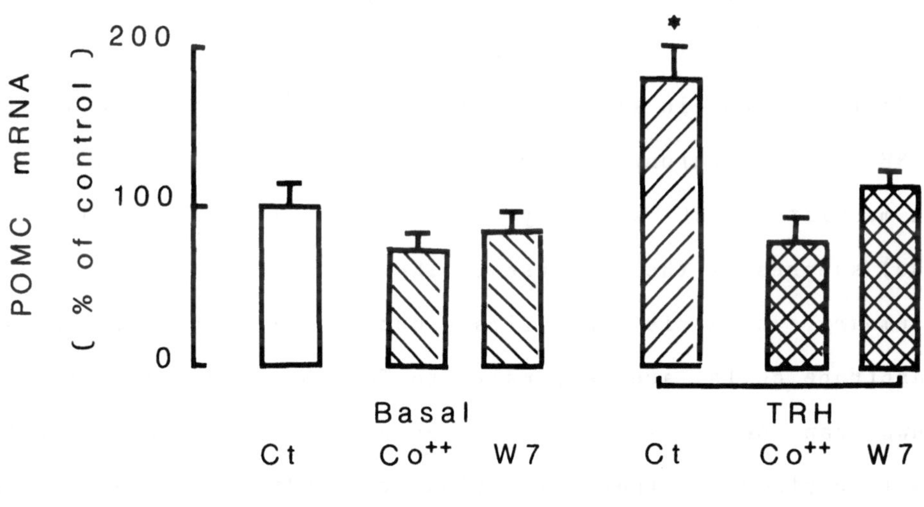

Fig 1 : IL cells were treated for 8 hours with either TRH (5.10^{-8} M) alone or in combination with Co^{++} (1 mM) or the calmodulin antagonist W7 (5.10^{-7} M). The mean values ± SEM of four cultures are given. * when p<0,05 as compared to untreated cultures.

Treatment of primary IL cells with TRH (10^{-8} M) for 8 hours resulted in an increase (25 to 125 %) of POMC mRNA levels (Fig 1). According to previous publications, Ca^{++} ion entry is required for maintenance of basal POMC mRNA levels (Loeffler et al., 1986 ; Taleb et al., 1986). Co^{++}, a potent Ca^{++} channel blocker, slyghtly decreased basal POMC mRNA levels and abolished the stimulatory effect of TRH (Fig 1). Since similar results were obtained with the dihydropyridine nifedipine (not shown) these experiment suggest that Ca^{++} entry through voltage dependent Ca^{++} channels is required for the stimulation of POMC gene expression induced by TRH. Similarly, inhibition by W7 of the Ca^{++} binding protein calmodulin (Cam) which is required for basal POMC gene expression (Loeffler et al., in press), suppressed the TRH-induced stimulation (Fig 1). This observation could suggest that Cam serves as an intracellular mediator of TRH effects on POMC mRNA levels.

In addition, protein kinase C (PKC) is implicated in the action of TRH, since chronic stimulation with phorbol esters, which down modulates PKC activity in IL cells (Loeffler et al., in press) suppresses the response to TRH.

In summary, this is the first demonstration that mammalian melanotrope cells are responsive to TRH. We show that the Ca^{++} pathway can be activated by a receptor-linked mechanism with this tripeptide and that this stimulation ultimatly leads to an increase in POMC biosynthesis. This stimulation of POMC gene expression requires the entry of Ca^{++}, probably through voltage dependent Ca^{++} channels, and is mediated by two Ca^{++}-dependent enzymes : Cam and PKC.

Dopamine and the Ca^{++} pathway : Treatment of porcine IL cells with dopamine (DA) and the D_2 receptor agonist bromocryptine (Bromo) decreased POMC mRNA levels by 50 to 70 % after two days of treatment. This inhibitory effect was abolished by an overnight pretreatment with pertussis toxin (PT) (100 ng/ml) indicating that the dopaminergic effect was mediated by a guanosine triphosphate (GTP) binding protein (Fig 2). The dopaminergic effect on IL cells is thought to be mediated by a decrease in cAMP levels which is indeed observed in porcine melanotrope cells (Fig 3). If the dopaminergic inhibition is mediated solely by a decrease in cAMP levels, one should be able to bypass the receptor mediated inhibition by stimulating directly the adenylate cyclase with forskolin. Although this drug efficiently increases cAMP levels in the presence of dopaminergic agonists in porcine IL cells, we could never observe a suppression of the dopaminergic inhibition. This experiment suggests that cAMP is not the main second messenger for this inhibition.

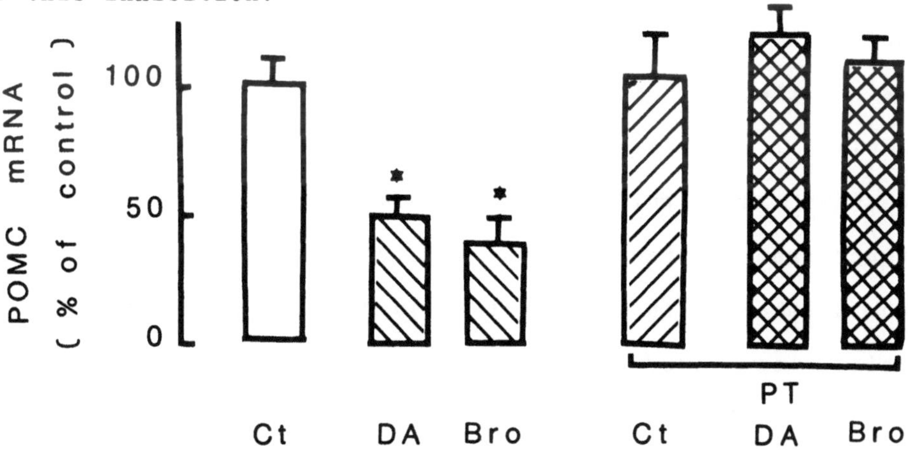

Fig 2 : IL cells were treated overnight with pertussis toxin (PT) (100 ng/ml). Cells were then cultured for another 2 days in serum free medium without toxin, supplemented with 1 % ascorbic acid and DA (10^{-8} M) or bromocryptine (10^{-8} M). Means ± SEM of four cultures are given. * when p<0,05.

In the anterior pituitary, DA, via D_2 receptors, clearly decreases internal free Ca^{++} by two mechanisms, one being cAMP independent (Malgaroli et al., 1987). In this light, the possibility of a Ca^{++} pathway mediated inhibition is also left open in IL cells. This hypothesis, that D_2 receptors are samehow linked to Ca^{++} metabolism rather than cAMP would be in line with the observation that the dopaminergic inhibition is lost when the normal Ca^{++} pathway is disturbed. Inhibition of Ca^{++} entry through voltage-dependent Ca^{++} channels with D600 and nifedipine or PKC down modulation with phorbol esters (Loeffler et al., in press) decreases POMC mRNA levels and this inhibition is not additive with the dopaminergic inhibition. This observation would again suggest that the inhibition induced by DA on POMC gene expression is largely cAMP independent.

The suppression of the DA effect by PT does not prove a cAMP mediated effect either since PT inactivates GTP binding proteins controling the adenylate cyclase Gi(s) but also Go (Sterweis et al., 1984) linked to Ca^{+} metabolism (see for example Miller, 1986). and Go is present in Il cells (Fig 3). In IL cells, D_2 might be coupled to Go and thus Ca^{++} metabolism : studies on D_2 receptor in the anterior pituitary have shown that this receptor is associated with a G protein distinct from Gi (Senogles et al., 1987).

Taken together, these results suggest that DA exerts an inhibition on POMC gene expression via GTP binding proteins controling cAMP and Ca^{++}. Although DA efficiently decreases cAMP levels in porcine IL cells, it does not appear to be the final trigger of POMC gene inhibition. In porcine melanotrope cells, this role seems to be devoted to the Ca^{++} pathway.

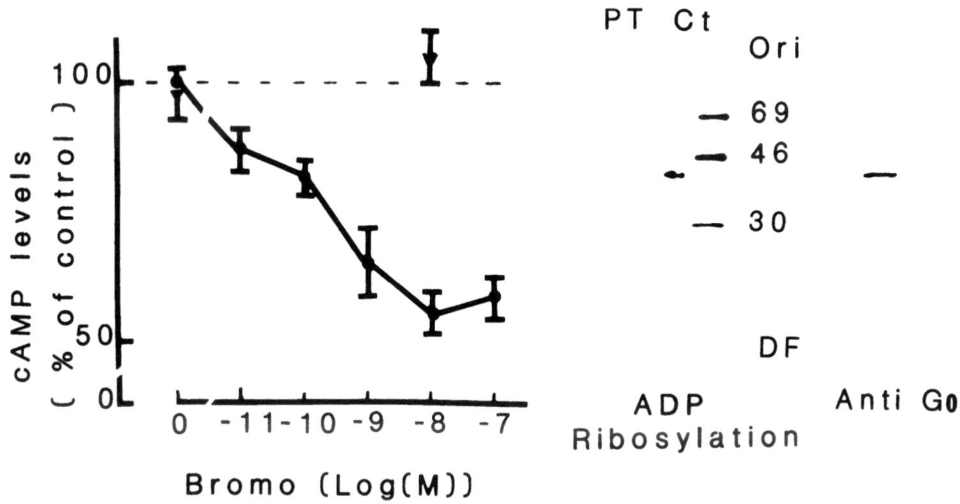

Fig 3 : Dose response curve showing the effect of bromocryptine on cAMP level in IL cells. Basal value (100 %) was 4,6 ± 0,15 pMole/well. n=6 for each value. Pertussis toxin (PT) (100 ng/ml) pretreatment (▼) alone had no effect on cAMP level but completely reversed the inhibitory effect of 10^{-8} M bromocryptine. ADP ribosylation : Molecular weight markers are indicated in K dalton. LI membranes from either control (Ct) or PT (100 ng/ml) pretreated cells were ribosylated and separated on a polyacrylamide gel according to the method of Laemli. Note the complete inactivation after PT pretreatment conditions which also suppress bromocryptin effects on CAMP and POMC mRNA levels. "Western" blot analysis of IL membranes. The antibody directed against Go recognises a single band in the 40 Kdalton region.

In conclusion, we report here that POMC gene expression can be stimulated via the Ca^{++} pathway (involving the Ca^{++} dependent enzymes Cam and PKC) and this pathway can be activated by TRH. Further, we show that basal POMC mRNA levels are dependent on normal Ca^{++} metabolism and that DA, in addition to its inhibitory effect on cAMP levels, might also inhibit the Ca^{++} pathway to exert its inhibitory effects on POMC gene expression. The topic of the current research is now a more precise targeting of intracellular relais (eg PKC versus PKA) or transacting factors involved in the mechanisms of each receptor type. To this aim we are now using liposomes to introduce macromolecules into melanotrope cells. This will allow to target specific proteins with antagonists or antibodies but also to express proteins (by introducing cloned DNA sequences) suspected to be putative transacting factors on the POMC gene.

Acknowledgment :

Antibody toward Go was a kind gift from Dr Tomaso Costa (NIH) and "Western" blot analysis was performed by Dr S. Ott at the Max Planck Institut for Psychyatrie in Munich. cAMP antibody was kindly provided by Prof. H. Glossmann (Insbrück, Austria). Plasmid containing the full lenght cDNA complementary to POMC was a kind gift from Dr J. Drouin (Canada). We thank Ms. Roth for expert help on cell culture work and to Ms E. Wendling for typing the manuscript. This work was done in the laboratory of Prof. P. Feltz. I would like to thank him for helpfull discussion and constructive criticism. This work was supported by INSERM (876013) and ARC (6089).

REFERENCES

Boileau G, Barbeau C, Jeanotte L, Chrétien M, Drouin J (1983) Complete structure of the porcine proopiomelanocortin mRNA derived from the nucleotide sequence of cloned DNA. Nucleic Acids Res 11:600-609

Demeneix BA, Taleb O, Loeffler JP, Feltz P (1986) $GABA_A$ and $GABA_B$ receptors on porcine pars intermedia cells in primary culture : functional role in modulating peptide release. Neuroscience 17:1275-1285

Eberwine JH, Jonassen JA, Evinger MJQ, Roberts JL (1987) Complex transcriptional regulation by glucocorticoids and corticotropin-releasing hormone of proopiomelanocortin gene expression in Rat pituitary cultures. D.N.A. 6:483-493

Eipper BA, Mains RE (1980) Structure and biosynthesis of pro-adrenocorticotropin/endorphin and related peptides. Endocr Rev 1:1-27

Harper JF, Brooker G (1975) Fentomole sensitive radioimmunoassay for cyclic AMP and cyclic GMP after 2'-O-acetylation by acetic anhydride in aqueous solution. J Cyclic Nucleotide Res 1:207-218

Lang J, Costa T (1987) Antisera against the 3-17 sequence of Rat G-alpha-i recognize only a 40 kDa G-protein in brain. Biochem Biophys Res Commun 148:838-848

Loeffler JP, Kley NA, Pittius CW, Höllt V (1986) Calcium ion and cyclic adenosine 3',5'-monophosphate regulate proopiomelanocortin messenger ribonucleic acid levels in rat intermediate and anterior pituitary lobe. Endocrinology 119:2840-2847

Loeffler JP, Demeneix BA, Kley NA, Höllt V (1988) Dopamine inhibition of proopiomelanocortin gene expression in the intermediate lobe of the pituitary. Neuroendocrinology 47:95-101

Loeffler JP, Kley NA, Louis JC, Demeneix BA (1988) Ca^{++} regulates hormone secretion and proopiomelanocortin gene expression in melanotrope cells via the calmodulin and the proteine kinase C pathways. J Neurochem (in press)

Malgaroli A, Vallar L, Elahi FR, Pozzan T, Spada A, Meldolesi J (1987) Dopamine inhibits cytosolic Ca^{++} increases in rat lactotroph cells. J Biol Chem 262:13920-13927

Miller RJ (1986) Protein kinase C : a key regulator of neuronal exitability. Trends Neurosci 9:538-541

Pittius CW, Kley NA, Loeffler JP, Höllt V (1985) Quantitation of proenkephalin A messenger RNA in bovine brain, pituitary and adrenal medulla: correlation between mRNA and peptide levels. Embo J 4:1257-1262

Senogles SE, Benovic JL, Amlaiky N, Unson C, Milligan G Vinitsky R, Spiegel AM, Caron MG (1987) The D_2-dopamine receptor of anterior pituitary is functionally associated with a pertussis toxin-sensitive guanine nucleotide binding protein. J Biol Chem 262:4860-4867

Sternweis PC, Robishow JD (1984) Isolation of two proteins with high affinity for guanine nucleotides from membranes of bovine brain. J Biol Chem 259:13806-13813

Taleb O, Loeffler JP, Trouslard J, Demeneix BA, Kley N, Höllt V, Feltz P (1986) Ionic conductances related to GABA action on secretory and biosynthetic activity of pars intermedia cells. Brain Res Bulletin 17:725-730

POSITIVE MODULATORS OF ACETYLCHOLINE RECEPTOR: DIFFERENCES BETWEEN SKELETAL MUSCLE AND ELECTRIC ORGAN

Vesna A. Eterović[1], Richard M. Hann[1], P.A. Ferchmin[1], Gladys Escalona de Motta[2], Jose del Castillo[2], Jose A. Prieto[3] and Jose A. Lasalde[3]

[1] Department of Biochemistry, Universidad Central del Caribe, School of Medicine, Cayey, Puerto Rico, USA

[2] Institute of Neurobiology and College of Pharmacy, University of Puerto Rico Medical Sciences Campus, San Juan, Puerto Rico, USA

[3] Department of Chemistry, University of Puerto Rico, Rio Piedras Campus, Rio Piedras, Puerto Rico, USA

INTRODUCTION

Synaptic receptors are regulated by allosteric modulators which may increase or decrease their response to agonists. For the nicotinic cholinergic receptor (AChR) very few positive modulators have been reported, although a number of allosteric inhibitors are known. This paper reviews work in which we have described two compounds that seem to act as positive modulators of muscle AChR, but have no effect on the AChRs present in membrane vesicles from Torpedo californica electric organ (Escalona de Motta and del Castillo, 1977; del Castillo et al., 1985; Eterović et al., 1981; Eterović et al. submitted to J Receptor Res).

NATO ASI Series, Vol. H 32
Molecular Biology of Neuroreceptors and Ion Channels. Edited by A. Maelicke

METHODS

Isolation, Characterization and Synthesis of TES Esters

When succinic anhydride (SA) is added to an aqueous solution of N-tris(hydroxymethyl)methyl-2-aminoethane sulphonic acid (TES), a succinylated derivative of this compound is formed with relatively low yield: for 1 to 4 mM SA and 5 to 20 mM TES, typically 5 to 30 % of the SA added was incorporated into the derivative. The yield increased with pH (from 6 to 8) and with reactant concentrations, within the mentioned range. Such a solution, called here the SA-TES solution, contains in addition to the main succinyl-TES derivative, traces of other succinylated derivatives, excess TES, and succinic acid formed by SA hydrolysis. The mayor derivative was isolated from the solution by ion-exchange HPLC, using a Synchropax AX-100 column eluted with 25 mM sodium phosphate buffer pH 6.5. The molecular structure of this compound was determined from IR, ^{1}HNMR and ^{13}CNMR spectra to be that shown in Fig 1a. This compound will be called here MST which stands for <u>M</u>ono<u>S</u>uccinyl<u>T</u>ES.

Radiolabeled MST was synthesized in aqueous solution, from [1,4-^{14}C]succinic anhydride (120 mCi/mmole) and TES. The product was separated from excess TES and succinic acid by ion-exchange HPLC as described above.

An organic synthesis was then developed: TES was dissolved in acetonitrile by reacting with triethylamine. The ester was formed by addition of equimolar amounts of succinic acid and dicyclohexylcarbodiimide as the dehydrating agent (Hassner et al., 1978). The product was purified by several precipitation and chromatographic steps. Its purity was assesed by HPLC as described above. Based on its chromatographic behavior and spectral characteristics, the synthetic compound was identical to that isolated from the SA-TES solution.

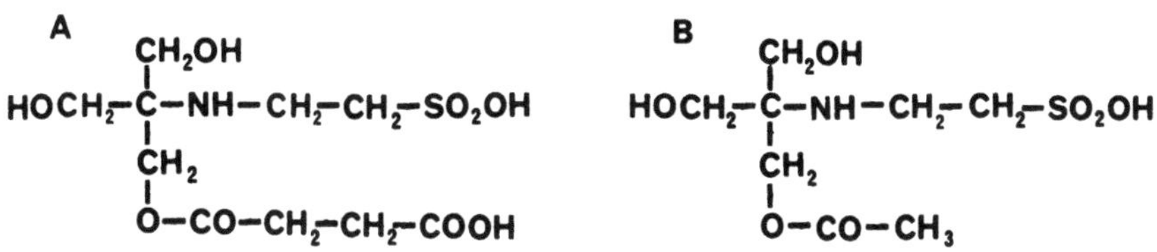

Fig 1: Molecular structure of MST (A) and MAT (B)

For the synthesis of the acetyl ester of TES, acetic anhydride and TES were reacted in acetonitrile, following the procedure of Durst and Gokel (1980). The remaining traces of acetic acid were removed by freeze-drying and the product was dissolved in methanol and precipitated with acetone. The structure was determined by IR and NMR spectroscopy to be that of Fig 1b. The acetyl derivative will be called MAT, for Mono-AcetylTES.

Physiological Methods

ACh-induced muscle contraction was measured in the frog rectus abdominus muscle. Muscle membrane depolarization, induced by bath-applied agonists, was measured in the frog sartorius muscle using the moving electrode technique (Fatt, 1950). Endplate potentials (epp's) and currents (epc's), and miniature endplate potentials and currents (mepp's and mepc's), were recorded from a sartorius-sciatic nerve preparation as described in del Castillo et al. (1985).

Biochemical Methods

AChR-rich membranes were prepared from frozen <u>Torpedo californica</u> electric organ (Lindstrom et al., 1980); alkali-washed membranes were used in some experiments, as indicated. Protein concentration was estimated by the method of Lowry et al. (1951), and receptor concentration from binding of [^{125}I]α-bungarotoxin (^{125}IαBgt)(Schmidt and Raftery, 1973). Equilibrium binding of [^{3}H]ACh and of [^{14}C]MST was measured by an ultracetrifugation method similar to that described by Krodel et al. (1979). Activity of acetylcholinesterase (AChE) was determined following the procedure of Fernandez et al. (1979). The affinity change induced by exposure to agonists was measured as described by Quast et al. (1978). AChR-regulated influx of ^{86}Rb into receptor-rich membrane vesicles was measured following the procedure of Lindstrom et al. (1980), with minor modifications.

RESULTS

ACTIVITY OF TES ESTERS ON MUSCLE MEMBRANES

Pure MST and SA-TES solution (MST concentration in this solution was aproximately 40 μM) acted similarly in all experiments tested and, therefore, it is safe to assume that all results obtained with the solution were due to MST. The other components of the SA-TES solution, TES and succinic acid, were physiologically inactive, and the trace amounts of other succinyl-TES derivatives did not have major effects.

ACh-induced Muscle Contraction

TES esters did not act as agonists on the frog rectus abdominus muscle: no effect was observed in the absence of ACh. But the strength of ACh-induced contraction was increased in their presence at all ACh concentrations tested. This effect was observed with the SA-TES solution as well as with pure MST (Fig 2) and MAT . The degree of increase was highly variable from muscle to muscle (5-300%). In these experiments, AChE was inhibited by neostigmine, excluding the possibility that TES esters potentiated ACh action by inhibiting the esterase.

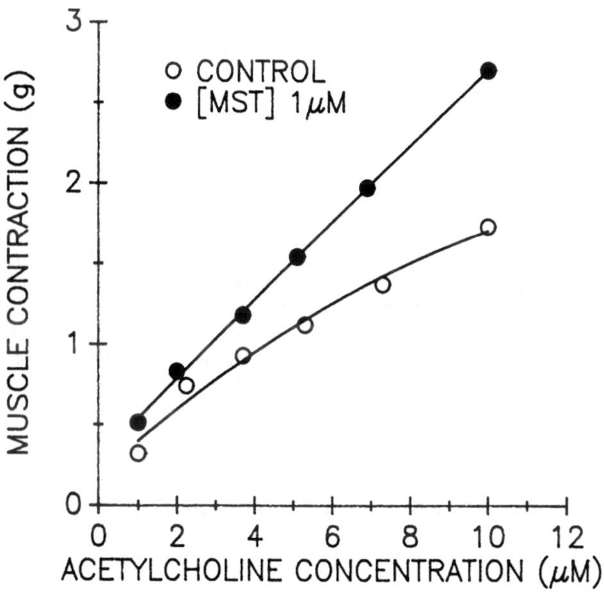

Fig 2: Effect of MST on muscle contraction. Maximum tension generated by frog rectus abdominus muscle in response to different concentrations of ACh was recorded in the absence of MST (control) and in the presence of 1μM MST. Control and experimental values were recorded from sister muscles. (From Eterović et al. submitted to J Receptor Res)

The dose-response curve for MST and MAT is shown in Fig 3. Both compounds increased the response to ACh at concentrations of 10^{-8} to 10^{-6} M. At higher concentrations they became antagonists, MST at lower concentrations that MAT.

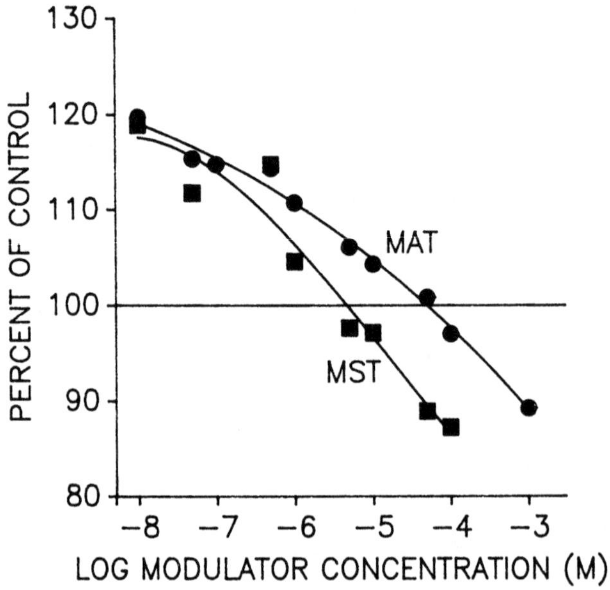

Fig 3: Modulator concentration-response curve for muscle contraction. Maximum tension generated by frog skeletal muscle in response to a fix amount of ACh was recorded in absence of TES esters (control) and in the presence of different concentrations of MST or MAT. Maximum tension values in the presence of TES esters are shown as percent of control values, recorded from sister muscles. ACh concentration was 0.1 μM for the MST curve and 0.01 μM for the MAT experiment. (From Eterović et al. submitted to J Receptor Res)

Muscle Membrane Depolarization by Externally Applied Agonists

The SA-TES solution did not affect the input resistance of the muscle fiber (control, $4.2 \times 10^5 \pm 1.5 \times 10^5$ Ω (s.d.) versus $4.1 \times 10^5 \pm 1.6 \times 10^5$ Ω in the presence of SA-TES; n=20 fibers).

On the other hand, the ACh-induced depolarization of sartorius muscle membrane was potentiated by the SA-TES solution. At saturating concentrations of ACh, the depolarization recorded in the presence of SA-TES solution was twice the size of control value (Fig 4).

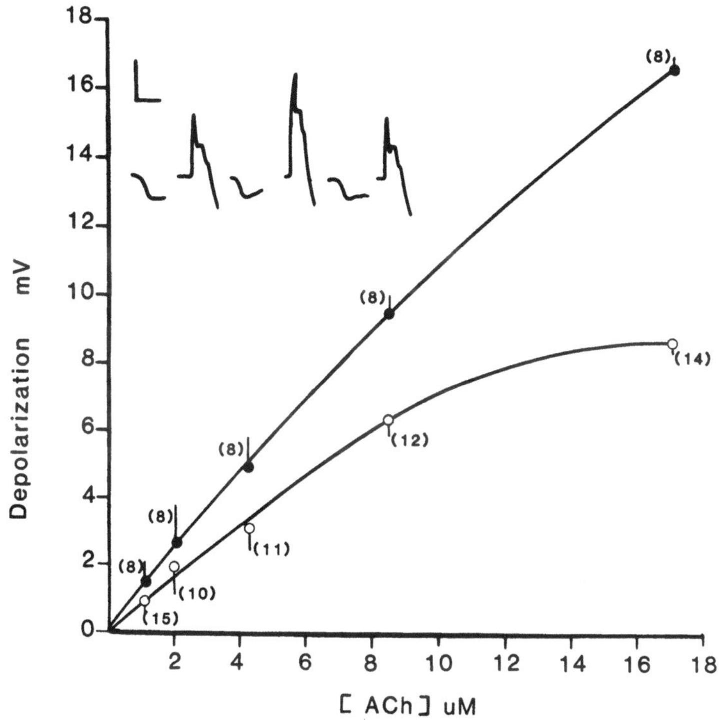

Fig 4: Depolariation of frog sartorius muscle by ACh. Points represent the average amplitude of the depolarization (in mV) elicited by 1 to 17 μM ACh in Ringer solution buffered with 5 mM TES (open circles) and after addition of 1 mM SA (solid circles). Number of measurements are in parenthesis. Vertical lines indicate s.d. for each point. Inset shows three records obtained from a muscle exposed to 10 μM ACh: The first in normal Ringer buffered with 5 mM TES, the second upon addition of 1 mM SA to the Ringer-TES (the SA-TES solution), and the third after return to normal Ringer-TES solution. The baseline record of the muscle surface potential is shown before each test scan. Inset calibration are: vertical, 1 mV; horizontal, 30 s. (Modified from del Castillo et al., 1985).

In addition to the increase in amplitude, the peak depolarization was achieved faster in the presence of the SA-TES solution, as was the repolarization after ACh was washed out of the bath (data not shown). To explain these results it was postulated that under normal conditions ACh binds to an inhibitory site on the AChR and this interaction delays its diffusion in the synaptic cleft. When the inhibitory site is occupied by MST, the inhibitory action is prevented (thus the amplitude of ACh potentials is increased) and ACh diffusion toward deeper receptors is accelerated.

All these effects of the SA-TES solutions were specific for choline esters. That is, they were observed with ACh and carbamylcholine (CCh), but not with nicotine, tetramethylammonium, benzyltrimethylammonium or decamethonium. It was therefore necessary to postulate that the proposed inhibitory site is an esteratic site binding ACh through its carbonyl moiety.

The Response to Neurally-released ACh

The amplitude of epp's was increased slightly in the presence of the SA-TES solution (Fig 5). This result showed that MST (the active component of this solution) affects the response to neurally released as well as externally applied agonist.

MAT and to a lesser degree MST increased the amplitude of mepp's. Here, as with muscle contraction, low concentrations of TES esters induced higher potentiations, while higher concentrations were inhibitory (Table 1).

Contrasting with this clearcut effect on the amplitude of mepp's, the time course of epc's or mepc's were not affected by either compound (data not shown). These results suggested that TES esters do not affect the lifetime of the AChR ion channel.

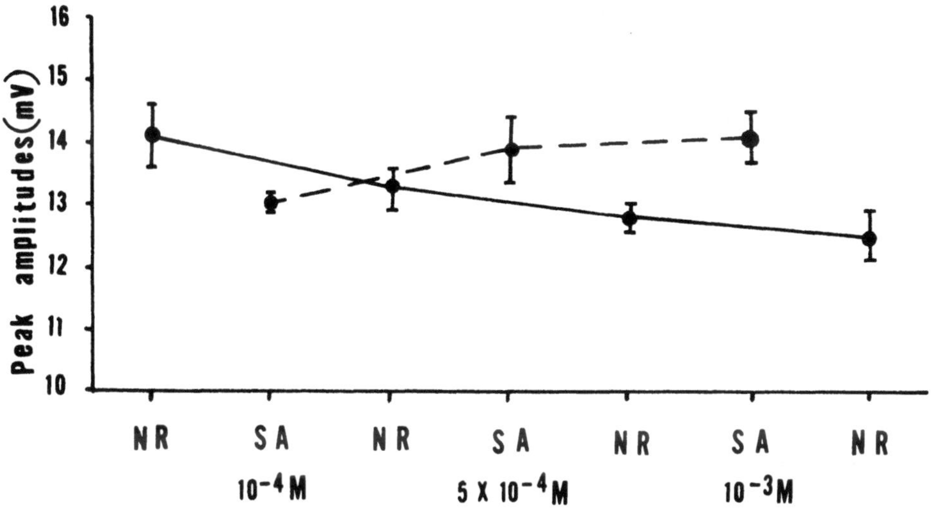

Fig 5: Peak amplitudes of epp's evoked by nerve stimulation. Average peak values of epp's obtained in 3 experiments, stimulating constantly at 0.1 Hz the nerve of sciatic-sartorius preparations, using a modification of the moving electrode technique (Fatt, 1950). Each point represent the average of 20 measurements; vertical bar indicates the s.d. Solid line corresponds to values obtained in normal Ringer buffered with 5 mM TES (NR); broken line indicates results obtained with SA-TES solution prepared with the indicated concentrations of SA and 5 mM TES.

In muscle-nerve preparations treated with neostigmine and stimulated at 200 Hz, the decay time of epc's increases progresively with the length of stimulation. This increase in decay time was postulated by Katz and Miledi (1973) to result from repetitive binding and activation of AChRs by ACh molecules, which are therefore delayed in their diffusion out of the synaptic cleft. This increase in epc's decay time was partially reverted in the presence of the SA-TES solution (Table 2 and Escalona de Motta and del Castillo, 1977). Neither this solution nor its purified active component, MST, appear to interact with the AChR active site. Thus the observed reduction in these epc's decay time was interpreted as indicating that ACh also binds to an additional site that delays its diffusion under normal conditions. In the presence of the SA-TES solution, this site is occupied by MST and the diffusion of ACh is facilitated.

TABLE 1

TES esters affect the amplitude mepp's

Exp. #	Compound	mepp's, mv		
		Control 1	Experimental	Control 2
A1	MST (5 μM)	0.50 (29)	0.69 (28)	--
A2		0.54 (48)	0.57 (30)	--
A3		0.38 (29)	0.44 (12)	--
B	MST (50 μM)	0.68 (30)	0.71 (32)	0.66 (29)
C1	MST (1 mM)	0.69 (48)	0.58 (42)	0.67 (38)
C2		1.01 (55)	0.79 (46)	0.96 (41)
C3		0.68 (30)	0.49 (36)	0.79 (46)
D	MAT (5 μM)	0.44 (101)	0.98 (150)	0.51 (123)
E	MAT (50 μM)	0.56 (96)	0.84 (235)	0.63 (66)

Values are averages of mepp's recorded intracellularly before (Control 1), during, and after (Control 2) the application of MST or MAT at the indicated concentrations. Only fibers having stable resting potential values were included. The values were collected and automatically averaged by a Nicolet digital oscilloscope which lacked the capability to calculate a measure of the dispersion of the values. Numbers in parentheses indicate total number of mepp's

Table 2

Average half-decay times of epc's evoked in antiesterase treated muscles by four consecutive pulses

Pulse #	Average half-decay time, ms ± sem		p^*
	Control	SA-TES solution	
1	3.25 ± 0.35 (10)	3.91 ± 0.40 (10)	n.s.
2	4.52 ± 0.40 (10)	4.98 ± 0.47 (10)	n.s.
3	6.87 ± 0.56 (9)	6.46 ± 0.54 (9)	n.s.
4	10.90 ± 0.85 (9)	7.02 ± 0.56 (9)	$p<0.025$

The averaged half-decay times were obtained from a total of 10 fibers in Ringer solution buffered with 5 mM TES, before and after the addition of 1 mM SA. Neostigmine (3 μM) was present throughout the experiment. *: t-Test for significance of the difference were made for each pair of observations.

BINDING STUDIES IN MUSCLE MEMBRANES

Amphibian rectus abdominus muscle contains multiple synaptic contacts per muscle fiber, and therefore is richer in AChR than other, singly innervated skeletal muscles. This fact has allowed us to study equilibrium binding of $[^3H]$ACh to membrane preparations from frog and toad rectus abdominus muscle (Eterovic et al., 1981). Two binding sites were detected by equilibrium dialysis, centrifugation and filtration assays: one of high affinity in the nanomolar range, and the other of lower affinity, in the 0.1 μmolar range (Fig 6). The high-affinity site was sensitive to αBgt, CCh, d-tubocurarine and nicotine, and so it most probably was the active site of the AChR. The concentration of this site was approximatelly 1 pmole per mg of protein (about 40 pmoles per g wet tissue) which is similar to AChR concentration in denervated diaphragm (Colquhoum and Rang, 1976). [^{3}H]ACh binding to the low-affinity site was not affected by αBgt; it was decreased by CCh (10^{-3}M) and decamethonium at 10^{-4} M, but low concentrations of decamethonium (10^{-8} to 10^{-7} M) stimulated this binding. The SA-TES solution consistently decreased [^{3}H]ACh binding to the low-affinity site.

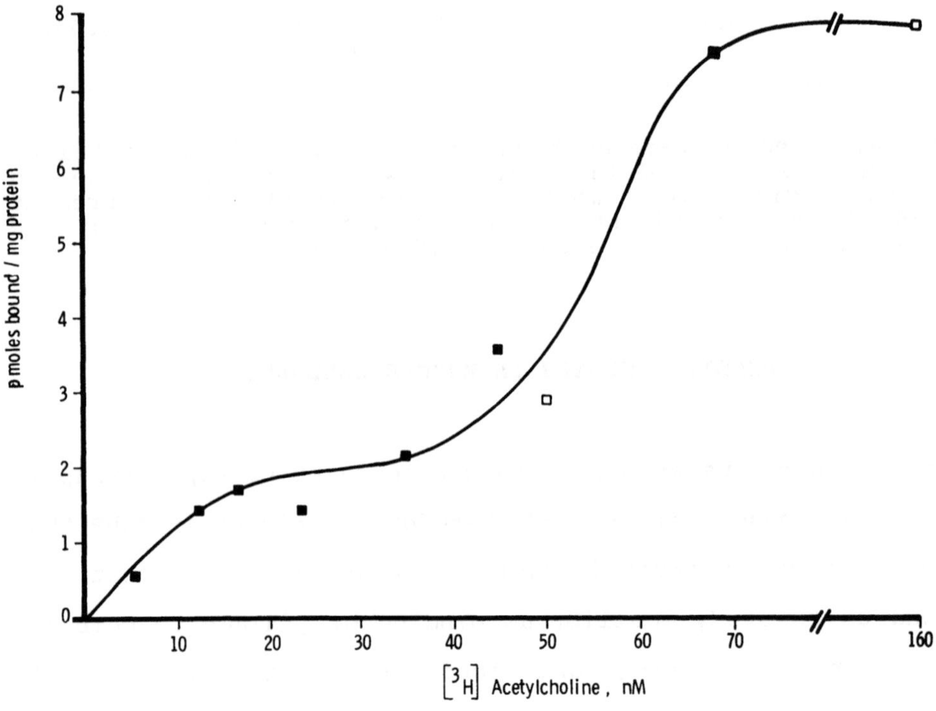

Fig 6: [3H]ACh binding to toad rectus abdominus muscle. This is a composite graph of two experiments (filled and empty squares respectively). Muscle particulate fraction was incubated at ice temperature, with [^{3}H]ACh or [^{3}H]ACh plus 10^{-4} M decamethonium (blank). For filled squares, protein concentration was 1.3 mg/ml and the incubation time 30 s. For empty squares, protein concentration was 0.53 mg/ml and the incubation time 10 min. The values were corrected for the binding to the filter displaceable by decamethonium.

BINDING AND ION-FLUX STUDIES IN MEMBRANES FROM ELECTRIC ORGAN

Due to the high concentration of AChR, membrane preparations derived from electric organs of fishes are more convenient to study ligand binding and other properties of AChR. Therefore, to clarify the mechanism of action of TES esters, and assuming that the receptors from the electric organ would behave similarly to those AChR from skeletal muscle, we studied ligand binding and ion-flux in receptor-rich membrane vesicles from electric organ. As will be seen by the following results, the assumption of the receptors' similarity was not warranted.

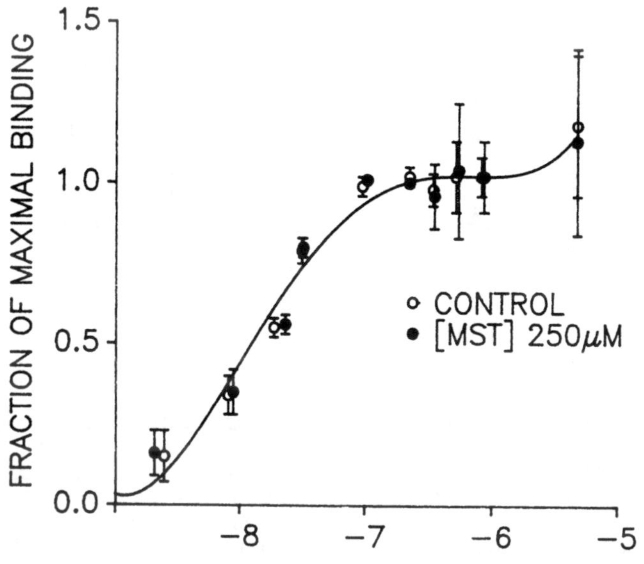

Fig 7: No effect of MST on [^{3}H]Ach equilibrium binding to membrane-bound AChR. Alkaline-washed AChR-rich membranes were preincubated either in the absence or presence of MST and then incubated in the indicated concentration of [^{3}H]Ach. The amount of receptor-bound [^{3}H]Ach was determined by an ultracentrifugation assay. B_{max} was approximately 350 nM in Bgt binding sites. Results are shown as the fraction of maximal binding ± 1 s.d. Points without error bars have s.d. too small to be plotted. (From Eterović et al., submitted to J Receptor Res)

MST and MAT did not affect the equilibrium binding of [^{3}H]ACh (Fig 7), at any concentrations of the compounds. They did not change the rate or extent of receptor desensitization, as measured by the change from a state of low affinity for agonists to a state of high affinity for agonists, which is induced by exposure to agonists (Fig 8).

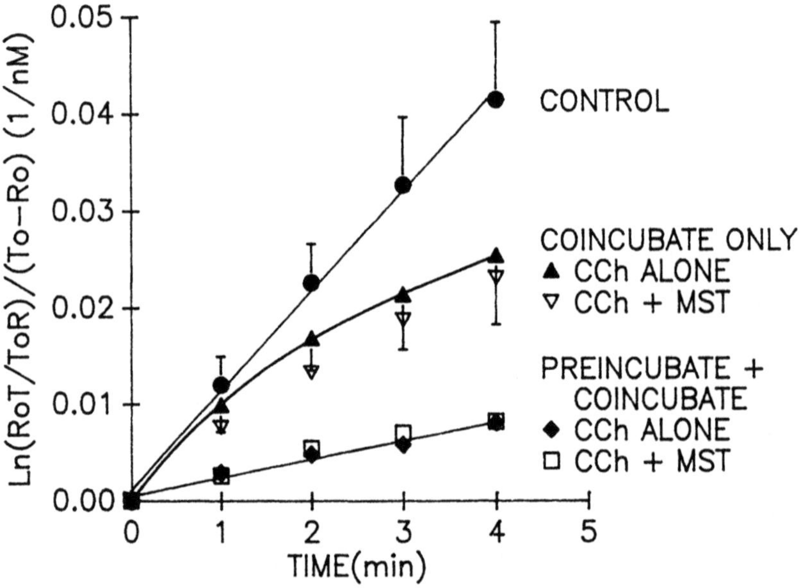

Fig 8: No effect of MST on transition of AChR from low to high agonist affinity. [^{125}I]αBgt binding after several preincubation conditions was determined as described in METHODS. Control samples (solid circles) contained neither CCh nor MST. Coincubation samples were preincubated without CCh or MST, but were then incubated with CCh either without MST (solid triangles) or with MST (open triangles). Preincubation samples were both preincubated and incubated with CCh either without MST (solid diamonds) or with MST (open squares). The values for control samples and for the preincubated samples were fitted by the following equation:

$$kt=\ln(R_oT/T_oR)/(T_o-R_o)$$

Where k is the initial rate constant for [^{125}I]αBgt binding, Ro is the total concentration of receptor binding sites, R is the concentration of unoccupied sites, To is the total toxin concentration and T is the unbound toxin concentration. Results are shown as the means of triplicate determination ± 1 s.d. Points without error bars have s.d. too small to be plotted. (From Eterović et al. submitted to J Receptor Res).

In addition, MAT (10^{-8} to 10^{-3} M) did not affect the receptor-regulated ^{86}Rb ion influx into membrane vesicles (Fig. 9). Neither compound inhibited electroplax AChE. Finally, [^{14}C]MST did not bind with high affinity to receptor-rich membranes, in a low-ionic strength, Ca++ free buffer.

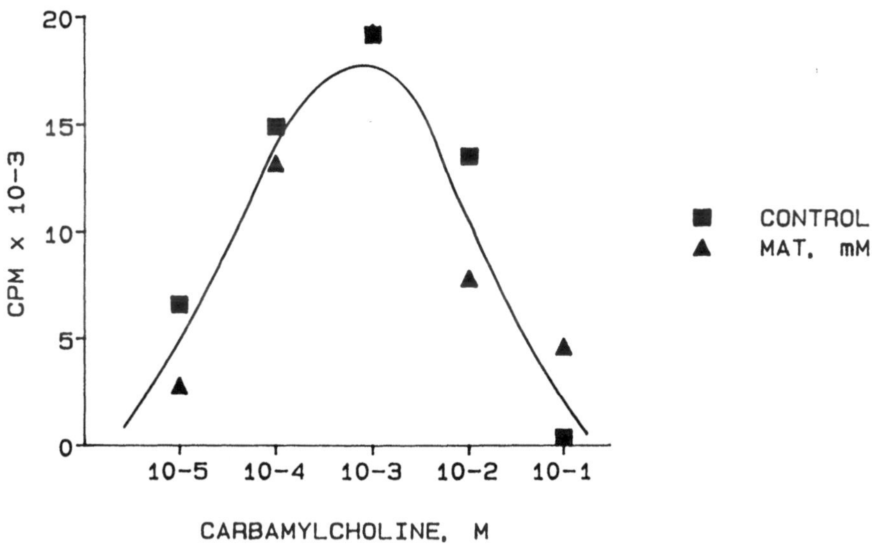

Fig 9: MAT has no effect on AChR-regulated ^{86}Rb influx into membrane vesicles from electric organ. Each point represents triplicate determinations. MAT did not affect the receptor-regulated ion flux, nor the pronounced inhibition of this flux which is observed at high agonist concentrations.

DISCUSSION

The succinyl and acetyl esters of TES potentiate the ACh-induced muscle depolarization and contraction. These compounds are not agonists per se, they do not change the input resistance of muscle membrane, nor inhibit the enzyme AChE. In addition to their effect on the amplitude of ACh potentials, these compounds increase the rate of depolarization when ACh is added to the bath and the rate of repolarization upon washing out of the agonist. The response to non-esteratic agonists is not affected.

TES esters are also active on the effect of neurally-released ACh. They increase the amplitude of mepp's and decrease the decay time of epc's elicited by multiple stimulation of neostigmine-treated preparations. However, the time course of single epc's and mepc's is not affected.

The simplest way to explain the above results is to postulate the existence of a low-affinity, regulatory (inhibitory) binding site for ACh on the AChR from skeletal muscle, which only binds esteratic agonists. By binding to this site, ACh would prevent the ion channel from opening, thus decreasing the response, especially at high ACh concentrations. If MST and MAT are acting as antagonists of the inhibitory site, then when this site is occupied by one of these compounds:

1. The amplitude of membrane depolarization (ACh potentials, epp's, or mepp's) will be increased because ACh inhibition will have been removed. For the same reason, the ACh-induced muscle contraction will be increased.
2. The diffusion of externally-applied ACh in the synaptic cleft to or from deeper receptors will be accelerated, because an additional interaction with the receptor has been eliminated. The diffusion of neurally-released ACh will be equally accelerated (when AChE is inhibited and ACh concentration in the synaptic cleft has been built up by repetitive stimulation).
3. The ion-channel lifetime will not be modified, but the

frequency of channel openings will be increased.

4. The input resistance of muscle membrane will not change.
5. The response to non-esteratic agonists will not be affected, because they will not be able to bind to the inhibitory site.

Our preliminary biochemical studies with muscle membranes did reveal an additional binding site for ACh, different from the receptor activating site. The fact the SA-TES solution displaced ACh from this site argues in favor of its being the postulated regulatory site. However, further experiments are necessary to clarify this matter.

The surprising difference between the effect of TES esters on AChRs from skeletal muscle and from the electrocyte deserves some comments. While the response to ACh was potentiated in muscle, no effect on ion flux was observed in Torpedo vesicles. In addition, the low-affinity binding site for ACh observed in muscle membranes in the 50-100 nM range, was not observed in electric organ membranes up to 5 μM ACh. It is interesting to note that Hess and collaborators have postulated a voltage-dependent, inhibitory site for ACh in the AChR from electric organ; the affinity of this site for ACh drops to about mM in vitro, where membrane potential is zero (Takeyasu et al., 1983). It is then possible that the inhibitory site for ACh exists in Torpedo receptor, but its affinity for ACh is low in vitro (K_D less than 10 μM judging from our binding experiments), and its affinity for TES esters is even lower (K_D less than 1 mM judging from our ion flux experiments). Alternatively, it is also possible that the electrocyte receptor, whose only function is to produce a massive electrical discharge, does not possess the capacity for subtle modulation which is present in muscle.

A thorough search of the literature for other positive modulators of muscle AChR rendered only four candidates: ATP, phospholipase C, ethanol and bisulfite. In frog muscle and sympathetic ganglia, ATP increased the amplitude of ACh and carbamylcholine potentials to a new maximum; agonist dissociation constant and the half-decay time of endplate currents were not affected (Saji et al., 1975; Akasu and Koketsu, 1985; Akasu et al., 1981). Treatment with phospholipase C also increased the amplitude of ACh-induced muscle contraction and of mepp's, without affecting the time course of mepc's (Harborne et al., 1987). In frog skeletal muscle, ethanol at low concentrations shifted the concentration -response curve for ACh to the left, without changing the maximum response. Channel lifetime was prolonged but mean conductance remained the same (Bradley et al., 1980 and 1984). On lizard endplates sodium bisulfite increased time-to-peak and decay-time of mepc's recorded under voltage clamp, but did not affect single channel conductance or lifetime. It was suggested that bisulfite affects ACh binding and unbinding rates (Steinacker and Zuazaga, 1987). Of these four modulators, only ATP and phospholipase resemble TES esters in their effects. We do not know whether these compounds act by related mechanisms, but this is an intriguing possibility.

This work was supported among others by the following grants: EPSCoR-NSF, RR08102 RCMI-NIH and NS07464 from NINDS to VAE; RR08102 from MBRS-NIH and NSF RIM-16333 to GEM

REFERENCES

Akasu T, Koketsu K (1985) Effect of adenosine triphosphate on the sensitivity of the nicotinic acetylcholine-receptor in the bullfrog sympathetic ganglion cell. Br J Pharmacol 84: 525-531

Akasu T, Hirai K, Koketsu K (1981) Increase of acetylcholine receptor sensitivity by adenosine triphosphate: a novel action of ATP on ACh-sensitivity. Br J Pharmacol 74: 505-507

Bradley RJ, Sterz R, Peper K (1984) The effect of alcohols and diols at the nicotinic acetylcholine receptor of the neuromuscular junction. Brain Res 295:101-112

Bradley RJ, Peper K, Sterz R (1980) Postsynaptic effects of ethanol at the frog neuromuscular junction. Nature (London) 284:60-62

del Castillo J, Escalona de Motta G, Eterovic VA, Ferchmin PA (1985) Succinyl derivatives of N-tris(hydroxymethyl)methyl-2-aminoethane sulfonic acid: their effects on the frog neuromuscular junction, Br J Pharmacol 84:275-288

Durst HD, Gokel GW (1980) Benzoyl esters from the acid chlorides, In: Durst HD, Gokel GW (eds) Experimental Organic Chemistry, 1st edn. McGraw Hill, inc., New York, N.Y., pp. 597-598

Escalona de Motta G, del Castillo J (1977) Diffusion barrier for acetylcholine at the frog neuromuscular junction. Nature (London) 270: 178-180

Eterovic VA, Escalona de Motta G, Hann RM, Lasalde JA, Prieto JA, Ferchmin PA (submitted) Positive modulators of muscle acetylcholine receptor, J. Receptor Res.

Eterovic VA, Ferchmin PA, Hann R, Escalona de Motta G (1981) Interaction of radiolabeled acetylcholine with muscle particulate fraction. Soc Neurosci Abstr 7: 496

Fatt P (1950) The electromotive action of acetylcholine at the motor endplate. J Physiol 111: 408-422

Fernandez HL, Duell MJ, Festoff BW (1979) Neurotrophic control of 16S acetylcholinesterase at the vertebrate neuro-muscular junction. J Neurobiol 10:441-454

Harborne AJ, Turner AW, Murphy RLW, Smith M (1987) Action of phospholipase C at the neuromuscular junction in rodent skeletal muscles. Eur J Pharmacol 143:163-170

Hassner A, Alexanion V (1978) Direct room temperature esterification of carboxylic acids. Tetrahedron Lett 46:4475-4478

Katz B, Miledi R (1973) The binding of acetylcholine to receptors and its removal from the synaptic cleft. J Physiol 231: 549-574

Kroedel ER, Beckman RA, Cohen JB (1979) Identification of a local anesthetic binding site in nicotinic postsynaptic membranes isolated from Torpedo marmorata electric tissue. Molec Pharmacol 15:294-312

Lindstrom J, Anholt R, Einarson B, Engle A, Osame M, Montal M (1980) Purification of acetylcholine receptors, reconstitution into lipid vesicles, and study of agonist-induced cation channel regulation, J Biol Chem 255:8340-8350

Lowry OH, Roseborough NJ, Farr AL, Randall RJ (1951) Protein measurement with the folin phenol reagent. J Biol Chem 193: 265-275

Quast U, Schimerlik M, Lee T, Witzemann V, Blanchard S, Raftery, MA (1978) Ligand-induced conformation changes in Torpedo californica membrane-bound acetylcholine receptor. Biochemistry 17:2405-2414

Saji Y, Escalona de Motta G, del Castillo J (1975) Depolarization and potentiation of responses to acetylcholine elicited by ATP on frog muscle, Life Sci 16:945-954

Schmidt J, Raftery MA (1973) A simple assay for the study of solubilized acetylcholine receptors. Biochemistry 12:852-856

Steinacker A, Zuazaga C (1987) Further kinetic analysis of the chemically modified acetylcholine receptor. Pflügers Arch 409: 555-560

Takeyasu K, Udgaonkar JB, Hess GP (1983) Acetylcholine receptor: evidence for a voltage-dependent regulatory site for acetylcholine. Chemical kinetic measurements in membrane vesicles using a voltage clamp, Biochemistry 22:5973-5978

LOCALISATION OF MUSCARINIC ACETYLCHOLINE RECEPTORS IN CAT VISUAL CORTEX DURING POSTNATAL DEVELOPMENT

F. van Huizen[1], C. Shaw[1], A.D. Strosberg[2] and M.S. Cynader[1]
[1] VGH/UBC Eye Care Centre
Department of Ophthalmology
University of British Columbia
2550 Willow Street
Vancouver, B.C.
Canada V5Z 3N9

[2] Laboratory of Molecular Biology of Receptors
Institut Pasteur
Paris, France

During the first months of postnatal life kittens undergo a critical period in which neurons in the visual cortex show a remarkable modifiability in response to changing visual stimuli (Hubel and Wiesel, 1970; Olson and Freeman, 1980; Cynader et al., 1980; Cynader and Shaw, 1986). Several reports suggest that this modifiability only occurs in cooperation with attentional mechanisms mediated by monoaminergic and/or cholinergic systems which substantially innervate the visual cortex (Kasamatsu et al., 1979; Bear and Singer, 1986). To learn more about the cellular mechanisms involved in visual cortex plasticity a systematic study of the development of the localisation and characteristics (B_{max} and K_D) of the different neurotransmitter receptors in the developing cortex. was carried out.

Changes in number, affinity and in laminar distribution of neurotransmitter receptors are now well-established features of the normal postnatal development of cat visual cortex (for a summary see Cynader and Shaw, 1986; Shaw et al., 1986)

Most receptor populations studied so far show some form of laminar redistribution during development (Shaw et al., 1984; Shaw et al., 1986), but a striking reversal of the early laminar binding pattern is seen for muscarinic acetylcholine receptors (mAChRs). In Figure 1A autoradiograms are shown which were produced by incubating coronal sections of the cat visual cortex with [3H]quinuclidinylbenzilate ([3H]QNB), a mAChR antagonist. Increased levels of darkness represent greater binding. In neonatal kittens (day 3) mAChR binding is densest in layer IV, but this pattern alters with age and from about 45 days

NATO ASI Series, Vol. H 32
Molecular Biology of Neuroreceptors and
Ion Channels. Edited by A. Maelicke

postnatal onward, binding is densest in cortical laminae I-III and VI. This period of rapid change in receptor distribution coincides with the peak of the physiologically-defined critical period for cortical plasticity (Hubel and Wiesel, 1970; Olson and Freeman, 1980). Since this critical period can be delayed by preventing the visual system from obtaining a normal pattern of input (Cynader and Mitchell, 1980), it is interesting that the redistribution of neurotransmitter receptors is also dependent on some form of extrinsic input. In Figure 1B the effect of undercutting a portion of the visual cortex in a young animal (24 days) is shown after allowing it to survive until 49 days of age when the adult distribution can be expected. In the control cortex (Fig. 1B left) mAChR distribution is typical for an animal of that age (see Fig. 1A), while in the undercut part of the cortex the distribution resembles that of a kitten of 24 days of age.

Indirect evidence regarding specific receptor distributions can be obtained by studying changes in the distribution of radiolabelled ligands after undercutting the cortex or after a selective lesion with toxins of neuronal or glial elements. Using these methods, Prusky et al., (1987) and Shaw et al. (1987,1988) were able to show that both muscarinic M_1 and M_2 AChR subtypes in adult cat visual cortex were located on intrinsic neuronal loci. They could not, however, determine whether receptors were located on dendrites, somata, or on local axon terminals.

The ultrastructural localisation of mAChRs in rat cortex has already been studied using electron microscopic autoradiography and the data indicate that a significant fraction of (but not all) autoradiographic grains (i.e., mAChR) were associated with synapses, but it was not possible to distinguish between pre- or postsynaptic locations using these methods (Kuhar et al., 1981).

More precise localisation data can be obtained at both the light and electron microscopic level by using mono- or polyclonal antibodies raised against the actual muscarinic receptors. We have studied the ultrastructural localisation of mAChR in cat visual cortex during postnatal development using M-35b, a mouse monoclonal antibody raised against purified muscarinic acetylcholine receptors (mAChR) from calf brain. M-35b recognizes active digitonin-solubilized muscarinic receptors (Andre et al., 1983, 1984) and also causes stimulation of the guinea pig myoterium (Leiber et al., 1984), a characteristic property of muscarinic agonists.

Methods

Cats varying in age from 28 days to adult were deeply anesthesized and and the brains were fixed by perfusion with cold fixative (2-4% paraformaldehyde and 0.05-0.1% glutaraldehyde in 0.1 M phosphate buffered saline (PBS), pH 7.3). The brains were removed and kept overnight in the fixative without glutaraldehyde. After washing in PBS, 40-100 μm slices were made from coronal blocks of the visual cortex using a Vibroslice. Individual slices were placed in the wells of a 24-well tissue culture plate. Alternate sections were used for Nissl staining to determine the exact dimensions and location of the various cortical layers, using criteria that we described elsewhere (Shaw et al., 1984, 1986)

For immunocytochemistry, the brain slices were processed with the avidin-biotin-HRP complex (Vector). All incubations were performed at room temperature, except when indicated, while under continuous agitation. Before starting the incubation steps, the slices were washed three times for 10 min in PBS. Slices were then incubated in 10% normal goat serum for 1-2 hr. Following this pre-incubation step, slices were incubated in the primary antibody (M-35b, 1:2000 in BSA-PBS plus 1% normal goat serum) for 2 hr at room temperature followed by 16 hr at 4 ^{o}C. After three washes of 10 min in PBS, slices were incubated in biotinylated goat anti-mouse IgM for 2-4 hr. This was followed by three washes of 10 min in PBS, 2 hr incubation in the ABC- complex (a mixture of avidin DH and biotinylated horseradish peroxidase H), a 10 min wash in PBS and two 10 min washes in 0.1 M Tris buffer (pH 7.2). Finally, the slices were reacted with 0.05% diaminobenzidine tetrahydrochloride (DAB) and 0.01% hydrogen peroxide made in 0.1 M Tris buffer (pH 7.2), followed by several washes in water. The slices were finally processed for light- or electron microscopy. Controls included replacement of the primary antibody by normal mouse serum (1:2000) and addition of aliquots of atropine, carbachol, quinuclidinyl benzilate or propyl benzyl choline mustard to the wells during the pre- and incubation steps in a final concentration of 10^{-4} M. The details of the immunocytochemical procedures are described in Van Huizen et al. (in press)

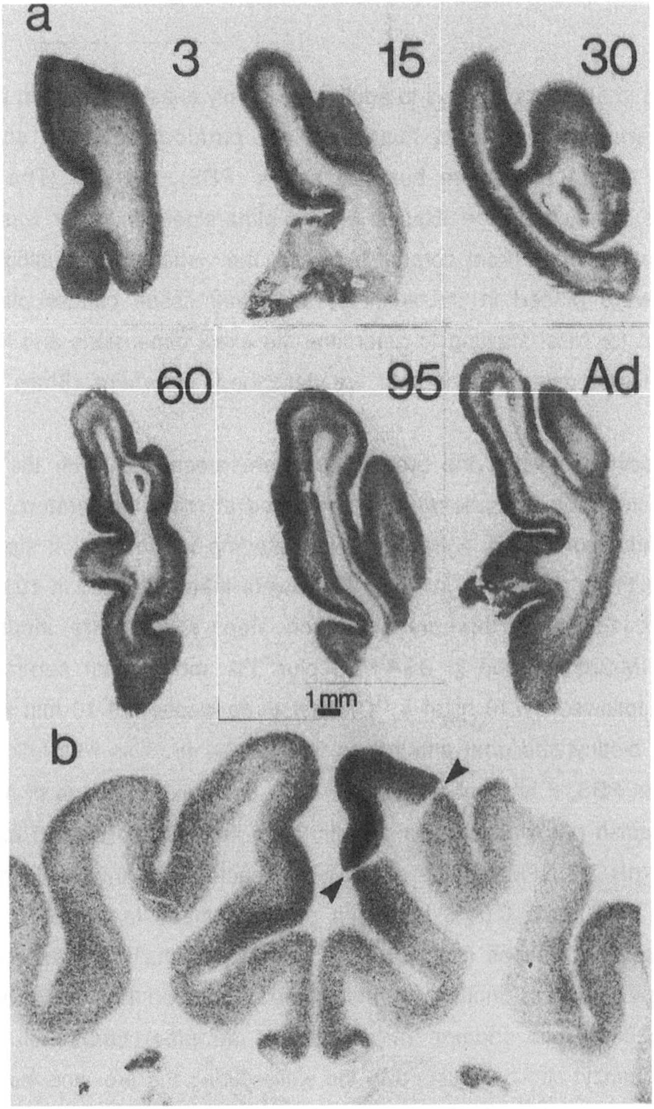

Fig. 1: (A) Laminar binding patterns for muscarinic acetylcholine receptors during postnatal development. Postnatal age is indicated in days. Dorsal is up and medial is left. Increased darkness in the figure indicates greater density of receptors.

(B) Autoradiogram illustrating the distribution of muscarinic ACh receptors in control and undercut visual cortex of a 49-day-old kitten after incubation with [3H]QNB. The kitten was operated at 24 days of age. In the control hemisphere (left-hand side) the receptors are located in layers I-III and VI. In the undercut zone of the opposite hemisphere receptors are instead concentrated in cortical layer IV, which closely resembles that of a normal kitten of that age.

Muscarinic receptors in both A and B were labelled with 5-7 nM [3H]QNB for 60 min at 20 °C. Nonspecific binding was determined by coincubation of alternate sections with 10^{-4} M atropine sulfate. Ultrofilm exposure: 14 days. All autoradiograms have been photographed for highest contrast. Calibration bar = 1 mm

Results and Discussion

None of the muscarinic receptor antagonists or agonists added during the pre-incubation and incubation steps could displace the antibody. This is likely due to the fact that the receptor domain interacting with the monoclonal antibody is different from the binding site for regulatory ligands Thus, agonist-like activity such as mAChR-mediated biochemical events was reported for M-35b in guinea pig myoterium, but this could only be partly inhibited by atropine sulphate (Leiber et al., 1984).

The peroxidase reaction product was readily identified in the slices. Replacement of the primary antibody by normal mouse serum resulted in non-stained or on some occasions, faintly stained slices.

Immunolabelling stained neuronal cell bodies and dendrites in a more or less punctate pattern. Sometimes staining could be seen as a rim outlining the cell bodies; in other cases the whole cell body was stained.

A striking change in laminar distribution of the immuno-reaction product was found during development (Fig. 2A, B). 28-day-old kittens showed a high level of immunoreactivity in non-pyramidal (stellate) cells in layer IV (Fig. 2A, C) and a small zone of cells located at the top of layer II . These latter neurons are the cells that have just finished migration (as the last group of visual cortex neurons) and have started to differentiate (Shatz and Luskin, 1986). Their classification into stellate or pyramidal cells is uncertain at this age. In adults the laminar distribution was markedly different with high density staining of fibers in layer I and pyramidal cells in layers II, III, V and VI (Fig. 2B). The large pyramidal cells in layer V were especially well-stained (Fig. 2D). The staining in layer I is apparently due to stained dendrites arising from pyramidal cells in layers II, III and V. These findings parallel the changes in laminar distribution which had been found using autoradiography and are described above. The only difference is the staining of layer II cells in the 28-day-old kitten found here. However, in studies using specific ligands for M1 and M2 mAChR separately (Prusky et al, 1987), it appears that [^{3}H]oxotremorine, which binds to M2 receptors gives a similar distribution to that observed with our monoclonal antibody. The developmental change in laminar distribution as seen with both immunocytochemical and radioautographic methods can not be accounted for by migration of cells from the middle of the cortex to the upper layers II and III, since migration of those cells is virtually complete after 3 weeks of postnatal age (Shatz and Luskin, 1986).

The observed laminar distribution was constant in the anterio-posterior direction in both kittens and adult cats as has been seen already with radioligand studies (Shaw et al, 1984).

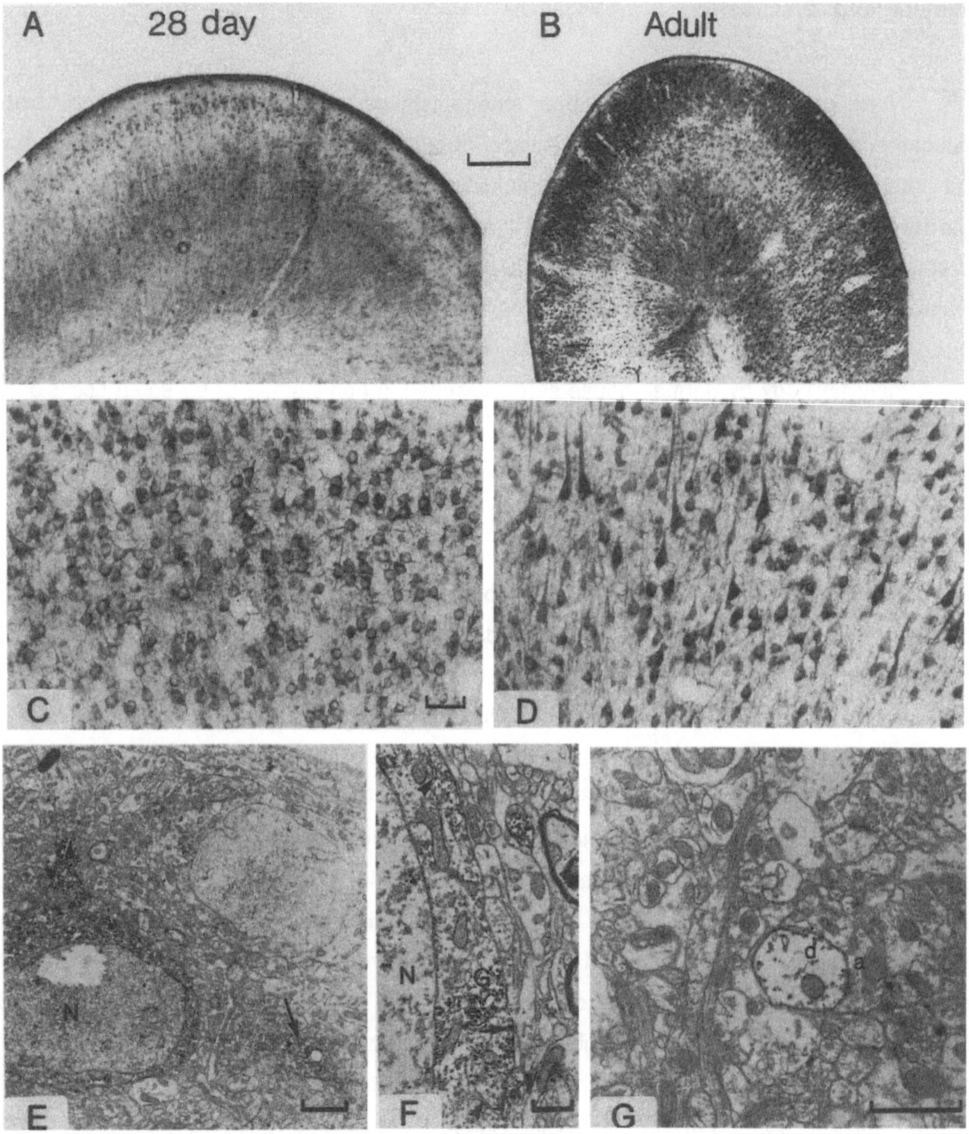

Fig. 2: Immunocytochemical labelling of the muscarinic acetylcholine receptor in 28-day-old (A, C) and adult (B, D) cat visual cortex using a monoclonal antibody.

(A, B) Laminar distribution in visual cortex of 28-day-old (A), and adult cat (B). Note that the immunostained cells are most dense in the middle cortical layer (layer IV) in the young animal and least dense in this layer of the adult animal. Bar = 0.5 mm.

(C, D) Higher power light micrographs of tissue showing the densely labelled non-pyramidal cells in layer IV of the young cat (C) and the pyramidal cells in layer V and VI in the adult cat (D). Bar = 50 µm.

Fig. 2: (E F, G) Electron micrographs illustrating distribution of anti-mAChR antibody labelling in cat visual cortex neurons.(E) Extensive intracellular staining in a layer V pyramidal neuron soma (N), in its apical dendrite (arrowhead), and in more distal dendrites (arrow) Bar = 2 µm.(F) High power micrograph of neuronal soma showing labelling of cell membrane, membranes of Golgi bodies (G), nucleus (N) and rough endoplasmatic reticulum (arrowhead), and vesicles (arrow). Bar = 0.25 µm.
(G) Staining of membranes of a distal dendrite (d) apposed to an axon terminal (a). No counterstaining with lead citrate. Bar = 0.5 µm.

Electron microscopic analysis shows (Fig. 2E) that the antibody can be found on membranes of cell bodies and large diameter dendritic profiles. In some cases also the membranes of small sized dendritic profiles were stained, often at loci apposed to an unlabelled axon terminal making a synaptic contact (Fig. 2G). Postsynaptic densities, however, display the same amount of electron density in unlabelled as in labelled dendritic profiles making it difficult to decide whether or not mACh receptors are concentrated in the densities. Besides the membrane staining, an abundance of reaction product was found <u>inside</u> the cell on membranes of Golgi bodies, rough endoplasmatic reticulum (RER), nucleus and in vesicles (Fig. 2F). The significance of these internal receptors is yet unclear. A large intracellular pool of mAChRs was found with the M-35b antibody in human fibroblasts in culture with EM-immunocytochemistry (Raposo et al., 1987), but this occurred only after stimulation of the receptors with carbachol. Internalization of mAChR into cortex cells followed by degradation was also found in living brain slices after stimulation of the slices with either veratridine, carbachol or high potassium (Shaw et al, in press and this volume) While internalized receptors should be associated with vesicles, labelling of nuclear membrane, RER and Golgi complexes may represent the sites of production and distribution of the receptor. Interestingly, an identical intracellular distribution of staining to that observed here was found with an antibody against the nicotinic AChR in chick ganglion neurons (Jacob et al., 1986).

Rough estimates of the percentage of immunolabelled to unlabelled cells indicated that about 20-30% cells in the labelled layers were immunopositive. The other, unlabelled cells may express one of the many other receptors (NMDA, GABA or CCK for example) that are found in the same cortical layers (Shaw et al., 1986), though double labelling studies would be required to address this issue.

Since we have found that receptors are found exclusively on or in cell bodies or dendrites, while the laminar distribution changes during development, we conclude that after a certain period, in this case between 45 and 60 days, non-pyramidal cells in layer IV stop producing mAChRs and subsequently eliminate the existing receptors, while pyramidal cells in layers III, V and VI begin to make mAChRs and express them at the membranes of cell bodies and dendrites. These data thus suggest that different populations of cortical neurons express the mAChR at different ages. Future studies using in situ hybridization can be used to determine whether this transient production of receptors is accompanied by alterations in the pattern of gene expression during the critical period for cortical plasticity.

This work was supported by a MRC of Canada grant to M.C. and C.S. and a Killam postdoctoral fellowship to F.v.H.

References

André C, De Backer JP, Guillet JG, Vanderheyden P, Vauquelin G, Strosberg AD (1983) Purification of muscarinic acetylcholine receptors by affinity chromatography. EMBO J 2:499-504

André C, Guillet JG, De Backer JP, Vanderheyden P, Hoebeke J, Strosberg AD (1984) Monoclonal antibodies against the native or denatured forms of muscarinic acetylcholine receptors. EMBO J 3:17-21

Bear MF, Singer W, (1986) Modulation of visual cortical plasticity by acetylcholine and noradrenaline. Nature (London) 320:172-176

Cynader M, Timney BN, Mitchell DE (1980)Period of susceptibility of kitten visual cortex to the effects of monocular deprivation extends beyond six months of age. Brain Res 191:545-550

Cynader MS, Mitchell D (1980) Prolonged sensitivity to monocular deprivation in dark-reared cats. J Neurophysiol 43:1026-1040

Cynader M, Shaw C (1986) Mechanisms underlying binocular competition in cat visual cortex. In: Kellar E, Zee D (eds.) Adaptive processes in visual and oculomotor systems. Pergamon Press, 1986, p 53

Hubel DH, Wiesel TN (1970) The period of susceptibility to the physiological effects of unilateral eye closure in kittens. J Physiol 206:419-436

Jacob MH, Lindstrom JM, Berg DK (1986) Surface and intracellular distribution of a putative neuronal nicotinic acetylcholine receptor. J Cell Biol 103;205-214

Kuhar MJ, Taylor N, Wamsley JK, Hulme EC, Birdsall NJM (1981) Muscarinic cholinergic receptor localization in brain by electron microscopic autoradiography. Brain Res 216:1-9

Leiber D, Harbon S, Guillet JG, André C, Strosberg AD (1984) Monoclonal antibodies to purified muscarinic receptor display agonist-like activity. Proc. Natl. Acad. Sci. U.S.A. 81:4331-4334

McCormick DA, Prince DA (1986) Mechanisms of action of acetylcholine in the guinea-pig cerebral cortex *in vitro* . J Physiol 375:169-194

Olson CR, Freeman RD (1980) Profile of the sensitive period for monocular deprivation in kittens. Exp Brain Res 39:17-21

Prusky G, Shaw C, Cynader M. (1987) The distribution and cellular localisation of muscarinic acetylcholine receptor subtypes in the developing cat striate cortex. Soc Neurosci Abstr 13:1045

Raposo G, Dunia I, Marullo S, André C, Guillet JG., Strosberg AD, Benedetti EL, Hoebeke J (1987) Redistribution of muscarinic acetylcholine receptors on human fibroblasts induced by regulatory ligands. Biology of the Cell 60:117-124

Shatz CJ, Luskin MB (1986) The relationship between geniculocortical afferents and their cortical target cells during development of the cat's primary visual cortex. J Neurosci 6:3655-3668

Shaw C, Needler MC, Cynader M (1984) Ontogenesis of muscarinic acetylcholine binding sites in cat visual cortex: reversal of laminar specific distribution during the critical period. Dev Brain Res 14:295-299

Shaw C, Wilkinson M, Cynader M, Needler MC, Aoki C, Hall SE. (1986) The laminar distributions and postnatal development of neurotransmitter and neuromodulator receptors in cat visual cortex. Brain Res Bull 16:661-671

Shaw C, Prusky G, Van Huizen F, Cynader M. (1987) Cellular localization of receptor populations in cat visual cortex using quinolinic acid. Soc Neurosci Abstr 13:1046

Shaw C, Prusky G, Cynader M (1988) Surgical undercutting prevents receptor redistribution in developing kitten visual cortex. Visual Neurosci 1:205-210

Shaw C, Van Huizen F, Cynader M, Wilkinson M.(to be published) A role for potassium channels in the regulation of cortical muscarinic ACh receptors in an in vitro slice preparation. Mol Brain Res

Shaw C, Van Huizen F, Cynader M, Wilkinson M.(to be published) Regulation of cortical muscarinic acetylcholine receptors: involvement of potassium channels (this volume)

Van Huizen F, Strosberg AD, Cynader MS (to be published) Cellular and subcellular localisation of muscarinic acetylcholine receptors during postnatal development of cat visual cortex using immunocytochemical procedures Dev Brain Res.

REGULATION OF CORTICAL MUSCARINIC ACETYLCHOLINE RECEPTORS: INVOLVEMENT OF POTASSIUM CHANNELS

Christopher Shaw[1], Frans van Huizen[1], Max S. Cynader[1] and Michael Wilkinson[2]
[1] VGH/UBC Eye Care Centre
Department of Ophthalmology
University of British Columbia
2550 Willow Street
Vancouver, B.C.
Canada V5Z 3N9

[2] Departments of Physiology and Biophysics
and Obsterics and Gynaecology,
Dalhousie University, Halifax, N.S., Canada

One characteristic feature of neurons in mammalian neocortex is their ability to modify their response properties following altered input activity. This neural "plasticity" may be confined to a "critical period" early in postnatal life, as it is for the cat visual cortex (Hubel and Wiesel, 1970). In other cortical areas it may occur in adult animals as well (Rasmusson, 1982; Merzenich et al., 1984). Changes in cortical properties may be induced with a time course of days to weeks (Hubel and Wiesel, 1970; Olson and Freeman, 1980; Rasmusson, 1982) or in some cases minutes to hours (Blakemore and Mitchell, 1973; Perkins and Teyler,1988).

Our attention has been focussed recently on the muscarinic acetylcholine receptor (mAChR) which shows pronounced changes in number, affinity, and in laminar distribution during postnatal development (Shaw et al., 1984, 1986a). The putative neurotransmitter, acetylcholine (ACh), appears to play a crucial role in regulating neural activity in the brain by altering potassium ion conductance (Brown, 1984; McCormick and Prince, 1985, 1986) following activation of mAChR M_1 and M_2 receptor subtypes (Watson et al., 1985; Spencer et al., 1986); it also has been implicated as playing a role in cortical plasticity mechanisms (Bear and Singer, 1986).

The developmental changes in mAChRs which we have studied occur with a time course of days to weeks and may contribute to some plasticity mechanisms in the visual cortex. These alterations take place over too long a time, however, to account for some of the shorter term alterations in cortical response properties which have been demonstrated (Bear and Singer, 1986; Marlin et al., 1988; Perkins and Teyler, 1988). Relatively rapid regulatory processes, the now extensively studied `up-' and `down-regulations' in receptor number,

NATO ASI Series, Vol. H 32
Molecular Biology of Neuroreceptors and
Ion Channels. Edited by A. Maelicke

would seem to provide a possible mechanism for rapid alterations in neural activity (for general reviews see Hollenberg, 1985a, 1985b; Jarett and Smith, 1985).

Studies of ACh receptors, both nicotinic and muscarinic (Nathanson, 1982) form a large part of the literature concerning receptor regulation, and include studies using a bewildering variety of preparations, ligands, and conditions. It is generally accepted that AChR down-regulation is a consequence of either cellular depolarization elicited by electrical stimulation (Luqmani et al., 1979; Abdul Ghani et al., 1981) various chemical agents or ions (Luqmani et al., 1979; Abdul Ghani et al., 1981; Bloch, 1986) (but see Liles and Nathanson, 1987) or agonist activation of the receptor (Klein et al., 1979; Luqmani et al., 1979; Siman and Klein, 1979; Shifrin and Klein, 1980;Maloteaux et al., 1983; Yamada et al., 1983; Bloch, 1986; El Fakahany and Lee, 1986; Hunter and Nathanson, 1986; Stollberg and Berg, 1987).

In order to investigate possible mAChR regulation in the cortex, we have employed an in vitro cortical slice preparation in which the majority of the neurons are still alive (Wilkinson et al., 1986; Van Huizen et al., 1987). The advantage of this preparation is that we are able to study cortical neurons in a situation in which they are maintained with relatively intact circuitry and activity (Shaw and Teyler, 1982; Perkins and Teyler, 1988), yet one in which we can effectively control a number of experimental parameters. We will show in the following results that the majority of neurons in the preparation are viable, that mAChR binding characteristics meet normal criteria, and that the distributions and characteristics of the mAChRs are quite similar for rat neocortex and cat visual cortex. Additionally, we will examine the effect of various alterations of bioelectric activity and/or receptor stimulation on mAChR regulation in adult rat neocortex.

Material and Methods

Animals Most experiments were performed with adult male Sprague-Dawley rats purchased from Canadian Hybrid Farms, Halls Harbour, Nova Scotia, Canada. Body weights were in the range of 150-250 g. The rats were housed under controlled lighting conditions (lights on 0700 hr - 1900 hr) and were given free access to food and water.

Radioligands and analogs L-[benzilic-4,4'-^{3}H]-quinuclidinyl benzilate ([^{3}H]QNB; 33.2 - 43.6 Ci/mmol) and [N-methyl-^{3}H]-scopolamine methyl chloride ([^{3}H]NMS; 80.0 - 87.0 Ci/mmol) were obtained from New England Nuclear (NEN, Boston, MA). Unlabelled QNB was obtained from Research Biochemicals Inc (Wayland, MA). All other chemicals were obtained from Sigma (St. Louis, MO)

Brain Slices and Assay Procedure Rats were killed by decapitation (between 10 - 12 a.m.) and the whole brain was rapidly removed and placed in ice cold Dulbecco's phosphate buffered saline (Gibco, Grand Island, NY) to which glucose (1 mg/ml) and Hepes (1 ml/100 ml) were added (Dulb+). Hydrogen peroxide (0.004%) was added as a source of molecular oxygen (Walton and Fulton, 1983). Blocks of cerebral cortex approximately 5 mm in width were rapidly dissected out and placed in cold medium. These blocks contained frontal cortical areas 4 and 6, parietal cortex areas 1, 2, 3 and 7, occipital cortex areas 17 and 18, and cingulate cortex areas 23, 24 and 29b,c (Krieg, 1946). Tissue slices (approx 400 μm thick) were obtained with a tissue slicer (model based on that described by Bennet et al., 1983). Normally about 30 slices could be obtained from one rat. Individual slices were placed in separate wells of a 24-well tissue culture plate in which each well contained 0.5 ml of Dulb+. In binding assays 4 slices were normally used to determine total binding while 2 slices were used for determination of nonspecific binding by adding 5 μl of a 10^{-2} M solution of the unlabelled ligand to the wells for a final concentration of 10^{-4} M. In addition, 8 to 10 representative fresh brain slices were weighed to obtain a mean wet weight of the slices. Protein content in these slices was measured by the method of Lowry et al. (1951). Specific binding is thus expressed conventionally as femtomoles bound per mg protein.

Immediately prior to the experiment, labelled ligand was diluted to the desired concentration in Dulb+, and 0.5 ml of this solution was added to the slices after the original Dulb+ was removed. Unlabelled ligands and other drugs were added in aliquots of 5 or 20 μl. Following incubation a sample of the buffer (20 μl) was removed for determination of 'free' ligand concentration. The remainder of the buffer was removed with a Pasteur pipette and the slices were washed (Jet pipette) in ice-cold Dulb medium or phosphate buffered saline (PBS) (2 x 0.5 ml; 2 x 5 min). The slices were then removed with a small piece of glass fiber filter paper (Whatman GFB) and placed directly in counting vials containing 2 ml Formula 963 (NEN). The vials were capped, shaken and counted in a LKB 1218 Rackbeta scintillation counter (efficiency: 41.5%) after a minimum interval of 12 hours.

Morphology Rat cortical slices were used to study the morphology of the tissue after incubation in Dulb+ at 4°C or 30°C for 0, 1, 2 and 6 hours. Directly after incubation, slices were placed in fixative (1.25% glutaraldehyde and 1% paraformaldehyde in 0.1 M PBS) for 2 hr, followed by 1% OsO_4 in PBS for 1 hr, dehydration in graded ethanols, and embedding in Epon 812.

Other slices were incubated in 0.4% trypan blue for five minutes before fixation to yield an estimate of the number of dead cells, since trypan blue is taken up only by cells whose membranes are not intact.

Autoradiography Cortical slices were frozen in liquid Freon after incubation for 2 hr in [^{3}H]NMS followed by two washes in PBS. Sections (16 μm thick) were cut on a cryostat and thaw-mounted onto subbed glass slides. After drying, the sections were exposed to LKB

Ultrofilm. The Ultrofilm was conventionally developed and fixed. The original sections were stained for Nissl substance with cresyl violet to identify the cortical laminae.

Preincubation conditions

Bioelectric activity of the slices was altered for various periods of time (2-4 hours) by various substances before addition of the labelled ligand (see concentrations in Table 1). Of these KCl, $CoCl_2$, $MgCl_2$ and $AuCl_2$ were obtained from Fisher Scientific. The remainder were purchased from Sigma Chemical Co. (For a general review of the effects of the various substances, see Narahashi, 1974; Catterall, 1980; Hille, 1984). These substances included those added to depolarize the neurons (e.g., veratridine and high external potassium (K^+_{out})). Glutamate and picrotoxin would be expected to increase general neural spontaneous activity, although the effects of glutamate may include activation of inhibitory circuits as well (Hess and Murata, 1974). Gamma-aminobutyric acid (GABA), a putative inhibitory neurotransmitter in visual cortex (Sillito, 1975) was used to depress neural activity by binding to GABA receptors, one class of which act through a chloride ionophore to hyperpolarize neurons (Simmonds, 1983). We used gold chloride ($AuCl_2$) to achieve the same effect: High Cl^-_{out} should increase Cl^- diffusion into the cell to produce a hyperpolarization which is not linked to activation of GABA receptors. Synaptic transmission was blocked by addition of cobalt chloride ($CoCl_2$) or magnesium chloride ($MgCl_2$) which also block calcium channels both pre and postsynaptically. Specific blockade of postsynaptic voltage-sensitive Ca^{++} (Shaw et al., 1986b) was achieved using nifedipine. Neural activity and sodium currents were suppressed by addition of tetrodotoxin (TTX), a sodium channel blocker. Different drugs were used to selectively block classes of potassium ion channels. The substances used included the quarternary anmonium ions tetraethylammonium chloride (TEA) and 4-aminopyridine (4-AP), cesium ions ($CsCl_2$) and apamin, a neurotoxin derived from bee venom. Each of these substances was used to block certain of the five types of channels thought to be present in hippocampus and neocortex (Prince and Huguenard, 1988) (Some of these K^+-channels appear to correspond to those described in other systems (Hille, 1984; Blatz and Magleby, 1987)). All of these substances were added 15 to 20 minutes before other substances such as veratridine, carbachol, etc. Agonist stimulation of mAChRs was accomplished by addition of carbamylcholine chloride (carbachol) (Bloch, 1986; El-Fakahany and Lee, 1986) and acetylcholine chloride (ACh). All preincubations were at 22-37°C and the incubation buffer containing the labelled ligand did not contain the preincubation substance.

Statistical Analysis

Differences in mAChR number following alterations of bioelectric activity and agonist stimulation were analysed using a one sample t-test (one tailed). The paired t-test was used for comparing the effects of an agent alone versus the effects of combinations of agents.

Table 1 Effects of changes in bioelectric activity and agonist stimulation on mAChRs

	Conc. (M)	Number mAChR: (percent difference from control ± S.E.M.)	
		$[^3H]NMS$	$[^3H]QNB$
Depolarizing Agents			
Veratridine	10^{-5}	-26.2±3.6***	-37.6±7.9***
K^+	5×10^{-2}	-11.3±3.0**	-8.1±4.0
L-glutamic acid (HCR)	10^{-5}	-11.6±4.9*	-4.5±3.9
Picrotoxin	10^{-5}	-10.4±6.2	-
Hyperpolarizing Agents			
GABA	10^{-5}	-5.8±7.1	-17.6±7.4
$AuCl_2$	10^{-5}	-8.5±9.7	-20.4±3.4
Synaptic Transmission/ Ca^{++}-channel Blockers			
$CoCl_2$	10^{-2}, 10^{-3}	+2.8, -11.2	
$MgCl_2$	10^{-2}	-10.7±5.6	-25.5±13.4
Nifedipin	10^{-5}	-8.8±1.0	+22.0
Ion-Channel Blockers			
TTX (Na^+)	10^{-5}	-8.4±5.6	-27.6±9.0
TEA (K^+)	10^{-3}	-2.4±5.1	+6.4±4.7
4-AP (K^+)	10^{-3}	-1.9±5.1	-16.2±11.2
Apamin(K^+)	10^{-5}, 5×10^{-5}	+16.8±16.0	+43.9±35.3
$CsCl_2$(K^+)	10^{-3}	-3.6±2.6	-7.1
mAChR Agonists			
ACh-chloride	10^{-5}	-6.3±8.7	-33.9±2.8*
Carbachol	10^{-3}	-26.1±1.7***	-35.9±3.8***
Miscellaneous			
Eth	1%	+16.2±8.0-	
Veratridine+Eth	10^{-5}, 1%	-15.0±4.1**-	

The effects of preincubation at 22-37°C for 4 hours with the different substances on $[^3H]$-NMS and $[^3H]$-QNB binding in cortical slices (in percent difference from control specific binding). Preincubation and incubation conditions are similar in this and all following figures: All preincubations were for 4 hrs at 22-37°C. All incubations with $[^3H]$-ligand (~5 nM) were for 2 hr at 4°C. In each experiment total binding was determined in 4 slices and non-specific binding, with 10^{-4} M of unlabelled ligand, was determined in 2 slices. The numbers inside each graph represent the number of separate experiments performed. Abbreviations in Tables and Figures: Ver - veratridine; K^+, - high potassium out; Glu - L-glutamatic acid; PTX - picrotoxin; GABA - γ-amino butyric acid; $AuCl_2$ - gold chloride; $CoCl_2$ - cobalt chloride; $MgCl_2$ - magnesium chloride; TTX - tetrodotoxin; TEA - tetraethylammonium chloride; 4-AP - 4-aminopyridine; $CsCl_2$ - cesium chloride; ACh - acetylcholine chloride; Carb - carbamylcholine chloride; Eth - ethanol (1% f.c.); +Eth - Veratridine dissolved in ethanol. In Table I and subsequent figures, error bars indicate S.E.M. Significance was analysed by one sample t-test: *$P<0.05$, **$P<0.01$, ***$P<0.001$ (one tailed).

Receptor binding values following various manipulations are usually expressed as percent difference from control mAChR number ± S.E.M.

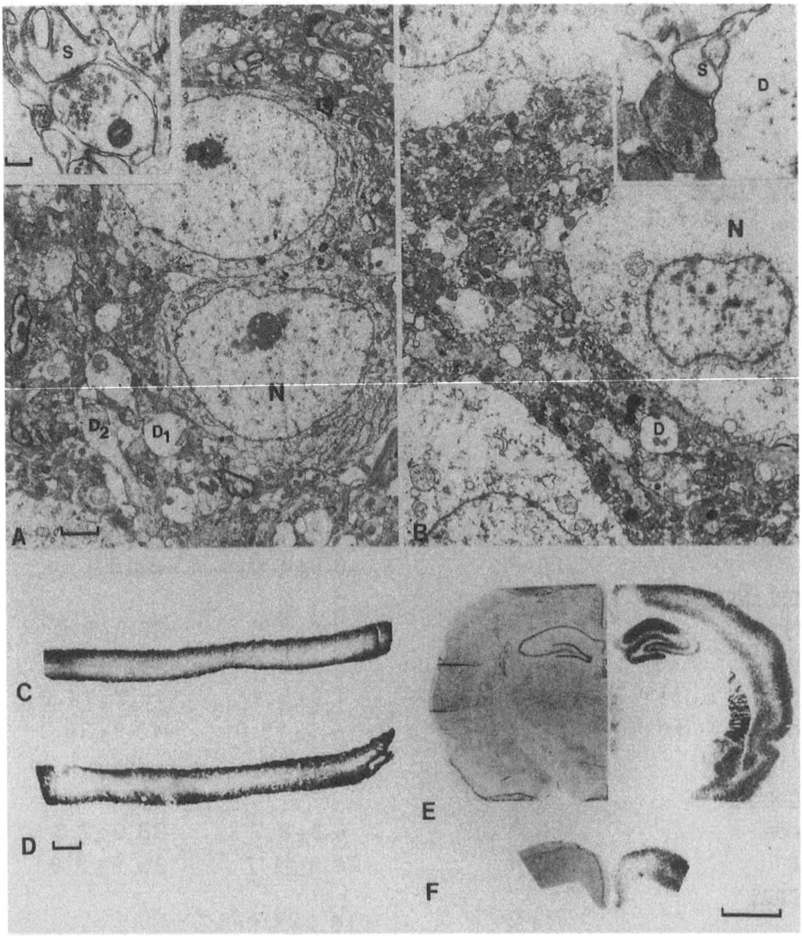

Fig. 1: Morphological evaluation of the viability of cerebral cortex slices.

(**A,B**):Electron micrographs of cerebral cortex slices. (**A**): Directly after cutting the slices (0 hrs) healthy looking neurons (N) are present. In the dense neuropil only a few dendritic profiles appear to be without cytoplasmic organelles (D_1), while other dendrites, running in the plane of of the section appear normal (D_2). The inset micrograph shows an axon terminal synapsing upon a dendritic spine (S). (**B**): After 6 hrs. in Dulb$^+$ medium at 30 ^oC.most neurons (N) and dendrites (D) are swollen and the cytoplasmic content is in different stages of degeneration. Membranes, however, are intact and axon terminals are still making synapses upon them as well as on dendritic spines (Inset). Bar = 2.5 μm. Bar inset = 0.25 μm.

(**C,D**): Trypan blue exclusion test performed on cortex slices after incubation for 0 (**C**) or 6 (**D**) hrs. in Dulb$^+$ medium at 30°C. Slices are viewed edge on from the pial surface. Dead cells are stained and there is only a small increase, from 15 - 30 %, between the 0 and 6 hr. period. Bar = 0.25 mm.

(**E,F**):Autoradiogram and Nissl stain of thin coronal sections (16 μm) of rat brain (**E**) and cortex slices (**F**). Sections or cortex slices were incubated in 5 nM [^{3}H]NMS and exposed to Ultrofilm. Both sections and slices show a laminar distribution of the muscarinic ACh receptor, with highest binding levels in L I - III and L V, VI. Bar = 2.5 mm.

Results

Morphology

Cross-sections of the trypan blue incubated slices showed that many dead cells could be found at the cut edges of the slice. Towards the center of the slice the number of dead cells gradually diminished so that in the inner part only intact cells were present (Fig. 1C). After 6 hrs. at 30 °C a more dense trypan blue staining area was found which included additional cells towards the center of the slice (Fig. 1D). Quantification of the percentage of dead cells was estimated to be 15% at 0 hr and 30% after 6 hr.

Electron microscopy also revealed dead cells and debris at the edges of the slice, with the center containing healthy tissue. Directly after cutting the slices and after 1 and 2 hours incubation at 30°C, a large number of swollen dendrites, without their cytoplasmic organelles, could be seen. Dendrites running in the plane of the section still contained cytoplasmic organelles, suggesting that only severed dendrites were degenerating. Axon terminals and cell bodies looked quite normal (Fig. 1A). After 6 hours, cell bodies and dendrites were swollen and their cytoplasmic content was in different stages of degeneration. Membranes, however, were intact and axon terminals were still making synapses upon them as well as on dendritic spines (Fig. 1B). At 4 °C the onset of the process of degeneration was delayed, so that at 6 hrs. after incubation more of the tissue was in a healthy condition. These degenerative changes observed at 30°C suggested the length of time that preincubation and incubation steps could be allowed to proceed for tissue to be considered living. We have consistently used a maximum treatment length of 6 hours in most of the following experiments.

Autoradiography of [^{3}H]NMS binding in rat cortex is illustrated in Fig. 1 D-E. In Fig. 1D, 16 μm thick coronal sections of rat brain were incubated and processed as described elsewhere (Shaw et al., 1984). Sections cut at 16 μm from a rat cortical slice previously incubated with [^{3}H]NMS are shown for comparison. Nissl stained sections of both conditions were compared to the autoradiogram and revealed that relative binding densities in both cases were highest in laminae I-III and VI. In both cases [^{3}H]NMS binding in the cingulate cortex (medial) is less dense than that seen in other cortical regions. The same binding pattern was observed in all sections throughout the cortical slice, indicating that the ligand penetrated to the inside of the slice.

Characterization of muscarinic binding sites

A plot of protein content versus amount of [^{3}H]NMS binding was made to determine if binding was linear with increasing concentrations of protein. Increasing numbers of cortex slices (1/2 - 5) per well, representing protein contents of 0.2 - 2 mg, resulted in a linear increase in total, specific and non- specific binding of [^{3}H]NMS. A linear relationship was also found between the wet weight of the slices and the protein content (results not shown). This relation allowed us to simply weigh subsequent slices in order to assign protein values to them.

Time course experiments at 4 °C showed that [^{3}H]NMS binding reached equilibrium at 8 hours. [^{3}H]QNB, however, reached an early equilibrium condition after 4 hr, but binding increased again to a new equilibrium after 8

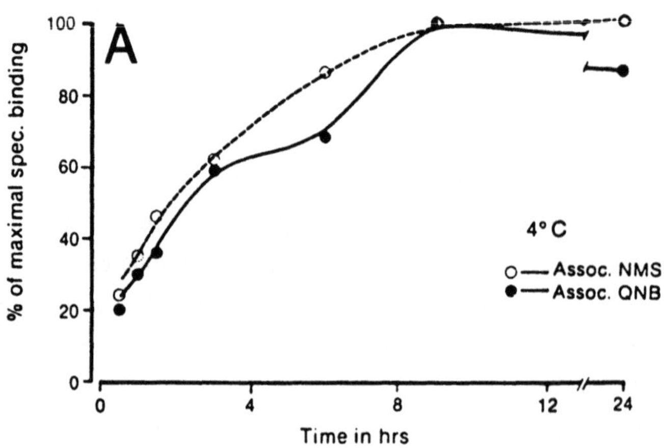

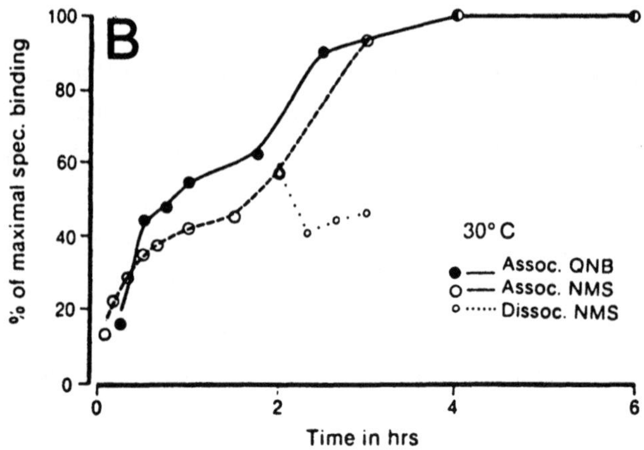

Fig. 2: (A,B) Association and dissociation time courses for [^{3}H]NMS and [^{3}H]QNB binding at 4°C (A) and 30°C (B). Values are given as percentage of maximal specific binding. Each graph represents data from two separate experiments. Slices were incubated in 2-10 nM [^{3}H]NMS or [^{3}H]QNB for times ranging from 5 min - 24 hr.

hrs (Fig. 2A). Binding at these long time points (>8 hrs.) does not, presumably, solely reflect binding to mAChRs in living cells (see section above).

At 30°C, [^{3}H]NMS and [^{3}H]QNB reached a first equilibrium after 90 minutes and a second final one after 3 hours (Fig. 2B). Preincubation of the slices in Dulb$^+$ at 30°C for 30 min. before the actual incubation did not affect the time course or amount of binding. The effects of longer Dulb$^+$ pre- incubations were not investigated. Post-incubation rinses of up to 60 minutes showed only a small decline in specific binding, in keeping with previous reports (Shaw et al., 1984). The highest specific to non-specific binding ratios were achieved with rinse times of less than 10 minutes. Consequently, in subsequent binding studies, incubations for 2 hr at 4°C with two washes of 5 min each, were routinely used.

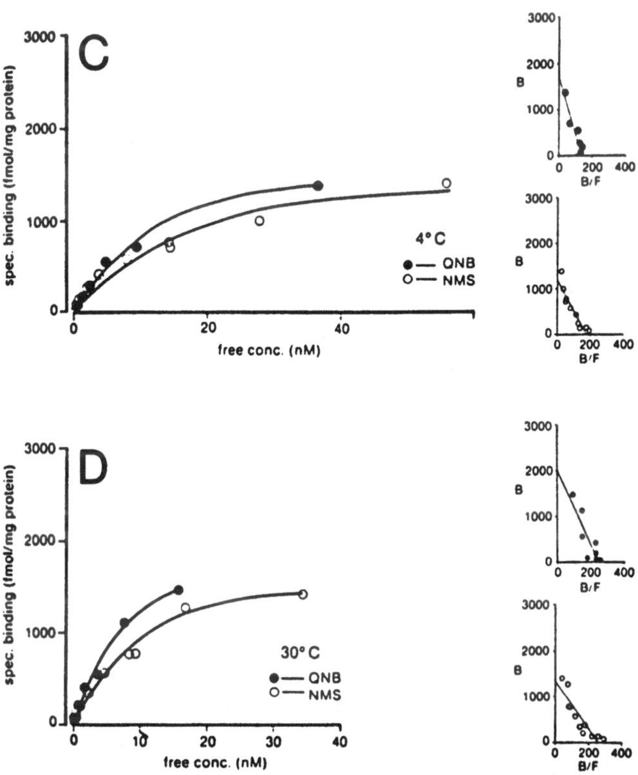

Fig.2: (C,D) Saturation binding of [^{3}H]NMS and [^{3}H]QNB in rat cerebral cortex slices at 4°C (C) and 30°C (D). Incubation was for 2 hr. with free concentrations ranging from 0.6 - 60 nM. Each point represents the mean of four determinations of specific binding (coefficient of variation always <20%). Non-specific binding is 5% of total binding for [^{3}H]NMS and 30% for [^{3}H]QNB. Each graph is a combination of 2-3 separate experiments.

Right panels show the Eadie-Hofstee plots of each binding curve. SD (E_{rad}) and Hill coefficient (see text for explanation) were respectively: [^{3}H]QNB, 4°C: 0.089 and 1.04; [^{3}H]NMS, 4°C: 0.15 and 0.95; [^{3}H]QNB, 30°C: 0.095 and 1.06; [^{3}H]NMS, 30°C: 0.106 and 1.15.

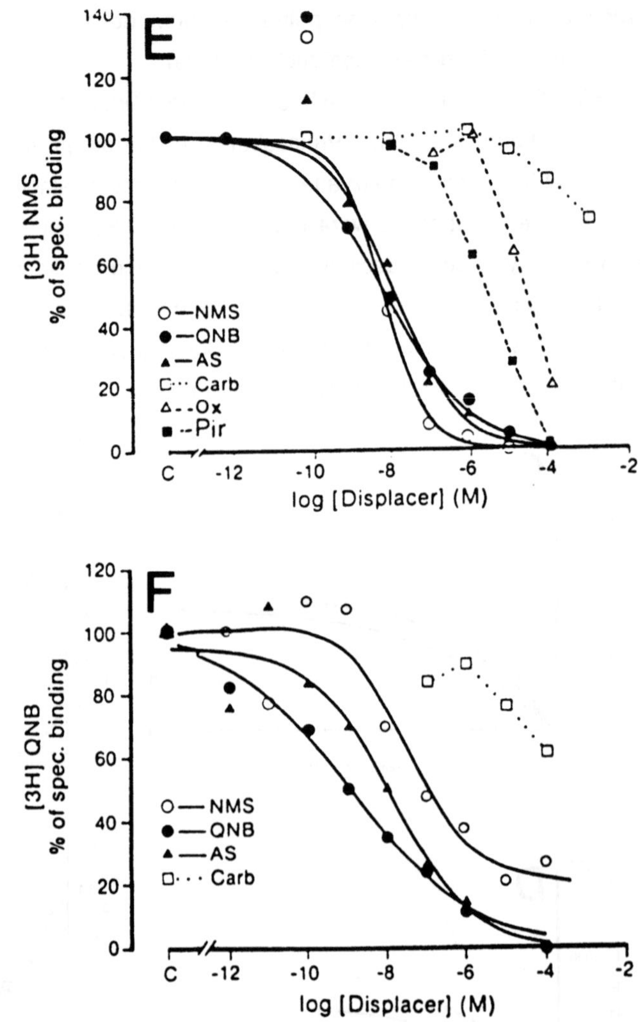

Fig. 2: (E,F) Competition curves for [^{3}H]NMS (E) and [^{3}H]QNB (F) binding from rat cerebral cortex slices. Slices were simultaneously incubated for 2 hr at 4°C in 5 nM [^{3}H]ligand and 10^{-3} to 10^{-12} M of the following displacers: NMS - N- methyl-scopolamine, QNB - quinuclinidyl-benzylate, AS - atropine sulphate, Carb - carbamyl choline chloride (carbachol), Ox - oxotremorine and Pir - pirenzipine. Values are given as percentage of control specific binding (C). Each graph combines data of two separate experiments.

Association (k_{+1}) and dissociation (k_{-1}) rate constants could be indirectly calculated from the observed rate constant (k_{obs}) (Bylund, 1980) and the equilibrium dissociation constant (K_D) derived from the saturation binding experiments (see below). For [^{3}H]QNB at 30°C, k_{-1} was 0.0085 $.min^{-1}$ and k_{+1} was 0.001 $nM^{-1}.min^{-1}$.

$[^3H]$NMS binding at 30°C yielded 0.0044 min^{-1} and 0.0008 $nM^{-1}.min^{-1}$ for k_{-1} and k_{+1}, respectively.

Saturation binding was performed at 4 °C and 30 °C for both $[^3H]$NMS and $[^3H]$QNB (Fig. 2C and D). Increasing concentration of ligands (0.6 - 60 nM, free concentration in the well) resulted in saturable binding. At 30 °C, a B_{max} of 2187 fmol/mg protein was found for $[^3H]$QNB (reflecting all mAChRs) and 1335 fmol/mg protein for $[^3H]$NMS (only mAChRs on the cell surface). K_D values were 8.2 and 5.2 nM for $[^3H]$QNB and $[^3H]$NMS, respectively. At 4 °C, a B_{max} of 1756 fmol/mg protein and a K_D of 12.1 nM was found for $[^3H]$QNB, while for $[^3H]$NMS 1191 fmol/mg protein and 6.7 nM were found for B_{max} and K_D, respectively. It is clear that at both temperatures the B_{max} of $[^3H]$ QNB was higher than the B_{max} of $[^3H]$NMS. The quality of these data can be assessed by the $SD(E_{rad})$ values and the Hill coefficient (Zivin and Waud,1982). $SD(E_{rad})$ values give an estimation of the bias in B_{max} and K_D values inherent in Eadie-Hofstee plots. The Hill coefficient is an indication of cooperativity: a value of 1.0 indicates that ligand-receptor interaction occurs via a biomolecular reaction obeying a simple mass action law. Representative values of Hill coefficient and $SD(E_{rad})$ are given in the legend to Fig. 2.

In a series of competition experiments we determined the IC_{50} (the concentration of cold ligand (M) needed to displace 50% of the specific binding) for QNB, NMS, atropine sulphate (AS), carbamylcholine chloride (carbachol), oxotremorine (Ox) and pirenzipine (Pir) on $[^3H]$NMS (Fig. 2E) and $[^3H]$QNB (Fig. 2F) binding. $[^3H]$QNB was displaced differentially by the respective analogs; the IC_{50}s for QNB, NMS and AS are respectively: 1.1 x 10^{-9}, 2.3 x 10^{-7} and 1.3 x 10^{-8} M. In contrast, close to identical IC_{50} values were found for $[^3H]$NMS using QNB, NMS and AS: 7.4 x 10^{-9}, 7.1 x 10^{-9} and 1.4 x 10^{-8} M, respectively. IC_{50} values for both Ox and Pir were higher: 2.0 x 10^{-6} and 2.4 x 10^{-6} respectively. Carbachol, even at a high concentration, had only a slight effect on specific binding of both ligands, as was found also in neuroblastoma cells (Gossuin et al., 1984) leading to IC_{50}s in the mM range. This can not be due to trapping (Gossuin et al., 1984) of the radiolabelled ligands since they are clearly displaced by their respective unlabelled compounds.

Table 1 summarizes the various agents, the concentrations used to alter bioelectrical activity and/or to stimulate mAChRs, and the effect on mAChR number. The various agents in each group had little or no effect on receptor number at 4°C under our conditions (data not shown).

Effects of Depolarization on AChRs

Increases in bioelectric activity induced by veratridine (10^{-5} M) led to a decrease in both $[^3H]$NMS and $[^3H]$QNB binding sites of 26.2% and 37.6%, respectively (Table 1).

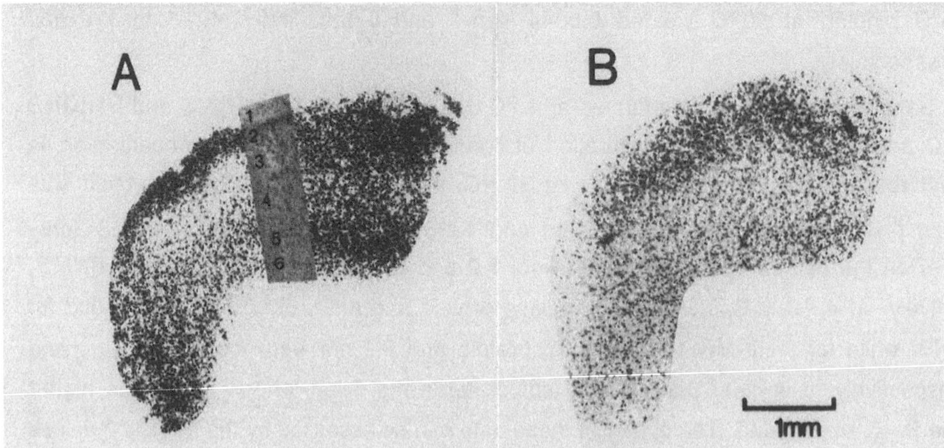

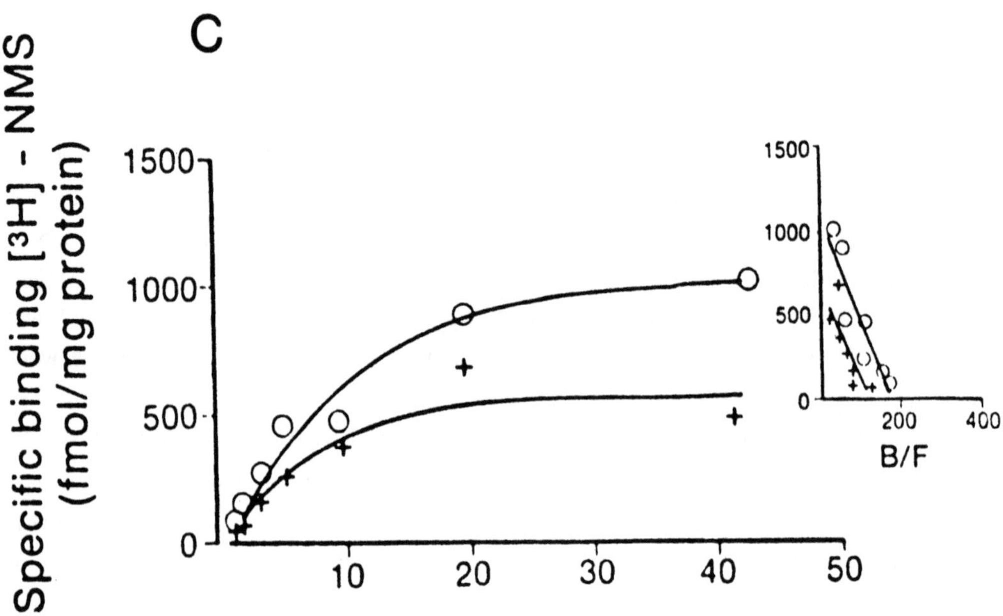

Fig. 3

(A) Effects of veratridine on mAChR distribution. Autoradiograms are from 16 μm thick sections cut from a cortex slice incubated with [^{3}H]NMS at 4°C for 2 hr. after a 2 hr preincubation at 22 °C in (A) control medium or (B) 10^{-5} M veratridine. Sections were apposed to LKB Ultrofilm for 21 days to generate autoradiograms. Following the development of the autoradiograms, the sections were stained with cresyl violet for Nissl substance. Both autoradiograms and stained sections were photographed individually for highest contrast, but to the same magnification. Sections of each print were aligned to identify the cortical laminae. Note the greater qualitative loss of [^{3}H]NMS binding sites in laminae IV-VI than in laminae I-III after veratridine preincubation. Dorsal is up, medial is to the left for both slices. Note that [^{3}H]NMS binding is less dense in the cingulate cortex (medial portion) than in the lateral neocortex . (Continued on next page)

The effects of veratridine, whose stock solution was first dissolved in 100% ethanol (Eth), depended on the final well concentration of Eth. Larger decreases with veratridine were seen for lower (0.01%) Eth concentrations. The observation that higher concentrations of Eth diminished the effects of veratridine provides evidence that veratridine's effects are not due solely to cellular damage induced by this substance. A 1% Eth well concentration alone increased mAChR number by 16.2% (Table 1).

The autoradiograms of sections from slices pre-incubated with veratridine (Fig. 3B) show a loss of mAChRs ([^{3}H]NMS binding) across all laminae in which mAChRs are normally present, although the loss in the middle and deep laminae is more pronounced.

Eadie-Hofstee analysis (Zivin and Waud, 1982) of saturation binding isotherms indicates that the loss of mAChRs reflects a lowered B_{max}, without a significant change in K_D for both [^{3}H]NMS (Fig. 3C) and [^{3}H]QNB (not shown).

In a separate series of experiments various influences on veratridine- induced down-regulation and the rate of down-regulation were examined. The amount of veratridine-induced down-regulation increased with increasing temperature to 37°C. At 37°C, [^{3}H]NMS labelled mAChR down-regulation in response to veratridine treatment showed an initial plateau followed by a decline to -27.4% of the control value (Fig. 4A, see Fig. C for [^{3}H]QNB labelled mAChRs). A similar loss following a 2 hr carbachol treatment was followed by an increase towards control values at longer preincubation times. Time to half maximal (at 4 hr.) down-regulation was about 30 min. for carbachol and about 2 hours for veratridine.

Muscarinic ACh receptor down-regulation was also observed following preincubation in high K^{+}_{out}, glutamate, or picrotoxin (Table 1). Of these, the effects of high K^{+}_{out} and glutamate on external mAChR labelled with [^{3}H]NMS were significant, showing a decrease in mAChRs of 11.3% and 11.6%, respectively.

Effects of Hyperpolarization on mAChRs

Neither GABA nor $AuCl_2$ treatment had a significant effect on mAChR number labelled by [^{3}H]NMS or [^{3}H]QNB (Table 1).

Fig. 3 (continued): (C)Saturation binding isotherms (at 4°C for 2 h) of [^{3}H]NMS after a 4 hr preincubation at 30°C in the presence (+) or absence (o) of veratridine. The right side of the figure shows Eadie-Hofstee plots of the data. B_{max} and K_D values for [^{3}H]NMS were 1042 ± 136 fmol/mg protein, and 6.2 ± 1.1 nM, respectively in the control condition and 604 ± 118 fmol/mg protein and 5.2 ± 1.8 nM following veratridine treatment. B_{max} and K_D values for [^{3}H]QNB were 1432 ± 301 fmol/mg protein and 8.7±2.8 nM, respectively in the control condition and 576 ± 182 fmol/mg protein and 7.2 ± 4.2 nM following veratridine treatment (graphs not shown).

Effects of Blocking Synaptic Transmission or Selected Ion Channels

Table I shows the effect on mAChRs of blocking Ca^{++}-, Na^{+}- and K^{+}- channels. $CoCl_2$ and $MgCl_2$ produced small, but insignificant, decreases of mAChRs. Nifedepine had a significant effect (-8.8%). Preincubation with TTX at 22-37°C led to a small decrease in mAChRs (Fig. 4 B,D), but had no effect on mAChRs when added during incubation at 4°C. The decrease in mAChRs following blockade of sodium channels by TTX led us to examine the blockade of other ion channels in particular K^{+}-channels, since ACh is believed to alter K^{+}-channel conductance. In a series of experiments we examined the effects of TEA, 4-AP, apamin, and Cs^{++} on mAChR number. TEA had no effect on [^{3}H]NMS binding to a

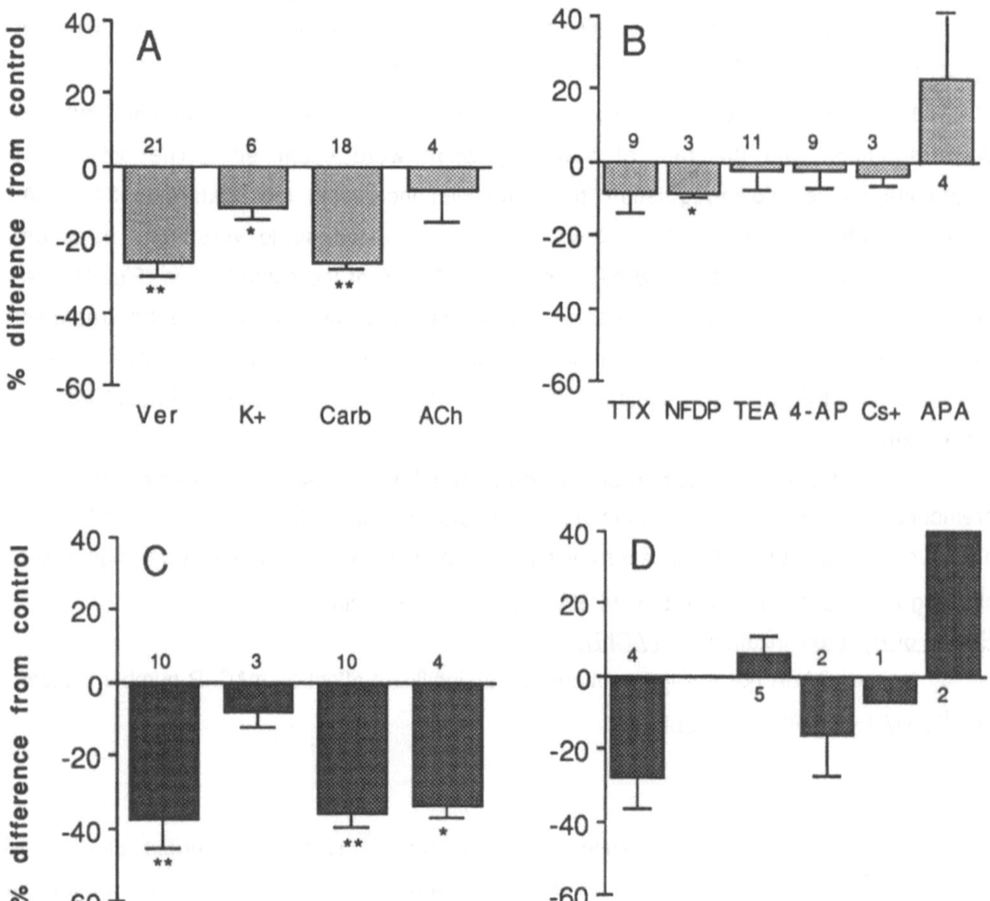

Fig. 4: (A,C) Effects of veratridine, high K^{+}_{out}, carbachol or ACh for 4 hours at 22- 37°C on [^{3}H]NMS (A) and [^{3}H]QNB (C) binding in cortex slices. Incubation conditions as in Table 1. (B,D) Effects of TTX, nifedipine, TEA, 4-AP, Cs^{++} or apamin on [^{3}H]NMS (B) and [^{3}H]QNB (D) binding in cortex slices. Preincubation and incubation conditions as above.* ($p<0.5$) and ** ($p<0.01$) significant difference from control (one-sample t- test)

concentration of 10^{-3} M at 4°C. After preincubation at 30-37 °C with 10^{-3} M TEA, mAChR number was also near control values (Fig. 4B). Similarly, preincubation with 4-AP and $CsCl_2$ had little effect on [^{3}H]NMS binding unlike apamin which gave a substantial increase (Fig. 4B). Similar results were obtained using these K^+-channel blockers with [^{3}H]QNB (Fig. 4D).

Effects of Agonist Stimulation on mAChR

Table 1 shows that the addition of a 1 mM concentration of the muscarinic agonist carbachol, led to a highly significant decrease (26.1%) in mAChR number following a 4 hour preincubation period at 22-37°C for receptors labelled with [^{3}H]NMS. The carbachol

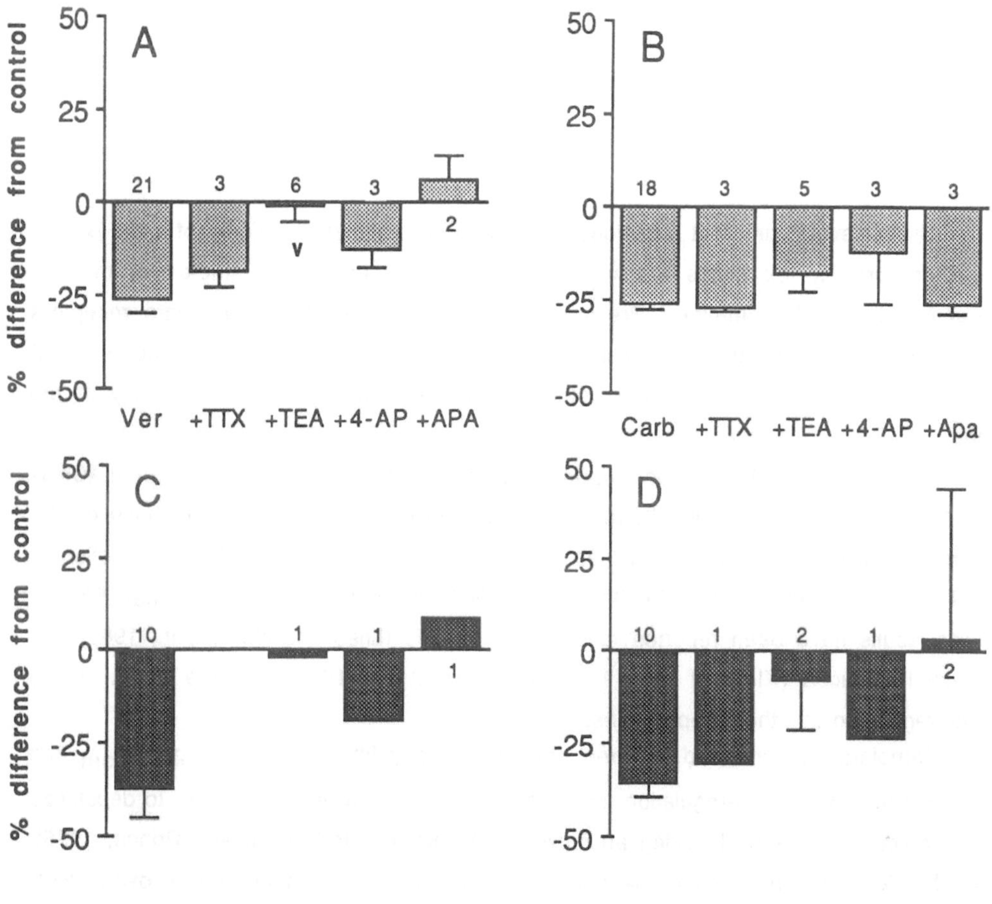

Fig. 5: Effects of TTX, TEA, 4-AP and apamin on veratridine (A, C) or carbachol(B, D) induced down regulation of [^{3}H]NMS (A, B) or [^{3}H]QNB (C, D) binding in cortex slices Conditions as in Fig.4. v significantly different from veratridine alone ($p<0.05$, paired t-test)

induced down-regulation was even larger for [^{3}H]QNB (33.9%) (Fig. 4 A,C). [^{3}H]NMS binding sites were not significantly altered by ACh, unlike those labelled by [^{3}H]QNB which showed a 35.9% decrease (Fig. 4 A,C).

Effects of Ion Channel Blockers on mAChR Down Regulation Induced by Veratridine, High K^+_{out} or Carbachol

Addition of TTX (10^{-5} M) totally failed to block carbachol-induced down-regulation of mAChRs but showed a partial block of the typical veratridine effects (Fig. 5 A,C). In contrast, addition of TEA, 4-AP and apamin) largely blocked down-regulation induced by veratridine (Fig. 5 A,C) and TEA and 4- AP, but not apamin, diminished that induced by carbachol (Fig. 5 B,D). In preliminary experiments effects of high K^+_{out} were not significantly affected by TEA.

Discussion

Using several different techniques, we have shown that a majority of cells (>70%) in rat cerebral cortex slices are viable and are thus useful for studies of mAChR characterization and regulation. Previous studies on β-adrenergic and opiate receptors (Wilkinson and Wilkinson, 1985; Wilkinson et al., 1986) had established the utility of this preparation for receptor binding studies, but had not explored the viability of the slices.

The results of the present report support the notion that regulation of mAChRs occurs by a process involving alterations in the number of receptors, both those on the external surface of neurons, as well as those in the internal milieu. Down-regulation of both surface and total mAChRs in response to veratridine involved a change in B_{max}, not K_D. Similar results have been reported following veratridine (Liles and Nathanson, 1987) and carbachol- induced (Klein et al., 1979; Maloteaux et al., 1983; Yamada et al., 1983) down- regulation in other preparations.

Depolarization induced in three different ways (with veratridine, high K^+_{out} and carbachol) all led to down-regulation for surface mAChRs. (Carbachol appears to depolarize some cortical neurons by blocking an outward K^+-current (McCormick and Prince, 1985). Note that for some neurons carbachol may increase K^+ inward current (Brown, 1984; McCormick and Prince, 1985).)

Depolarization of cortical slices by veratridine led to a reasonably rapid ($t_{1/2} \simeq 2$ hr), temperature dependent loss of 26.2% of surface mAChRs. Carbachol-induced down-regulation (26.1%) showed a faster rate of loss of surface mAChRs ($t_{1/2}$ = 30 minutes)

than that found using veratridine. The amount and rate of mAChR down-regulation is similar to that reported by other investigators (Klein et al., 1979; Siman and Klein, 1979; Shifrin and Klein, 1980; Maloteaux et al., 1983; El Fakahany and Lee, 1986; Hunter and Nathanson, 1986; Raposo et al., 1987)

Using [^{3}H]QNB, the loss of total mAChR was 37.6% and 35.9% for veratridine and carbachol, respectively. Since [^{3}H]QNB binds to both internal and surface mAChRs a loss of mAChRs thus labelled may reflect the loss of receptors from both the cell surface and the internal receptor pool. A secondary loss of internal mAChRs would indicate a possibly degradative (or phosphorylating) process and has been reported by other workers (Stollberg and Berg, 1987; Bouvier et al., 1988). Whether the loss of internal mAChRs in the present experiments is restricted to those mAChRs already inside the cell, or to those first removed from the cell surface awaits clarification.

In spite of this uncertainty, it seems to us valuable to be able to estimate the relative proportions of surface and internal mAChR down- regulation, and in the following discussion we will attempt such an estimate. We begin with our previous estimate of surface to internal mAChR distribution based on the differential penetration of the ligands [^{3}H]NMS and [^{3}H]QNB (Maloteaux et al, 1983; Gossuin et al., 1984). Since the former binds to surface sites only, while the latter binds all mAChRs, calculating internal mAChRs can be determined by a single relation: $R_t - R_s = R_i$, where R_t = total mAChRs labelled, R_s = surface mAChRs and R_i = internal mAChRs. Our measurements suggest that at 30°C approximately 61% of mAChRs were on the surface while 39% were internal. Estimating the number of mAChRs left after the various preincubation treatments ('experimental') is of the same form: $R_{t(exp)} - R_{s(exp)} = R_{i(exp)}$. Given the surface to internal ratios of 61:39 we can then calculate the percent loss of surface or internal mAChRs by the following equations: for R_i: $100 \times (R_{i(exp)} - R_{i(control)}) R_{i(control)}$; for R_s: $100 \times (R_{s(exp)} - R_{s(control)})/R_{s(control)}$.

The results of these calculations are given in Table 2 and show the percent loss of mAChRs from the surface or internal pools and their proportions after veratridine, high K^+_{out}, carbachol, and ACh preincubations. Further, the effects of the various K^+-channel blockers are shown. From Table 2 it is apparent that the different preincubation treatments have quite different effects on surface and internal mAChRs. For example, inwardly directed potassium current (high K^+_{out}) affects only surface mAChRs. In contrast, veratridine and carbachol affect both sites although the main effects are on the internal sites while ACh's actions are exclusively on internal mAChRs following preincubations of the concentrations and lengths used here. Other investigators have also noted differences in the effect of ACh and carbachol when both drugs were used at the same high concentrations (Klein et al., 1979).

Table 2 Percent loss*** mAChRs from surface or internal sites after different preincubation conditions

Treatment	Surface (s)*	Internal (i)**	s:i Ratio
Veratridine	-26.2	-55.3	1:2.1
High K^{+}_{out}	-11.3	- 3.1	1:0.3
Carbachol	-26.1	54.8	1:2.1
ACh	-6.3	77.2	1:12.2
TEA	-2.4	+20.3	1:22.7
4-AP	-1.9	-38.5	1:20.3
apamin	+16.8	+86.4	1:5.1

* Surface mAChRs (R_s) were measured using [^{3}H]NMS.

** Internal mAChRs were determined from the equation $R_i = R_t - R_s$, where R_t is the total number of mAChRs labelled by [^{3}H]QNB.

*** Percent loss of surface or internal mAChRs following experimental preincubation were determined by the equations: for R_i: $100 \times (R_{i(exp)} - R_{i(control)})/R_{i(control)}$;
for R_s: $100 \times (R_{s(exp)} - R_{s(control)})/R_{s(control)}$.

One reason why the ACh and carbachol effects differ may be that they bind to different membrane sites. In our preparation carbachol does not appear to have a high affinity for the mAChR since we have observed little displacement with this agonist except at very high concentrations (~28% at 10^{-3} M, ~5% at 10^{-5} M) either in the present study or in cat visual cortex (Shaw et al., 1984). These results may suggest that carbachol exerts much of its influence on mAChR regulation indirectly, perhaps at a site linked to, but not identical with, the mAChR.

One of the largest effects on mAChR down-regulation observed in the present study followed exposure to veratridine, a drug which decreases sodium channel inactivation leaving these channels open for more than 50 times longer than normal (Narahashi, 1974; Catterall, 1980; Hille, 1984). TTX failed to block veratridine's effect. While veratridine exerts its primary effect on Na^{+}- channels, an increased K^{+}-flux may be expected to occur as well if not all Na^{+}-channels have been blocked by the TTX, or if other ion channels, e.g. Ca^{++}, are still active. Additionally, veratridine increases the relative K^{+}- permeability (Hille, 1984). Depolarization induced in other ways, notably by increased K^{+}_{out}, also led to a down-regulation of mAChRs as observed in other studies (Betz, 1983) (Some reports, however, e.g. Liles and Nathanson, 1987 have shown the opposite effect). Our results suggest that the observed down- regulation is not strictly related to depolarization and that it is not Na^{+}- dependent. Furthermore, mAChR regulation in the cortical slice does not appear to be directly Ca^{++}-dependent, unlike in other preparations (Liles and Nathanson, 1983) since a general block of Ca^{++}-channels with either Mg^{++} or Co^{++} had little effect. More specific blocking of postsynaptic voltage dependent Ca^{++}-channels with nifedipine, however, gave a small decrease in mAChRs. Hyperpolarization of neurons by either GABA (which acts to increase inward Cl- conductance) or $AuCl_2$ had little effect on mAChR regulation.

Carbachol-induced down-regulation also appears to act through a Na^+-, Ca^{++}- or Cl^--independent process. Like ACh, carbachol has been shown to alter a K^+-current (Brown, 1984; McCormick and Prince, 1985) and its effects on the mAChR are clearly not blocked by TTX.

The K^+-channel blockers TEA and 4-AP showed little effect on surface mAChRs, but had a more pronounced effect on internal mAChRs. Incubation of the slices with TEA led to a large increase in internal mAChRs. 4-AP had the opposite effect. Apamin increased both surface and internal mAChRs. For all K^+-channel blockers the greatest effects observed were on internal mAChRs. K^+-channel blockers had a significant effect on down-regulation elicited with veratridine and carbachol. These results lead us to suggest that changes in K^+-current, in whatever direction, might result in mAChR down-regulation. The relative failure of alterations in Na^+, Ca^{++} or Cl^- ion channels to alter mAChR number in this preparation lends further support to this notion. The relative potencies of the three main K^+-channel blockers employed in these experiments (TEA, 4-AP and apamin) may further suggest the specific K^+- channels through which veratridine and carbachol exert their effects (Table I and Fig. 5). An increasing order of effectiveness in veratridine-induced down- regulation was found such that apamin>TEA>4-AP. Since TEA is known to block both Ca^{++}- and non-Ca^{++} dependent K^+-channels, while apamin is only effective in Ca^{++}-dependent K^+-channels, our results suggest that both types of K^+-channel normally participate in mAChR down-regulation, although to differing degrees. The less effective, but still substantial, effects of 4-AP on mAChR regulation support this view. Alternatively, the greater effectiveness of TEA than apamin may derive in part from TEA's known action on all Ca^{++}- dependent K^+-channels (i.e. large conductance Ca^{++}-dependent K^+-channels (Blatz and Magleby, 1987)) versus apamin's effects on only a subclass of these channels. In contrast to the effects of K^+-channel blockers on veratridine, only TEA and 4-AP were able to diminish carbachol-induced mAChR down- regulation, while apamin was without significant effect. These results suggest that the Ca^{++}-dependent K^+-channels are not involved in this type of agonist- induced down-regulation. Finally, the down-regulation observed following the various experimental manipulations may result from effects not only on the different K^+-channels, but also on their relationship to surface or internal mAChRs. This difference is shown clearly in Table 2. High K^+_{out} appears to act mostly on surface mAChRs. Carbachol and veratridine show an approximate 1:2 effect on surface vs internal receptors while ACh and K^+-channel blockers appear to act mainly on internal receptors. Based on these results we argue for separate mechanisms of regulation of surface and internal mAChR; all however, are based on changes in K^+-currents.

Given the above data, we speculate that specific receptor regulations might occur in response to changes in ionic current in specific ion channels induced by the relevant neurotransmitters. Thus, in the present study changes in K^+-channel currents led to decreases in mAChRs, while in muscle myotubes, in which ACh stimulation leads to increased Na^+-current, down- regulation of AChRs appears to be linked to alterations in Na^+-current (Bloch, 1986). If this principle is a general one, we speculate that the regulation of other receptor populations in the neocortex, e.g., the $GABA_A$ and glutaminergic (kainate or quisqualate) follow changes in GABA or glutamate-induced Cl- or Na^+ currents, respectively. Studies in progress are directed at this issue.

It is of interest that no manipulation in the present study, except for the effect of K^+-channel blockers on internal mAChRs, reliably gave a mAChR up- regulation. Since depolarization leads, however indirectly, to mAChR down regulation, it was surprising to find that TTX which blocks Na^+-induced depolarization, and in other preparations gives up-regulation (Betz,1983),had an insignificant effect on mAChR number, yet one similar in direction to that observed with veratridine. How can we account for this? One possibility is that up-regulation occurs only after longer periods of incubation than we employed here, as been suggested by other studies (Liles and Nathanson, 1987). If this is a correct interpretation, then up-regulation may be expected to occur at longer times following TTX. Another possibility is that mAChR up-regulation following a loss of neural activity occurs indirectly and is mediated through heterospecific receptor regulatory mechanisms (Hollenberg, 1985a,b). Further experiments are needed to clarify this issue.

The relatively rapid alterations of mAChRs which we have observed may underlie some of the rapid alterations of neural activity observed following various forms of synaptic stimulation (Blakemore and Mitchell, 1973; Marlin et al, 1988; Perkins and Teyler, 1988). Similar regulatory effects on other receptor populations may contribute further to dynamic alterations of cortical chemical circuitry. These regulatory effects on mAChRs and other receptor populations may further be dependent on a number of variables, including the regions of neocortex studied (we have shown elsewhere (Shaw et al, 1984) that mAChR binding differs in primary sensory from association cortex in the cat), the postnatal age of the animal, and the species under study (Shaw et al., 1986a). The latter are questions of ongoing investigations.

Acknowledgements

The authors thank D. March and G. Troop for assistance and Prof. J. Simmons for comments on the manuscript. This work was supported by a Killam Postdoctoral Fellowship to F.v.H. and a Medical Research Council of Canada grant (to M.W.,M.S.C and C.S.).

References

Adbul Ghani A-S, Boyar MM, Coutinho-Netto J, Bradford HF, Benie CP, Hulme EC, Birdsall NJM. (1981) Effect of tityus toxin and sensory stimulation on muscarinic cholinergic receptors in vivo. Biochem Pharmacol 30:2713-1714

Bear MF, Singer W (1986) Modulation of visual cortical plasticity by acetylcholine and noradrenaline. Nature 320:172-176

Bennett GW, Sharp T, Marsden CA, Parker TL (1983) A manually-operated brain tissue slicer suitable for neurotransmitter release studies. J Neurosci Methods 7:107-115

Betz H (1983) Regulation of -bungarotoxin receptor accumulation in chick retinal cultures: effects of membrane depolarization, cyclic nucleotide derivatives and Ca^{++}. J Neurosci 3:1333-1341

Blakemore C, Mitchell DE Environmental modification of the visual cortex and the neural basis of learning and memory Nature 241:467-468

Blatz AL, Magleby KL (1987) Potassium channels, calcium-activated. In: Adelman G (ed.) Encyclopedia of Neuroscience, Vol. 2. Birkhauser, Boston, p. 966

Bloch RJ (1986) Loss of acetylcholine receptor clusters induced by treatment of cultured rat myotubes with carbachol. J Neurosci 6:691-700

Bouvier M, Hausdorff WP, De Blasi A, O'Dowd BF, Kobilka BK, Caron MG, Lefkowitz RJ (1988) Removal of phosphorylation sites from B2-adrenergic receptor delays onset of agonist-promoted desensitization. Nature 333:370- 373

Brown DA (1984) Muscarinic excitation of sympathetic and central neurons. Trends in Pharm Sci (Suppl) 6:32-34.

Bylund DB (1980) Analysis of receptor binding data. In: Receptor Binding Techniques, Short Course Syllabus, Soc. for Neuroscience. p. 709

Catterall WA (1980) Neurotoxins that act on voltage-sensitive sodium channels in excitable membranes. Ann Rev Pharmacol Toxicol 20:15-43

El-Fakahany EE, Lee J-H (1986) Agonist-induced muscarinic acetylcholine receptor down-regulation in intact rat brain cells. Eur J Pharmacol 132:21- 30

Gilbert RFT, Hanley MR, Iversen LL (1979) [^{3}H]-Quinuclidinyl benzilate binding to muscarinic receptors in rat brain: comparison of results from intact brain slices and homogenates. Br J Pharmac 65:451-456

Gossuin A, Maloteaux J-M, Trouet A, Laduron P (1984) Differentiation between ligand trapping into intact cells and binding on muscarinic receptors. Biochem Biophys Acta 804:100-106

Harden TK, Petch LA, Treynels SF, Waldo GL (1985) Agonist-induced alteration in the membrane form of muscarinic cholinergic receptors. J Biol Chem 260:13060-13066

Hess R, Murata K (1974) Effects of glutamate and GABA on specific response properties of neurons in the visual cortex. Exp Brain Res 21:285-297

Hille B (1984) Potassium channels and chloride channels In: Ionic Channels of Excitable Membranes. Sinauer Sunderland,Mass., p. 99

Hollenberg MD (1985a) Biochemical mechanisms of receptor regulation. Trends in Pharm Sci 6:299-302

Hollenberg MD (1985b) Control of receptor function by homologous and heterologous ligands. In:

Poste G, Crooke ST (eds.) Mechanisms of Receptor Regulation. Plenum, New York, p. 295
Hubel DH, Wiesel TN (1970) The period of susceptibility to the physiological effect of unilateral eyelid closure in kittens. J Physiol (Lond) 206:419-436
Hunter DD, Nathanson NM (1986) Biochemical and physical analysis of newly synthesized muscarinic acetylcholine receptors on cultrued embryonic chicken cardiac cells. J Neurosci 6:3739-3748
Jarett L, Smith RM (1985) Receptor regulation, problems and perspectives. In: Poste G, Crooke ST (eds.) Mechanisms of Receptor Regulation. Plenum, New York, p. 1
Klein WL, Nathanson N, Nirenberg M (1979) Muscarinic acetylcholine receptor regulation by accelerated rate of receptor loss. Biochem Biophys Res Comm 90: 506-512
Krieg WJS (1946) Connections of the cerebral cortex. I. The albino rat. A. Topography of the cortical areas. J Comp Neurol 84: 221-275
Liles WC, Nathanson NM (1987) Regulation of acetylcholine receptor number in cultured neuronal cells by chronic membrane depolarization. J Neurosci 7(8):2556-2563
Lowry OH, Rosebrough NJ, Farr AL, Randall RJ (1951) Protein measurement with the Folin phenol reagent. J Biol Chem 193:265-275
Luqmani YA, Bradford HF, Birdsall NJM, Hulme EC (1979) Depolarization induced changes in muscarinic cholinergic receptors in synaptosomes. Nature 277:481-483
Maloteaux J-M, Gossuin A, Pauwels PJ, Laduron PM (1983) Short-term disappeara_nce of muscarinic cell surface receptors in carbachol-induced desensitizat_ion. FEBS Letters 156:103-107
Marlin SG, Hasan SJ, Cynader MS (1988) Direction selective adaptation in simple and complex cells in cat striate cortex. J Neurophysiol 59:1314-1330
McCormick DA, Prince DA (1985) Two types of muscarinic response to acetylcholine in mammalian cortical neurons. Proc Natl Acad Sci (USA) 82:6344-6348
McCormick DA, Prince DA (1986) Acetylcholine induces burst firing in thalamic reticular neurons by activating a potassium conductance. Nature 319:402-405
Merzenich MM, Nelson RJ, Stryker MP, Cynader MS, Schoppmann A, Zook JM (1984) Somatosensory cortical map changes following digit amputation in adult monkeys J Comp Neurol 224:591-605
Narahashi T (1974) Chemicals as tools in the study of excitable membranes Physiol Rev 54:813-889
Nathanson NM (1982) Regulation and development of muscarinic acetylcholine receptors Trends in Neurosci 5:401-404
Olson CR, Freeman RD (1980) Profile of the sensitive period for monocular deprivation in kittens Exp Brain Res 39:17-21
Perkins AT, Teyler TJ (1988) A critical period for long-term potentiation in the developing rat visual cortex Brain Res 439:222-229
Prince DA, Huguenard JR (1988) Functional properties of neocortical neurons. In: Rakic P, Singer W (eds.) Neurobiology of the Neocortex. Wiley, New York, p. 153
Raposo G, Dunia I, Marullo S, Andre C, Guillet JG, Strosberg AD, Benedetti EL, Hoebeke J (1987) Redistribution of muscarinic acetylcholine receptors on human fibroblasts induced by regulatory ligands Biology of the Cell 60:117-124
Rasmusson DD (1982) Reorganization of raccoon somatosensory cortex following removal of the fifth digit J Comp Neurol 205:313-326
Shaw C, Needler MC, Cynader M (1984) Ontogenesis of muscarinic acetylcholine bind_ing sites in cat visual cortex: reversal of specific laminar distributions during the critical period Develop Brain Res 14:295- 299
Shaw C, Teyler TJ (1982) The neural circuitry of the neocortex examined in the in vitro brain slice preparation Brain Res 243:35-47
Shaw C, Wilkinson M, Cynader M, Needler MC, Aoki C, Hall SE (1986a) The laminar distributions and postnatal development of neurotransmitter and neuromodulator receptors in cat visual cortex Brain Res Bull 16:661-671
Shaw C, Wilkinson M, Cynader MS (1986b) Calcium channel binding site distribution in cat and kitten visual cortex Soc Neurosci Abstr 12:1229

Shifrin GS, Klein WL (1980) Regulation of muscarinic acetylcholine receptor concentration in cloned neuroblastoma cells J Neurochem 34:993-999

Sillito AM (1975) The contribution of inhibitory mechanisms to the receptive field properties of neurons in the striate cortex of the cat J Physiol (Lond.) 250:305-329

Siman RG, Klein WL (1979) Cholinergic activity regulates muscarinic receptors in cerebral nervous system cultures Proc Natl Acad Sci USA 76:4141-4145

Simmonds M (1983) Multiple GABA receptors and associated regulatory sites Trends in Neurosci 6:279-281

Spencer DG, Horvath E, Traber J (1986) Direct autoradiographic determination of M_1 and M_2 muscarinic acetylcholine receptor distribution in the rat brain: relation to cholinergic nuclei and projections Brain Res 380:59-68

Stollberg J, Berg DK (1987) Neuronal acetylcholine receptors: fate of surface and internal pools in cell culture J Neurosci 7:1809-1815

Van Huizen F, Shaw C, Wilkinson M, Cynader M (1987) Muscarinic receptor regulation in slices of rat cerebral cortex Soc Neurosci Abstr 13:727

Walton K, Fulton B (1983) Hydrogen peroxide as a source of molecular oxygen for in vitro mammalian CNS preparations Brain Res 278:387-393

Watson M, Roeske WR, Vickroy TW, Smith TL, Akiyama K, Gulya K, Duckles SP, Serra M, Adem A, Nordberg A, Gehlert DR, Walmsley JK, Yamamura HI (1985) Biochemical and functional basis of putative muscarinic receptor subtypes and its implications Trends in Pharm Sci (Suppl.) 6:46-55

Wilkinson M, Wilkinson DA (1985) Beta-adrenergic ([^{3}H]CGP-12177) binding to brain slices and single intact pineal glands Neurochemical Res 10:829-839

Wilkinson M, Jacobson W, Watson-Wright W (1986) Tissue slices in radioligand binding assays: Studies in brain, pineal and muscle Life Sci 39:2037-2048

Yamada S, Isogai M, Okudaira H, Ash EH (1983) Regional adaptation of muscarinic receptors and choline uptake in brain following repeated administration of diisopropylfluorophosphate and atropine Brain Res 268:315-320

Zivin JA, Waud DR (1982) How to analyse binding, enzyme and uptake data: the simplest case, a single phase Life Sci 30:1407-1422

REGULATION OF CALCIUM CHANNELS AND ADENYLATE CYCLASE BY OPIATES

Z. Vogel, S.Y. Nah, B. Attali

Department of Neurobiology
Weizmann Institute of Science
76100 Rehovot
Israel

INTRODUCTION

The spinal cord (SC) is an important site for the mediation of the antinociceptive action of opiates (Yaksh, 1983), as well as for the induction of tolerance to these drugs following prolonged opiate exposure (Delander and Takemori, 1983).

Several opioid receptor subtypes μ, δ and two κ (κ_1 and κ_2) are present in the SC (Attali et al., 1982; Quirion, 1984; Gouarderes et al., 1985). The κ receptor subtypes represent the majority of spinal opiate receptors and have an important role in the opiate-induced spinal analgesia (Attali et al., 1982; Czlonkowski et al., 1983; Wood et al., 1981).

The molecular mechanisms underlying the opiate action in the SC are not well understood. Cyclic nucleotides seem to be involved in spinal opiate analgesia. The spinal antinociceptive effect of morphine is prevented by the administration of aminophylline, a cAMP phosphodiesterase inhibitor (Jurna, 1984). cAMP analogs as well as forskolin, a selective activator of adenylate cyclase, reduce the depressant effect of opiates on the sensory-evoked dorsal-horn electrophysiological responses in mouse

NATO ASI Series, Vol. H 32
Molecular Biology of Neuroreceptors and Ion Channels. Edited by A. Maelicke

SC dorsal root ganglion (DRG) cocultured explants (Crain et al., 1986). Moreover, we and others have shown that opiate agonist binding in SC is regulated by GTP and sodium ions (Attali et al., 1986; Kelly et al., 1982). All these findings suggest that opiate receptors in the SC are coupled to the adenylate cyclase system.

Opiates have also been implicated in inhibition of neurotransmitter release by mechanisms which involve reduced entry of Ca^{2+} through voltage-sensitive Ca^{2+} channels (Mudge et al., 1979). Recent electrophysiological studies indicate that in DRG neurons, the κ opioid peptide dynorphin depresses Ca^{2+} currents suggesting an inhibitory interaction of κ opiate receptors with voltage-sensitive Ca^{2+} channels (MacDonald and Werz, 1986). Three different types of voltage-dependent Ca^{2+} channels have recently been classified (L, N and T), according to their electrophysiological properties (Miller, 1987; Nowycky et al., 1985). They also differ in their response to dihydropyridine (DHP) agonists and antagonists. The L type channels in many tissues are modulated by these drugs, while T and N channels are not (Miller, 1987).

The SC-DRG coculture system contains both SC and primary sensory neurons and serves as a convenient system for analyzing the mechanisms underlying opiate signal transduction and tolerance (Crain et al., 1979; Crain et al., 1986). We have used mixed cultures of dissociated rat SC and DRG neurons and showed that κ opiate agonists decrease the activity of the adenylate cyclase and of the DHP-sensitive voltage-dependent calcium channels. Moreover, prolonged exposure of the cells to κ agonists decreases the capacity of these drugs to modulate these activities (see also Attali and Vogel, 1986a,b; Vogel et al., 1988; Attali et al., 1988a).

METHODS

Cell cultures

Neuronal cultures were prepared according to Attali *et al.* (1988a). SC with attached DRG were dissected from 12 to 14 day old fetal rats, the tissue minced and trypsinized. Cells were centrifuged (160 g; 5 min) and resuspended in culture medium which consisted of Dulbecco's modified Eagle's medium (DMEM) supplemented with 10% fetal calf serum, 10% heat inactivated horse serum and 7 ng/ml nerve growth factor (to promote survival of DRG neurons). Cells were plated onto poly-L-lysine precoated 60 mm tissue culture dishes ($5x10^6$ cells/dish). Five days after plating, culture medium was replaced and cells treated for 2 days with 15 μg/ml 5'-fluoro-2'-deoxyuridine and 35 μg/ml uridine to suppress growth of non-neuronal background cells. Cultures were then returned to culture medium without fetal calf serum and used for adenylate cyclase assays. SC-DRG cocultures prepared for $^{45}Ca^{2+}$ flux assays were similarly prepared except that the cells were plated on 35 mm dishes ($1.5x10^6$ cells/dish). One day following plating, the culture medium was replaced with serum-free defined medium which consisted of DMEM supplemented with 5 μg/ml insulin, 10 μg/ml transferrin, 20 nM progesterone, 100 μM putrescine, 30 nM selenium dioxide and 7 ng/ml nerve growth factor.

Adenylate cyclase assay

SC-DRG cocultures were maintained for 3 weeks *in vitro*, the last 4 days in the presence or absence of opiate agonists. Cells were washed twice with Dulbecco's phosphate buffered saline (Ca^{2+} and Mg^{2+} free) and homogenized in 10 mM Tris-HCl buffer, pH 7.4, containing 1 mM dithiothreitol and 5 mM $MgCl_2$ to give a protein concentration of 0.5-0.7 mg/ml.

Homogenates were assayed for adenylate cyclase activity as previously described (Attali et al., 1988a). Cyclase assay mixtures also contained 10 μM GTP, 100 mM NaCl, and the opiate agonist or antagonist as stated.

$^{45}Ca^{2+}$ influx assay

Procedure was modified from Yano et al. (1984). Cells were washed and preincubated at 37°C in Tris-buffered saline, which consisted of 20 mM Tris-HCl, pH 7.4, 150 mM NaCl, 5.4 mM KCl, 1.8 mM $CaCl_2$, 0.8 mM $MgSO_4$ and 25 mM glucose. Reaction was started by addition of 3 μCi $^{45}CaCl_2$ in the presence of 5.4 or 50 mM KCl. After incubation for 5 min, cells were quickly washed with cold buffer containing 150 mM NaCl, 5 mM $CaCl_2$, 0.1 mM $LaCl_3$ and 20 mM Tris-HCl, pH 7.4, and counted for radioactivity.

RESULTS

κ opiates regulate cyclase activity

The experiments described were performed with mixed cultures of rat SC and DRG cells. Such a coculture is shown in Figure 1. Most of the neurons present have aggregated into small cell clusters. However, many individual cells are also present. The number of nonneuronal background cells is kept low through the use of 5'-fluoro-2'-deoxyuridine or alternatively by maintaining the cells in serum-free defined medium.

We have previously shown that these cocultures express opiate receptors. More than 70% of the opiate receptors appearing in these cocultures are of the κ type. The amount of μ receptors is ca. 20% and that of δ is negligible (Attali et al., 1986). Similar results have been obtained for adult SC (Attali et al., 1982; Traynor and Wood, 1987).

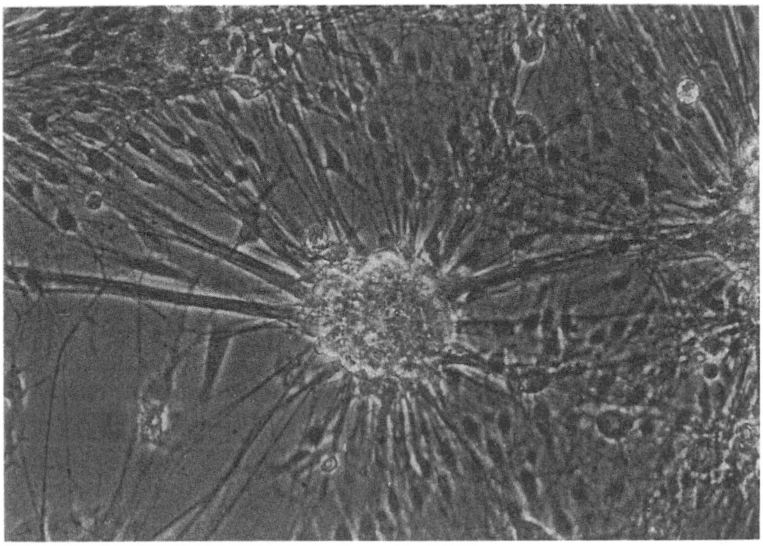

Figure 1. Phase contrast photomicrograph of cocultured rat SC-DRG cells, 3 weeks after plating.

Interaction of opiate agonists with δ type opiate receptor was shown to inhibit adenylate cyclase activity and to decrease cAMP levels in NG108-15 neuroblastoma X glioma hybrid cells as well as in CNS tissue (Sharma et al., 1975; Girardot et al., 1983; Law et al., 1981; Smith and Harden, 1984). We found that μ and δ selective opiate agonists, such as [D-Ala^2, N-Me-Phe^4, Gly^5-ol]enkephalin (DAGO) and [D-Ala^2, D-Leu^5]enkephalin (DADL), had no effect on the cyclase activity in membranes prepared from rat SC as well as in membranes of SC-DRG cocultured cells (see Table 1 and Attali et al., 1988a). On the other hand, the κ specific opiate agonist U50488 significantly inhibited the basal cyclase activity by 31±5% ($p<0.01$) at 10 μM. This inhibition was dose-dependent with an apparent IC_{50} value of 0.1 μM (Attali et al., 1988a). Etorphine (1 μM) and levorphanol (1 μM), two nonselective opiate agonists known to interact with κ opiate receptors in the SC (Attali et al., 1982), inhibited cyclase activity by 20% and 21%, respectively. Naloxone, a nonselective opiate antagonist, reversed the U50488 and etorphine inhibitory effects. The κ agonist regulation of adenylate cyclase showed the stereospecificity criterion established for opiate receptors, since levorphanol was far more effective than its respective enantiomer dextrorphan.

Table 1: Effect of opiates on basal cyclase activity in rat spinal cord membranes

Drugs added	Adenylate cyclase activity (% of control)
Control basal activity	100(3.9±0.04)
+U50488 10 μM	69±5
+U50488 10 μM + naloxone 10 μM	87±4
+Naloxone 10 μM	102±3
+Etorphine 1 μM	80±3
+Levorphanol 1 μM	79±2
+Dextrorphan 10 μM	103±5
+DADL 10 μM	99±8
+DAGO 10 μM	95±5

Drugs were added to the adenylate cyclase assay. Activity, in parentheses, is expressed in nmol cAMP/15 min/mg protein.

Table 2: Effect of chronic opiate treatment on adenylate cyclase activity in rat SC-DRG cocultures

	Adenylate cyclase activity		
Drugs added	Untreated (% of control)	Etorphine-treated (% of control)	U50488-treated (% of control)
Basal activity	100(4.97±0.14)	100(4.87±0.33)	100(5.06±0.04)
+U50488	76±2	102±7	94±1
+U50488+Naloxone	98±1	100±1	96±2
+Naloxone	99±2	102±1	103±3

Cocultures were maintained for 3 weeks, the last 4 days in the presence or absence of etorphine (1 μM) or U50488 (10 μM). Drugs added to cyclase assay were U50488 (10 μM) and naloxone (100 μM). Numbers in parentheses correspond to adenylate cyclase activity in nmol cAMP/15 min/mg protein.

The SC is involved in the development of tolerance and dependence resulting from chronic opiate administration (Delander and Takemori, 1983). To study the regulation of adenylate cyclase following chronic opiate treatment, we treated the SC-DRG cocultures for 4 days with 1 μM etorphine or 10 μM U50488 and compared the cyclase activity with that of control cultures. Table 2 shows that the basal adenylate cyclase activity was not affected by the chronic opiate treatment. However, the κ agonist U50488 lost its ability to inhibit the basal adenylate cyclase. Similar desensitization was also obtained with other κ ligands (Attali et al., 1988a).

κ opiates regulate Ca^{2+} channel activity

Recent electrophysiological experiments have shown that the κ opiate peptide dynorphin depresses Ca^{2+} currents in DRG neurons (MacDonald and Werz, 1986). Utilizing the SC-DRG cocultures, we were able to show that κ agonists inhibit $^{45}Ca^{2+}$ uptake through the DHP sensitive voltage-dependent Ca^{2+} channels (Vogel et al., 1988; Attali et al., 1988b). The rate of $^{45}Ca^{2+}$ uptake into the cells was increased by the addition of KCl (to depolarize the cell) in a dose-dependent manner (Fig. 2). Concentration of 50 mM KCl gave the optimum rate of uptake and was routinely used in our experiments. The DHP Ca^{2+} channel agonist Bay K8644, which is known to prolong the opening time of the voltage-dependent Ca^{2+} channels (Fox et al., 1987) further enhanced by approximately 2-fold, the voltage-dependent $^{45}Ca^{2+}$ uptake. On the other hand, the DHP Ca^{2+} channel antagonist nitrendipine reduced the stimulated Ca^{2+} uptake produced by 50 mM KCl and 1 μM Bay K8644 (K^+/Bay K-stimulated $^{45}Ca^{2+}$ uptake) by 95% (Table 3). Table 3 also shows that κ agonists such as U50488 and EKC profoundly

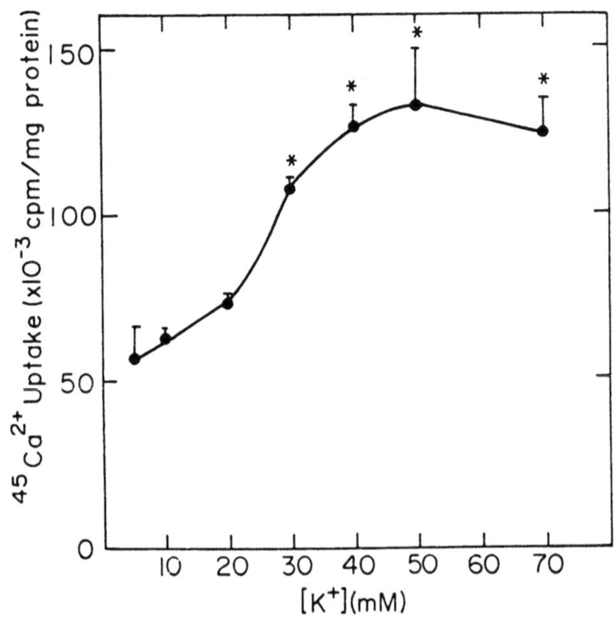

Figure 2. Effect of increasing K^+ concentration on $^{45}Ca^{2+}$ uptake into SC-DRG cultures. Depolarization-induced Ca^{2+} influx was produced by increasing K^+ concentrations. Data are presented as mean±SEM of 3 experiments. Significance (*p<0.01).

Table 3: Effect of voltage-dependent Ca^{2+} channel antagonist and opiate ligands on $^{45}Ca^{2+}$ uptake

Drugs added	Inhibition of K^+/Bay K-stimulated $^{45}Ca^{2+}$ uptake (%)
Nitrendipine	95±14
U50488	53±12
EKC	78±6
Naloxone	9±1
U50488+naloxone	15±6
Levorphanol	47±3
Dextrorphan	2±1
DAGO	1±1
DADL	3±4

All drugs were at 10 μM, except EKC (ethylketocyclazocine) at 100 μM. Results are expressed as the percentage of inhibition of the net K^+/Bay K-stimulated $^{45}Ca^{2+}$ uptake.

depressed the K^+/Bay K-stimulated $^{45}Ca^{2+}$ uptake (86% and 78% inhibition, respectively), while μ (DAGO) and δ (DADL) agonists had no effects. As shown above for adenylate cyclase, the κ opiate inhibition was stereospecific since the nonselective opiate agonist levorphanol exhibited significant inhibition, while dextrorphan failed to affect the $^{45}Ca^{2+}$ uptake. Moreover, the antagonist naloxone, which by itself did not have any significant effect, suppressed almost completely the effect of U50488 on Ca^{2+} influx.

Following chronic exposure of SC-DRG cocultures to U50488 (10 μM, 24 h) and after an extensive wash of the cells, the basal $^{45}Ca^{2+}$ uptake remained unaffected and the K^+/Bay K-stimulated Ca^{2+} uptake was only slightly reduced when compared with control untreated cells. However, as shown in Table 4, the capacity of the opiate agonists U50488 and EKC to inhibit the K^+/Bay K-stimulated Ca^{2+} influx was significantly reduced. For example, EKC, which inhibited the stimulated Ca^{2+} uptake by 78% in untreated cells, inhibited this uptake by only 38% following the chronic U50488 treatment. The extent of inhibition by nitrendipine was not significantly changed.

Table 4: Effect of chronic κ agonist treatment on Ca^{2+} influx in rat SC-DRG cocultures

Drugs added	Inhibition of stimulated Ca^{2+} uptake (%) Untreated	U50488-treated
U50488 10 μM	53±12	7±4
EKC 100 μM	78± 6	38±2
Nitrendipine 10 μM	95±14	88±6

Cells were treated with 10 μM U50488 for 24 h, washed and assayed for K^+/Bay K-stimulated $^{45}Ca^{2+}$ uptake, in the presence of the drugs described above.

Possible mechanism

GTP binding proteins are known to serve as a coupling system between adenylate cyclase and the activated neurohormone receptor (Gilman, 1987). Moreover, it seems very likely that GTP binding proteins are also involved in the regulation of several ionic channels, including the voltage-dependent Ca^{2+} channel (Dunlap et al., 1987). GTP binding proteins seem to be involved in mediating the κ opiate response in the SC-DRG cells. We have previously shown that the affinity of the κ agonist U50488 to κ receptors in these cells is decreased in the presence of GTP and NaCl (Attali et al., 1986). In Table 5, we show that the levels of inhibition of adenylate cyclase activity and of $^{45}Ca^{2+}$ influx by the κ agonist U50488 are markedly reduced following pretreatment of the SC-DRG cultures with pertussis toxin. The inhibition of cyclase activity by U50488 was completely eliminated and that of Ca^{2+} influx was reduced by 60%. Similar results were obtained with other κ agonists.

Table 5: Effect of pertussis toxin on the sensitivity of adenylate cyclase and Ca^{2+} channels to U50488

	Percent inhibition by U50488 of Adenylate cyclase	Ca^{2+} influx
Untreated	22±3	53±12
Pertussis-treated	0±3	21± 3

Effect of U50488 is expressed as percentage of inhibition of adenylate cyclase and K^+/Bay K-stimulated $^{45}Ca^{2+}$ uptake in control and pertussis toxin-treated (100 ng/ml, 18 h) cells. Values are means±SEM from 4 different experiments.

As described above, chronic exposure of the SC-DRG cells to κ agonists leads to desensitization of the κ opiate action on adenylate cyclase, as well as on Ca^{2+} channel activity. Additional experiments (Attali et al., 1988a; Vogel et al., 1988) demonstrated that the κ agonist treatment also induced desensitization to α_2 adrenergic and muscarinic agonists, suggesting a postreceptor modification. The chronic opiate exposure does not seem to be accompanied by alterations in the catalytic unit of adenylate cyclase, since its basal and forskolin-stimulated levels of activity remained the same in control and chronic opiate-treated cultures (Table 2 and Attali et al., 1988a). On the other hand, we have recently found that chronic treatment with κ, α_2 or muscarinic agonists decreases the amount of the α subunit of the inhibitory G protein (α_i). More detailed analysis revealed that chronic κ agonist exposure leads to a dramatic (>50%) reduction in the α_i of type 1 (α_i-1), with no significant change in α_s, α_o and β subunits (Vogel et al., 1988; Attali et al., in preparation).

REFERENCES

Attali B, Gouarderes C, Mazarguil H, Audigier Y, Cros J (1982) Evidence of multiple "kappa" binding sites by use of opioid peptides in the guinea pig lumbo-sacral spinal cord. Neuropeptides 3: 53-64.

Attali B, Vogel Z (1986a) Inhibition of adenylate cyclase and induction of heterologous desensitization by kappa agonists in rat spinal cord. In: Holaday JW, Law PY, Herz A (eds) Progress in Opioid Research, NIDA Research Monograph, vol 75, US Govt Printing Office, Washington DC, pp. 141-144.

Attali B, Vogel Z (1986b) Kappa opiate receptors in rat spinal cord-dorsal root ganglion cocultures: regulation and modulation of adenylate cyclase. Soc Neurosci Abstr 12: 1012.

Attali B, Saya D, Vogel Z (1986) Expression and regulation of κ opiate receptors in rat spinal cord-dorsal root ganglion cocultures. In: Holaday JW, Law PY, Herz A (eds) Progress in Opioid Research, NIDA Research Monograph, vol 75, US Govt Printing Office, Washington, DC, pp. 434-437.

Attali B, Saya D, Vogel Z (1988a) Kappa opiate agonists inhibit adenylate cyclase and produce heterologous desensitization in rat spinal cord. J Neurochem (in press).

Attali B, Saya D, Nah SY, Vogel Z (1988b) κ opiate agonists inhibit Ca^{2+} influx in rat spinal cord-dorsal root ganglion cocultures: involvement of a GTP-binding protein. J Biol Chem (in press).

Crain SM, Crain B, Finnigan T, Simon EJ (1979) Development of tolerance to opiates and opioid peptides in organotypic cultures of mouse spinal cord. Life Sci 25: 1797-1802.

Crain SM, Crain B, Peterson ER (1986) Cyclic AMP or forskolin rapidly attenuates the depressant effects of opioids on sensory-evoked dorsal-horn responses in mouse spinal cord ganglion explants. Brain Res 370: 61-72.

Czlonkowski A, Costa T, Przewlocki R, Pasi A, Herz A (1983) Opiate receptor binding sites in human spinal cord. Brain Res 267: 392-396.

Delander GE, Takemori AE (1983) Spinal antagonism of tolerance and dependence induced by systemically administered morphine. Eur J Pharmacol 94: 35-42.

Dunlap K, Holz G, Rane SG (1987) G proteins as regulators of ion channel function. Trends Neurosci 10: 241-244.

Fox AP, Nowycky MC, Tsien RW (1987) Single-channel recordings of three types of calcium channels in chick sensory neurons. J Physiol (Lond) 394: 173-200.

Gilman AG (1987) G proteins: transducers of receptor-generated signals. Ann Rev Biochem 56: 615-649.

Girardot JM, Kempf J, Cooper DMF (1983) Role of calmodulin in the effect of guanyl nucleotides on rat hippocampal adenylate cyclase: involvement of adenosine and opiates. J Neurochem 41: 848-859.

Jurna I (1984) Cyclic nucleotides and aminophylline produce different effects on nociceptive motor and sensory responses in the rat spinal cord. Naunyn-Schmiedeberg's Arch Pharmacol 327: 23-30.

Kelly PD, Rance MJ, Traynor JR (1982) Properties of opiate binding in the rat spinal cord. Neuropeptides 2: 319-324.

Law PY, Wu J, Koehler JE, Loh HH (1981) Demonstration and characterization of opiate inhibition of the striatal adenylate cyclase. J Neurochem 36: 1834-1846.

MacDonald RL, Werz MA (1986) Dynorphin A decreases voltage-dependent calcium conductance of mouse dorsal root ganglion neurons. J Physiol (Lond) 377: 237-249.

Miller RJ (1987) Multiple calcium channels and neuronal function. Science 235: 46-52.

Mudge AW, Leeman SE, Fischbach GD (1979) Enkephalin inhibits release of substance P from sensory neurons in culture and decreases action potential duration. Proc Natl Acad Sci USA 76: 526-530.

Nowycky MC, Fox AP, Tsien RW (1985) Three types of neuronal calcium channel with different calcium agonist sensitivity. Nature 316: 440-443.

Quirion R (1984) Pain, nociception and spinal opioid receptors. Prog Neuropsychopharmacol Biol Psychiat 8: 571-579.

Sharma SK, Nirenberg M, Klee WA (1975) Morphine receptors as regulators of adenylate cyclase activity. Proc Natl Acad Sci USA 72: 590-594.

Smith MM, Harden TK (1984) Modification of receptor-mediated inhibition of adenylate cyclase in NG108-15 neuroblastoma x glioma cells by N-ethylmaleimide. J Pharmacol Exp Ther 228: 425-433.

Traynor JR, Wood MS (1987) Distribution of opioid binding sites in spinal cord. Neuropeptides 10: 313-320.

Vogel Z, Saya D, Attali B (1988) Kappa agonists inhibit Ca^{2+} influx and induce heterologous desensitization in rat spinal cord-dorsal root ganglion cocultures: involvement of G proteins. Soc Neurosci Abstr 14: 83.

Wood PL, Rackham A, Richard J (1981) Spinal analgesia: comparison of the μ agonist morphine and the κ agonist ethylketocyclazocine. Life Sci 28: 2119-2125.

Yaksh TL (1983) In vivo studies on spinal opiate receptor systems mediating antinociception: I. Mu and delta receptor profiles in the primate. J Pharmacol Exp Ther 226: 303-316.

Yano K, Higashida H, Inoue R, Nozawa Y (1984) Bradykinin-induced rapid breakdown of phosphatidylinositol 4,5-bisphosphate in neuroblastoma x glioma hybrid NG108-15 cells. J Biol Chem 259: 10201-10207.

TRANSDIFFERENTIATION TO A NEURONAL PHENOTYPE IN ADULT BOVINE CHROMAFFIN CELLS: EFFECTS OF αMSH, bFGF AND HISTAMINE

Barbara A. Demeneix* & Nikolai Kley
Department of Neuropharmacology
Max-Planck-Institut für Psychiatrie
Am Klopferspitz 18a
8033 Planegg-Martinsried, FRG

ABSTRACT

The effects of histamine, basic fibroblast growth factor (bFGF) and α melanocyte stimulating hormone (αMSH) on neurite outgrowth, proenkephalin A (pENK) and tyrosine hydroxylase (TH) gene expression have been studied in primary cultures of chromaffin cells from the adult bovine adrenal medulla. All three factors promote transdifferentiation to the neuronal phenotype (assessed by neurite outgrowth) but histamine acts with a much shorter latency than bFGF and αMSH. Moreover, only histamine produces any modification of pENK gene expression at the times examined (8,24,48 and 72h). Possible intracellular signalling mechanisms linked to transdifferentiation in bovine chromaffin cells are discussed.

INTRODUCTION

The molecular mechanisms involved in differentiation are an integral response to extracellular factors that impinge on gene expression through intracellular signals. To biochemically dissect the signals that transduce membrane effects into modulation of gene expression one needs a homogeneous population of responsive cells that can be obtained in high yields. A model that fulfils these criteria for neuronal differentiation is the adult bovine adrenal medulla chromaffin cell in primary culture. Chromaffin cells are phenotypically plastic as they can "transdifferentiate" from an endocrine to a sympathetic neurone-like phenotype. This transdifferentiation may thus reflect the com-

* on leave from Institut de Physiologie Général, UACNRS 309, Université Louis Pasteur, 67000 Strasbourg, France

NATO ASI Series, Vol. H 32
Molecular Biology of Neuroreceptors and Ion Channels. Edited by A. Maelicke

mon origin of adrenal chromaffin cells and sympathetic neurones from a shared neural crest precursor.

Many studies on plasticity in sympathoadrenal cells have used embryological tissues or the clonal rat pheochromocytoma cell line, PC12 cells. In these cell types neuronal phenotype can be induced by a number of factors, the most well documented being nerve growth factor (NGF) (Unsicker et al. 1978, Doupe et al. 1985, Heidemann et al. 1985, Anderson & Axel 1986). Glucocorticoids, on the other hand inhibit neurite outgrowth and induce the endocrine phenotype in rat embryo adrenal cells (Unsicker et al. 1978, Doupe et al. 1985, Anderson & Axel 1986). NGF does not induce neurite extension in adult bovine chromaffin cells (Naujoks et al. 1978). So to take advantage of the high yields afforded by cultures from adult bovine sources one needs a competent stimulus for transdifferentiation. We induced neurite outgrowth in these cells using conditioned medium from pituitary intermediate lobe cultures (ILCM) or with αMSH, a peptide secreted by intermediate lobe (IL) cells. These treatments are significant promoters of stable neurite formation (Demeneix & Grant 1988) and the neuronal phenotype induced is confirmed by the presence of phosphorylated neurofilaments in the neurites (Grant et al. in press).

As neurite outgrowth can be a long-term response (over days) we looked for precocious changes in gene expression, that could more easily be correlated with intracellular signals. We measured specific mRNAs coding for proenkephalin (pENK) and tyrosine hydroxylase (TH). Changes in gene expression were correlated with the morphological changes in cells challenged with αMSH, ILCM, basic FGF and histamine. Basic FGF (bFGF) was chosen as it is known to be produced in the pituitary (Ferrara et al. 1987) and thus it could be an active component of ILCM. Histamine was included as it is known modulator of pENK mRNA in bovine chromaffin cells (Kley et al. 1987a, Kley 1988) and moreover it proved to be the most rapid and efficient inductor of neuronal phenotype. The histamine stimulus has the added advantage that its receptor type in the chromaffin cell is well defined, as are certain of the intracellular signals linked to it (Kley et al. 1987, Kley 1988).

EXPERIMENTAL METHODS

Cell culture

Isolated adrenal medullary chromaffin cells were obtained by enzymatic digestion of bovine adrenal glands as previously described (Kley et al. 1986). Cells were allowed to attach to collagen-coated plastic dishes (or glass coverslips for morphology) for 24h in Dulbecco's modified Eagle medium (DMEM) and Ham F12 (1:1, v/v) supplemented with 10% foetal calf serum, antibiotics (penicillin 100 µ/ml, kanamycin 50 µg/ml) and antimitotics (fluorodeoxyuridine and cytosine arabinoside, each at 10 µM). At the onset of the experiment cells were switched to defined serum-free medium (Bottenstein and Sato 1979) with antibiotics and antimitotics as above.

Immunofluorescence and evaluation of neurite outgrowth

Staining for TH was carried out using the biotin-streptavidin method. Cells were fixed for 15 min with 4% paraformaldehyde in PBS, washed, permeabilized for 15 min with PBT1 (PBS supplemented with 1% BSA and 0.1% Triton-X100) and then incubated for 30 min with mouse anti-TH (1:200) in the same buffer. After washing (4 times) biotinylated goat anti-mouse antibodies (1:100, Amersham) dissolved in PBT1 were added for 30 min. After washing (2 x with PBT1 and PBT2 each) Texas Red Streptavidin (1:100, Amersham) dissolved in PBT2 (PBS supplemented with 0.1% BSA and 0.1% Triton X100) was added for 20 min. Then the cultures were washed and mounted in PBS/glycerol. Chromaffin cells were examined using a 40 X Neofluor objective and cell counts accomplished by sampling all cells in 3 randomly selected transects covering the entire width of the coverslip (100-150 cells per transect; cell processes extending twice the diameter of the cell body (>40 µm) were considered to be neurites).

Proenkephalin mRNA and tyrosine hydroxylase mRNA measurement

Quantification of mRNA for pENK and TH was performed by Northern blot hybridization. Total nucleic acids were extracted by phenol/chloroform extraction as previously described (Kley et al. 1986), denatured in glyoxal and size fractionated on a 1.2% agarose gel, transferred to intracellulose paper

(Schleicher & Schuell), then prehybridized and hybridized with the appropriate cDNA at 42°C. A single stranded ^{32}P labelled pENK cDNA was prepared as described in Pittius et al. (1985) and the TH cDNA according to Saadat et al. (1987).

Transmitter release

To study histamine, αMSH and bFGF effects on secretion, cells were plated on collagen-coated plastic multiwell tissue culture plates at a density of 5×10^5 per well (in 1 ml plating medium, see above). Two days later cells were loaded with $[^3H]$norepinephrine (Amersham) in DMEM and 0.01% ascorbic acid, washed 3times in DMEM and incubated 30' with test substances (all procedures at 37°C). Medium was collected, centrifuged and radioactivity determined. The cells were lysed in 10% TCA, scraped from the dish and cellular radioactivity determined. Released $[^3H]$NE was calculated as a % of total content.

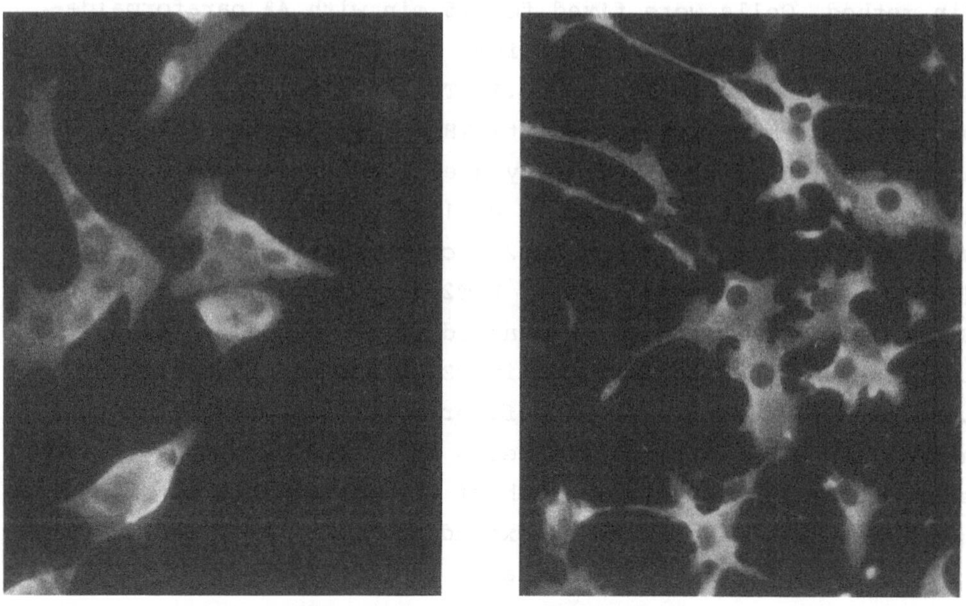

Figure 1 Bovine chromaffin cells can be induced to adopt a neuronal phenotype. Immunofluorescent identification of TH positive cells. A: control cells, grown in defined medium 4d *in vitro*. B: Neurite extension in cells (at 4d *in vitro*) treated with histamine (20 μM) for 40h prior to fixation (x 400).

RESULTS

Neuronal differentiation

The differentiating capacity of adult bovine chromaffin cells is illustrated in Fig. 1a+b. Fig. 1a shows an immunofluorescent TH marking of a chromaffin cell culture grown in defined medium for 4 days. The cells are either spherical, or slightly flattened but with few extensions. However, treatment with histamine (20 µM) for 40h induces extensive cell flattening, a decrease in cell body size and neurite outgrowth (Fig. 1b). Histamine treatment gave the most rapid responses of all the factors tested, and was the only one producing significant neurite extension at 40h (Fig. 2). However, bFGF (10 ng/ml), like αMSH (50 ng/ml) and ILCM (Demeneix & Grant 1988) promoted significant neurite outgrowth after 7 days of treatment (Fig. 2b). The kinetics of the induction with these latter three treatments are therefore significantly slower than with histamine treatment.

pENK mRNA and TH mRNA in response to neurotropic factors.

The effect of histamine, bFGF and MSH on pENK mRNA and TH mRNA levels were tested after 7,24 and 48h of treatment. Fig. 3a shows the Northern blot analysis of pENK mRNA.

Histamine treatment stimulated pENK mRNA levels giving an increase that was 2fold at 8 hours and 5fold at 24 hours (values measured by densitometric scanning of the photographic negative of the Nothern blot). None of the other treatments produced any variation in pRNK mRNA at 24h (Fig. 3a) nor at any of the other times examined.

TH mRNA, levels were modified in parallel to pENK mRNA (Fig. 3b), only histamine producing a significant increase.

Transmitter release

We tested the effects of bFGF (10 ng/ml) and αMSH (50 ng/ml) on a calcium-dependent mechanism, basal [^{3}H]NE secretion. In contrast to histamine (20 µM), which doubled basal secretion (Fig. 4), neither bFGF nor αMSH modified secretion when used at concentrations shown to induce transdifferentiation.

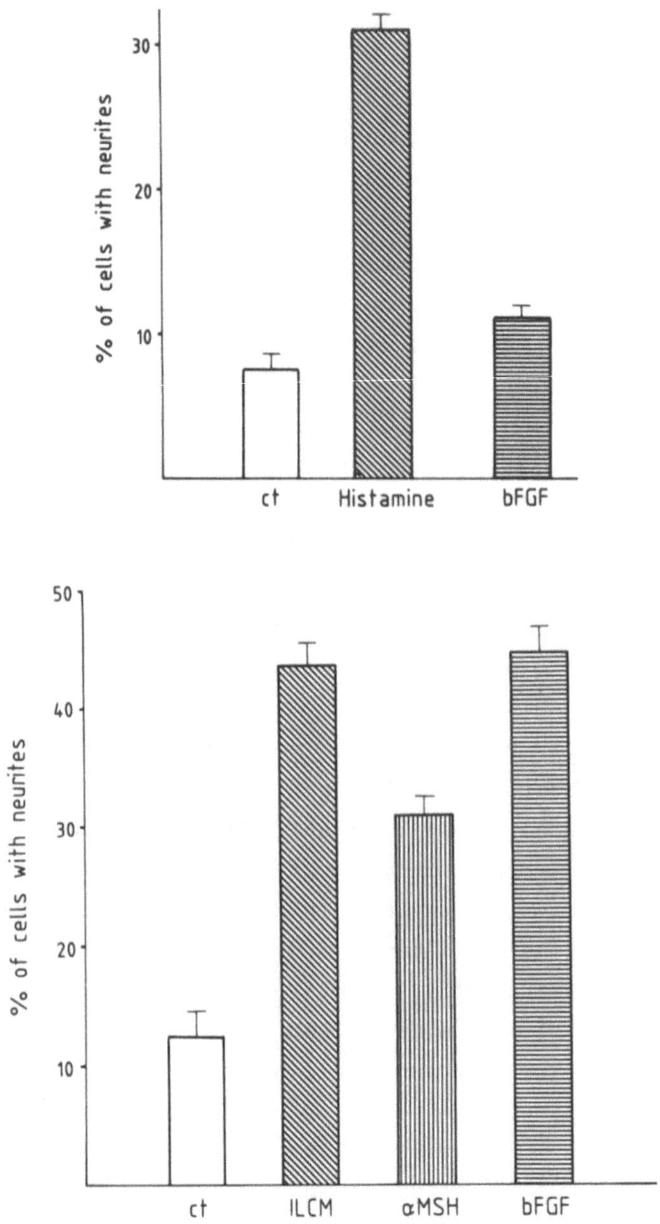

Figure 2 Histamine (20μM), ILCM (prepared as in Demeneix & Grant 1988), αMSH (50 ng/ml) and bFGF (10 ng/ml) have different latencies for the induction of neurite outgrowth in bovine chromaffin cells. The percent of chromaffin cells with neurites is shown after 40h treatment (A) and after 7 days of treatment (B). A minimum of 300 cells was counted for each of 4 slides per treatment. (Means ± SEM are given.)

DISCUSSION

The data presented show that the neuronal phenotype can be induced in adult bovine chromaffin cells by ILCM, αMSH (Demeneix, & Grant 1988), bFGF (Demeneix, Grant & Loeffler, in preparation) and histamine. Indeed it was the initial demonstration that adult bovine chromaffin cells could be induced to undergo transdifferentiation to a neuronal phenotype in the presence of IL cells (Demeneix & Grant 1988) that suggested this model could be used to analyse neurotrophic effects of pituitary products and the signalling mechanisms implicated in transdifferentiation itself.

The intermediate lobe of the pituitary is known to produce an array of peptides (αMSH, ßendorphin, ACTH, etc.) derived from one precursor molecule: pro-opiomelanocortin (POMC). We began our analysis of possible neurotrophic properties of these peptides by testing ACTH and the two forms of αMSH (N-acetylated and desacetylated) produced after cleavage of the ACTH molecule. Of the three peptides, tested at concentrations found in ILCM, ACTH was without effect, whilst both forms of αMSH were equipotent in stimulating neurite outgrowth (Demeneix & Grant 1988). Interestingly, desacetylated αMSH is less effective than N-acetylated αMSH on the best-documented target of αMSH, melanocytes, which like chromaffin cells are derived from the neural crest. However, desacetylated αMSH is the major form present in the developing hypothalamus and may play a neurotrophic role in the brain.

The pituitary is also a major source of bFGF (Ferrara et al. 1987) a factor known to induce differentiation in PC12 cells (Togari et al. 1985) and rat embryo chromaffin cells (Stemple et al. 1988). Here we show that bFGF does in fact induce neurite extension in adult bovine chromaffin cells, so it could indeed be an active component of ILCM (Demeneix, Grant & Loeffler, in preparation).

Thus three treatments ILCM, αMSH (50 ng/ml) and bFGF (10 ng/ml) were found to induce neuronal differentiation over 5 to 7 days in bovine chromaffin cells, a time course similar to NGF induction of neurite outgrowth in PC12 cells (Greene 1984). To

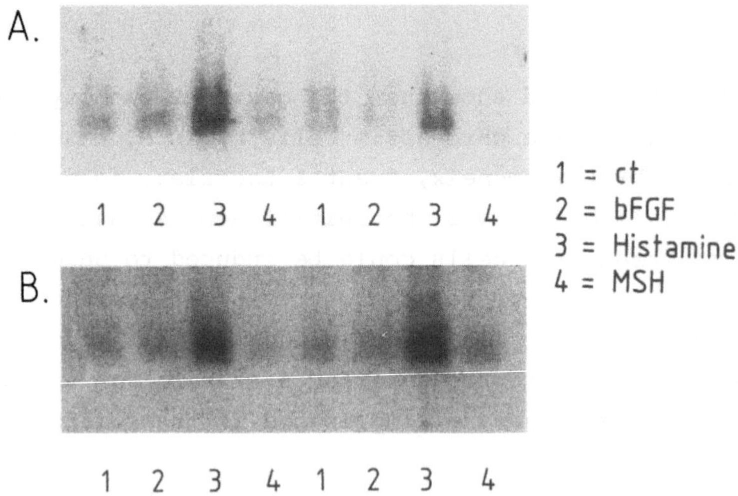

Figure 3 Northern hybridization of pENK mRNA (A) and TH mRNA (B) in chromaffin cells after 24 h treatment with control defined medium (ct), αMSH (50 ng/ml), histamine (20 μM) and bFGF (10 ng/ml). 6 individual cultures were extracted and equal aliquots pooled and subjected to analysis. Total RNA applied in each sample was determined by scanning densitometrically negative photographs of UV-transilluminated, ethidium bromide stained gels run in parallel to gels used for Northern analysis No increases in total RNA were seen with any treatments.

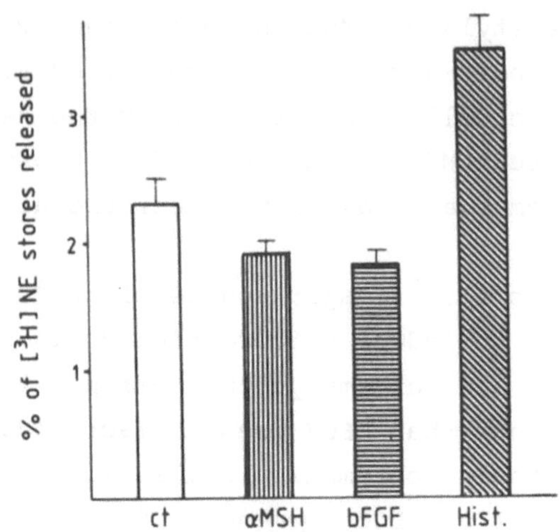

Figure 4 Percent total [^{3}H]norepinephrine (NE) release over 30 min in control (ct) conditions, or in the presence of histamine (20 μM), αMSH (50 ng/ml) and bFGF (10 ng/ml). Only the histamine response was significantly different from controls (student's t'test $p<0.05$). Means ± SEM are given (n=6 in each group).

see if neurite extension was preceded by changes in peptide or catecholamine synthesis we followed pENK and TH mRNA levels in cells treated with ILCM (data not shown), αMSH and bFGF for 8,24 and 48h. However, none of these treatments produced any modification of pENK nor TH mRNA levels at the times examined. These findings contrast with NGF induction of TH in cultures of rat symathetic neurones (Rohrer et al. 1987) and calf chromaffin cells (Acheson et al. 1984).

Histamine (20 μM), on the other hand produced increases in both pENK and TH mRNA levels and induced rapid neurite outgrowth (within 24-40h). So adult bovine chromaffin cell responses to neurotropic factors fall into three categories. First, neurite outgrowth can be induced in parallel with pENK and TH gene expression (histamine, and to a lesser extent K^+ depolarisation). Second, morphological transdifferentiation can be induced without any precocious changes in the expression of these genes (as with ILCM, bFGF and αMSH). Third, certain treatments, such as cAMP, known to increase neurite outgrowth in PC12 cells (Heidemann et al. 1985) and sympathetic neurones (Rydel & Greene 1988) induce increases in pENK and TH gene expression in bovine chromaffin cells (Kley et al. 1987b, Kley unpublished observations) but have no neurite outgrowth promoting effects (Unsicker et al. 1983).

Two possible explanations can account for these findings. One is that the culture conditions used uncouple certain receptors from the signalling mechanisms linking them to one or other of the responses. Practically, this is a difficult hypothesis to test. A more attractive alternative is that sub-populations of cells are present that do or do not possess certain receptors or inducible genes involved in initiating differentiation. It is known that bovine adrenal medulla has a heterogenous cell population, for instance pENK is only expressed in the peripheral area of the medulla <u>in vivo</u>. One means of ascertaining whether those cells undergoing transdifferentiation in response to histamine are also expressing pENK would be to correlate morphological changes with immunochemical localisation of pENK-derived peptides. However, even if one found that all the cells responding to histamine with neurite outgrowth were also those expressing pENK one would still have to consider whether one signal was responsible for generating both responses or whether independent mechanisms accounted for each. At the moment, however, little is known about the intracellular signals mediating effects of growth factors. Indeed it is in this respect that the rapid induction of neurite outgrowth in bovine chromaffin cells by histamine will be useful as a model for analysing the intracellular triggers in neuronal differentiation.

Our current knowledge of histamine receptor activation on bovine chromaffin cells does permit some consideration of the signal mechanisms implicated in effects on neurite outgrowth and gene expression. Histamine acts on these cells through the H_1 receptor, which activates phospholipase C (PLC) generating inositol triphosphate (IP_3) and diacylglycerol. IP_3 production in turn mobilises internal Ca^{++} (Ca^{++}_i) and diacylglycerol production results in activation of protein kinase C (PKC). Increased Ca^{++}_i levels, whether from mobilisation of internal stores or by entry of extracellular Ca^{++}, is a potent stimulator of pENK gene expression (Kley et al. 1986, 1987a), so the questions arise as to whether the same signal transduces the histamine-induced neuronal differentiation and whether other agents that induce elevations in Ca^{++}_i also induce neurite outgrowth. One treatment that should increase Ca^{++}_i is K^+ depolarisation. However, even 50 mM K^+ produces only moderate neurite outgrowth (Unsicker et al. 1983), compared to histamine. This in itself suggests that the histamine effect on neurite outgrowth might involve a Ca^{++}-independent pathway. Evidence that Ca^{++}-independent mechanisms might be implicated in transdifferentiation also come from the αMSH and bFGF experiments. Apparently neither treatment modifies Ca^{++}_i, as secretion and pENK gene expression were unaffected by their presence.

As stated above histamine activation of PLC should also result in activation of PKC. However, it is unlikely that this pathway is implicated in the transdifferentiation observed here as inhibiting PKC activity with H7 (20 μM) does not block, but rather enhances both histamine and bFGF effects on neurite outgrowth (Kley & Demeneix, unpublished observations). This is in agreement with data from PC12 cells where direct activation of

PKC with phorbol esters does not induce differentiation (Kruijer et al. 1985).

In conclusion, induction of neuronal differentiation in bovine chromaffin cells can be induced by various neurotrophic substances, with different latencies and that do or do not affect TH and pENK gene expression. Thus we now have a model that will allow biochemical dissection of the different pathways involved in induction of phenotypic transdifferentiation.

REFERENCES

Acheson A, Naujoks K, Thoenen H (1984) Nerve growth factor mediated enzyme induction in primary cultures of bovine adrenal chromaffin cells: specificity and level of regulation. J Neurosci 4:1771-1780.

Anderson JA, Axel R (1986) A bipotential neuroendocrine precursor whose choice of cell fate is determined by NGF and glucocorticoids. Cell 47:1079-1090

Bottenstein JE, Sato GH (1979) Growth of a rat neuroblastoma cell line in serum-free supplemented medium. Proc Natl Acad Sci USA 76:514-517

Demeneix BA, Grant NJ (1988) Alpha-melanocyte stimulating hormone promotes neurite outgrowth in chromaffin cells. FEBS Lett 226:337-342

Doupe AJ, Landis SC, Patterson PH (1985) Environmental influences in the development of neural crest derivatives: glucocorticoids, growth factors, and chromaffin cell plasticity. J Neurosci 5:2119-2142

Ferrara N, Schweiger L, Gosporadowicz D (1987) Pituitary follicular cells produce basic fibroblast growth factor. Proc Nat Acad Sci 84:5773-5777.

Grant NJ, Demeneix BA, Aunis D, Langley KO, Induction of neurofilament phosphorylation in cultured chromaffin cells. Neuroscience (in press).

Greene LA (1984) The importance of both early and delayed responses in the biological actions of nerve growth factor. Trends in Neurosci 7:91-94

Heidemann SR, Joshi HC, Schechter A, Fletcher JR, Bothwell M (1985) Synergistic effects of cyclic AMP and Nerve Growth Factor on neurite outgrowth and microtubule stability of PC12 cells. J C Biol 100:916-927

Kley N (1988) Multiple regulation of proenkephalin gene expression by protein kinase C. J Biol Chem 243: 2003-2008

Kley N, Loeffler JP, Pittius CW, Höllt V (1986) Proenkephalin A gene expression in bovine adrenal chromaffin cells is regulated by changes in electrical activity. EMBO J 5:967-970

Kley N, Loeffler JP, Höllt V (1987a) Ca^{++}-dependent histaminergic regulation of proenkephalin mRNA levels in cultured adrenal chromaffin cells. Neuroendocrinology 46:89-92

Kley N, Loeffler JP, Pittius CW, Höllt V (1987b) Involvement of ion channels in the induction of proenkephalin A gene expression by nicotine and cAMP in bovine chromaffin cells. J Biol Chem 262:4083-4089

Kruijer W, Schubert D, Verma IM (1985) Induction of the proto-oncogene fos by nerve growth factor. Proc Natl Acad Sci 82: 7330-7334

Naujoks KW, Korsching S, Rohrer H, Thoenen H (1982) Nerve growth factor-mediated induction of tyrosine hydroxylase and of neurite outgrowth in cultures of bovine adrenal chromaffin cells: dependence on developmental stage. Dev Biol 92:365-379

Pittius CW, Kley N, Loeffler JP, Höllt V (1985) Quantitation of proenkephalin A messenger RNA in bovine brain, pituitary and adrenal medulla: correlation between mRNA and peptide levels. EMBO J 4:1257-1260

Rohrer H, Otten U, Thoenen H (1978) On the role of RNA synthesis in the selective induction of tyrosine hydroxylase by nerve growth factor. Brain Res 159:436-439

Rohrer H, Acheson AL, Thibault J, Thoenen H (1986) Development potential of quail dorsal root ganglion cells analyzed in vitro and in vivo. J Neurosci 6:2616-2624

Rydel RE, Greene LA (1988) cAMP analogs promote survival and neurite outgrowth in cultures of rat sympathetic and sensory neurones independently of nerve growth factor. Proc Natl Acad Sci USA 85:1257-1261

Saadat S, Stehle AD, Lamouroux A, Mallet J, Thoenen H (1987) Influence of cell-cell contact on levels of tyrosine hydroxylase in cultured bovine adrenal chromaffin cells. J Biol Chem 262:13007-13014

Stemple DL, Mahanthappa NK, Anderson DJ (1988) Basic FGF induces neuronal differentiation, cell division, and NGF dependence in chromaffin cells: a sequence of events in sympathetic development. Neuron 1:517-525

Togari A, Dickens G, Kuzuya H, Guroff G (1985) The effect of fibroblast growth factor on PC12 cells. J Neurosci 5:307-316

Unsicker K, Krisch B, Otten U, Thoenen H (1978) Nerve growth factor-induced fiber outgrowth from isolated rat adrenal chromaffin cells: impairment by glucocorticoids. Proc Natl Acad Sci 75:3498-3502

Zeytin FN, Rusk SF, Raymond V, Mandell A (1988) Fibroblast growth factor stabilizes ribonucleic acid and regulates differential functions in a multipeptide-secreting neuroendocrine cell line. Endocrinology 122:1121-1128.

ACKNOWLEDGEMENTS

We thank Dr. H. Rohrer (Martinsried) for the gift of monoclonal TH antibody, Prof. S. Numa (Kyoto) for the proenkephalin A clone, pENK5 and Dr. J. Mallet (Paris), for permission to use the bovine TH cDNA clone. Basic FGF was generously provided by Drs. M. Sensenbrenner and G. Labourdette (Strasbourg).

We thank Prof. A. Herz and Prof. P. Feltz for all their encouragement and Ms. H. Roth for excellent secretarial assistance. Supported by INSERM CRE grant 846010 and the DFG.

MOLECULAR BIOLOGY OF NEURORECEPTORS: IMPLICATIONS FOR CLINICAL NEUROSCIENCE

M. Baron
Departments of Medical Genetics and Psychiatry
New York State Psychiatric Institute and
Columbia University
722 West 168th Street
New York, New York 10032
USA

Recent advances in molecular biology hold great promise for clinical neuroscience. Recombinant DNA technology can be used to localize, clone and characterize the genes responsible for neurogenetic disorders, and to identify the defective gene products. This, in turn, could have major implications for etiology, molecular pathology and, possibly, prevention and treatment of these disorders.

Applications of molecular biology techniques, in particular the use of DNA probes that reveal restriction-fragment-length- polymorphisms (RFLPs) or DNA markers, have already yielded important new insights concerning genetic diseases affecting the central nervous systems. In conjunction with linkage analysis, the genetic marker strategy has led to the chromosomal localization of the genes for several neurological and neuropsychiatric disorders (see Table 1) including Huntington's disease (Gusella et al, 1983), Lesch-Nyhan syndrome (Nussbaum et al, 1983), Duchenne dystrophy (Murray et al, 1982), Wilson's disease (Frydman et

NATO ASI Series, Vol. H 32
Molecular Biology of Neuroreceptors and
Ion Channels. Edited by A. Maelicke

al, 1985), von Recklinghausen Neurofibromatosis (Barker et al, 1987; Seizinger et al, 1987), manic-depressive illness (Baron et al, 1987; Mendlewicz et al, 1987; Egeland et al, 1987), and Alzheimer's disease (St. George-Hyslop et al, 1987). In addition, methods of somatic cell genetics and cytogenetics have been used to identify and define chromosomal aberrations, such as deletions and translocations, associated with some of these disorders.

This brief review summarizes some current notions in molecular biology as they bear upon neuroreceptors and their possible role in clinical neuroscience. Some of these issues are discussed in greater detail elsewhere (Baron, 1988; Baron and Rainer, 1987; Gurling, 1986).

Until recently, studies with linkage markers were limited to the classical polymorphisms such as leukocyte (HLA) and erythrocyte antigens (ABO, Rh, MNS) that may be detected serologically, and for erythrocyte and serum proteins and isozymes that may be detected electrophoretically. However, because they are small in number, the prior probability of detecting linkage with any of these markers, even if randomly distributed throughout the human genome, is less than 30%. The dramatic expansion in the number of polymorphic DNA sequences brought about by molecular biology techniques has all but eliminated this methodological limitation. At the recent Human Gene Mapping Workshop 9 (1987) nearly 1,200 mapped DNA polymorphisms were listed. Of these over 200 were cloned genes and nearly 1,000 were arbitrary, or random DNA seqments. These RFLPs span the

Table 1. Genetic Mapping of Central Nervous System Disorder

Disorder	Chromosomal location	Gene defect	Reference
Tay-Sachs	15q22-25	Hexosaminidase (α chain)	Gilbert et al (1975)
Duchenne	Xp21	Deletions	Murray et al (1982)
		Dystrophin	Hoffman et al (1987)
Huntington's	4p16	Unknown	Gusella et al (1983)
Lesch-Nyhan	Xq27	HPRT	Nussbaum et al (1983)
Wilson's	13q14	Unknown	Frydman et al (1985)
Neurofibro-matosis (VR)	17cen	Unknown	Barker et al (1987) Seizinger et al (1987)
Manic Depression	Xq27-28	Unknown	Baron et al (1987) Mendlewicz et al(1987)
	11p	Unknown	Egeland et al (1987)
Alzheimer's	21q21	Unknown	St. George - Hyslop et al (1987)

entire length of the human genome. There will soon be enough DNA markers to permit the detection of linkage anywhere in the human genome.

Once linkage between a mapped DNA and a putative disease gene has been confirmed, a fine genetic map can be constructed in the region of interest beginning, initially, with flanking markers on both sides of the disease locus, and generating additional markers for a fine marker-to-marker resolution. The positioning of the various marker loci in relation to one another and to the disease gene can be attained by multipoint linkage analysis. Physical mapping strategies, such as the use of cosmid libraries and pulsed field or inverse-field gel electrophroesis, can then be harnessed in orde to identify and, ultimately, to clone the gene itself. Finally, studies of transcribed DNA sequences and protein products in conjunction with animal models could contribute to the elucidation of molecular pathology. One example of the successful application of these methods is Duchenne muscular dystrophy where both the disease gene and the associated protein product have been identified (Monaco et al, 1986; Hoffman et al, 1987).

Two general strategies can be employed in genetic marker studies of neurological and neuropsychiatric disorders: 1. linkage studies with arbitrary or random RFLPs; 2. studies with candidate genes. The notion of arbitrary RFLPs refers to random DNA variations spread throughout the genome whose biological function, if any, need not be related to the molecular pathology of the disorder in question. By contrast, candidate genes have known neurobiological functions of possible relevance for the pathophysiology of the disease.

Candidate Genes: The Protein-Gene Approach

The various classes of candidate proteins are listed in Table 2. They include neurotransmitter receptors such as dompamine, norepinephrine, acetylcholine, serotonin and γ-aminobutyric acid (GABA); enzymes involved in the metabolism of neurotransmitters such as tyrosine hydroxylase (TH), dopamine-beta-hydroxylase (DBH), monoamine oxidase (MAO), catechol-o-methyltransferase (COMT), and tryptophan hydroxylase; neuropeptide receptors such as enkephalins, neurotensin, substance P, bradykinin, somatostatin, and neuropeptide Y; endocrine agents such as thyroid releasing hormone (TRH), thyroid stimulation hormone (TSH), and adrenocorticotropin (ACTH); neurotrophic factors such as nerve growth factor (NGF); intracellular recognition factors such as nerve cell adhesion molecules; neuro-specific factors such as granule cells, antigens and phosphoproteins; components of second messenger systems such as guanine nucleotide-binding proteins (G proteins); pharmacoreceptors, or receptors for psychoactive drugs such as major and minor tranquillizers, antidepressants, lithium, and stimulants; and ion channels, or membrane proteins mediating transmission of information in the brain.

The human genes for some of these candidate proteins have already been cloned (Human Gene Mapping 9, 1987). These include the genes for the alpha and beta adrenergic

Table 2. Candidate Genes in Clinical Neuroscience

Class	Protein
Neurotransmitter receptors	Dopamine
	Norepinephrine
	Acetylcholine
	Serotonin
	γ-aminobutyric acid (GABA)
Neurotransmitter enzymes	Tyrosine hydroxylase (TH)
	Dopamine-beta-hydroxylase(DBH)
	Monoamine oxidase (MAO)
	Catechol-O-methyltransferase
	Tryptophan hydroxylase
Neuropeptide receptors	Enkephalins
	Neurotensin
	Substance P
	Somatostatin
	Neuropeptide Y (NPY)
Hormones	Thyroid-releasing-hormone
	Thyroid-stimulating-hormone
	Adrenocorticotropin (ACTH)

Table 2. Cont'd

Neurotrophic factors	Nerve growth factor (NGF)
Intracellular recognition factors	Nerve cell adhesion molcules
Neuron-specific factors	Granule cell antigens Phosphoproteins
G proteins	Gs, Gi, Go
Pharmacoreceptors (Psychoactive drugs)	Major tranquillizers Antidepressants Lithium Stimulants Minor tranquillizers
Ion channels receptors	ligand-gated (GABA, Acetylcholine) Voltage-gated (Sodium/calcium, potassium)

receptors; several acetylcholine muscarinic receptors; serotonin; the enzymes TH, DBH, MAO-A, MAO-B, and tryptophan hydroxylase;somatostatin; NPY; the gene coding for ACTH and beta-endorphin (pro-opiomelanocortin, or POMC); NGF; TSH; TRH; several G proteins; and the benzodiazepine receptor. Some of these genes have been mapped to specific chromosomal locations. A selective list appears in Table 3. Other genes have been cloned in other organisms, but the likely homology with the human sequences renders them suitable for genetic probing in the human. The gene-specific probes could be hybridized with digested DNA (using suitable restriction enzymes) from affected individuals and normal subjects. A different pattern of restriction fragments would suggest that the abnormal version of the gene is present in the affected individuals and is therefore associated with the disease gene. The probe could also be used in a pedigree linkage study as described earlier.

The successful application of the candidate gene strategy, namely, selecting the 'right' gene for the 'right' disorder, has the distinct advantage of circumventing the aforementioned labrorious and, often, lengthy genetic and physical mapping involved in bridging the distance between a random RFLP and the disease gene. However, should this approach fail in a given disorder, the particular genomic region could be excluded from further consideration, and so the overall search for the putative gene would be somewhat narrowed.

Table 3. Chromosomal Localization of Selected Candidate Genes

Gene	Chromosomal location
Alpha-2 adrenergic receptor	10q
Beta-2 adrenergic receptor	5q
Acetylcholine muscarinic receptors	1q
	7p
	11p
	11q
	15q
Tyrosine hydroxylase	11p
Dopamine-beta-hydroxylase	9
Monoamine oxidase A	Xq22-11
Monoamine oxidase B	X
Somatostatin	7
Neuropeptide Y	3
Thyroid stimulating hormone receptor	22q
Thyroid stimulating hormone beta subunit	1p22
Nerve growth factor	17q12-22

Applications

Studies with neuroreceptor genes and other candidate genes in neurological and neuropsychiatric disorders have already begun. The elucidation of molecular pathology in Lesch-Nyhan syndrome and Tay-Sachs disease has been facilitated by knowledge of the candidate proteins, HPRT (Nussbaum et al, 1983) and hexosaminidase (Gilbert et al, 1975), respectively. Other applications of the candidate gene approach have not been as successful. The beta-amyloid gene and the NGF receptor gene, the candidate genes for Alzheimer's disease and von Recklinghausen's neurofibromatosis, are not closely linked to these disorders, making it unlikely that the disease genes and the candidate genes coincide (Van Broeckhoven et al, 1987; Barker et al, 1987). Three potential candidate genes for schizophrenia and/or manic-depressive illness, POMC (Feder el at, 1985), NPY and somatostatin (Detera-Wadleigh et al, 1985), and beta adrenergic receptors (Berrettini et al, 1988) did not show clear association or linkage with these disorders. The reported linkage between an hereditary form of manic-depressive illness and RFLPs for the Ha-ras-1 (HRAS1) and insulin loci (Egeland et al, 1987) is of potential relevance because the HRAS locus is closely linked to the gene encoding TH, the rate limiting enzyme in catecholamine synthesis. However, further study is needed to resolve this issue in view of the likely heterogeneity of manic depression (Baron et al, 1987; Detera-Wadleigh 1987; Hodkinson et al, 1987) and the uncertain map position of the TH locus vis-a-vis the manic depression putative gene (Human Gene Mapping 9, 1987).

Limitations and Prospects

The candidate gene approach is predicated on the assumption that the protein being considered plays a major role in the biology of the disorder. However, several different neuroreceptors or other candidate proteins, encoded by different genes, may be involved in a given neurological or neuropsychiatric disorder. Another likely scenario is that certain genes cause non-specific genetic predisposition to disease and that other etiological factors are required for phenotypic expression.

Candidate genes that mediate several neuroreceptor systems appear more promising than those with a more limited range of activity. In that regard G proteins constitute an appealing candidate. G Proteins are involved in a variety of receptor-mediated signal transduction systems (for review see Gilman, 1987) including a large array of biogenic amines, polypeptide hormones, neurotransmitters, and other pivotal neurobiological mechanisms and ion channels. Of special interest is the recent finding that lithium, an effective therapeutic agent in manic-depressive illness, blocks both adrenergic and cholinergic-induced increases in G protein binding to cerebral membranes (Avissar et al, 1988). The findings suggest G proteins as the possible molecular site for both the antimanic and antidepressant effects of lithium, and are consistent with the adrenergic-cholinergic balance hypothesis of some types of psychosis. If true, some G proteins with central neuronal activity could be considered as candidates for molecular genetic studies of neuropsychiatric conditions such as manic depression. cDNA clones are available for several G proteins, and in some instances the human genes have been isolated (Gilman, 1987).

Ion channels constitute a related class of candidate proteins. By virtue of their multi-purpose mode of operation, ion channels occupy a special place in molecular neurobiology. For instance, different neuroreceptors converge onto the some ion channel, and, conversely, most neurotransmitter receptors are coupled to different ion channels on the same cell (Nicoll, 1988). This remarkable convergence and divergence in neurotransmitter action allows for considerable diversity in neuronal signaling. It is conceivable that mutations or polymorphisms in the DNA encoding some channel regions may have greater relevance for the molecular biology of complex neurological and neuropsychiatric disorders than candidate genes whose function is more circumscribed. In conjunction with G proteins, ion channels in the brain can add an important dimension to studies of both normal and pathological information processing in the central nervous system.

Finally, genes coding exclusively for brain outnumber available candidate genes by several orders of magnitude. Thus, important coding sequences in the central nervous system can be easily missed. Current efforts at expanding the inventory of candidate genes address this issue.

The localization, characterzation and cloning of abnormal genes, and the discovery of a protein abnormality will likely be followed by the introduction of the abnormal genes into transgenic animals in order to examine disease expression and to elucidate and confirm pathological function. The induction of neurofibromatosis-like condition in tansgenic mice bearing the HTLV-1 gene is an illustration of this approach. (Hinrichs et al, 1987). Ultimately, the correction of the neurogenetic defect can be attempted by transgenic experiments, as in the shiverer mouse (Readhead et al, 1987), although gene therapy of central nervous system disorders will depend, among other considerations, on the development of site-specific gene insertion techniques applicable to postmitotic brain neurons.

Acknowledgement

Supported by NIMH Research Scientist Development Award MH00176 and by NIMH grants MH42535 and MH44115.

References

Avissar S, Schreiber G, Danon A, Belmaker RH (1988) Lithium inhibits adrenergic and cholinergic increases in GTP binding in rat cortex. Nature 331:440-442.

Barker D, Wright, E, Nguyen L, Cannon L, Fain P et al (1987) Gene for von Recklinghausen neurofibromatosis is in the pericentromeric region of chromosome 17. Science 236:1100-1102.

Baron M (1988) Molecular genetic studies in affective disorders. In: Neuropsychopharmacology. Bunney WE, Hippius H, Laakmann G, Schmauss M (Eds.) Proceeding of the XVI CINP Congress. Springer-Verlag. In press.

Baron M, Rainer JD (1988) Molecular genetics and human disease: Implications for modern psychiatric research and practice. Brit J Psychiat 152:741-753.

Baron M, Risch N, Hamburger R, Mandel B, Kushner S, Newman M, Drumer D, Belmaker RH (1987) Genetic linkage between X-chromosome markers and bipolar affective illness. Nature 326:289-292.

Berrettini WE, Hoehe MR, Lentes K-V, Goldin LR, Vente JC, Gershon ES (1987) Beta adrenergic receptors in manic depressive illness: Functional and molecular genetic studies with peripheral cells. In: 16th Congr. Intl Neuropsychopharmacol. 15 Aug - 19 Aug 1988. Springer International.

Detera-Wadleigh SD, Delisi LE, Berrettini WE, Nurnberger JL, Boorman DG, Gershon ES, Minth CD, Dixon JE (1985) DNA polymorphisms in schizophrenia and affective disorders. In: Shagass C, Josiassen RC, Bridger WH, Weiss KJ, Stoff D, Simpson GM (Eds.) Biological Psychiatry 1985. Elsevier, New York, p 67-69.

Detera-Wadleigh S, Berrettini W, Goldin L, Boorman D, Anderson S, Gershon ES (1987) Close linkage of c-Harvey-ras-1 and the insulin gene to affective disorder is ruled out in three North Anerican pedigrees. Nature 325:806-808.

Egeland JA, Gerhard DS, Pauls DL, Sussex JN, Kidd KK, Allen CR, Hostetter AM, Housman DE (1987) Bipolar affective disorders linked to DNA markers on chromosome 11. Nature 325:783-787.

Feder J, Gurling HMD, Darby J, Cavalli-Sforza LL (1985) DNA retriction fragment analysis of the Proopiomelanocortin gene in schizophrenia and bipolar disorder. Amer J Hum Genet 37:286-294.

Frydman M, Bonne-Tamir B, Farrer L et al (1985) Assignment of the gene for Wilson's disease to chromosome 13: linkage to the esterase D locus. Proc Natl Acad Sci USA 82:1819-1821.

Gilbert F, Kucherlapati R, Creagan RP et al (1975) Tay-Sachs and Sandhoff's diseases: the assignment of genes for hexosaminidases A and B to individual human chromosomes. Proc Natl Acad Sci USA 72:263-267.

Gilman AG (1987) G proteins: Transducers of receptor-generated signals. Ann Rev Biochem 56:615-649.
Gurling H (1986) Candidate genes and favoured loci: strategies for molecular genetic research into schizophrenia, manic depression, autism, alcoholism and Alzheimer's disease. Psychiat Develop 4:289-309.
Gusella JF, Wexler NS, Conneally PM et al (1983) A polymorphic DNA marker genetically linked to Huntington's disease. Nature 306:234-238.
Hinrichs SH, Nerenberg M, Reynolds RK, Khoury G, Jay G (1987) A transgenic mouse model for human neurofibromatosis. Science 237:1340-1343.
Hoffman EP, Brown RH, Kunkel LM (1987) Dystrophin: the protein product of the Duchenne muscular dystrophy lucus. Cell 919-928.
Hodkinson S, Sherrington R, Gurling H et al (1987) Molecular genetic evidence for heterogeneity in manic depression. Nature 325:805-806.
Human Gene Mapping 9 (1987) Cytogenet Cell Genet 46:No.1-4
Mendlewicz J, Simon P, Sevy S et al (1987) Polymorphic DNA marker on X-chromosome and manic-depression. Lancet 1:1230-1232.
Monaco AP, Neve RL, Colletti-Feener C, Bertelson CJ, Kurnit DM, Kunkel LM (1986) Isolation of candidate cDNAs for portions of the Duchenne muscular gene. Nature 323:646-650.
Murray JM, Davies KE, Harper PS, Meredith L, Mueller CR, Williamson R (1982) Linkage relationship of a cloned DNA sequence on the short arm of the X chromosome to Duchenne muscular dystrophy. Nature 300:69-71
Nicoll RA (1988) The coupling of neurotransmitter receptors to ion channels in the brain. Science 241:545-551.
Nussbaum R, Brennand J, Chinault C et al (1983) Molecular analysis of the hypoxanthine phosphoribosyltransferase locus. In: Caskey CT, White RL (Eds.). Recombinant DNA Application to Human Disease. Cold Spring Harbor Laboratory, New York, p 81-90.
Readhead C, Popko B, Takahashi N et al (1987) Expression of a Myelin basic protein gene in transgenic shiverer mice: Correction of the dysmeylinating phenotype. Cell 48:703-710.
Seizinger BR, Rouleaux GA, Ozelius LJ et al (1987) Genetic linkage of von Recklinghausen neurofibromatosis to the nerve growth factor receptor gene. Cell 49:589-594.
St. George-Hyslop PH, Tanzi RE, Polinsky RJ et al (1987) The genetic defect causing familial Alzheimer's disease maps on chromosome 4. Science 235:885-889.
Van Broeckhoven C, Genthe AM, Vandenberghe A et al (1987) Failure of familial Alzheimer's disease to segregate with the A4-amyloid gene in several European families. Nature 329:153-155.

Authors Index

Subject Index

NATO ASI Series H

Vol. 1: **Biology and Molecular Biology of Plant-Pathogen Interactions.**
Edited by J.A. Bailey. 415 pages. 1986.

Vol. 2: **Glial-Neuronal Communication in Development and Regeneration.**
Edited by H.H. Althaus and W. Seifert. 865 pages. 1987.

Vol. 3: **Nicotinic Acetylcholine Receptor: Structure and Function.**
Edited by A. Maelicke. 489 pages. 1986.

Vol. 4: **Recognition in Microbe-Plant Symbiotic and Pathogenic Interactions.**
Edited by B. Lugtenberg. 449 pages. 1986.

Vol. 5: **Mesenchymal-Epithelial Interactions in Neural Development.**
Edited by J.R. Wolff, J. Sievers, and M. Berry. 428 pages. 1987.

Vol. 6: **Molecular Mechanisms of Desensitization to Signal Molecules.**
Edited by T.M. Konijn, P.J.M. Van Haastert, H. Van der Starre, H. Van der Wel, and M.D. Houslay. 336 pages. 1987.

Vol. 7: **Gangliosides and Modulation of Neuronal Functions.**
Edited by H. Rahmann. 647 pages. 1987.

Vol. 8: **Molecular and Cellular Aspects of Erythropoietin and Erythropoiesis.**
Edited by I.N. Rich. 460 pages. 1987.

Vol. 9: **Modification of Cell to Cell Signals During Normal and Pathological Aging.**
Edited by S. Govoni and F. Battaini. 297 pages. 1987.

Vol. 10: **Plant Hormone Receptors.** Edited by D. Klämbt. 319 pages. 1987.

Vol. 11: **Host-Parasite Cellular and Molecular Interactions in Protozoal Infections.**
Edited by K.-P. Chang and D. Snary. 425 pages. 1987.

Vol. 12: **The Cell Surface in Signal Transduction.**
Edited by E. Wagner, H. Greppin, and B. Millet. 243 pages. 1987.

Vol. 13: **Toxicology of Pesticides: Experimental, Clinical and Regulatory Perspectives.**
Edited by L.G. Costa, C.L. Galli, and S.D. Murphy. 320 pages. 1987.

Vol. 14: **Genetics of Translation. New Approaches.**
Edited by M.F. Tuite, M. Picard, and M. Bolotin-Fukuhara. 524 pages. 1988.

Vol. 15: **Photosensitisation. Molecular, Cellular and Medical Aspects.**
Edited by G. Moreno, R.H. Pottier, and T.G. Truscott. 521 pages. 1988.

Vol. 16: **Membrane Biogenesis.** Edited by J.A.F. Op den Kamp. 477 pages. 1988.

Vol. 17: **Cell to Cell Signals in Plant, Animal and Microbial Symbiosis.**
Edited by S. Scannerini, D. Smith, P. Bonfante-Fasolo, and V. Gianinazzi-Pearson. 414 pages. 1988.

Vol. 18: **Plant Cell Biotechnology.**
Edited by M.S.S. Pais, F. Mavituna, and J.M. Novais. 500 pages. 1988.

Vol. 19: **Modulation of Synaptic Transmission and Plasticity in Nervous Systems.**
Edited by G. Hertting and H.-C. Spatz. 457 pages. 1988.

Vol. 20: **Amino Acid Availability and Brain Function in Health and Disease.**
Edited by G. Huether. 487 pages. 1988.

NATO ASI Series H

Vol. 21: **Cellular and Molecular Basis of Synaptic Transmission.**
Edited by H. Zimmermann. 547 pages. 1988.

Vol. 22: **Neural Development and Regeneration. Cellular and Molecular Aspects.**
Edited by A. Gorio, J.R. Perez-Polo, J. de Vellis, and B. Haber. 711 pages. 1988.

Vol. 23: **The Semiotics of Cellular Communication in the Immune System.**
Edited by E.E. Sercarz, F. Celada, N.A. Mitchison, and T. Tada. 326 pages. 1988.

Vol. 24: **Bacteria, Complement and the Phagocytic Cell.**
Edited by F.C. Cabello und C. Pruzzo. 372 pages. 1988.

Vol. 25: **Nicotinic Acetylcholine Receptors in the Nervous System.**
Edited by F. Clementi, C. Gotti, and E. Sher. 424 pages. 1988.

Vol. 26: **Cell to Cell Signals in Mammalian Development.**
Edited by S.W. de Laat, J.G. Bluemink, and C.L. Mummery. 322 pages. 1989.

Vol. 27: **Phytotoxins and Plant Pathogenesis.**
Edited by A. Graniti, R.D. Durbin, and A. Ballio. 508 pages. 1989.

Vol. 28: **Vascular Wilt Diseases of Plants. Basic Studies and Control.**
Edited by E.C. Tjamos and C.H. Beckman. 590 pages. 1989.

Vol. 29: **Receptors, Membrane Transport and Signal Transduction.**
Edited by A.E. Evangelopoulos, J.P. Changeux, L. Packer, T.G. Sotiroudis, and K.W.A. Wirtz. 387 pages. 1989.

Vol. 30: **Effects of Mineral Dusts on Cells.**
Edited by B.T. Mossman and R.O. Bégin. 470 pages. 1989.

Vol. 31: **Neurobiology of the Inner Retina.**
Edited by R. Weiler and N.N. Osborne. 529 pages. 1989.

Vol. 32: **Molecular Biology of Neuroreceptors and Ion Channels.**
Edited by A. Maelicke. 675 pages. 1989.

MIX
Papier aus verantwortungsvollen Quellen
Paper from responsible sources
FSC® C105338

If you have any concerns about our products,
you can contact us on
ProductSafety@springernature.com

In case Publisher is established outside the EU,
the EU authorized representative is:
Springer Nature Customer Service Center GmbH
Europaplatz 3, 69115 Heidelberg, Germany

Printed by Libri Plureos GmbH
in Hamburg, Germany